Praise for *Visio*

"As a guide to the inner workings of the mathematical jungle, Stewart provides an engaging and informative experience."
—*Science News*

"A designated math popularizer, Stewart writes books that are always enlightening and enjoyable. . . . Again, Stewart provides another interesting read for anyone intrigued by mathematics."
—*Choice*

"Anyone who has always loved math for its own sake or for the way it provides new perspectives on important real-world phenomena will find hours of brain-teasing and mind-challenging delight in the British professor's survey of recently answered or still open mathematical questions."
—*Dallas Morning News*

"Entertaining and accessible. . . . Ian Stewart belongs to a very small, very exclusive club of popular science and mathematics writers who are worth reading today."
—*New York Journal of Books*

"Few of us share Stewart's mathematical skills. But we relish the intellectual stimulation of joining him in exploring mathematical problems that have pushed even genius to the limit. . . . Stewart repeatedly shows how a trivial mathematical curiosity can open up vital new conceptual insights."
—*Booklist*, starred review

"An entertaining history of mathematics and a fresh look at some of the most challenging problems and puzzles in the history of the field. . . . Once again, Stewart delivers an intriguing book that rewards random reading as much as dedicated study."
—*Publishers Weekly*

"Stewart's imaginative, often-witty anecdotes, analogies and diagrams succeed in illuminating . . . some very difficult ideas. It will enchant math enthusiasts as well as general readers who pay close attention."
—*Kirkus*

VISIONS OF INFINITY

The Great Mathematical Problems

IAN STEWART

BASIC BOOKS

A Member of the Perseus Books Group
New York

Copyright © 2013 by Joat Enterprises

Hardcover first published in the United States in 2013 by Basic Books,
A Member of the Perseus Books Group
Paperback first published in the United States in 2014 by Basic Books

All rights reserved. Printed in the United States of America. No part of this book
may be reproduced in any manner whatsoever without written permission except
in the case of brief quotations embodied in critical articles and reviews. For information, address Basic Books, 250 West 57th Street, New York, NY 10107.

Books published by Basic Books are available at special discounts for bulk pur-
chases in the United States by corporations, institutions, and other organizations.
For more information, please contact the Special Markets Department at the
Perseus Books Group, 2300 Chestnut Street, Suite 200, Philadelphia, PA 19103,
or call (800) 810-4145, ext. 5000, or e-mail special.markets@perseusbooks.com.

Library of Congress Control Number: 2012924095

978-0-465-02240-3 (hardcover ISBN)
978-0-465-06599-8 (e-book ISBN)
978-0-465-06489-2 (paperback ISBN)

British ISBN: 978-1-84668-1998

Contents

We must know. We shall know.
David Hilbert

Speech about mathematical problems in 1930, on the occasion of his honorary citizenship of Königsberg.[1]

Preface

Mathematics is a vast, ever-growing, ever-changing subject. Among the innumerable questions that mathematicians ask, and mostly answer, some stand out from the rest: prominent peaks that tower over the lowly foothills. These are the really big questions, the difficult and challenging problems that any mathematician would give his or her right arm to solve. Some remained unanswered for decades, some for centuries, a few for millennia. Some have yet to be conquered. Fermat's last theorem was an enigma for 350 years until Andrew Wiles dispatched it after seven years of toil. The Poincaré conjecture stayed open for over a century until it was solved by the eccentric genius Grigori Perelman, who declined all academic honours and a million-dollar prize for his work. The Riemann hypothesis continues to baffle the world's mathematicians, impenetrable as ever after 150 years.

Visions of Infinity contains a selection of the really big questions that have driven the mathematical enterprise in radically new directions. It describes their origins, explains why they are important, and places them in the context of mathematics and science as a whole. It includes both solved and unsolved problems, which range over more than two thousand years of mathematical development, but its main focus is on questions that either remain open today, or have been solved within the past fifty years.

A basic aim of mathematics is to uncover the underlying simplicity of apparently complicated questions. This may not always be apparent, however, because the mathematician's conception of 'simple' relies on many technical and difficult concepts. An important feature of this book is to emphasise the deep simplicities, and avoid – or at the very least explain in straightforward terms – the complexities.

Mathematics is newer, and more diverse, than most of us imagine. At a rough estimate, the world's research mathematicians number about a hundred thousand, and they produce more than *two million* pages of new mathematics every year. Not 'new numbers', which are not what the enterprise is really about. Not 'new sums' like existing ones, but bigger – though we do work out some pretty big sums. One recent piece of algebra, carried out by a team of some 25 mathematicians, was described as 'a calculation the size of Manhattan'. That wasn't quite true, but it erred on the side of conservatism. The *answer* was the size of Manhattan; the calculation was a lot bigger. That's impressive, but what matters is quality, not quantity. The Manhattan-sized calculation qualifies on both counts, because it provides valuable basic information about a symmetry group that seems to be important in quantum physics, and is definitely important in mathematics. Brilliant mathematics can occupy one line, or an encyclopaedia – whatever the problem demands.

When we think of mathematics, what springs to mind is endless pages of dense symbols and formulas. However, those two million pages generally contain more words than symbols. The words are there to explain the background to the problem, the flow of the argument, the meaning of the calculations, and how it all fits into the ever-growing edifice of mathematics. As the great Carl Friedrich Gauss remarked around 1800, the essence of mathematics is 'notions, not notations'. Ideas, not symbols. Even so, the usual language for expressing mathematical ideas is symbolic. Many published research papers do contain more symbols than words. Formulas have a precision that words cannot always match.

However, it is often possible to explain the ideas while leaving out most of the symbols. *Visions of Infinity* takes this as its guiding principle. It illuminates what mathematicians do, how they think, and why their subject is interesting and important. Significantly, it shows how today's mathematicians are rising to the challenges set by their predecessors, as one by one the great enigmas of the past surrender to the powerful techniques of the present, which changes the mathematics and science of the future. Mathematics ranks among humanity's greatest achievements, and its great problems – solved and unsolved – have guided and stimulated its astonishing power for millennia, both past and yet to come.

Coventry, June 2012

Figure Credits

Great problems

TELEVISION PROGRAMMES ABOUT MATHEMATICS are rare, good ones rarer. One of the best, in terms of audience involvement and interest as well as content, was Fermat's last theorem. The programme was produced by John Lynch for the British Broadcasting Corporation's flagship popular science series *Horizon* in 1996. Simon Singh, who was also involved in its making, turned the story into a spectacular bestselling book.[2] On a website, he pointed out that the programme's stunning success was a surprise:

> It was 50 minutes of mathematicians talking about mathematics, which is not the obvious recipe for a TV blockbuster, but the result was a programme that captured the public imagination and which received critical acclaim. The programme won the BAFTA for best documentary, a Priz Italia, other international prizes and an Emmy nomination – this proves that mathematics can be as emotional and as gripping as any other subject on the planet.

I think that there are several reasons for the success of both the television programme and the book and they have implications for the stories I want to tell here. To keep the discussion focused, I'll concentrate on the television documentary.

Fermat's last theorem is one of the truly great mathematical problems, arising from an apparently innocuous remark which one of the leading mathematicians of the seventeenth century wrote in the margin of a classic textbook. The problem became notorious because no one could prove what Pierre de Fermat's marginal note claimed, and this state of affairs continued for more than 300 years despite strenuous efforts by extraordinarily clever people. So when the British mathematician Andrew Wiles finally cracked the problem in 1995, the magnitude of his achievement was obvious to anyone. You didn't even need to know what the problem was, let alone how he had solved it. It was the mathematical equivalent of the first ascent of Mount Everest.

In addition to its significance for mathematics, Wiles's solution also involved a massive human-interest story. At the age of ten, he had become so intrigued by the problem that he decided to become a mathematician and solve it. He carried out the first part of the plan, and got as far as specialising in number theory, the general area to which Fermat's last theorem belongs. But the more he learned about real mathematics, the more impossible the whole enterprise seemed. Fermat's last theorem was a baffling curiosity, an isolated question of the kind that any number theorist could dream up without a shred of convincing evidence. It didn't fit into any powerful body of technique. In a letter to Heinrich Olbers, the great Gauss had dismissed it out of hand, saying that the problem had 'little interest for me, since a multitude of such propositions, which one can neither prove nor refute, can easily be formulated'.[3] Wiles decided that his childhood dream had been unrealistic and put Fermat on the back burner. But then, miraculously, other mathematicians suddenly made a breakthrough that linked the problem to a core topic in number theory, one on which Wiles was already an expert. Gauss, uncharacteristically, had underestimated the problem's significance, and was unaware that it could be linked to a deep, though apparently unrelated, area of mathematics.

With this link established, Wiles could now work on Fermat's enigma *and* do credible research in modern number theory at the same time. Better still, if Fermat didn't work out, anything significant that he discovered while trying to prove it would be publishable in its own right. So off the back burner it came, and Wiles began to think about Fermat's problem in earnest. After seven years of obsessive research, carried on in private and in secret – an unusual precaution in mathematics – he became convinced that he had found a solution. He delivered a series of lectures at a prestigious number theory conference, under an obscure title that fooled no one.[4] The exciting news broke, in the media as well as the halls of academe: Fermat's last theorem had been proved.

The proof was impressive and elegant, full of good ideas. Unfortunately, experts quickly discovered a serious gap in its logic. In attempts to demolish great unsolved problems of mathematics, this kind of development is depressingly common, and it almost always proves fatal. However, for once the Fates were kind. With assistance from his former student Richard Taylor, Wiles managed to bridge the gap, repair the proof, and complete his solution. The emotional burden involved became vividly clear in the television programme: it must have been the only occasion when a mathematician has burst into tears on screen, just recalling the traumatic events and the eventual triumph.

You may have noticed that I haven't told you what Fermat's last theorem *is*. That's deliberate; it will be dealt with in its proper place. As far as the success of the television programme goes, it doesn't actually matter. In fact, mathematicians have never greatly cared whether the theorem that Fermat scribbled in his margin is true or false, because nothing of great import hangs on the answer. So why all the fuss? Because a huge amount hangs on the inability of the mathematical community to *find* the answer. It's not just a blow to our self-esteem: it means that existing mathematical theories are missing something vital. In addition, the theorem is very easy to state; this adds to its air of mystery. How can something that seems so simple turn out to be so hard?

Although mathematicians didn't really care about the answer, they cared deeply that they didn't know what it was. And they cared even more about finding a method that could solve it, because that must surely shed light not just on Fermat's question, but on a host of others. This is often the case with great mathematical problems: it is the methods used to solve them, rather than the results themselves, that matter most. Of course, sometimes the actual result matters too: it depends on what its consequences are.

Wiles's solution is much too complicated and technical for television; in fact, the details are accessible only to specialists.[5] The proof does involve a nice mathematical story, as we'll see in due course, but any attempt to explain that on television would have lost most of the audience immediately. Instead, the programme sensibly concentrated on a more personal question: what is it like to tackle a notoriously difficult mathematical problem that carries a lot of historical baggage? Viewers were shown that there existed a small but dedicated band of mathematicians, scattered across the globe, who cared deeply about their research area, talked to each other, took note of each other's work, and devoted a large part of their lives to advancing mathematical knowledge. Their emotional investment and social interaction came over vividly. These were not clever automata, but real people, engaged with their subject. That was the message.

Those are three big reasons why the programme was such a success: a major problem, a hero with a wonderful human story, and a supporting cast of emotionally involved people. But I suspect there was a fourth, not quite so worthy. The majority of non-mathematicians seldom hear about new developments in the subject, for a variety of perfectly sensible reasons: they're not terribly interested anyway; newspapers hardly ever mention anything mathematical; when they do, it's often facetious or trivial; and nothing much in daily life seems to be affected by whatever it is that mathematicians are doing behind the scenes. All too often, school mathematics is presented as a closed book in which every question has an answer. Students can easily come to imagine that new mathematics is as rare as hen's teeth.

From this point of view, the big news was not that Fermat's last theorem had been proved. It was that at last *someone had done some new mathematics*. Since it had taken mathematicians more than 300 years to find a solution, many viewers subconsciously concluded that the breakthrough was the first important new mathematics discovered in the last 300 years. I'm not suggesting that they *explicitly* believed that. It ceases to be a sustainable position as soon as you ask some obvious questions, such as 'Why does the Government spend good money on university mathematics departments?' But subconsciously it was a common default assumption, unquestioned and unexamined. It made the magnitude of Wiles's achievement seem even greater.

One of the aims of this book is to show that mathematical research is thriving, with new discoveries being made all the time. You don't hear much about this activity because most of it is too technical for non-specialists, because most of the media are wary of anything intellectually more challenging than *The X Factor*, and because the applications of mathematics are deliberately hidden away to avoid causing alarm. 'What? My iPhone depends on advanced mathematics? How will I log in to Facebook when I failed my maths exams?'

Historically, new mathematics often arises from discoveries in other areas. When Isaac Newton worked out his laws of motion and his law of gravity, which together describe the motion of the planets, he did not polish off the problem of understanding the solar system. On the contrary, mathematicians had to grapple with a whole new range of questions: yes, we know the laws, but what do they imply? Newton invented calculus to answer that question, but his new method also has limitations. Often it rephrases the question instead of providing the answer. It turns the problem into a special kind of formula, called a differential equation, whose *solution* is the answer. But you still have to solve the equation. Nevertheless, calculus was a brilliant start. It showed us that answers were possible, and it provided one effective way to seek them, which continues to provide major insights more than 300 years later.

As humanity's collective mathematical knowledge grew, a second source of inspiration started to play an increasing role in the creation of even more: the internal demands of mathematics itself. If, for example, you know how to solve algebraic equations of the first, second, third, and fourth degree, then you don't need much imagination to ask about the fifth degree. (The degree is basically a measure of complexity, but you don't even need to know what it is to ask the obvious question.) If a solution proves elusive, as it did, that fact *alone* makes mathematicians even more determined to find an answer, whether or not the result has useful applications.

I'm not suggesting applications don't matter. But if a particular piece of mathematics keeps appearing in questions about the physics of waves – ocean waves, vibrations, sound, light – then it surely makes sense to investigate the gadget concerned in its own right. You don't need to know ahead of time exactly how any new idea will be used: the topic of waves is common to so many important areas that significant new insights are bound to be useful for something. In this case, those somethings included radio, television, and radar.[6] If somebody thinks up a new way to understand heat flow, and comes up with a brilliant new technique that unfortunately lacks proper mathematical support, then it makes sense to sort the whole thing out *as a piece of mathematics*. Even if you don't give a fig about how heat flows, the results might well be applicable elsewhere. Fourier analysis, which emerged from this particular line of investigation, is arguably the most useful single mathematical idea ever found. It underpins modern telecommunications, makes digital cameras possible, helps to clean up old movies and recordings, and a modern extension is used by the FBI to store fingerprint records.[7]

After a few thousand years of this kind of interchange between the external uses of mathematics and its internal structure, these two aspects of the subject have become so densely interwoven that picking them apart is almost impossible. The mental attitudes involved are more readily distinguishable, though, leading to a broad classification of mathematics into two kinds: pure and applied. This is defensible as a rough-and-ready way to locate mathematical ideas in the intellectual landscape, but it's not a terribly accurate description of the subject itself. At best it distinguishes two ends of a continuous spectrum of mathematical styles. At worst, it misrepresents which parts of the subject are useful and where the ideas come from. As with all branches of science, what gives mathematics its power is the *combination* of abstract reasoning and inspiration from the outside world, each feeding off the other. Not only is it impossible to pick the two strands apart: it's pointless.

Most of the really important mathematical problems, the great problems that this book is about, have arisen within the subject through a kind of intellectual navel-gazing. The reason is simple: they are *mathematical* problems. Mathematics often looks like a collection of isolated areas, each with its own special techniques: algebra, geometry, trigonometry, analysis, combinatorics, probability. It tends to be taught that way, with good reason: locating each separate topic in a single well-defined area helps students to organise the material in their minds. It's a reasonable first approximation to the structure of mathematics, especially long-established mathematics. At the research frontiers, however, this tidy delineation often breaks down. It's not just that the boundaries between the major areas of mathematics are blurred. It's that they don't really exist.

Every research mathematician is aware that, at any moment, suddenly and unpredictably, the problem they are working on may turn out to require ideas from some apparently unrelated area. Indeed, new research often combines areas. For instance, my own research mostly centres on pattern formation in dynamical systems, systems that change over time according to specific rules. A typical example is the way animals move. A trotting horse repeats the same sequence of leg movements over and over again, and there is a clear pattern: the legs hit the ground together in diagonally related pairs. That is, first the front left and back right legs hit, then the other two. Is this a problem about patterns, in which case the appropriate methods come from group theory, the algebra of symmetry? Or is it a problem about dynamics, in which case the appropriate area is Newtonian-style differential equations?

The answer is that, by definition, it has to be both. It is not their intersection, which would be the material they have in common – basically, nothing. Instead, it is a new 'area', which straddles two of the traditional divisions of mathematics. It is like a bridge across a river that separates two countries; it links the two, but belongs to neither. But this bridge is not a thin strip of roadway; it is comparable in size to each of the countries. Even more vitally, the methods involved are not limited to those two areas. In fact, virtually every course in mathematics that I have ever studied has played a role somewhere in my research. My Galois theory course as an undergraduate at Cambridge was about how to solve (more precisely, why we can't solve) an algebraic equation of the fifth degree. My graph theory course was about networks, dots joined by lines. I never took a course in dynamical systems, because my PhD was in algebra, but over the years I picked up the basics, from steady states to chaos. Galois theory, graph theory, dynamical systems: three separate areas. Or so I assumed until 2011, when I wanted to understand how to detect chaotic dynamics in a network of dynamical systems, and a crucial step depended on things I'd learned 45 years earlier in my Galois theory course.

Mathematics, then, is not like a political map of the world, with each speciality neatly surrounded by a clear boundary, each country tidily distinguished from its neighbours by being coloured pink, green, or pale blue. It is more like a natural landscape, where you can never really say where the valley ends and the foothills begin, where the forest merges into woodland, scrub, and grassy plains, where lakes insert regions of water into every other kind of terrain, where rivers link the snow-clad slopes of the mountains to the distant, low-lying oceans. But this ever-changing mathematical landscape consists not of rocks, water, and plants, but of ideas; it is tied together not by geography, but by logic. And it is a dynamic landscape, which changes as new ideas and methods are discovered or invented. Important concepts with extensive implications are like mountain peaks, techniques with lots of uses are like broad rivers that carry travellers across the fertile plains. The more clearly defined the landscape becomes, the easier it is to spot unscaled peaks, or unexplored terrain that creates unwanted obstacles. Over time, some of the peaks and obstacles acquire iconic status. These are the great problems.

What makes a great mathematical problem great? Intellectual depth, combined with simplicity and elegance. Plus: it has to be *hard*. Anyone can climb a hillock; Everest is another matter entirely. A great problem is usually simple to state, although the terms required may be elementary or highly technical. The statements of Fermat's last theorem and the four colour problem make immediate sense to anyone familiar with school mathematics. In contrast, it is impossible even to state the Hodge conjecture or the mass gap hypothesis without invoking deep concepts at the research frontiers – the latter, after all, comes from quantum field theory. However, to those versed in such areas, the statement of the question concerned is simple and natural. It does not involve pages and pages of dense, impenetrable text. In between are problems that require something at the level of undergraduate mathematics, if you want to understand them in complete detail. A more general feeling for the essentials of the problem – where it came from, why it's important, what you could do if you possessed a solution – is usually accessible to any interested person, and that's what I will be attempting to provide. I admit that the Hodge conjecture is a hard nut to crack in that respect, because it is very technical and very abstract. However, it is one of the seven Clay Institute millennium mathematics problems, with a million-dollar prize attached, and it absolutely must be included.

Great problems are creative: they help to bring new mathematics into being. In 1900 David Hilbert delivered a lecture at the

International Congress of Mathematicians in Paris, in which he listed 23 of the most important problems in mathematics. He didn't include Fermat's last theorem, but he mentioned it in his introduction. When a distinguished mathematician lists what he thinks are some of the great problems, other mathematicians pay attention. The problems wouldn't be on the list unless they were important, and hard. It is natural to rise to the challenge, and try to answer them. Ever since, solving one of Hilbert's problems has been a good way to win your mathematical spurs. Many of these problems are too technical to include here, many are open-ended programmes rather than specific problems, and several appear later in their own right. But they deserve to be mentioned, so I've put a brief summary in the notes.[8]

That's what makes a great mathematical problem great. What makes it problematic is seldom deciding what the answer should be. For virtually all great problems, mathematicians have a very clear idea of what the answer ought to be – or had one, if a solution is now known. Indeed, the statement of the problem often includes the expected answer. Anything described as a conjecture is like that: a plausible guess, based on a variety of evidence. Most well-studied conjectures eventually turn out to be correct, though not all. Older terms like hypothesis carry the same meaning, and in the Fermat case the word 'theorem' is (more precisely, was) abused – a theorem requires a proof, but that was precisely what was missing until Wiles came along.

Proof, in fact, is the requirement that makes great problems problematic. Anyone moderately competent can carry out a few calculations, spot an apparent pattern, and distil its essence into a pithy statement. Mathematicians demand more evidence than that: they insist on a complete, logically impeccable proof. Or, if the answer turns out to be negative, a disproof. It isn't really possible to appreciate the seductive allure of a great problem without appreciating the vital role of proof in the mathematical enterprise. Anyone can make an educated guess. What's hard is to prove it's right. Or wrong.

The concept of mathematical proof has changed over the course of history, with the logical requirements generally becoming more stringent. There have been many highbrow philosophical discussions of the nature of proof, and these have raised some important issues. Precise logical definitions of 'proof' have been proposed and implemented. The one we teach to undergraduates is that a proof begins with a collection of explicit assumptions called axioms. The axioms are, so to speak, the rules of the game. Other axioms are possible, but they lead to different games. It was Euclid, the ancient Greek geometer, who introduced this approach to mathematics, and it is still valid today. Having agreed on the axioms, a proof of some statement is a series of steps, each of which is a logical consequence of either the axioms, or previously proved statements, or both. In effect, the mathematician is exploring a logical maze, whose junctions are statements and whose passages are valid deductions. A proof is a path through the maze, starting from the axioms. What it proves is the statement at which it terminates.

However, this tidy concept of proof is not the whole story. It's not even the most important part of the story. It's like saying that a symphony is a sequence of musical notes, subject to the rules of harmony. It misses out all of the creativity. It doesn't tell us how to find proofs, or even how to validate other people's proofs. It doesn't tell us which locations in the maze are significant. It doesn't tell us which paths are elegant and which are ugly, which are important and which are irrelevant. It is a formal, mechanical description of a process that has many other aspects, notably a human dimension. Proofs are discovered by people, and research in mathematics is not just a matter of step-by-step logic.

Taking the formal definition of proof literally can lead to proofs that are virtually unreadable, because most of the time is spent dotting logical i's and crossing logical t's in circumstances where the outcome already stares you in the face. So practising mathematicians cut to the chase, and leave out anything that is routine or obvious. They make it clear that there's a gap by using stock phrases like 'it is easy to verify that' or 'routine calculations imply'. What they don't do, at least not consciously, is to slither past a logical difficulty and to try to pretend it's not there. In fact, a competent mathematician will go out of his or her way to point out exactly those parts of the argument that are logically fragile, and they will devote most of their time to explaining how to make them sufficiently robust. The upshot is that a proof, in practice, is a mathematical story with its own narrative flow. It has a beginning, a middle, and an end. It often has subplots, growing out of the main plot, each with its own resolution. The British mathematician

Christopher Zeeman once remarked that a theorem is an intellectual resting point. You can stop, get your breath back, and feel you've got somewhere definite. The subplot ties off a loose end in the main story. Proofs resemble narratives in other ways: they often have one or more central characters – ideas rather than people, of course – whose complex interactions lead to the final revelation.

As the undergraduate definition indicates, a proof starts with some clearly stated assumptions, derives logical consequences in a coherent and structured way, and ends with whatever it is you want to prove. But a proof is not just a list of deductions, and logic is not the sole criterion. A proof is a story told to and dissected by people who have spent much of their life learning how to read such stories and find mistakes or inconsistencies: people whose main aim is to prove the storyteller *wrong*, and who possess the uncanny knack of spotting weaknesses and hammering away at them until they collapse in a cloud of dust. If any mathematician claims to have solved a significant problem, be it a great one or something worthy but less exalted, the professional reflex is not to shout 'hurray!' and sink a bottle of champagne, but to try to shoot it down.

That may sound negative, but proof is the only reliable tool that mathematicians have for making sure that what they say is correct. Anticipating this kind of response, researchers spend a lot of their effort trying to shoot their own ideas and proofs down. It's less embarrassing that way. When the story has survived this kind of critical appraisal, the consensus soon switches to agreement that it is correct, and at that point the inventor of the proof receives appropriate praise, credit, and reward. At any rate, that's how it usually works out, though it may not always seem that way to the people involved. If you're close to the action, your picture of what's going on may be different from that of a more detached observer.

How do mathematicians solve problems? There have been few rigorous scientific studies of this question. Modern educational research, based on cognitive science, largely focuses on education up to high school level. Some studies address the teaching of undergraduate mathematics, but those are relatively few. There are significant differences between learning and teaching existing mathematics and creating new mathematics. Many of us can play a musical instrument, but far fewer can compose a concerto or even write a pop song.

When it comes to creativity at the highest levels, much of what we know – or think we know – comes from introspection. We ask mathematicians to explain their thought processes, and seek general principles. One of the first serious attempts to find out how mathematicians think was Jacques Hadamard's *The Psychology of Invention in the Mathematical Field*, first published in 1945.[9] Hadamard interviewed leading mathematicians and scientists of his day and asked them to describe how they thought when working on difficult problems. What emerged, very strongly, was the vital role of what for lack of a better term must be described as intuition. Some feature of the subconscious mind guided their thoughts. Their most creative insights did not arise through step by step logic, but by sudden, wild leaps.

One of the most detailed descriptions of this apparently illogical approach to logical questions was provided by the French mathematician Henri Poincaré, one of the leading figures of the late nineteenth and early twentieth centuries. Poincaré ranged across most of mathematics, founding several new areas and radically changing many others. He plays a prominent role in several later chapters. He also wrote popular science books, and this breadth of experience may have helped him to gain a deeper understanding of his own thought processes. At any rate, Poincaré was adamant that conscious logic was only part of the creative process. Yes, there were times when it was indispensable: deciding what the problem really was, systematically verifying the answer. But in between, Poincaré felt that his brain was often working on the problem without telling him, in ways that he simply could not fathom.

His outline of the creative process distinguished three key stages: preparation, incubation, and illumination. Preparation consists of conscious logical efforts to pin the problem down, make it precise, and attack it by conventional methods. This stage Poincaré considered essential: it gets the subconscious going and provides raw materials for it to work with. Incubation takes place when you stop thinking about the problem and go off and do something else. The subconscious now starts combining ideas with each other, often quite wild ideas, until light starts to dawn. With luck, this leads to illumination: your subconscious taps you on the shoulder and the proverbial light bulb goes off in your mind.

This kind of creativity is like walking a tightrope. On the one hand, you won't solve a difficult problem unless you make yourself familiar with the area to which it seems to belong – along with many other areas, which may or may not be related, just in case they are. On the other hand, if all you do is get trapped into standard ways of thinking, which others have already tried, fruitlessly, then you will be stuck in a mental rut and discover nothing new. So the trick is to know a lot, integrate it consciously, put your brain in gear for weeks ... and then set the question aside. The intuitive part of your mind then goes to work, rubs ideas against each other to see whether the sparks fly, and notifies you when it has found something. This can happen at any moment: Poincaré suddenly saw how to solve a problem that had been bugging him for months when he was stepping off a bus. Srinivasa Ramanujan, a self-taught Indian mathematician with a talent for remarkable formulas, often got his ideas in dreams. Archimedes famously worked out how to test metal to see if it were gold when he was having a bath.

Poincaré took pains to point out that without the initial period of preparation, progress is unlikely. The subconscious, he insisted, needs to be given plenty to think about, otherwise the fortuitous combinations of ideas that will eventually lead to a solution cannot form. Perspiration begets inspiration. He must also have known – because any creative mathematician does – that this simple three-stage process seldom occurs just once. Solving a problem often requires more than one breakthrough. The incubation stage for one idea may be interrupted by a subsidiary process of preparation, incubation, and illumination for something that is needed to make the first idea work. The solution to any problem worth its salt, be it great or not, typically involves many such sequences, nested inside each other like one of Benoît Mandelbrot's intricate fractals. You solve a problem by breaking it down into subproblems. You convince yourself that if you can solve these subproblems, then you can assemble the results to solve the whole thing. Then you work on the subproblems. Sometimes you solve one; sometimes you fail, and a rethink is in order. Sometimes a subproblem itself breaks up into more pieces. It can be quite a task just to keep track of the plan.

I described the workings of the subconscious as 'intuition'. This is one of those seductive words like 'instinct', which is widely used even though it is devoid of any real meaning. It's a name for something whose presence we recognise, but which we do not understand. Mathematical intuition is the mind's ability to sense form and structure, to detect patterns that we cannot consciously perceive. Intuition lacks the crystal clarity of conscious logic, but it makes up for that by drawing attention to things we would never have consciously considered. Neuroscientists are barely starting to understand how the brain carries out much simpler tasks. But however intuition works, it must be a consequence of the structure of the brain and how it interacts with the external world.

Often the key contribution of intuition is to make us aware of weak points in a problem, places where it may be vulnerable to attack. A mathematical proof is like a battle, or if you prefer a less warlike metaphor, a game of chess. Once a potential weak point has been identified, the mathematician's technical grasp of the machinery of mathematics can be brought to bear to exploit it. Like Archimedes, who wanted a firm place to stand so that he could move the Earth, the research mathematician needs some way to exert leverage on the problem. One key idea can open it up, making it vulnerable to standard methods. After that, it's just a matter of technique.

My favourite example of this kind of leverage is a puzzle that has no intrinsic mathematical significance, but drives home an important message. Suppose you have a chessboard, with 64 squares, and a supply of dominoes just the right size to cover two adjacent squares of the board. Then it's easy to cover the entire board with 32 dominoes. But now suppose that two diagonally opposite corners of the board have been removed, as in Figure 1. Can the remaining 62 squares be covered using 31 dominoes? If you experiment, nothing seems to work. On the other hand, it's hard to see any obvious reason for the task to be impossible. Until you realise that however the dominoes are arranged, each of them must cover one black square and one white square. This is your lever; all you have to do now is to wield it. It implies that any region covered by dominoes contains the same number of black squares as it does white squares. But diagonally opposite

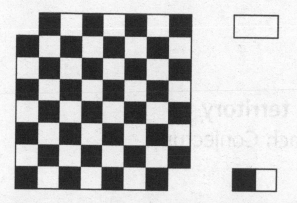

Fig 1 Can you cover the hacked chessboard with dominoes, each covering two squares (top right)? If you colour the domino (bottom right) and count how many black and white squares there are, the answer is clear.

squares have the same colour, so removing two of them (here white ones) leads to a shape with two more black squares than white. So no such shape can be covered. The observation about the combination of colours that *any* domino covers is the weak point in the puzzle. It gives you a place to plant your logical lever, and push. If you were a medieval baron assaulting a castle, this would be the weak point in the wall – the place where you should concentrate the firepower of your trebuchets, or dig a tunnel to undermine it.

Mathematical research differs from a battle in one important way. Any territory you once occupy remains yours for ever. You may decide to concentrate your efforts somewhere else, but once a theorem is proved, it doesn't disappear again. This is how mathematicians make progress on a problem, even when they fail to solve it. They establish a new fact, which is then available for anyone else to use, in any context whatsoever. Often the launchpad for a fresh assault on an age-old problem emerges from a previously unnoticed jewel half-buried in a shapeless heap of assorted facts. And that's one reason why new mathematics can be important for its own sake, even if its uses are not immediately apparent. It is one more piece of territory occupied, one more weapon in the armoury. Its time may yet come – but it certainly won't if it is deemed 'useless' and forgotten, or never allowed to come into existence because no one can see what it is *for*.

2

Prime territory
Goldbach Conjecture

SOME OF THE GREAT PROBLEMS show up very early in our mathematical education, although we may not notice. Soon after we are taught multiplication, we run into the concept of a prime number. Some numbers can be obtained by multiplying two smaller numbers together; for example, $6 = 2 \times 3$. Others, such as 5, cannot be broken up in this manner; the best we can do is $5 = 1 \times 5$, which doesn't involve two *smaller* numbers. Numbers that can be broken up are said to be composite; those that can't are prime. Prime numbers seem such simple things. As soon as you can multiply whole numbers together you can understand what a prime number is. Primes are the basic building blocks for whole numbers, and they turn up all over mathematics. They are also deeply mysterious, and they seem to be scattered almost at random. There's no doubting it: primes are an enigma. Perhaps this is a consequence of their definition – not so much what they are as what they are not. On the other hand, they are fundamental to mathematics, so we can't just throw up our hands in horror and give up. We need to come to terms with primes, and ferret out their innermost secrets.

A few features are obvious. With the exception of the smallest prime, 2, all primes are odd. With the exception of 3, the sum of their digits can't be a multiple of 3. With the exception of 5, they can't end in the digit 5. Aside from these rules, and a few subtler ones, you can't look at a number and immediately spot whether it is prime. There do exist formulas for primes, but to a great extent they are cheats: they don't provide useful new information about primes; they are just clever ways to encode the definition of 'prime' in a formula. Primes are like people: they are individuals, and they don't conform to standard rules.

Over the millennia, mathematicians have gradually increased their understanding of prime numbers, and every so often another big problem about them is solved. However, many questions still remain unanswered. Some are basic and easy to state; others are more esoteric. This chapter discusses what we do and don't know about these infuriating, yet fundamental, numbers. It begins by setting up some of the basic concepts, in particular, prime factorisation – how to express a given number by multiplying primes together. Even this familiar process leads into deep waters as soon as we start asking for genuinely effective methods for finding a number's prime factors. One surprise is that it seems to be relatively easy to test a number to determine whether it is prime, but if it's composite, finding its prime factors is often much harder.

Having sorted out the basics, we move on to the most famous unsolved problem about primes, the 250-year-old Goldbach conjecture. Recent progress on this question has been dramatic, but not yet decisive. A few other problems provide a brief sample of what is still to be discovered about this rich but unruly area of mathematics.

Prime numbers and factorisation are familiar from school arithmetic, but most of the interesting features of primes are seldom taught at that level, and virtually nothing is proved. There are sound reasons for that: the proofs, even of apparently obvious properties, are surprisingly hard. Instead, pupils are taught some simple methods for working with primes, and the emphasis is on calculations with relatively small numbers. As a result, our early experience of primes is a bit misleading.

The ancient Greeks knew some of the basic properties of primes, and they knew how to prove them. Primes and factors are the main topic of Book VII of Euclid's *Elements*, the great geometry classic. This particular book contains a geometric presentation of division and multiplication in arithmetic. The Greeks preferred to work with lengths of lines, rather than numbers as such, but it is easy to reformulate their results in the language of numbers. Euclid takes care to prove statements that may seem obvious: for example, Proposition

16 of Book VII proves that when two numbers are multiplied together, the result is independent of the order in which they are taken. That is, $ab = ba$, a basic law of algebra.

In school arithmetic, prime factors are used to find the greatest common divisor (or highest common factor) of two numbers. For instance, to find the greatest common divisor of 135 and 630, we factorise them into primes:

$$135 = 3^3 \times 5 \qquad 630 = 2 \times 3^2 \times 5 \times 7$$

Then, for each prime, we take the largest power that occurs in both factorisations, obtaining $3^2 \times 5$. Multiply out to get 45: this is the greatest common divisor. This procedure gives the impression that prime factorisation is needed to find greatest common divisors. Actually, the logical relationship goes the other way. Book VII Proposition 2 of the *Elements* presents a method for finding the greatest common divisor of two whole numbers without factorising them. It works by repeatedly subtracting the smaller number from the larger one, then applying a similar process to the resulting remainder and the smaller number, and continuing until there is no remainder. For 135 and 630, a typical example using smallish numbers, the process goes like this. Subtract 135 repeatedly from 630:

$$630 - 135 = 495$$
$$495 - 135 = 360$$
$$360 - 135 = 225$$
$$225 - 135 = 90$$

Since 90 is smaller than 135, switch to the two numbers 90 and 135:

$$135 - 90 = 45$$

Since 45 is smaller than 90, switch to 45 and 90:

$$90 - 45 = 45$$
$$45 - 45 = 0$$

Therefore the greatest common divisor of 135 and 630 is 45.

This procedure works because at each stage it replaces the original pair of numbers by a simpler pair (one of the numbers is smaller) that has the same greatest common divisor. Eventually one of the numbers divides the other exactly, and at that stage we stop. Today's term for an explicit computational method that is guaranteed to find an answer to a given problem is 'algorithm'. So Euclid's procedure is now called the Euclidean algorithm. It is logically prior to prime factorisation. Indeed, Euclid uses his algorithm to prove basic properties about prime factors, and so do university courses in mathematics today.

Euclid's Proposition 30 is vital to the whole enterprise. In modern terms, it states that if a prime divides the product of two numbers – what you get by multiplying them together – then it must divide one of them. Proposition 32 states that either a number is prime or it has a prime factor. Putting the two together, it is easy to deduce that every number is a product of prime factors, and that this expression is unique apart from the order in which the factors are written. For example,

$$60 = 2 \times 2 \times 3 \times 5 = 2 \times 3 \times 2 \times 5 = 5 \times 3 \times 2 \times 2$$

and so on, but the only way to get 60 is to rearrange the first factorisation. There is no factorisation, for example, looking like $60 = 7 \times$ *something*. The existence of the factorisation comes from Proposition 32. If the number is prime, stop. If not, find a prime factor, divide to get a smaller number, and repeat. Uniqueness comes from Proposition 30. For example, if there were a factorisation $60 = 7 \times$ *something*, then 7 must divide one of the numbers 2, 3, or 5, but it doesn't.

At this point I need to clear up a small but important point: the exceptional status of the number 1. According to the definition as stated so far, 1 is clearly prime: if we try to break it up, the best we can do is $1 = 1 \times 1$, which does not involve smaller numbers. However, this interpretation causes problems later in the theory, so for the last century or two, mathematicians have added an extra restriction. The number 1 is so special that it should be considered as neither prime nor composite. Instead, it is a third manner of beast, a unit. One reason for treating 1 as a special case, rather than a genuine prime, is that if we call 1 a prime then uniqueness fails. In fact, $1 \times 1 = 1$ already exhibits the failure, and $1 \times 1 \times 1 \times 1 \times 1 \times 1 \times 1 \times 1 = 1$ rubs our noses in it. We could modify uniqueness to say 'unique except for extra 1s', but that's just another way to admit that 1 is special.

Much later, in Proposition 20 of Book IX, Euclid proves another key fact: 'Prime numbers are more than any assigned multitude of prime numbers.' That is, the number of primes is infinite. It's a wonderful theorem with a clever proof, but it opened up a huge can of worms. If the primes go on for ever, yet seem to have no pattern, how can we describe what they look like?

We have to face up to that question because we can't ignore the primes. They are essential features of the mathematical landscape. They are especially common, and useful, in number theory. This area of mathematics studies properties of whole numbers. That may sound a bit elementary, but actually number theory is one of the deepest and most difficult areas of mathematics. We will see plenty of evidence for that statement later. In 1801 Gauss, the leading number theorist of his age – arguably one of the leading mathematicians of all time, perhaps even the greatest of them all – wrote an advanced textbook of number theory, the *Disquisitiones Arithmeticae* ('Investigations in arithmetic'). In among the high-level topics, he pointed out that we should not lose sight of two very basic issues: 'The problem of distinguishing prime numbers from composite numbers and of resolving the latter into their prime factors is known to be one of the most important and useful in arithmetic.'

At school, we are usually taught exactly one way to find the prime factors of a number: try all possible factors in turn until you find something that goes exactly. If you haven't found a factor by the time you reach the square root of the original number – more precisely, the largest whole number that is less than or equal to that square root – then the number is prime. Otherwise you find a factor, divide out by that, and repeat. It's more efficient to try just prime factors, which requires having a list of primes. You stop at the square root because the smallest factor of any composite number is no greater than its square root. However, this procedure is hopelessly inefficient when the numbers become large. For example, if the number is

$$1,080,813,321,843,836,712,253$$

then its prime factorisation is

$$13,929,010,429 \times 77,594,408,257$$

and you would have to try the first 624,401,249 primes in turn to find the smaller of the two factors. Of course, with a computer this is fairly easy, but if we start with a 100-digit number that happens to be the product of two 50-digit numbers, and employ a systematic search through successive primes, the universe will end before the computer finds the answer.

In fact, today's computers can generally factorise 100-digit numbers. My computer takes less than a second to find the prime factors of $10^{99} + 1$, which looks like 1000 ... 001 with 98 zeros. It is a product of 13 primes (one of them occurs twice), of which the smallest is 7 and the largest is

$$141,122,524,877,886,182,282,233,539,317,796,144,938,305,$$
$$111,168,717$$

But if I tell the computer to factorise $10^{199} + 1$, with 200 digits, it churns away for ages and gets nowhere. Even so, the 100-digit calculation is impressive. What's the secret? Find more efficient methods than trying all potential prime factors in turn.

We now know a lot more than Gauss did about the first of his problems (testing for primes) and a lot less than we'd like to about the second (factorisation). The conventional wisdom is that primality testing is far simpler than factorisation. This generally comes as a surprise to non-mathematicians, who were taught at school to test whether a number is prime by the same method used for factorisation: try all possible divisors. It turns out that there are slick ways to prove that a number is prime without doing that. They also allow us to prove that a number is composite, without finding any of its factors. Just show that it fails a primality test.

The great grand-daddy of all modern primality tests is Fermat's theorem, not to be confused with the celebrated Fermat's last theorem, chapter 7. This theorem is based on modular arithmetic, sometimes known as 'clock arithmetic' because the numbers wrap round like those on a clock face. Pick a number – for a 12-hour analogue clock it is 12 – and call it the modulus. In any arithmetical calculation with whole numbers, you now allow yourself to replace any multiple of 12 by zero. For example, $5 \times 5 = 25$, but 24 is twice 12, so subtracting 24 we obtain $5 \times 5 = 1$ to the modulus 12. Modular arithmetic is very pretty, because nearly all of the usual rules of arithmetic still work. The main difference is that you can't always divide one number by another, even when it's not zero. Modular arithmetic is also useful, because it provides a tidy way to deal with questions about divisibility: which numbers are divisible by the chosen modulus, and what is the remainder when they're not? Gauss introduced modular arithmetic in the *Disquisitiones Arithmeticae*, and today it is widely used in computer science, physics, and engineering, as well as mathematics.

Fermat's theorem states that if we choose a prime modulus p, and take any number a that is not a multiple of p, then the $(p-1)$ th power of a is equal to 1 in arithmetic to the modulus p. Suppose, for example, that $p = 17$ and $a = 3$. Then the theorem predicts that when we divide 3^{16} by 17, the remainder is 1. As a check,

$$3^{16} = 43,046,721 = 2,532,160 \times 17 + 1$$

No one in their right mind would want to do the sums that way for, say, 100-digit primes. Fortunately, there is a clever, quick way to carry out this kind of calculation. The point is that if the answer is not equal to 1 then the modulus we started with is composite. So Fermat's theorem forms the basis of an efficient test that provides a necessary condition for a number to be prime.

Unfortunately, the test is not sufficient. Many composite numbers, known as Carmichael numbers, pass the test. The smallest is 561, and in 2003 Red Alford, Andrew Granville, and Carl Pomerance proved, to general amazement, that there are infinitely many. The amazement was because they found a proof; the actual result was less of a surprise. In fact, they showed that there are at least $x^{2/7}$ Carmichael numbers less than or equal to x if x is large enough.

However, more sophisticated variants of Fermat's theorem can be turned into genuine tests for primality, such as one published in 1976 by Gary Miller. Unfortunately, the proof of the validity of Miller's test depends on an unsolved great problem, the generalised Riemann hypothesis, chapter 9. In 1980 Michael Rabin turned Miller's test into a probabilistic one, a test that might occasionally give the wrong answer. The exceptions, if they exist, are very rare, but they can't be ruled out altogether. The most efficient deterministic (that is, guaranteed correct) test to date is the Adleman-Pomerance-Rumely test, named for Leonard Adleman, Pomerance, and Robert Rumely. It uses ideas from number theory that are more sophisticated than Fermat's theorem, but in a similar spirit.

I still vividly recall a letter from one hopeful amateur, who proposed a variant of trial division. Try all possible divisors, but start at the square root and work *downwards*. This method sometimes gets the answer more quickly than doing things in the usual order, but as the numbers get bigger it runs into the same kind of trouble as the usual method. If you try it on my example above, the 22-digit number 1,080,813,321,843,836,712,253, then the square root is about 32,875,725,419. You have to try 794,582,971 prime divisors before you find one that works. This is *worse* than searching in the usual direction.

In 1956 The famous logician Kurt Gödel, writing to John von Neumann, echoed Gauss's plea. He asked whether trial division could be improved, and if so, by how much. Von Neumann didn't pursue the question, but over the years others answered Gödel by discovering practical methods for finding primes with up to 100 digits, sometimes more. These methods, of which the best known is called the quadratic sieve, have been known since about 1980. However, nearly all of them are either probabilistic, or they are inefficient in the following sense.

How does the running time of a computer algorithm grow as the input size increases? For primality testing, the input size is not the number concerned, but how many digits it has. The core distinction in such questions is between two classes of algorithms called P and not-P. If the running time grows like some fixed power of the input size, then the algorithm is class P; otherwise, it's not-P. Roughly speaking, class P algorithms are useful, whereas not-P algorithms are impractical, but there's a stretch of no-man's-land in between where other considerations come into play. Here P stands for 'polynomial time', a fancy way to talk about powers, and we return to the topic of efficient algorithms in chapter 11.

By the class P standard, trial division performs very badly. It's all right in the classroom, where the numbers that occur have two or three digits, but it's completely hopeless for 100-digit numbers. Trial division is firmly in the not-P class. In fact, the running time is roughly $10^{n/2}$ for an n-digit number, which grows faster than any fixed power of n. This type of growth, called exponential, is *really* bad, computational cloud-cuckoo-land.

Until the 1980s all known algorithms for primality testing, excluding probabilistic ones or those whose validity was unproved, had exponential growth rate. However, in 1983 an algorithm was found that lies tantalisingly in the no-man's-land adjacent to P territory: the aforementioned Adleman-Pomerance-Rumely test. An improved version by Henri Cohen and Hendrik Lenstra has running time n raised to the power log log n, where log denotes the logarithm. Technically, log log n can be as large as we wish, so this algorithm is not in class P. But that doesn't prevent it being practical: if n is a googolplex, 1 followed by 10^{100} zeros, then log log n is about 230. An old joke goes: 'It has been proved that log log n tends to infinity, but it has never been observed doing it.'

The first primality test in class P was discovered in 2002 by Manindra Agrawal and his students Neeraj Kayal and Nitin Saxena, who were undergraduates at the time. I've put some details in the Notes.[10] They proved that their algorithm has running time proportional to at most n^{12}; this was quickly improved to $n^{7.5}$. However, even though their algorithm is class P, hence classed as 'efficient', its advantages don't show up until the number n becomes very large indeed. It should beat the Adleman-Pomerance-Rumely test when the number of *digits* in n is about 10^{1000}. There isn't room to fit a number that big into a computer's memory, or, indeed, into the known universe. However, now that we *know* that a class P algorithm for primality testing exists, it becomes worthwhile to look for better ones. Lenstra and Pomerance reduced the power from 7.5 to 6. If various other conjectures about primes are true, then the power can be reduced to 3, which starts to look practical.

The most exciting aspect of the Agrawal-Kayal-Saxena algorithm, however, is not the result, but the method. It is simple – to mathematicians, anyway – and novel. The underlying idea is a variant of Fermat's theorem, but instead of working with numbers, Agrawal's team used a polynomial. This is a combination of powers of a variable x, such as $5x^3 + 4x - 1$. You can add, subtract, and multiply polynomials, and the usual algebraic laws remain valid. Chapter 3 explains polynomials in more detail.

This is a truly lovely idea: expand the domain of discourse and transport the problem into a new realm of thought. It is one of those ideas that are so simple you have to be a genius to spot them. It developed from a 1999 paper by Agrawal and his PhD supervisor Somenath Biswas, giving a probabilistic primality test based on an analogue of Fermat's theorem in the world of polynomials. Agrawal was convinced that the probabilistic element could be removed. In 2001 his students came up with a crucial, rather technical, observation. Pursuing that led the team into deep number-theoretic waters, but eventually everything was reduced to a single obstacle, the existence of a prime p such that $p - 1$ has a sufficiently large prime divisor. A bit of asking around and searching the Internet led to a theorem proved by Etienne Fouvry in 1985 using deep and technical methods. This was exactly what they needed to prove that their algorithm worked, and the final piece of the jigsaw slotted neatly into place.

In the days when number theory was safely tucked away inside its own little ivory tower, none of this would have mattered to the rest of the world. But over the last 20 years, prime numbers have become important in cryptography, the science of secret codes. Codes aren't just important for military use; commercial companies have secrets too. In this Internet age, we all do: we don't want criminals to gain access to our bank accounts, credit card numbers, or, with the growth of identity theft, the name of our cat. But the Internet is such a convenient way to pay bills, insure cars, and book holidays, that we have to accept some risk that our sensitive, private information might fall into the wrong hands.

Computer manufacturers and Internet service providers try to reduce that risk by making various encryption systems available. The involvement of computers has changed both cryptography and cryptanalysis, the dark art of code-breaking. Many novel codes have been devised, and one of the most famous, invented by Ron Rivest, Adi Shamir, and Leonard Adleman in 1978, uses prime numbers. Big ones, about a hundred digits long. The Rivest-Shamir-Adleman system is employed in many computer operating systems, is built into the main protocols for secure Internet communication, and is widely used by governments, corporations, and universities. That doesn't mean that every new result about primes is significant for the security of your Internet bank account, but it adds a definite frisson of excitement to any discovery that relates primes to computation. The Agrawal-Kayal-Saxena test is a case in point. Mathematically, it is elegant and important, but it has no direct practical significance.

It does, however, cast the general issue of Rivest-Shamir-Adleman cryptography in a new and slightly disturbing light. There is still no class P algorithm to solve Gauss's second problem, factorisation. Most experts think nothing of the kind exists, but they're not quite as sure as they used to be. Since new discoveries like the Agrawal-Kayal-Saxena test can lurk unsuspected in the wings, based on such simple ideas as polynomial versions of Fermat's theorem, cryptosystems based on prime factorisation might not be quite as secure as we fondly imagine. Don't reveal your cat's name on the Internet just yet.

Even the basic mathematics of primes quickly leads to more advanced concepts. The mystery becomes even deeper when we ask subtler questions. Euclid proved that the primes go on for ever, so we can't just list them all and stop. Neither can we give a simple, useful algebraic formula for successive primes, in the way that x^2 specifies squares. (There do exist simple formulas, but they 'cheat' by building the primes into the formula in disguise, and don't tell us anything new.[11]) To grasp the nature of these elusive, erratic numbers, we can carry out experiments, look for hints of structure, and try to prove that these apparent patterns persist no matter how large the primes become. For instance, we can ask how the primes are distributed among all whole numbers. Tables of primes strongly suggest that they tend to thin out as they get bigger. Table 1 shows how many primes there are in various ranges of 1000 consecutive numbers.

The numbers in the second column mostly decrease as we move down the rows, though sometimes there are brief periods when they go the other way: 114 is followed by 117, for instance. This is a symptom of the irregularity of the primes, but despite that, there is a clear general tendency for primes to become rarer as their size increases. The

range	number of primes
1–1000	168
1001–2000	135
2001–3000	127
3001–4000	119
4001–5000	118
5001–6000	114
6001–7000	117
7001–8000	106
8001–9000	110
9001–10,000	111

Table 1 The number of primes in successive intervals of 1000 numbers.

reason is not far to seek: the bigger a number becomes, the more potential factors there are. Primes have to avoid all of these factors. It's like fishing for non-primes with a net: the finer the net becomes, the fewer primes slip through.

The 'net' even has a name: the sieve of Eratosthenes. Eratosthenes of Cyrene was an ancient Greek mathematician who lived around 250 BC. He was also an athlete with interests in poetry, geography, astronomy, and music. He made the first reasonable estimate of the size of the Earth by observing the position of the Sun at noon in two different locations, Alexandria and Syene – present-day Aswan. At noon, the Sun was directly overhead at Syene, but about 7 degrees from the vertical at Alexandria. Since this angle is one fiftieth of a circle, the Earth's circumference must be 50 times the distance from Alexandria to Syene. Eratosthenes couldn't measure that distance directly, so he asked traders how long it took to make the journey by camel, and estimated how far a camel typically went in a day. He gave an explicit figure in a unit known as a *stadium*, but we don't know how long that unit was. Historians generally think that Eratosthenes's estimate was reasonably accurate.

His sieve is an algorithm to find all primes by successively eliminating all multiples of numbers already known to be prime. Figure 2 illustrates the method on the numbers up to 102, arranged to

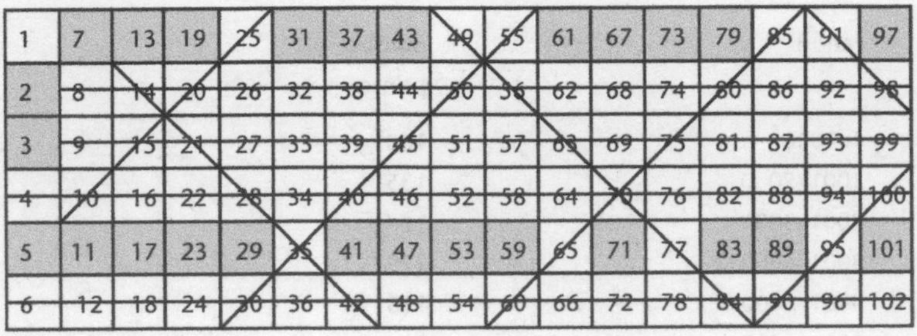

Fig 2 The sieve of Eratosthenes.

make the elimination process easy to follow. To see what's going on, I suggest you construct the diagram for yourself. Start with just the grid, omitting the lines that cross numbers out. Then you can add those lines one by one. Omit 1 because it's a unit. The next number is 2, so that's prime. Cross out all multiples of 2: these lie on the horizontal lines starting from 4, 6, and 8. The next number not crossed out is 3, so that's prime. Cross out all multiples of 3: these lie on the horizontal lines starting from 6, already crossed out, and 9. The next number not crossed out is 5, so that's prime. Cross out all multiples of 5: these lie on the diagonal lines sloping up and to the right, starting at 10. The next number not crossed out is 7, so that's prime. Cross out all multiples of 7: these lie on the diagonal lines sloping down and to the right, starting at 14. The next number not crossed out is 11, so that's prime. The first multiple of 11 that has not already been crossed out because it has a smaller divisor is 121, which is outside the picture, so stop. The remaining numbers, shaded, are the primes.

The sieve of Eratosthenes is not just a historical curiosity; it is still one of the most efficient methods known for making extensive lists of primes. And related methods have led to significant progress on what is probably the most famous unsolved great problem about primes: the Goldbach conjecture. The German amateur mathematician Christian Goldbach corresponded with many of the famous figures of his time. In 1742 he stated a number of curious conjectures about primes in a letter to Leonhard Euler. Historians later noticed that René Descartes had said much the same a few years before. The first of Goldbach's statements was: 'Every integer which can be written as the sum of two primes, can also be written as the sum of as many primes as one wishes, until all terms are units.' The second, added in the margin of his letter, was: 'Every integer greater than 2 can be written as the sum of three primes.' With today's definition of 'prime' there are obvious exceptions to these statements. For example, 4 is not the sum of three primes, because the smallest prime is 2, so the sum of three primes must be at least 6. But in Goldbach's day, the number 1 was considered to be prime. It is straightforward to rephrase his conjectures using the modern convention.

In his reply, Euler recalled a previous conversation with Goldbach, when Goldbach had pointed out that his first conjecture followed from a simpler one, his third conjecture: 'Every even integer is the sum of two primes.' With the prevailing convention that 1 is prime, this statement also implies the second conjecture, because any number can be written as either $n + 1$ or $n + 2$ where n is even. If n is the sum of two primes, the original number is the sum of three primes. Euler's opinion of the third conjecture was unequivocal: 'I regard this as a completely certain theorem, although I cannot prove it.' That pretty much sums up its status today.

The modern convention, in which 1 is not prime, splits Goldbach's conjectures into two different ones. The even Goldbach conjecture states:

Every even integer greater than 2 is the sum of two primes.

The odd Goldbach conjecture is:

Every odd integer greater than 5 is the sum of three primes.

The even conjecture implies the odd one, but not conversely.[12] It is useful to consider both conjectures separately because we still don't know whether either of them is true. The odd conjecture seems to be slightly easier than the even one, in the sense that more progress has been made.

Some quick calculations verify the even Goldbach conjecture for small numbers:

$$4 = 2 + 2$$
$$6 = 3 + 3$$
$$8 = 5 + 3$$
$$10 = 7 + 3 = 5 + 5$$
$$12 = 7 + 5$$
$$14 = 11 + 3 = 7 + 7$$
$$16 = 13 + 3 = 11 + 5$$
$$18 = 13 + 5 = 11 + 7$$
$$20 = 17 + 3 = 13 + 7$$

It is easy to continue by hand up to, say, 1000 or so – more if you're persistent. For example $1000 = 3 + 997$, and $1,000,000 = 17 + 999,993$. In 1938 Nils Pipping verified the even Goldbach conjecture for all even numbers up to 100,000.

It also became apparent that as the number concerned gets bigger, there tend to be more and more ways to write it as a sum of primes. This makes sense. If you take a big even number, and keep subtracting primes in turn, how likely is it that *all* of the results will be composite? It takes just one prime to turn up among the resulting list of differences and the conjecture is verified for that number. Using statistical features of primes, we can assess the probability of such an outcome. The analysts Godfrey Harold Hardy and John Littlewood performed such a calculation in 1923, and derived a plausible but non-rigorous formula for the number of different ways to express a given even number n as a sum of two primes: approximately $n/[2(\log n)^2]$. This number increases as n becomes larger, and it also agrees with numerical evidence. But even if this calculation could be made rigorous, there might just be an occasional rare exception, so it doesn't greatly help.

The main obstacle to a proof of Goldbach's conjecture is that it combines two very different properties. Primes are defined in terms of multiplication, but the conjectures are about addition. So it is extraordinarily difficult to relate the desired conclusion to any reasonable features of primes. There seems to be nowhere to insert a lever. This must have been music to the ears of the publisher Faber & Faber in 2000, when it offered a million-dollar prize for a proof of the conjecture to promote the novel *Uncle Petros and Goldbach's Conjecture* by Apostolos Doxiadis. The deadline was tight: a solution had to be submitted before April 2002. No one made a successful claim to the prize, which is hardly surprising given that the problem has remained unsolved for over 250 years.

The Goldbach conjecture is often reformulated as a question about adding sets of integers together. The even Goldbach conjecture is the simplest example for this particular way of thinking, because we add just *two* sets of integers together. To do this, take any number from the first set, add any number from the second set, and then take the set of all such sums. For instance, the sum of {1, 2, 3} and {4, 5} contains $1 + 4, 2 + 4, 3 + 4, 1 + 5, 2 + 5, 3 + 5$, which is {5, 6, 7, 8}. Some numbers occur more than once, for instance $6 = 2 + 4 = 1 + 5$. I'll call this kind of repetition 'overlap'.

The even Goldbach conjecture can now be restated: if we add the set of primes to itself, the result contains every even number greater than 2. This reformulation may sound a bit trite – and is – but it moves the problem into an area where there are some powerful general theorems. The number 2 is a bit of a nuisance, but we can easily get rid of it. It is the only even prime, and if we add it to any other prime the result is odd. So as far as the even Goldbach conjecture is concerned, we can forget about 2. However, we need $2 + 2$ to represent 4, so we must also restrict attention to even numbers that are at least 6.

As a simple experiment, consider the even numbers up to and including 30. There are nine odd primes in this range: {3, 5, 7, 11, 13, 17, 19, 23, 29}. Adding them gives Figure 3: I've marked the sums that are less than or equal to 30 (a range of even numbers that includes all primes up to 29) in bold. Two simple patterns appear. The whole table is symmetric about its main diagonal because $a + b = b + a$. The bold numbers occupy roughly the top left half of the table, above the thick (diagonal) line. If anything, they tend to bulge out beyond it in the middle. This happens because on the whole, large primes are rarer than small ones. The extra region of the bulge more than compensates for the two 32s at top right and bottom left.

Now we make some rough estimates. I could be more precise, but these are good enough. The number of slots in the table is $9 \times 9 = 81$.

| 3 | 5 | 7 | 11 | 13 | 17 | 19 | 23 | 29 |

3	6	8	10	14	16	20	22	26	32
5	8	10	12	16	18	22	24	28	34
7	10	12	14	18	20	24	26	30	36
11	14	16	18	22	24	28	30	34	40
13	16	18	20	24	26	30	32	36	42
17	20	22	24	28	30	34	36	40	46
19	22	24	26	30	32	36	38	42	48
23	26	28	30	34	36	40	42	46	52
29	32	34	36	40	42	46	48	52	58

Fig 3 Sums of pairs of primes up to 30. Boldface: sums that are 30 or smaller. Thick line: diagonal. Shaded region: eliminating symmetrically related pairs. The shaded region is slightly more than one quarter of the square.

About half of the numbers in those slots are in the top left triangle. Because of the symmetry, these arise in pairs except along the diagonal, so the number of unrelated slots is about 81/4, roughly 20. The number of even integers in the range from 6 to 30 is 13. So the 20 (and more) boldface sums have to hit only 13 even numbers. There are more potential sums of two primes in the right range than there are even numbers. It's like throwing 20 balls at 13 coconuts at the fair. You have a reasonable chance of hitting a lot of them. Even so, you could miss a few coconuts. Some even numbers might still be missing.

In this case they're not, but this kind of counting argument can't eliminate that possibility. However, it does tell us that there must be quite a bit of overlap, where the same boldface number occurs several times in the relevant quarter of the table. Why? Because 20 sums have to fit into a set with only 13 members. So on average each boldface number appears about 1.5 times. (The actual number of sums is 27, so a better estimate shows that each boldface number appears twice.) If any even numbers are missing, the overlap must be bigger still.

We can play the same game with a larger upper limit – say 1

million. A formula called the prime number theorem, chapter 9, provides a simple estimate for the number of primes up to any given size x. The formula is $x/\log x$. Here, the estimate is about 72,380. (The exact figure is 78,497.) The corresponding shaded region occupies about one quarter of the table, so it provides about $n^2/4 = 250$ billion boldface numbers: sums of two primes in this range. This is vastly larger than the number of even numbers in the range, which is half a million. Now the amount of overlap has to be gigantic, with each sum occurring on average 500,000 times. So the chance of any particular even number escaping is greatly reduced.

With more effort, we can turn this approach into an estimate of the probability that some even number in a given range is not the sum of two primes, assuming that the primes are distributed at random and with frequencies given by the prime number theorem – that is, about $x/\log x$ primes less than any given x. This is what Hardy and Littlewood did. They knew that their approach wasn't rigorous, because primes are defined by a specific process and they're not actually random. Nevertheless, it's sensible to expect the actual results to be consistent with this probabilistic model, because the defining property of primes seems to have very little connection with what happens when we add two of them together.

Several standard methods in this area adopt a similar point of view, but taking extra care to make the argument rigorous. Sieve methods, which build on the sieve of Eratosthenes, are examples. General theorems about the density of numbers in sums of two sets – the proportion of numbers that occur, as the sets become very large – provide other useful tools.

When a mathematical conjecture eventually turns out to be correct, its history often follows a standard pattern. Over a period of time, various people prove the conjecture to be true provided special restrictions apply. Each such result improves on the previous one by relaxing some restrictions, but eventually this process runs out of steam. Finally, a new and much cleverer idea completes the proof.

For example, a conjecture in number theory may state that every positive integer can be represented in some manner using, say, six special numbers (prime, square, cube, whatever). Here the key features are *every* positive integer and *six* special numbers. Initial advances lead to much weaker results, but successive stages in the process slowly improve them.

The first step is often a proof along these lines: every positive integer that is not divisible by 3 or 11, except for some finite number of them, can be represented in terms of some gigantic number of special numbers – say 10^{666}. The theorem typically does not specify how many exceptions there are, so the result cannot be applied directly to any specific integer. The next step is to make the bound effective: that is, to prove that every integer greater than $10^{10^{42}}$ can be so represented. Then the restriction on divisibility by 3 is eliminated, followed by a similar advance for 11. After that, successive authors reduce one of the numbers 10^{666} or $10^{10^{42}}$, often both. A typical improvement might be that every integer greater than 5.8×10^{17} can be represented using at most 4298 special numbers, for instance.

Meanwhile, other researchers are working upwards from small numbers, often with computer assistance, proving that, say, every number less than or equal to 10^{12} can be represented using at most six special numbers. Within a year, 10^{12} has been improved in five stages, by different researchers or groups, to 11.0337×10^{29}. These improvements are neither routine nor easy, but the way they are achieved involves intricate special methods that provide no hint of a more general approach, and each successive contribution is more complicated and longer. After a few years of this kind of incremental improvement, applying the same general ideas but with more powerful computers and new tweaks, this number has risen to 10^{43}. But now the method grinds to a halt, and everyone agrees that however much tweaking is done, it will never lead to the full conjecture.

At that point the conjecture disappears from view, because no one is working on it any more. Sometimes, progress pretty much stops. Sometimes, twenty years pass with nothing new ... and then, apparently from nowhere, Cheesberger and Fries announce that by reformulating the conjecture in terms of complex meta-ergodic quasiheaps and applying byzantine quisling theory, they have obtained a complete proof. After several years arguing about fine points of logic, and plugging a few gaps, the mathematical community accepts that the proof is correct, and immediately asks if there's a better way to achieve the same result, or to push it further.

You will see this pattern work itself out many times in later chapters. Because such accounts become tedious, no matter how proud Buggins and Krumm are of their latest improvement of the exponent in the Jekyll-Hyde conjecture from 1.773 to $1.771 + \varepsilon$ for any positive ε, I will describe a few representative contributions and leave out the rest. This is not to deny the importance of the work of Buggins and Krumm. It may even have paved the way to the great Cheesberger-Fries breakthrough. But only experts, following the developing story, are likely to await the next tiny improvement with bated breath.

In future I'll provide less detail, but let's see how it goes for Goldbach.

Theorems that go some way towards establishing Goldbach's conjecture have been proved. The first big breakthrough came in 1923, when Hardy and Littlewood used their analytic techniques to prove the odd Goldbach conjecture for all sufficiently large odd numbers. However, their proof relied on another big conjecture, the generalised Riemann hypothesis, which we discuss in chapter 9. This problem is still open, so their approach had a significant gap. In 1930 Lev Schnirelmann bridged the gap using a fancy version of their reasoning, based on sieve methods. He proved that a nonzero proportion of all numbers can be represented as a sum of two primes. By combining this result with some generalities about adding sequences together, he proved that there is some number C such that every integer greater than 1 is a sum of at most C prime numbers. This number became known as Schnirelmann's constant. Ivan Matveyevich Vinogradov obtained similar results in 1937, but his method also did not specify how big 'significantly large' is. In 1939 K. Borozdin proved that it is no greater than $3^{14,348,907}$. By 2002 Liu Ming-Chit and Wang Tian-Ze had reduced this 'upper bound' to e^{3100}, which is about 2×10^{1346}. This is a lot smaller, but it is still too big for the intermediate numbers to be checked by computer.

In 1969 N.I. Klimov obtained the first specific estimate for Schnirelmann's constant: it is at most 6 billion. Other mathematicians reduced that number considerably, and by 1982 Hans Riesel and Robert Vaughan had brought it down to 19. Although 19 is a lot better than 6 billion, the evidence pointed to

Schnirelmann's constant being a mere 3. In 1995 Leszek Kaniecki reduced the upper bound to 6, with five primes for any odd number, but he had to assume the truth of the Riemann hypothesis. His results, combined with J. Richstein's numerical verification of the Riemann hypothesis up to 4×10^{14}, would prove that Schnirelmann's constant is at most 4, again assuming the Riemann hypothesis. In 1997 Jean-Marc Deshouillers, Gove Effinger, Herman te Riele, and Dmitrii Zinoviev showed that the generalised Riemann hypothesis (chapter 9) implies the odd Goldbach conjecture. That is, every odd number except 1, 3, and 5 is the sum of three primes.

Since the Riemann hypothesis is currently not proved, it is worth trying to remove this assumption. In 1995 the French mathematician Olivier Ramaré reduced the upper estimate for representing odd numbers to 7, without using the Riemann hypothesis. In fact, he proved something stronger: every even number is a sum of at most six primes. (To deal with odd numbers, subtract 3: the result is even, so it is a sum of six or fewer primes. The original number is this sum plus the prime 3, requiring seven or fewer primes.) The main breakthrough was to improve existing estimates for the proportion of numbers, in some specified range, that are the sum of two primes. Ramaré's key result is that for any number n greater than e^{67} (about 1.25×10^{29}), at least one fifth of the numbers between n and $2n$ are the sum of two primes. Using sieve methods, in conjunction with a theorem of Hans-Heinrich Ostmann about sums of sequences, refined by Deshouillers, this leads to a proof that every even number greater than 10^{30} is a sum of at most six primes.

The remaining obstacle is to deal with the gap between 4×10^{14}, where Jörg Richstein had checked the theorem by computer, and 10^{30}. As is common, the numbers are too big for a direct computer search, so Ramaré proved a series of specialised theorems about the number of primes in small intervals. These theorems depend on the truth of the Riemann hypothesis up to specific limits, which can be verified by computer. So the proof consists mainly of conceptual pencil-and-paper deductions, with computer assistance in this particular respect. Ramaré ended his paper by pointing out that in principle a similar approach could reduce the number of primes from 7 to 5. However, there were huge practical obstacles, and he wrote that such a proof 'can not be reached by today's computers'.

In 2012 Terence Tao overcame those difficulties with some new and very different ideas. He posted a paper on the Internet, which as I write is under review for publication. Its main theorem is: every odd number is a sum of at most five primes. This reduces Schnirelmann's constant to 6. Tao is renowned for his ability to solve difficult problems in many areas of mathematics. His proof throws several powerful techniques at the problem, and requires computer assistance. If the number 5 in Tao's theorem could be reduced to 3, the odd Goldbach conjecture would be proved, and the bound on Schnirelmann's constant reduced to 4. Tao suspects that it should be possible to do this, although further new ideas will be needed.

The even Goldbach conjecture seems harder still. In 1998 Deshouillers, Saouter, and te Riele verified it for all even numbers up to 10^{14}. By 2007, Tomás Oliveira e Silva had improved that to 10^{18}, and his computations continue. We know that every even integer is the sum of at most six primes – proved by Ramaré in 1995. In 1973 Chen Jing-Run proved that every sufficiently large even integer is the sum of a prime and a semiprime (either a prime or a product of two primes). Close, but no cigar. Tao has stated that the even Goldbach conjecture is beyond the reach of his methods. Adding three primes together creates far more overlap in the resulting numbers – in the sense discussed in connection with Figure 3 – than the two primes needed for the even Goldbach conjecture, and Tao's and Ramaré's methods exploit this feature repeatedly.

In a few years' time, then, we may have a complete proof of the odd Goldbach conjecture, in particular implying that every even number is the sum of at most four primes. But the even Goldbach conjecture will probably still be just as baffling as it was for Euler and Goldbach.

In the 2300 years since Euclid proved several basic theorems about primes, we have learned a great deal more about these elusive, yet vitally important, numbers. But what we now know puts into stark perspective the long list of what we don't know.

We know, for instance, that there are infinitely many primes of the form $4k + 1$ and $4k + 3$; more generally, that any arithmetic sequence[13] $ak + b$ for fixed a and b contains infinitely many primes provided a and b have no common factor. For instance, suppose that $a = 18$. Then

$b = 1, 5, 7, 11, 13$, or 17. Therefore there exist infinitely many primes of each of the forms $18k + 1$, $18k + 5$, $18k + 7$, $18k + 11$, $18k + 13$, or $18k + 17$. This is not true for, say, $18k + 6$, because this is a multiple of 6. No arithmetic sequence can contain *only* primes, but a recent major breakthrough, the Green-Tao theorem, shows that the set of primes contains arbitrarily long arithmetic sequences. The proof, obtained in 2004 by Ben Green and Tao, is deep and difficult. It gives us hope: difficult open questions, however impenetrable they may appear, can sometimes be answered.

Putting on our algebraist's hat we immediately wonder about more complicated formulas involving k. There are no primes of the form k^2, and none except 3 for the form $k^2 - 1$, because these expressions factorise. However, the expression $k^2 + 1$ does not have obvious factors, and here we can find plenty of primes:

$$2 = 1^2 + 1 \quad 5 = 2^2 + 1 \quad 17 = 4^2 + 1 \quad 37 = 6^2 + 1$$

and so on. A larger example of no special significance is

$$18,672,907,718,657 = (4,321,216)^2 + 1$$

It is conjectured that infinitely many such primes exist, but no such statement has yet been proved for any specific polynomial in which k occurs to a higher power than the first. A very plausible conjecture is the one made by V. Bouniakowsky in 1857: any polynomial in k that does not have obvious divisors represents infinitely many primes. The exceptions here include not only reducible polynomials, but ones like $k^2 + k + 2$ which is always divisible by 2, despite having no algebraic factors.

Some polynomials seem to have special properties. The classic case is $k^2 + k + 41$, which is prime for $k = 0, 1, 2, ..., 40$, and indeed also for $k = -1, -2, ..., -40$. Long runs of primes for consecutive values of k are rare, and a certain amount is known about them. But the whole area is very mysterious.

Almost as famous as the Goldbach conjecture, and apparently just as hard, is the twin primes conjecture: there are infinitely many pairs of primes that differ by 2. Examples are

3, 5   5, 7   11, 13   17, 19

The largest known twin primes (as of January 2012) are

$$3,756,801,695,685 \times 2^{666,669} \pm 1$$

which have 200,700 decimal digits. They were found by the PrimeGrid distributed computing project in 2011. In 1915, Viggo Brun used a variant of the sieve of Eratosthenes to prove that the sum of reciprocals of all twin primes converges, unlike the sum of the reciprocals of all primes. So in this sense, twin primes are relatively rare. He also proved, using similar methods, that there exist infinitely many integers n such that n and $n+2$ have at most nine prime factors. Hardy and Littlewood used their heuristic methods to argue that the number of twin prime pairs less than x should be asymptotic to

$$2a \frac{n}{(\log n)^2}$$

where a is a constant whose value is about 0.660161. The underlying idea is that for this purpose primes can be assumed to arise at random, at a rate that makes the number of primes up to x approximately equal to $x/\log x$. There are many similar conjectures and heuristic formulas, but again, no rigorous proofs.

Indeed, there are hundreds of open questions about primes. Some are just curios, some are deep and significant. We will meet some of the latter in chapter 9. Despite all of the advances mathematicians have made over the last two and a half millennia, the humble primes have lost none of their allure and none of their mystery.

3

The puzzle of pi
Squaring the Circle

PRIMES ARE AN OLD IDEA, but circles are even older. Circles led to a great problem that took more than 2000 years to solve. It is one of several related geometric problems that have come down to us from antiquity. The central character in the story is the number π (Greek 'pi') which we meet at school in connection with circles and spheres. Numerically it is 3.14159 and a bit; often the approximation 22/7 is used. The digits of π never stop, and they never repeat the same sequence over and over again. The current record for calculating digits of π is 10 trillion digits, by Alexander Yee and Chigeru Kondo in October 2011.[14] Computations like this are significant as ways to test fast computers, or to inspire and test new methods to calculate π, but very little hinges on the numerical results. The reason for being interested in π is not to calculate the circumference of a circle. The same strange number appears all over mathematics, not just in formulas related to circles and spheres, and it leads into very deep waters indeed. The school formulas are important, even so, and they reflect π's origins in Greek geometry.

There, one of the great problems was the unsolved task of squaring the circle. This phrase is often employed colloquially to indicate a wrong-headed approach to something, rather like trying to fit a square peg into a round hole. Like many common phrases extracted from science, this one's meaning has changed over the centuries.[15] In Greek times, trying to square the circle was a perfectly reasonable idea. The difference in the two shapes – straight or curved – is totally irrelevant:

similar problems have valid solutions.[16] However, it eventually turned out that this particular problem cannot be solved using the specified methods. The proof is ingenious and technical, but its general nature is comprehensible.

In mathematics, squaring the circle means constructing a square whose *area* is the same as that of a given circle, using the traditional methods of Euclid. Greek geometry actually permitted other methods, so one aspect of the problem is to pin down which methods are to be used. The impossibility of solving the problem is then a statement about the limitations of those methods; it doesn't imply that we can't work out the area of a circle. We just have to find another approach. The impossibility proof explains why the Greek geometers and their successors failed to find a construction of the required kind: there isn't one. In retrospect, that explains why they had to introduce more esoteric methods. So the solution, despite being negative, clears up what would otherwise be a big historical puzzle. It also stops people wasting time in a continuing search for a construction that doesn't exist – except for a few hardy souls who regrettably seem unable to get the message, no matter how carefully it is explained.[17]

In Euclid's *Elements* the traditional methods for constructing geometric figures are idealised versions of two mathematical instruments: the ruler and the compass. To be pedantic, compass*es*, for the same reason that you cut paper with scissor*s*, not with a scissor – but I will follow common parlance and avoid the plural. These instruments are used to 'draw' diagrams on a notional sheet of paper, the Euclidean plane.

Their form determines what they can draw. A compass comprises two rigid rods, hinged together. One has a sharp point, the other holds a sharp pencil. The instrument is used to draw a circle, or part of one, with a specific centre and a specific radius. A ruler is simpler: it has a straight edge, and is used to draw a straight line. Unlike the rulers you buy in stationery shops, Euclid's rulers have no marks on them, and this is an important restriction for the mathematical analysis of what they can create.

The sense in which the geometer's ruler and compass are idealisations is straightforward: they are assumed to draw infinitely thin lines. Moreover, the straight lines are exactly straight and the circles are perfectly round. The paper is perfectly flat and even. The other key ingredient of Euclid's geometry is the notion of a point, another ideal. A point is a dot on the paper, but it is a physical impossibility: it has no size. 'A point', said Euclid, in the first sentence of the *Elements*, 'is that which has no part.' This sounds a bit like an atom, or if you're clued into modern physics, a subatomic particle, but compared to a geometric point, those are gigantic. From an everyday human perspective, however, Euclid's ideal point, an atom, and a pencil dot on a sheet of paper, are similar enough for the purposes of geometry.

These ideals are not attainable in the real world, however carefully you make the instruments and sharpen the pencil, and however smooth you make the paper. But idealism can be a virtue, because these requirements make the mathematics much simpler. For instance, two pencil lines cross in a small fuzzy region shaped like a parallelogram, but mathematical lines meet at a single point. Insights gained from ideal circles and lines can often be transferred to real, imperfect ones. This is how mathematics works its magic.

Two points determine a (straight) line, the unique line that passes through them. To construct the line, place your ideal ruler so that it passes through the two points, and run your ideal pencil along it. Two points also determine a circle: choose one as the centre, and place the compass point there; then adjust it so that the tip of the pencil lies on the other point. Now swing the pencil round in an arc, keeping the central point fixed. Two lines determine a unique point, where they cross, unless they are parallel, in which case they don't cross, but a Pandora's box of logical issues yawns wide. A line and a circle determine two points, if they cross; one point, if the line cuts the circle at a tangent; nothing at all if the circle is too small to meet the line. Similarly two circles either meet in two points, one, or none.

Distance is a fundamental concept in the modern treatment of Euclidean geometry. The distance between any two points is measured along the line that joins them. Euclid managed to get his geometry working without an explicit concept of distance, by finding a way to say that two line segments have the *same* length without defining length itself. In fact, this is easy: just stretch a compass between the ends of one segment, transfer it to the second, and see if the ends fit. If they do, the lengths are equal; if they don't, they're not. At no stage do you measure an actual length.

From these basic ingredients, geometers can build up more interesting shapes and configurations. Three points determine a triangle unless they all lie on the same line. When two lines cross, they form an angle. A right angle is especially significant; a straight line corresponds to two right angles joined together. And so on, and so on, and so on. Euclid's *Elements* consists of 13 books, delving ever deeper into the consequences of these simple beginnings.

The bulk of the *Elements* consists of theorems – valid features of geometry. But Euclid also explains how to solve geometric problems, using 'constructions' based on ruler and compass. Given two points joined by a segment of a line, construct their midpoint. Or trisect the segment: construct a point exactly one third of the way along it. Given an angle, construct one that bisects it – is half the size. But some simple constructions proved elusive. Given an angle, construct one that trisects it – is one third the size. You can do that for line segments, but no one could find a method for angles. Approximations, as close as you wish, yes. Exact constructions using only an unmarked ruler and a compass: no. However, no one really needs to trisect angles exactly anyway, so this particular issue didn't cause much trouble.

More embarrassing was a construction that could not be ignored: given a circle, construct a square that has the same area. This is the problem of squaring the circle. From the Greek point of view, if you couldn't solve that, you weren't entitled to claim that a circle *had* an area. Even though it visibly encloses a well-defined space, and intuitively the area is *how much* space. Euclid and his successors, notably Archimedes, settled for a pragmatic solution: assume circles have areas, but don't expect to be able to construct squares with the same area. You can still say a lot; for instance, you can prove, in full logical rigour, that the area of a circle is proportional to the square of its diameter. What you can't do, without squaring the circle, is to construct a line whose length is the constant of proportionality.

The Greeks couldn't square the circle using ruler and compass, so they settled for other methods. One used a curve called a quadratrix.[18] The importance they attached to using only ruler and compass was exaggerated by some later commentators, and it's not even clear that the Greeks considered squaring the circle to be a vital issue. By the nineteenth century, however, the problem was becoming a major nuisance. Mathematics that was unable to answer such a straightforward question was like a cordon bleu cook who didn't know how to boil an egg.

Squaring the circle sounds like a problem in geometry. That's because it is a problem in geometry. But its solution turned out to lie not in geometry at all, but in algebra. Making unexpected connections between apparently unrelated areas of mathematics often lies at the heart of solving a great problem. Here, the connection was not entirely unprecedented, but its link to squaring the circle was not at first appreciated. Even when it was, there was a technical difficulty, and dealing with that required yet another area of mathematics: analysis, the rigorous version of calculus. Ironically, the first breakthrough came from a fourth area: number theory. And it solved a geometric problem that the Greeks would never in their wildest dreams have believed to possess a solution, and as far as we can tell never thought about: how to construct, with ruler and compass, a regular polygon with 17 sides.

It sounds mad, especially if I add that no such construction exists for regular polygons with 7, 9, 11, 13, or 14 sides, but one does for 3, 4, 5, 6, 8, 10, and 12. However, there is method behind the madness, and it is the method that enriched mathematics.

First: what is a regular polygon? A polygon is a shape bounded by straight lines. It is regular if those lines have equal length and meet at equal angles. The most familiar example is the square: all four sides have the same length and all four angles are right angles. There are other shapes with four equal sides or four equal angles: the rhombus and rectangle respectively. Only a square has both features. A regular 3-sided polygon is an equilateral triangle, a regular 5-sided polygon is a regular pentagon, and so on, Figure 4. Euclid provides ruler-and-compass constructions for regular polygons with 3, 4, and 5 sides. The Greeks also knew how to repeatedly double the number of sides, giving 6, 8, 10, 12, 16, 20, and so on. By combining the constructions for 3- and 5-sided regular polygons they could obtain a 15-sided one. But there, their knowledge stopped. And for about 2000 years that's how it remained. No one imagined that any other numbers were feasible.

They didn't even ask, it just seemed obvious that nothing more could be done.

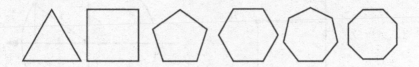

Fig 4 The first few regular polygons. *From left to right*: equilateral triangle, square, pentagon, hexagon, heptagon, octagon.

It took one of the greatest mathematicians who have ever lived to think the unthinkable, ask the unaskable, and discover a truly astonishing answer. Namely, Gauss Carl Friedrich. Gauss was born into a poor, working-class family in the city of Braunschweig (Brunswick) in Germany. His mother Dorothea could not read or write, and failed to write down the date of his birth, but she did remember that it was on a Wednesday, eight days before the feast of the ascension, in 1777. Gauss later worked out the exact date from a mathematical formula he devised for the date of Easter. His father Gebhard came from a farming family, but made a living in a series of low-level jobs: gardener, canal labourer, street butcher, funeral parlour accountant. Their son was a child prodigy who is reputed to have corrected his father's arithmetic at the age of three, and his abilities, which extended to languages as well as mathematics, led the Duke of Braunschweig to fund his university studies at the Collegium Carolinum. While an undergraduate Gauss independently rediscovered several important mathematical theorems that had been proved by illustrious people such as Euler. But his theorem about the regular 17-sided polygon came as a bolt from the blue.

By then, the close link between geometry and algebra had been understood for 140 years. In an appendix to *Discours de la Méthode* ('Discourse on the method') René Descartes formalised an idea that had been floating around in rudimentary form for some time: the notion of a coordinate system. In effect, this takes Euclid's barren plane, a blank sheet of paper, and turns it into paper ruled into squares, which engineers and scientists call graph paper. Draw two straight lines on the paper, one horizontal, the other vertical: these are

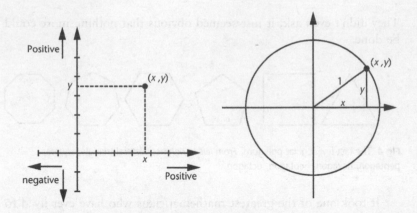

Fig 5 *Left*: Coordinates in the plane. *Right*: How to derive the equation for the unit circle.

called axes. Now you can pin down the location of any point of the plane by asking how far it lies in the direction along the horizontal axis, and how far up the vertical axis, Figure 5 (left). These two numbers, which may be positive or negative, provide a complete description of the point, and they are called its coordinates.

All geometric properties of points, lines, circles, and so on can be translated into algebraic statements about the corresponding coordinates. It's very difficult to talk meaningfully about these connections without using some actual algebra – just as it's hard to talk sensibly about football without mentioning the word 'goal'. So the next few pages will include some formulas. They are there to ensure that the main players in the drama have names and the relationship between them is clear. 'Romeo' is so much simpler to follow than 'the son of an Italian patriarch who falls in love with his father's sworn enemy's beautiful daughter'. Our Romeo will bear the prosaic name x, and his Juliet will be y.

As an example of how geometry converts into algebra, Figure 5 (right) shows how to find the equation for a circle of unit radius centred at the origin, where the two axes cross. The marked point has coordinates (x, y), so the right-angled triangle in the figure has horizontal side of length x and vertical side of length y. The longest side of the triangle is the radius of the circle, which is 1. Pythagoras's theorem now tells us that the sum of the squares of the two coordinates is 1. In symbols, a point with coordinates x and y lies on the circle if (and only if) it satisfies the condition $x^2 + y^2 = 1$. This symbolic characterisation of the circle is brief and precise, and it shows that we really are talking algebra. Conversely, any algebraic property of pairs of numbers, any equation involving x and y, can be reinterpreted as a geometric statement about points, lines, circles, or more elaborate curves.[19]

The basic equations of algebra involve polynomials, combinations of powers of an unknown quantity x, where each power is multiplied by some number, called a coefficient. The largest power of x that occurs is the degree of the polynomial. For example, the equation

$$x^4 - 3x^3 - 3x^2 + 15x - 10 = 0$$

involves a polynomial starting with x^4, so its degree is 4. The coefficients are 1, -3, -3, 15, and -10. There are four distinct solutions: $x = 1, 2, \sqrt{5}$, and $\sqrt{5}$. For these numbers the left-hand side of the equation is equal to zero – the right-hand side. Polynomials of degree 1, like $7x + 2$, are said to be linear, and they involve only the first power of the unknown. Equations of degree 2, like $x^2 - 3x + 2$, are said to be quadratic, and they involve the second power – the square. The equation for a circle involves a second variable, y. However, if we know a second equation relating x and y, for example the equation defining some straight line, then we can solve for y in terms of x and reduce the equation for a circle to one that involves only x. This new equation tells us where the line meets the circle. In this case the new equation is quadratic, with two solutions; this is how the algebra reflects the geometry, in which a line meets the circle at two distinct points.

This feature of the algebra has an important implication for ruler-and-compass constructions. Such a construction, however complicated, breaks up into a sequence of simple steps. Each step produces new points at places where two lines, two circles, or a line and a circle, meet. Those lines and circles are determined by previously constructed points. By translating geometry into algebra, it can be proved that the algebraic equation that corresponds to the intersection of two lines is always linear, while that for a line and a circle, or two circles, is quadratic. Ultimately this happens because the equation for a circle involves x^2 but no higher power of x. So every individual step in a construction corresponds to solving an equation of degree 1 or 2 only.

More complex constructions are sequences of these basic operations, and a certain amount of algebraic technique lets us deduce that each coordinate of any point that can be constructed by ruler and compass is a solution of a polynomial equation, with integer coefficients, whose degree is a power of 2. That is, the degree has to be one of the numbers 1, 2, 4, 8, 16, and so on.[20] This condition is necessary for a construction to exist, but it can be beefed up into a precise characterisation of which regular polygons are constructible. Suddenly a tidy algebraic condition emerges from a complicated geometric muddle – and it applies to *any construction whatsoever*. You don't even need to know what the construction is: just that it uses only ruler and compass.

Gauss was aware of this elegant idea. He also knew (indeed, any competent mathematician would quickly realise) that the question of which regular polygons can be constructed by ruler and compass boils down to a special case, when the polygon has a prime number of sides. To see why, think of a composite number like 15, which is 3×5. Any hypothetical construction of a 15-sided regular polygon automatically yields a 3-sided one (consider every fifth vertex) and a 5-sided one (consider every third vertex), Figure 6. With a bit more effort you can combine constructions for a 3-gon and a 5-gon to get a 15-gon.[21] The numbers 3 and 5 are prime, and the same idea applies in general. So Gauss focused on polygons with a prime number of sides, and asked what the relevant equation looked like. The answer was surprisingly neat. Constructing a regular 5-sided polygon, for example, is equivalent to solving the equation $x^5 - 1 = 0$. Replace 5 by any other prime, and the corresponding statement is true.

The degree of this polynomial is 5, which is *not* one of the powers of 2 that I listed; even so, a construction exists. Gauss quickly figured out why: the equation splits into two pieces, one of degree 1 and the other of degree 4. Both 1 and 4 are powers of 2, and it turns out that the degree-4 equation is the crucial one. To see why, we need to connect the equation to the geometry. That involves a new kind of number, one that is largely ignored in school mathematics but is indispensable for

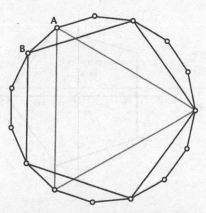

Fig 6 Constructing an equilateral triangle and a regular pentagon from a regular 15-gon. For the reverse, observe that A and B are consecutive points on the regular 15-gon.

anything beyond that. They are called complex numbers, and their defining feature is that in the complex number system −1 has a square root.[22]

An ordinary 'real' number is either positive or negative, and either way, its square is positive, so −1 can't be the square of any real number. This is such a nuisance that mathematicians invented a new kind of 'imaginary' number whose square is −1. They needed a new symbol for it, so they called it i (for 'imaginary'). The usual operations of algebra – adding, subtracting, multiplying, dividing – lead to combinations of real and imaginary numbers such as 3 + 2i. These are said to be complex, which doesn't mean 'complicated', but indicates that they come in two parts: 3 and 2i. Real numbers lie on the famous number line, like the numbers on a ruler. Complex numbers lie in a number plane, in which an imaginary ruler is placed at right angles to a real one, and the two together form a system of coordinates, Figure 7 (left).

For the last 200 years, mathematicians have considered complex numbers to be fundamental to their subject. We now recognise that logically they are on the same footing as the more familiar 'real' numbers – which, like all mathematical structures, are abstract concepts, not real physical things. Complex numbers were in widespread use before the time of Gauss, but their status was still

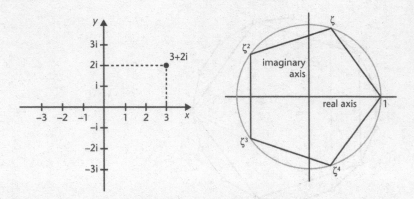

Fig 7 *Left:* The complex plane. *Right:* The complex fifth roots of unity.

mysterious until Gauss and several others demystified them. The source of their attraction was paradoxical: despite the mystery surrounding their meaning, complex numbers were much better behaved than real numbers. They supplied a missing ingredient that the real numbers lacked. They provided a complete set of solutions for an algebraic equation.

Quadratic equations are the simplest example. Some quadratics have two real solutions, while others have none. For example $x^2 - 1 = 0$ has the solutions 1 and −1, but $x^2 + 1 = 0$ has no solutions. In between is $x^2 = 0$, whose sole solution is 0, but there is a sense in which this is the same solution 'repeated twice'.[23] If we allow complex solutions, however, then $x^2 + 1 = 0$ also has two solutions: i and −i. Gauss had no qualms about using complex numbers; in fact, his doctoral thesis provided the first logically sound proof of the fundamental theorem of algebra: the number of complex solutions to any polynomial equation (with multiplicities counted correctly) is equal to the degree of the equation. So quadratics (degree 2) always have two complex solutions, cubics (degree 3) always have three complex solutions, and so on.

The equation $x^5 - 1 = 0$, which I claimed defines a regular pentagon, has degree 5. Therefore it has five complex solutions. There is just one real solution: $x = 1$. What about the other four? They provide four vertexes of a perfect regular pentagon in the complex plane, with $x = 1$ being the fifth, Figure 7 (right). This correspondence is an example of mathematical beauty: an elegant geometric shape becomes an elegant equation.

Now, the equation whose solutions are these five points has degree 5, which is not a power of 2. But, as mentioned earlier, the degree-5 equation splits into two pieces with degrees 1 and 4, called its irreducible factors:

$$x^5 - 1 = (x - 1)(x^4 + x^3 + x^2 + x + 1)$$

('Irreducible' means that no further factors exist, just like prime numbers.) The first factor yields the real solution $x = 1$. The other factor yields the four complex solutions and the other four vertexes of the pentagon. So everything makes much more sense, and is far more elegant, when we use complex numbers.

It is often difficult to reconstruct how the mathematicians of the past arrived at new discoveries, because they had a habit of presenting only the final outcome of their deliberations, not the many false steps they took along the way. This problem is often compounded, because the natural thought patterns in past ages were different from today's. Gauss in particular was notorious for covering his tracks and publishing only his final, highly polished, analysis. But when it comes to Gauss's research on the 17-sided polygon, we are on fairly safe ground; the analysis that he eventually published provides several useful clues.

His starting-point was not new. Several earlier mathematicians were well aware that the above analysis of regular pentagons works in complete generality. Constructing a polygon with any number n of sides is equivalent to solving the equation $x^n - 1 = 0$ in complex numbers. Moreover, this polynomial factorises as

$$(x - 1)(x^{n-1} + x^{n-2} + \cdots + x^2 + x + 1)$$

Again the first factor gives the real solution $x = 1$ and the remaining $n - 1$ solutions come from the second factor. When n is odd, these are all complex; when n is even, one of them is a second real solution $x = -1$.

What Gauss noticed, and everyone else had missed, is that

sometimes the second factor can be expressed using a series of quadratic equations. Not by representing it as a product of simpler factors, because that's not possible, but by using equations whose coefficients solve other equations. The key fact here – the weak point in the problem – is an elegant property of algebraic equations, which arises when we solve several of them in turn in this manner. The calculation is always equivalent to solving a single equation, but the degree generally gets bigger. So the price we pay for having fewer equations is an increase in the degree. It can be messy, but there is one feature that we can predict: how big the degree becomes. Just multiply together the degrees of the successive polynomials.

If they are all quadratics, the result is $2 \times 2 \times \cdots \times 2$, a power of 2. So $n - 1$ must be a power of 2 if a construction exists. However, this condition is not always sufficient. When $n = 9$, $n - 1 = 8$, which is a power of 2. But Gauss discovered that no construction exists for the regular 9-gon. The reason is that 9 is not prime.[24] What about the next case, in which we solve a series of four quadratic equations? Now the degree $n - 1$ of the corresponding single equation is $2 \times 2 \times 2 \times 2 = 16$. So $n = 17$, and this is prime.

By this point Gauss must have known he was on to something, but there is a further technical point, possibly fatal. Gauss had convinced himself that in order for construction for a regular pentagon with a prime number of sides to exist, that prime must be a power of 2, plus 1. So this condition is necessary for a construction to exist: if it fails, there is no such construction. However, the condition might not be sufficient: in fact there are plenty of equations of degree 16 that do not reduce to a series of four quadratics.

There was a reason to be optimistic, however: the Greek constructions. Which primes occurred there? Only three: 2, 3, and 5. These are all 1 more than a power of 2, namely $2^0 + 1$, $2^1 + 1$, and $2^2 + 1$. The algebra associated with the pentagon provides further clues. Thinking it all through, Gauss proved that the degree-16 polynomial associated with the 17-sided polygon can indeed be reduced to a series of quadratics. Therefore a ruler-and-compass construction must exist. A similar method proved that the same is true whenever the number of sides is a prime that is 1 greater than some power of 2. The ideas are a tribute to Gauss's ability to understand mathematical patterns. At their heart are some general theorems in number theory, which I won't go into here. The point is, none of this was accidental. There were solid structural reasons for it to work. You just had to be a Gauss to notice them.

Gauss didn't provide an explicit construction, but he did give a formula for the solutions of the degree-16 equation that can be turned into such a construction if you really want one.[25] When he wrote down his ideas in the *Disquisitiones Arithmeticae*, he omitted quite a few of the details, but he did assert that he possessed complete proofs. His epic discovery convinced him that he should devote his life to mathematics rather than languages. The Duke continued to support Gauss financially, but Gauss wanted something more permanent and reliable. When the astronomer Giuseppe Piazzi discovered the first asteroid, Ceres, only a few observations could be made before this new world became invisible against the glare of the Sun. Astronomers were worried that they wouldn't be able to find it again. In a *tour de force* that involved new techniques for calculating orbits, Gauss predicted where it would reappear – and he was right. This led to him being appointed Professor of Astronomy and Director of the Göttingen Observatory. He continued to hold the post for the rest of his life.

It turns out that 17 is not the only new number of this kind. Two more are known: $2^8 + 1 = 257$ and $2^{16} + 1 = 65,537$. (A bit of algebra shows that the power of 2 that occurs must itself be a power of 2; if not, then the number cannot be prime.) However, the pattern stops at that point, because $2^{32} + 1 = 4,294,967,297$ is equal to $641 \times 6,700,417$, hence is not prime. The so-called Fermat numbers $2^{2^n} + 1$ are known not to be prime for $n = 5, 6, 7, \ldots$ up to 32. Many larger Fermat numbers are also known not to be prime. No further prime Fermat numbers have been found, but their existence is by no means impossible.[26] A construction for the 257-sided polygon is known. One mathematician devoted many years to the 65,537-sided polygon, a somewhat pointless task, and his results contain errors anyway.[27]

The upshot of Gauss's analysis is that a regular polygon can be constructed with ruler and compass if and only if the number of sides is a product of a power of 2 and *distinct* odd prime Fermat numbers. In particular, a regular 9-sided polygon cannot be constructed in this manner. This immediately implies that at least one angle cannot be trisected, because the angle in an equilateral triangle is 60 degrees, and one third of that is 20 degrees. Given this angle, it is easy to construct a regular 9-sided polygon. Since that is impossible, there is no general ruler-and-compass construction to trisect an angle.

Gauss omitted many details of the proofs when he wrote up his results, and mathematicians couldn't simply take his word for it. In 1837 the French mathematician Pierre Wantzel published a complete proof of Gauss's characterisation of constructible regular polygons, and deduced the impossibility of trisecting a general angle by ruler-and-compass construction. He also proved that it is impossible to construct a cube whose volume is twice that of a given cube, another ancient Greek problem known as 'duplicating the cube'.

Both angle-trisection and cube-duplication turn out to be impossible because the lengths involved satisfy irreducible *cubic* equations – degree 3. Since 3 is not a power of 2, this knocks the question on the head. However, this method didn't seem to work for the problem of squaring the circle, for interesting reasons. A circle of unit radius has area π, and square of that area has side $\sqrt{\pi}$. Geometric constructions for square roots exist, and so do constructions for squares, so squaring the circle boils down to starting with a line of length 1 and constructing one of length π. If π happened to satisfy an irreducible cubic equation – or any irreducible equation whose degree is not a power of 2 – then Wantzel's methods would prove that it's impossible to square the circle.

However, no one knew any algebraic equation that was satisfied exactly by π, let alone one whose degree is not a power of 2. The school value 22/7 satisfies $7x - 22 = 0$, but that's just an approximation to π, a tiny bit too large, so it doesn't help. If it could be proved that no such equation exists – and many suspected this on the grounds that it would have been found if it did – the impossibility of squaring the circle would follow. Unfortunately, no one could prove that there is no such equation. The algebraic status of π was in limbo. The eventual solution employed methods that did not just go beyond geometry: they went beyond algebra as well.

To appreciate the main issue here, we need to start with a simpler idea. There is an important distinction in mathematics between numbers that can be expressed as exact fractions p/q, where p and q

are whole numbers, and those that can't be so expressed. The former are said to be rational (they are ratios of whole numbers), and the latter are irrational. The approximation 22/7 to π is rational, for example. There are better approximations; a famous one is 355/113, correct to six decimal places. However, it is known that no fraction can represent π exactly: it is irrational. This long-suspected property was first proved by the Swiss mathematician Johann Heinrich Lambert in 1768. His proof is based on a clever formula for the tangent function in trigonometry, which he expressed as a continued fraction: an infinite stack of ordinary fractions.[28] In 1873 Charles Hermite found a simpler proof, based on formulas in calculus, which went further: it proved that $π^2$ is irrational. Therefore π is not the square root of a rational number either.

Lambert suspected something much stronger. In the article that proved π to be irrational, he conjectured that π is transcendental; that is, π does not satisfy any polynomial equation with integer coefficients. It transcends algebraic expression. Subsequent discoveries proved him right. The breakthrough came in two stages. Hermite's new method for proving irrationality set the scene by hinting that calculus – more precisely, its rigorous version, analysis – might be a useful strategy. By pushing that idea further, Hermite found a wonderful proof that the other famous curious number in mathematics, the base e of natural logarithms, is transcendental. Numerically, e is roughly 2.71828, and if anything it is even more important than π. Hermite's transcendence proof is magical, a rabbit extracted with a flourish from the top hat of analysis. The rabbit is a complicated formula related to a hypothetical algebraic equation that e is assumed to satisfy. Using algebra, Hermite proves that this formula is equal to some nonzero integer. Using analysis, he proves that it must lie between $-\frac{1}{2}$ and $\frac{1}{2}$. Since the only integer in this range is zero, these results are contradictory. Therefore the assumption that e satisfies an algebraic equation must be false, so e is transcendental.

In 1882 Ferdinand Lindemann added some bells and whistles to Hermite's method, and proved that if a nonzero number satisfies an algebraic equation, then e raised to the power of that number does not satisfy an algebraic equation. He then took advantage of a relationship that was known to Euler involving π, e, and the imaginary number i: the famous formula $e^{i\pi} = -1$. Suppose that π satisfies some algebraic equation. Then so does $i\pi$, and Lindemann's theorem implies that -1 does *not* satisfy an algebraic equation. However, it visibly does: it is the solution of $x + 1 = 0$. The only way out of this logical contradiction is that π does not satisfy an algebraic equation; that is, it is transcendental. And that means you can't square the circle.

It was a long and indirect journey from Euclid's geometry to Lindemann's proof, and it took more than 2000 years, but mathematicians finally got there. The story doesn't just tell us that the circle can't be squared. It is an object lesson in how great mathematical problems get solved. It required mathematicians to formulate carefully what they meant by 'geometric construction'. They had to pin down general features of such constructions that might place limits on what they could achieve. Finding those features required making connections with another area of mathematics: algebra. Solving the algebraic problem, even in simpler cases such as the construction of regular polygons, also involved number theory. Dealing with the difficult case of π required further innovations, and the problem had to be transported into yet another area of mathematics: analysis.

None of these steps was simple or obvious. It took about a century to complete the proof, even when the main ideas were in place. The mathematicians involved were among the best of their age, and at least one was among the best of *any* age. Solving great problems requires a deep understanding of mathematics, plus persistence and ingenuity. It can involve years of concentrated effort, most of it apparently fruitless. But imagine how it must feel when your persistence pays off, and you crack wide open something that has baffled the rest of humanity for centuries. As President John F. Kennedy said in 1962 when announcing the Moon landing project: 'We choose to ... do [these] ... things, not because they are easy, but because they are hard.'

Few stories in mathematics end, and π is no exception. Every so often, amazing new discoveries about π appear. In 1997 Fabrice Bellard announced that the trillionth digit of π, in binary notation, is 1.[29] What made the statement remarkable was not the answer. The amazing feature was that he didn't calculate any of the earlier digits. He just plucked one particular digit from the air.

The calculation was made possible by a curious formula for π discovered by David Bailey, Peter Borwein, and Simon Plouffe in 1996. It may seem a bit complicated, but let's take a look anyway:

$$\pi = \sum_{n=0}^{\infty} \frac{1}{2^{4n}} \left(\frac{4}{8n+1} - \frac{2}{8n+4} - \frac{1}{8n+5} - \frac{1}{8n+6} \right)$$

The big Σ means 'add up', over the range specified. Here n runs from 0 to infinity (∞). Bellard actually used a formula that he had derived using similar methods, which is marginally faster for computations:

$$\pi = \frac{1}{64} \sum_{n=0}^{\infty} \frac{(-1)^n}{2^{10n}} \left(-\frac{32}{4n+1} - \frac{1}{4n+3} + \frac{256}{10n+1} - \frac{64}{10n+3} \right.$$
$$\left. - \frac{4}{10n+5} - \frac{4}{10n+7} + \frac{1}{10n+9} \right)$$

The key point is that many of the numbers that occur here – 1, 4, 32, 64, 256, and also 2^{4n} and 2^{10n} – are powers of 2, which of course are very simple in the binary system used for the internal workings of computers. This discovery stimulated a flood of new formulas for π, and for several other interesting numbers. The record for finding a single binary digit of π is broken regularly: in 2010 Yahoo's Nicholas Sze computed the two-quadrillionth binary digit of π, which turns out to be 0.

The same formulas can be used to find isolated digits of π in arithmetic to the bases 4, 8, and 16. Nothing of the kind is known for any other base; in particular, we can't compute decimal digits in isolation. Do such formulas exist? Until the Bailey-Borwein-Plouffe formula was found, no one imagined it could be done in binary.

4

Mapmaking mysteries
Four Colour Theorem

MANY OF THE GREAT MATHEMATICAL problems stem from deep and difficult questions in well-established areas of the subject. They are the big challenges that emerge when a major area has been thoroughly explored. They tend to be quite technical, and everyone in the area knows they're hard to answer, because many experts have tried and failed. The area concerned will already possess many powerful techniques, massive mathematical machines whose handles can be cranked if you've done your homework – but if the problem is still open, then all of the plausible ways to use those techniques have already been tried, and *they didn't work*. So either there is a less plausible way to use the tried-and-tested techniques of the area, or you need new techniques.

Both have happened.

Other great problems are very different. They appear from nowhere – a scribble in the sand, a scrawl in a margin, a passing whim. Their statements are simple, but because they do not already have an extensive mathematical background, there are no established methods for thinking about them. It may take many years before their difficulty becomes apparent: for all anyone knows, there might be some clever but straightforward trick that solves them in half a page. The four colour problem is of this second kind. It took decades before mathematicians began to grasp how difficult the question was, and for a large part of that time they thought that it *had* been solved, in a few pages. It seemed to be a fringe issue, so few people bothered to take it seriously. When they did, the alleged solution turned out to be flawed. The final solution fixed the flaws, but by then the argument had become so complicated that massive computer assistance was required.

In the long run, both types of problem converge, despite their different backgrounds, because resolving them requires new ways of thinking. Problems of the first type may be embedded in a well-understood area, but the traditional methods of that area are inadequate. Problems of the second type don't belong to any established area – in fact, they motivate the creation of new ones – so there are no traditional methods that can be brought to bear. In both cases, solving the problem demands inventing new methods and forging new links with the existing body of mathematics.

We know exactly where the four colour problem came from, and it wasn't mathematics. In 1852 Francis Guthrie, a young South African mathematician and botanist working for a degree in law, was attempting to colour the counties in a map of England. He wanted to ensure that any two adjacent counties were assigned different colours, so that the borders were clear to the eye. Guthrie discovered that he needed only four different colours to complete the task, and after some experimentation convinced himself that this statement would be true for any map whatsoever. By 'adjacent' he meant that the counties concerned shared a border of nonzero length: if two counties touched at a point, or several isolated points, they could if necessary have the same colour. Without this proviso, there is no limit to the number of colours, because any number of regions can meet at a point, Figure 8 (left).

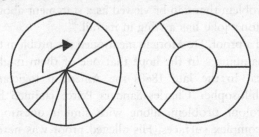

Fig 8 *Left*: Any number of regions can meet at a point. *Right*: At least four colours are necessary.

Wondering whether this statement was a known mathematical theorem, he asked his brother Frederick, who was studying mathematics under the distinguished but eccentric Augustus De Morgan at University College, London. De Morgan didn't know, so he wrote to an even more distinguished mathematican, the Irishman Sir William Rowan Hamilton:

> A student of mine [identified later as Frederick Guthrie] asked me today to give him a reason for a fact which I did not know was a fact – and do not yet. He says that if a figure be any how divided and the compartments differently coloured so that figures with any portion of common boundary *line* are differently coloured – four colours may be wanted but not more... Query cannot a necessity for five or more be invented... What do you say? And has it, if true, been noticed?

Frederick later referred to a 'proof' that his brother had suggested, but he also said that the key idea was a drawing equivalent to Figure 8, which proves only that fewer than four colours won't work.

Hamilton's reply was brief and unhelpful. 'I am unlikely to attempt your "quaternion" of colours very soon,' he wrote. At the time he was working on an algebraic system that became a lifelong obsession, analogous to the complex numbers but involving four types of number rather than the two (real and imaginary) of the complex numbers. This he called 'quaternions'. The system remains important in mathematics; indeed, it is probably more important now than it was in Hamilton's day. But it has never really attained the heights that Hamilton hoped for. Hamilton was just making an academic joke when he used the word, and for a long time there seemed to be no link between quaternions and the four colour problem. However, there is a reformulation of the problem that can be viewed as a statement about quaternions, so Hamilton's joke has a sting in its tail.[30]

Being unable to find a proof, De Morgan mentioned the problem to his mathematical acquaintances in the hope that one of them might come up with an idea. In the late 1860s the American logician, mathematician, and philosopher Charles Sanders Peirce claimed he had solved the four colour problem, along with similar questions about maps on more complex surfaces. His alleged proof was never published, and it is doubtful that the methods available to him would have been adequate.

Although the four colour problem is ostensibly about maps, it has no useful applications to cartography. Practical criteria for colouring maps mainly reflect political differences, and if that means that adjacent regions must have the same colour, so be it. The problem's interest lay entirely within pure mathematics, in a new area that had only just started to develop: topology. This is 'rubber-sheet geometry' in which shapes can be deformed in any continuous manner. But even there, the four colour problem didn't belong to the mainstream. It seemed to be no more than a minor curiosity.

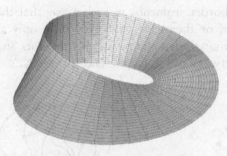

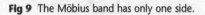

Fig 9 The Möbius band has only one side.

One of the pioneers of topology was August Möbius, famous today for his one-sided band, Figure 9. You can make a model by taking a strip of paper, bending it round into a ring like a short fat cylinder, twisting one end through 180 degrees, and gluing the ends together. A friend of his, the linguist Benjamin Weiske, set Möbius a puzzle: could an Indian king with five sons, all princes, divide up his kingdom so that the region belonging to each prince shared a border of nonzero length with the regions belonging to the other four princes? Möbius passed the puzzle on to his students as an exercise. But in the next lecture, he apologised for asking them to perform the impossible. By this he meant that he could *prove* it was impossible.[31]

It's hard to tackle this puzzle geometrically, because the shapes of the regions and how they are arranged might in principle be very complicated. Progress depends on a big simplification: all that really matters is which regions are adjacent to which, and how the common boundaries are arranged relative to each other. This is topological information, independent of the precise shapes. It can be represented in a clean, simple way, known as a graph – or, nowadays, a network, which is a more evocative term.

A network is a devastatingly simple concept: a set of vertexes, represented by dots, some of which are linked together by edges, drawn as lines. Take any map, such as Figure 10 (left). To convert it into a network, place a dot inside each region, Figure 10 (middle). Whenever two regions have a common length of border, draw a line between the corresponding dots, passing through that segment. If there are several separate common border segments, each gets its own line. Do this for all regions and all common border segments, in such a way that the lines do not cross each other, or themselves, and they meet only at dots. Then throw away the original map and retain only the dots and lines. These form the dual network of the map, Figure 10 (right).[32]

Fig 10 *Left*: A map. *Middle*: Place a point in each region. *Right*: Connect dots across borders to form the dual network (black lines and dots only).

The word 'dual' is used because the procedure takes regions, lines, and dots (junctions between regions of the map) and turns them into dots, lines, and regions. A region in the map corresponds to a dot in the dual network. A border segment in the map corresponds to a line in the dual network; not the same line, but one that crosses the border and links the corresponding dots. A point in the map where three or more regions meet corresponds to a region in the dual network bounded by a closed loop of lines. So the dual network is itself a map, because the lines enclose regions, and it turns out that the dual of the dual is the original map, give or take a few technicalities that exclude unnecessary dots and lines.

The problem of the five princes can be reinterpreted using the dual network: Is it possible to join five points in the plane by lines, with no crossings? The answer is 'no', and the key is Euler's formula, which states that if map on the plane consists of F faces (regions), E edges (lines), and V vertexes (dots), then $F + V - E = 2$. Here we count the rest of the plane, outside the network, as one big region. This formula was one of the first hints that topological considerations could be worth investigating, and it reappears in chapter 10.

The proof that the Indian princes puzzle is impossible starts by assuming that a solution exists, and deduces a contradiction. Any solution will have $V = 5$, the number of dots. Since each pair of dots is joined by a line, and there are 10 pairs, $E = 10$. Euler's theorem implies that $F = E - V + 2 = 7$. The regions of the dual network are surrounded by closed loops of lines, and only one line joins any pair of points; therefore these loops must contain at least 3 lines. Since there are 7 regions, that makes at least 21 lines … except that every line is being counted twice because it separates 2 regions. So there are at least $10\frac{1}{2}$ lines. The number of lines is an integer, so in fact there must be at least 11 lines. However, we already know there are 10 lines. This is a logical contradiction, and it proves that no such network exists. The king cannot divide his land in the prescribed way.

The encouraging aspect of this argument is that elegant topological methods enable us to prove something specific and interesting about maps. However, contrary to a common misconception, which De Morgan seems to have shared, the impossibility of solving the puzzle of the five Indian princes does *not* prove the four colour theorem. A proof may be wrong even if its conclusion is correct, or not known to be incorrect. If somewhere in an alleged proof I encounter a triangle with four sides, I can stop reading, because the proof is wrong. It doesn't matter what happens after that, or what the conclusion is. Our answer to the Indian princes puzzle shows that one particular way to disprove the four colour theorem doesn't work. However, that doesn't imply that no *other* way to disprove it could work. Potentially, there might be many obstacles to 4-colouring a map (from now on I will use this term rather than the clumsy 'colour the map with four colours'). The existence of five regions all adjacent to each other is merely one such obstacle. For all we know, there might be a very complicated map with 703 regions, such that however you 4-colour 702 of them, the final region always needs a fifth colour. That region would have to adjoin at least four others, but that's entirely feasible, and it doesn't require an Indian prince arrangement. If a map like that existed, it would prove that four colours are not enough. Any proof has to rule out that kind of obstacle. And that statement is valid even if I don't show you – can't show you – an explicit example of such an obstacle.

For a time the four colour problem seemed to have sunk without trace, but it resurfaced in 1878 when Arthur Cayley mentioned it at a meeting of the London Mathematical Society. Despite its name, this organisation represented the whole of British (or at least English) mathematics, and its founder was De Morgan. Cayley asked whether anyone had obtained a solution. His query was published soon after in the science journal *Nature*. A year later he wrote a more extended article for the *Proceedings of the Royal Geographical Society*.[33] It presumably seemed a logical place to put the paper, because the problem is ostensibly about maps. He may even have been asked to submit it. But it wasn't really a sensible choice, because no mapmaker would have any reason to want to know the answer, other than idle curiosity. Unfortunately, the choice of journal meant that few mathematicians would be aware of the article's existence. That was a pity, because Cayley explained why the problem might be tricky.

In chapter 1, I said that a proof is a bit like a battle. The military recognises a difference between tactics and strategy. Tactics are how you win local skirmishes; strategy sets the broad structure of the campaign. Tactics involve detailed troop movements; strategy involves broad plans, with room for many different tactical decisions at any stage. Cayley's article was short on tactics, but it contained the vaguest hint of a strategy which, in the fullness of time, cracked the four colour problem wide open. He observed that adding regions one at a time didn't work if you followed the obvious line of reasoning. But maybe it would work if you found a less obvious line of reasoning.

Suppose you take a map, remove one region – say by merging it with a neighbour, or shrinking it to a point. Suppose that the resulting map can be 4-coloured. Now put the original region back again. If you're lucky, its neighbours might use only three colours. Then all you have to do is colour it using the fourth. Cayley's point was that this procedure might not work, because the neighbours of the final region might use four distinct colours. But that doesn't mean you're stuck. There are two ways to get round this obstacle: you may have chosen the wrong region, or you may have chosen the wrong way to colour the smaller map.

Still running on unsubstantiated suppositions (this is a very effective way to get research ideas, although at some point you have to substantiate them), assume that something of this kind can always be fixed up. Then it tells you that a map can always be 4-coloured provided some smaller map can be 4-coloured. This may not seem like progress: how do we know the smaller map can be 4-coloured? The answer is that the same procedure applies to the smaller map, leading to an even smaller map ... and so on. Eventually you get to a map so small that it has only four regions, and then you know it can be 4-coloured. Now you reverse your steps, colouring slightly larger maps at each stage ... and eventually climb back to your original map.

This line of reasoning is called 'proof by mathematical induction'. It's a standard method with a more technical formulation, and the logic behind it can be made rigorous. Cayley's proposed proof strategy becomes more transparent if the method is reformulated using a logically equivalent concept: that of a minimal criminal. In this context, a criminal is any hypothetical map that can't be 4-coloured. Such a map is minimal if any map with a smaller number of countries *can* be 4-coloured. If a criminal exists, there must be a minimal one: just choose a criminal with the smallest possible number of regions. Therefore, if minimal criminals do not exist, then criminals do not exist. And if there are no criminals, the four colour theorem must be true.

The induction procedure boils down to this. Suppose we can prove that 4-colouring a minimal criminal is always possible, provided some related smaller map can be 4-coloured. Then the minimal criminal can't actually be a criminal. Because the map is minimal, *all* smaller maps can be 4-coloured, so by what we've supposed can be proved, the same is true of the original map. Therefore there are no minimal criminals, so there are no criminals. This idea shifts the focus of the problem from all maps to just the hypothetical minimal criminals, and specifying a reduction procedure – a systematic way to turn a 4-colouring of some related smaller map into a 4-colouring of the original map.

Why fuss about with minimal criminals, rather than plain criminals? It's a matter of technique. Even though we initially don't know whether criminals exist, one of the paradoxical but useful features of this strategy is that we can say quite a lot about what minimal ones would look like if they did exist.

This requires the ability to think logically about hypotheticals, a vital skill for any mathematician. To give a flavour of the process, I'll prove the *six* colour theorem. To do so, we borrow a trick from the five princes puzzle, and reformulate everything in terms of the dual network, in which regions become dots. The four colour problem is then equivalent to a different question: Given a network in the plane whose lines do not cross, is it possible to 4-colour the *dots*, so that two dots joined by a line always have different colours? The same reformulation applies with any number of colours.

To illustrate the power of minimal criminals, I'm going to use them to prove that any planar network can be 6-coloured. Again the main technical tool is Euler's formula. Given a dot in the dual network, define its neighbours to be those dots that are linked to it by a line. A dot may have many neighbours, or just a few. It can be shown that Euler's formula implies that some dots must have few neighbours. More precisely, in a planar network it is impossible for all dots to have six or more neighbours. I've put a proof of this in the notes to avoid interrupting the flow of ideas.[34] This fact provides the necessary lever to start pulling the problem to pieces. Consider a hypothetical minimal criminal for the 6-colour theorem. This is a network that can't be 6-coloured, but every smaller network can be 6-coloured. I now prove this map can't exist. By the above consequence of Euler's formula, it contains at least one dot with five or fewer neighbours. Temporarily delete this dot and the lines that link it to its neighbours. The resulting network has fewer dots, so by minimality it can be 6-coloured. (Here is where we get stuck unless our hypothetical criminal is minimal.) Now put the deleted dot and lines back. That dot has at most five neighbours, so there is always a sixth colour. Use this to colour the deleted dot. Now we have successfully 6-coloured our minimal criminal – but that contradicts its criminality. So minimal criminals for the 6-colour theorem don't exist, and that implies that the six colour theorem is true.

This is encouraging. Until now, for all we knew, some maps might need 20 colours, or 703, or millions. Now we know that maps like that are no more real than the pot of gold at the end of the rainbow. A specific, limited number of colours definitely works for *any* map. This is a genuine triumph for minimal criminals, and it encouraged mathematicians to tighten up the argument in the hope of replacing six colours by five, or if you were really clever, four.

All criminals need lawyers. A barrister named Alfred Kempe was at the meeting in which Cayley mentioned the four colour problem. He had studied mathematics under Cayley as an undergraduate at Cambridge, and his interest in the subject remained undiminished. Within a year, Kempe convinced himself that he had cracked the problem, and he published his solution in 1879 in the newly founded *American Journal of Mathematics*. A year later he published a simplified proof, which corrected some errors in the first one. He pointed out that

> A very small alteration in one part of a map may render it necessary to recolour it throughout. After a somewhat arduous search, I have succeeded ... in hitting upon the weak point, which proved an easy one to attack.

I'll reinterpret Kempe's ideas in terms of the dual network. Once again he started from Euler's formula and the consequent existence of a dot with either three, four, or five neighbours. (A dot with two neighbours lies in the middle of a line, and contributes nothing to the network or the map: it can safely be omitted.)

If there is a dot with three neighbours, the procedure I used in proving the six colour theorem applies when there are only four colours. Remove the dot and the lines that meet it, 4-colour the result, put the dot and lines back, use a spare colour for the dot. We may therefore assume that no dot has three neighbours.

If there is a dot with four neighbours the above tactic fails, because a spare colour might not be available. Kempe devised a clever way to deal with this obstacle: delete that dot anyway, but after doing so, change the colouring of the resulting smaller map so that two of those four neighbours have the same colour. After this change, the neighbours of the deleted dot use at most three colours, leaving a spare one for the deleted dot. The basic idea of Kempe's recolouring scheme is that two of the neighbouring dots must have different colours – say red and blue, with the other colours being green or yellow. If both are green or both yellow, the other colour is available for the deleted dot. So we can assume one is green and one yellow. Now find all of the dots that can be connected to the blue one by a sequence of lines, using only blue and red dots. Call this a blue-red Kempe chain.[35] By definition, every neighbour of any dot in the Kempe chain, that is not itself in the Kempe chain, is either green or yellow, because a blue or red neighbour would already be in the chain. Having found such a chain, observe that swapping the two colours blue and red for all dots within the chain produces another colouring of the network, still satisfying the key condition that adjacent dots get different colours, Figure 11.

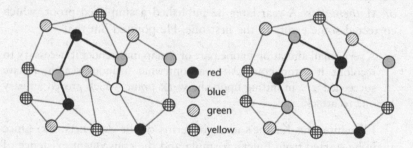

Fig 11 Swapping colours in a Kempe chain (thick black lines) associated with a dot of degree 4 (white) that has neighbours of all four colours. *Left*: Original colours. *Right*: With swapped colours, blue is available for the white dot.

If the red neighbour of our original dot is not in this blue-red chain, make such a change. The blue neighbour of the original dots turns red; the red neighbour remains red. Now the neighbours of the original dot use at most three different colours: red, green, and yellow. That leaves blue for the original dot, and we're done. However, the blue-red chain might loop round and join up with the blue neighbour. If so, leave the blue and red chain alone, and use the same trick for the yellow and green neighbours of the original dot instead. Start with the green one and form a green-yellow Kempe chain. This chain *can't* join up with the yellow neighbour, because the previous blue-red chain gets in the way. Swap yellow and green, and you're done.

That leaves one final case, when there are no dots with three or four neighbours, but at least one has five neighbours. Kempe proposed a similar but more complicated recolouring rule, which seemed to sort out that case as well. Conclusion: the four colour theorem is true, and Kempe had proved it. It even hit the media: *The Nation*, an American magazine, mentioned the solution in its review section.

Kempe's proof seemed to have laid the problem to rest. For most mathematicians it was a done deal. Peter Guthrie Tait continued publishing papers on the problem, seeking a simpler proof; this led him to some useful discoveries, but the simpler proof eluded him.

Enter Percy Heawood, a lecturer in mathematics at Durham University known by the nickname 'Pussy', thanks to his magnificent moustache. As an undergraduate at Oxford he had learned of the four colour problem from Henry Smith, the Professor of Geometry. Smith told him that the theorem, though probably true, was unproved, so Heawood had a go. Along the way he came across Kempe's paper, and tried to understand it. He published the outcome in 1889 as 'Map colour theorem', regretting that the aim of his article was more 'destructive than constructive, for it will be shown that there is a defect in the now apparently recognized proof.' Kempe had made a mistake.

It was a subtle mistake, and it occurred in the recolouring method when the dot being deleted had five neighbours. Kempe's scheme could occasionally change the colour of some dot as a knock-on effect of later changes. But Kempe had assumed that once the colour of a dot had been changed, it didn't change again. Heawood found a network for which Kempe's recolouring scheme went wrong, so his proof was flawed. Kempe was quick to acknowledge the error, and added that he had 'not succeeded in remedying the defect'. The four colour theorem was up for grabs again.

Heawood extracted some crumbs of comfort for Kempe from the débâcle: his method successfully proved the five colour theorem. Heawood also worked on two generalisations of the problem: empires, in which the regions can consist of several disconnected pieces, all requiring the same colour; and maps on more complicated surfaces. The analogous question on a sphere has the same answer as it does for the plane. Imagine a map on a sphere, and rotate it until the north pole is somewhere inside one region. If you delete the north pole you can open up the punctured sphere to obtain a space that is topologically equivalent to the infinite plane. The region that contains the pole becomes the infinitely large one surrounding the rest of the map. But there are other, more interesting, surfaces. Among them is the torus, shaped like a doughnut with a hole, Figure 12 (left).

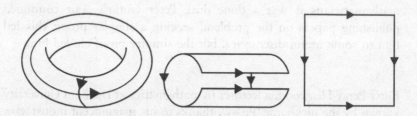

Fig 12 Cutting a torus open and unwrapping it to make a square.

There is a useful way to visualise the torus, which often makes life simpler. If we cut the torus along two closed curves, Figure 12 (middle), we can open it out into a square, Figure 12 (right). This transformation changes the topology of the torus, but we can get round that by 'identifying' opposite edges of the square. In effect (and a rigorous definition makes this idea precise) we agree to treat corresponding points on these edges as though they were identical. To see how this goes, reverse the sequence of pictures. The square gets rolled up, and opposite edges really do get glued together. Now comes the clever part. You don't actually need to roll up the square and join corresponding edges. You can just work with the flat square, provided you bear in mind the rule for identifying the edges. Everything you do to the torus, such as drawing curves on it, has a precise corresponding construction on the square.

Heawood proved that seven colours are both necessary and sufficient to colour any map on a torus. Figure 13 (left) shows that seven are necessary, using a square to represent the torus as just described. Observe how the regions match at opposite edges. There are surfaces like a torus, but with more holes, Figure 13 (right). The number of holes is called the genus, and denoted by the letter g. Heawood conjectured a formula for the number of colours required on

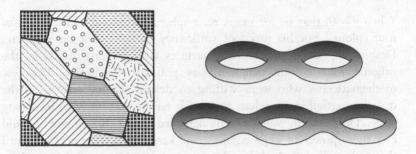

Fig 13 *Left*: Seven-colour map on a torus. The torus is represented as a square whose opposite sides are conceptually 'wrapped round' so that they glue together. Regions of the map are required to match up across corresponding edges. *Right*: Toruses with two and three holes.

a torus with g holes when $g \geqslant 1$: it is the smallest whole number less than or equal to

$$\frac{7 + \sqrt{48g + 1}}{2}$$

When g ranges from 1 to 10, this formula yields the numbers

7 8 9 10 11 12 12 13 13 14

The number of colours specified by the formula grows more slowly than the genus, and often it makes no difference if you add an extra hole to the torus. This is a surprise, because every extra hole provides more freedom to invent complicated maps.

Heawood didn't just pluck this formula from thin air. It arose by generalising the way I proved the six-colour theorem in the plane. He could prove that this number of colours is always sufficient. The big question, for many years, was whether this number can be made smaller. Examples for small values of the genus suggested that Heawood's estimate is the best possible. In 1968, after a lengthy investigation, Gerhard Ringel and John W.T. (Ted) Youngs filled in the final details in a proof that this is correct, building on their own work and that of several others. Their methods are combinatorial, based on special kinds of networks, and complicated enough to fill an entire book.[36]

When $g = 0$, that is, for maps on a sphere, Heawood's formula gives four colours, but his proof of sufficiency doesn't work on a sphere. Despite impressive progress for surfaces with at least one hole, the original four colour problem was still up for grabs. The few mathematicians who were willing to devote serious efforts to the question settled in for what, in warrior terms, was likely to be a long siege. The problem was a heavily defended castle; they hoped to build ever more powerful siege engines, and keep knocking pieces off until the castle walls fell. And they did, and it didn't. However, the attackers slowly accumulated a lot of information about how not to solve the problem, and the types of obstacle that seemed unavoidable. Out of these failures an ambitious strategy began to emerge. It was a natural extension of Kempe's and Heawood's methods, and it came in three parts. I'll state them using the dual network, the standard viewpoint nowadays:

1 Consider a minimal criminal.
2 Find a list of unavoidable configurations: smaller networks, with the property that any minimal criminal must contain something in the list.
3 Prove that each of the unavoidable configurations is reducible. That is: if a smaller network, obtained by deleting the unavoidable configuration, can be 4-coloured, then these colours can be redistributed so that when the unavoidable configuration is restored, the 4-colouring of the smaller network extends to the whole network.

Putting these three steps together, we can prove that a minimal criminal doesn't exist. If it did, it would contain an unavoidable configuration. But the rest of the network is smaller, so minimality implies that it can be 4-coloured. Reducibility now implies that the original network can be 4-coloured. This is a contradiction.

In these terms, Kempe had correctly found a list of unavoidable configurations: a dot with three lines sticking out, one with four, and one with five, Figure 14. He had also correctly proved that the first two are reducible. His mistake lay in his proof that the third configuration is reducible. It's not. Proposal: replace this bad configuration by a longer list, making sure that the list remains unavoidable. Do this in

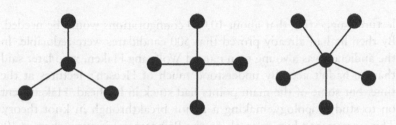

Fig 14 Kempe's list of unavoidable configurations.

such a manner that each configuration on the new list is reducible. That is: look for an unavoidable list of reducible configurations. If you succeed, you've proved the four colour theorem.

There might not be such a list, but this strategy is worth a try, and no one had any better ideas. It has one awkward inner tension, though. On the one hand, the longer the list, the more chance it has of being unavoidable, which is good. On the other hand, the longer list, the less likely it is that every configuration in it will be reducible. If even a single one of them is not, the whole proof collapses, and this danger becomes more acute as the list grows. Which is bad. On the *third* hand ... a longer list provides more opportunities to choose reducible configurations, which is good. On the fourth hand, it increases the work needed to prove reducibility, which is bad. And on the fifth hand, there were no good methods for doing that anyway, which was worse.

This sort of thing is what makes great problems great.

So for a time the odd bit of the castle occasionally got chipped off, and its loss made not one whit of difference to the stronghold's solidity; meanwhile mainstream mathematics yawned, if it noticed at all. But someone was building a better battering-ram, and his name was Heinrich Heesch. His big contribution was a systematic way to prove that a configuration is reducible. He called it 'discharging', and it is roughly analogous to imagining that the dots in the network carry electric charges, and allowing the electricity to flow from one dot to another.

Even with this method, finding an unavoidable set of reducible configurations by hand would be a daunting task. The individual configurations would probably be fairly small, but there would have to be a lot of them. Heesch persevered, and in 1948 he gave a course of lectures suggesting that about 10,000 configurations would be needed. By then he had already proved that 500 candidates were reducible. In the audience was a young man named Wolfgang Haken, who later said that he hadn't actually understood much of Heesch's lectures at the time, but some of the main points had stuck in his head. Haken went on to study topology, making a major breakthrough in knot theory. This encouraged him to work on the Poincaré conjecture, chapter 10. For a particular line of attack, he classified the possibilities into 200 cases, solved 198 of them, and grappled with the remaining two for 13 years. At that point he gave up, and started working on the four colour problem instead. Haken clearly liked tough problems, but he was worried that something similar might happen with Heesch's 10,000 configurations. Imagine dealing successfully with 9,998, and getting stuck on the final two. So in 1967 he invited Heesch to visit the University of Illinois, where Haken was based, to ask his advice.

In those days computers were starting to become useful for real mathematics, but they were huge machines located in some central building, not things that sat on your desk or inside your briefcase. Haken wondered whether they could help. Heesch had already had the same idea and made a rough estimate of the complexity of the problem. This indicated that the best computer available to him wasn't up to the task. Illinois had a far more powerful supercomputer, ILLIAC-IV, so Haken requested time on it. But the supercomputer wasn't ready, so he was told to try the Cray 6600 in the Brookhaven Laboratory on Long Island. The director of the lab's computer centre was Yoshio Shimamoto, who had long been fascinated by the four colour problem – a stroke of good luck that gave Heesch and Haken access to the machine.

The computer lived up to expectations, but Haken started to wonder whether it could be used more efficiently. They were generating lots of reducible configurations and hoping to assemble an unavoidable list, but that strategy wasted a lot of time on potential configurations that turned out not to be reducible. Why not do it the other way round: make unavoidability the main objective, and check reducibility later? Of course, you'd need to use configurations that had a good chance of being reducible, but it seemed a better way to go. By then, however, the

Cray at Brookhaven was being used for more important things. Worse, several experts told Haken that the methods he wanted to use couldn't be turned into computer programs at all. He believed them, and gave a lecture saying that the problem couldn't be solved without computers, but now it seemed it couldn't be solved with computers either. He had decided to give up.

In the audience was an expert programmer, Kenneth Appel, who told Haken that the alleged experts were probably just trying to put him off because the programs would take a lot of work and the outcome was very uncertain. In Appel's view, there was no mathematical problem that could not be programmed. The crucial issue was whether the program would get anywhere in a reasonable time. They joined forces. The strategy evolved as improvements to the discharging method caused changes to the program and improvements to the program caused changes to the discharging method. This led them to a new concept: 'geographically good' configurations, which didn't contain certain nasty configurations that prevented reducibility. The chance of such a configuration being reducible was much improved, and the defining property was easy to check. Appel and Haken decided to prove theoretically, rather than by computer, that there was an unavoidable list of geographically good configurations. By 1974 they had succeeded.

This was encouraging, but they knew what was likely to happen. Some of their geographically good configurations would turn out not to be reducible, so they would have to remove those and replace them by a longer, more complicated, list. The calculation would be chasing its tail, and would succeed only if it caught it. Rather than wasting years on a fruitless pursuit, they did some rough calculations to estimate how long the process might take. The results were mildly encouraging, so the work went on. Theory and computation fed off and changed each other. At times, the computer seemed to have a mind of its own, 'discovering' useful features of configurations. Then the university administration bought itself a new, very powerful computer – more powerful than those available to the university's scientists. After some protest, and pointed questions, half of the machine's time was made available for scientific use. Appel and Haken's ever-changing list of unavoidable configurations stabilised at around 2000 of them. In June 1976, the computer ground out its last reducibility check, and the proof was complete. The story hit the media, starting in *The Times* and rapidly spreading worldwide.

They still had to make sure there were no silly mistakes, and by then several other teams were in hot pursuit. By July, Appel and Haken were confident their method worked, and they announced their proof officially to the mathematical community by circulating a preprint – a cheaply duplicated draft version of a paper, intended for later publication. At the time, it typically took between one and two years to get a mathematical paper into print. To avoid holding up progress the profession had to find a quicker way to get important results out to the community, and preprints were how they did it. Nowadays, the preprint goes on the web. Preprints are always provisional; full publication requires peer review. Preprints assist this process, because anyone can read them, look for mistakes or improvements, and tell the authors. In fact, the published version often differs considerably from the preprint, for just that reason.

The final proof took a thousand hours of computer time and involved 487 discharging rules; the results were published as two papers with a 450-page supplement showing all 1482 configurations. At the time, it was a *tour de force*.

The main reaction from the wider mathematical community, however, was vague disappointment. Not at the result; not at the remarkable computational achievement. What was disappointing was the method. In the 1970s, mathematical proofs were things that you put together by hand and checked by hand. As I said in Chapter 1, a proof is a story whose plot convinces you the statement is true. But this story didn't have a plot. Or if it did, there was a big hole in the middle:

> Once upon a time there was a beautiful conjecture. Her mother told her never to enter the dark, dangerous forest. But one day Little Four Colour Conjecture slipped away and wandered into the unavoidable forest. She knew that if each configuration in the forest were reducible, she would have a proof, become Little Four Colour Theorem, and be published in a journal run by Prince Ton. She came across a candy-coated computer, deep in the forest, and inside was a Wolf disguised as a programmer. And the Wolf said 'Yes, they are all reducible', and they all lived happily ever after.

No, it doesn't work. I'm being flippant, but the hole in this fairytale is the same as the hole in the Appel-Haken proof, or at least, what most mathematicians considered to be the hole in the proof. *How do we know the Wolf is right?*

We run our own computer program and find out whether it agrees. But however many times we do that, it won't have the same ring of authenticity as, say, my proof that you can't cover a hacked chessboard with dominoes. You can't grasp it as a whole. You couldn't check all the calculations by hand if you lived for a billion years. Worse, you wouldn't believe the answer if you could. Humans make mistakes. In a billion years, they make a lot of mistakes.

Computers, by and large, don't. If a computer and a human both do a really complicated piece of arithmetic and disagree, the smart money is on the computer. But that's not certain. A computer that is functioning exactly as designed can make an error; for example, a cosmic ray can zip through its memory and change a 0 to a 1. You can guard against that by doing the computation again, but more seriously, designers can make mistakes. The Intel P5 Pentium chip had an error in its routines for floating-point arithmetic: if asked to divide 4195835 by 3145727, it responded with 1.33373, when the correct answer is 1.33382. Apparently, four entries in a table had been left out.[37] Other things that can go wrong include the computer's operating system and bugs in the user's program.

A lot of philosophical hot air has been expended on the proposition that the Appel-Haken computer-assisted proof changed the nature of 'proof'. I can see what the philosophers are getting at, but the concept of proof that working mathematicians use is not the one we teach to undergraduates in mathematical logic classes. And even when that more formal concept applies, nothing requires the logic of each step to be checked by a human. For centuries, mathematicians have used machines for routine arithmetic. And even if a human does go through a proof line by line, finding no mistakes, how do we know they haven't missed one? Perfect, unassailable logic is an ideal at which we aim. Imperfect humans do the best they can, but they can never remove every element of uncertainty.

In *Four Colours Suffice*, Robin Wilson put his finger on a key sociological aspect of the community's reaction:

The audience split into two groups: the over-forties could not be convinced that a proof by computer was correct, while the under-forties could not be convinced that a proof containing 700 pages of hand calculations could be correct.

If our machines are better at some things than we are, it makes sense to use machines. Proof *techniques* may change, but they do that all the time anyway: it's called 'research'. The concept of proof does not radically alter if some steps are done by a computer. A proof is a story; a computer-assisted proof is a story that's too long to be told in full, so you have to settle for the executive summary and a huge automated appendix.

Since Appel and Haken's pioneering work, mathematicians have become accustomed to computer assistance. They still *prefer* proofs that rely solely on human brainpower, but most of them don't make that a requirement any more. In the 1990s, though, there was still a certain amount of justifiable unease about the Appel-Haken proof. So instead of rechecking the work, some mathematicians decided to redo the whole proof, taking advantage of new theoretical advances and much improved computers. In 1994 Neil Robertson, Daniel Sanders, Paul Seymour, and Robin Thomas threw away everything in the Appel-Haken paper except the basic strategy. Within a year they had found an unavoidable set of 633 configurations, each of which could be proved reducible using just 32 discharging rules. This was much simpler than Appel and Haken's 1482 configurations and 487 discharging rules. Today's computers are so fast that the entire proof can now be verified on a home computer in a few hours.

That's all very well, but the computer is still king. Can we get rid of it? There is a growing feeling that in this particular instance a story that humans can grasp in its entirety may not be totally inconceivable. Perhaps new insights into the four colour problem will eventually lead to a simpler proof, with little or no computer assistance, so that mathematicians can read it, think about it, and say 'Yes!' We don't know such a proof yet, and it may not exist, but there's a feeling in the air...

Mathematicians are learning a lot about networks. Topologists and geometers are finding deep relations between networks and utterly different areas of mathematics, including some that apply to mathematical physics. One of the concepts that turns up, from time to time, is curvature. The name is apt: the curvature of a space tells you how bent it is. If it's flat like the plane, its curvature is zero. If it bends in the same direction, the way a hilltop curves downwards on all sides, it has positive curvature. If it's like a mountain pass, curving up in some directions but down in others, it has negative curvature. There are geometric theorems, descendants of Euler's formula, that relate networks drawn in a space to the space's own curvature. Heawood's formula for a g-holed torus hints at this. A sphere has positive curvature, a torus represented as a square with opposite edges identified, Figure 12 (right), has zero curvature, and a torus with two or more holes has negative curvature. So there is some sort of link between curvature and map-colouring.

Behind this link is a useful feature of curvature: it's hard to get rid of it. It's like a cat under a carpet. If the carpet is flat, there's no cat, but if you see a bump, there's a cat underneath. You can chase the cat around the carpet, but all that does is move the bump from one place to another. Similarly, curvature can be moved, but not removed. Unless the cat gets to the edge of the carpet, in which case it can escape, taking its curvature with it. Heesch's discharging rules look a bit like curvature in disguise. They shift electrical charge around, but don't destroy it. Might there exist some concept of curvature for a network, and some cunning discharging rule that, in effect, pushes curvature around?

If so, you might be able to persuade a network to colour itself automatically. Assign curvature to its dots (and perhaps lines); then let the network redistribute the curvature more evenly. Perhaps 'evenly' here implies that four colours suffice, if we set everything up correctly. It's only an idea, it's not mine, and I haven't explained it in enough detail to make much sense. But it reflects some mathematicians' intuition, and it offers hope that a more conceptual proof of the four colour theorem – a ripping yarn rather than a summary with a billion telephone books as an appendix – might yet be found. We will encounter a similar idea, in a far more sophisticated context, in chapter 10, and it solved an even greater problem in topology.

5

Sphereful symmetry
Kepler Conjecture

IT ALL STARTED WITH A SNOWFLAKE.

Snow has a strange beauty. It falls from the sky in fluffy white flakes, it blows in the wind to create soft humps and hillocks that cover the landscape, it forms spontaneously into outlandish shapes. It's cold. You can ski on it, ride a sleigh over it, make snowballs and snowmen out of it ... and if you're unlucky, you can be buried by thousands of tonnes of it. When it goes away, it doesn't go back into the sky – not directly as white flakes. It turns into plain ordinary water. Which can evaporate and go back in to the sky, of course, but may travel down rivers to the sea along the way, and spend a very long time in the oceans. Snow is a form of ice, and ice is frozen water.

This is not news. It must have been obvious to Neanderthals.

Snowflakes are by no means formless lumps. When they are pristine, before they start to melt, many of them are tiny, intricate stars: flat, six-sided, and symmetric. Others are simple hexagons. Some have less symmetry, some have a substantial third dimension, but sixfold snowflakes are iconic and widespread. Snowflakes are ice crystals. That's not news either: you just have to recognise a crystal when you see one. But these are no ordinary crystals, with flat, polygonal facets. Their most puzzling feature adds a dash of chaos: despite having the same symmetry, the detailed structure differs from each snowflake to the next. No two snowflakes are alike, they say. I've always wondered how they know, but the numbers favour that view if you are sufficiently pedantic about what counts as being alike.

Why are snowflakes six-sided? Four hundred years ago, one of the great mathematicians and astronomers of the seventeenth century asked himself that question, and thought his way towards an answer. It turned out to be an amazingly good answer, all the more so because he carried out no special experiments. He just put together some simple ideas that were known to everyone. Like the way pomegranate seeds are packed inside the fruit.

His name was Johannes Kepler, and he had a very good reason to think about snowflakes. His livelihood depended on a rich sponsor, John Wacker of Wackenfels. At that time Kepler was court mathematician to the Holy Roman Emperor Rudolf II, and Wacker, a diplomat, was a counsellor to the emperor. Kepler wanted to give his sponsor a New Year present. Ideally it should be cheap, unusual, and stimulating. It should give Wacker an insight into the remarkable discoveries that his money was making possible. So Kepler collected his thoughts about snowflakes in a small book, and that was the present. Its title was *De Nive Sexangula* ('On the six-cornered snowflake'). The date was 1611. Tucked away inside it, one of the main steps in Kepler's thinking, was a brief remark: a mathematical puzzle that would not be solved for 387 years.

Kepler was an inveterate pattern-seeker. His most influential scientific work was the discovery of three basic laws of planetary motion, the first and best known being that the orbit is an ellipse. He was also a mystic, thoroughly immersed in the Pythagorean world-view that the universe is based on numbers, patterns, and mathematical shapes. He did astrology as well as astronomy: mathematicians often moonlighted as astrologers in those days, because they could actually do the sums to work out when Aquarius was in the ascendant. Wealthy patrons, even royalty, paid them to cast horoscopes.

In his book, Kepler pointed out that snow begins as water vapour, which is formless, yet somehow the vapour turns into six-sided solid flakes. Some agent must cause this transition, Kepler insisted:

Did [this agent] stamp the six-cornered shape on the stuff as the stuff demanded, or out of its own nature – a nature, for instance, in which there is inborn either the idea of the beauty inherent in the hexagon or knowledge of the purpose which that form subserves?

In search of the answer, he considered other examples of hexagonal forms in nature. Honeycombs in beehives sprang to mind. They are made from two layers of hexagonal cells, back to back, and their common ends are formed by three rhombuses – parallelograms with all sides equal. This shape reminded Kepler of a solid called a rhombic dodecahedron, Figure 15. It's not one of the five regular solids that the Pythagoreans knew about and Euclid classified, but it has one distinctive property: identical copies can pack together exactly to fill space, with no gaps. The same shape occurs in pomegranates, where small round seeds grow, get squeezed together, and are therefore forced to create an efficient packing.

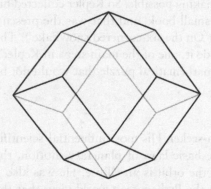

Fig 15 The rhombic dodecahedron, a solid with 12 rhombic faces.

Like any sensible mathematician, Kepler begins with the simplest case in which the spheres form a single plane layer. This is equivalent to packing identical circles in the plane. Here he finds just two regular arrangements. In one, the spheres are arranged in squares, Figure 16 (left); in the other, they are arranged in equilateral triangles, Figure 16 (right). These arrangements, repeated over the entire infinite plane, are the square lattice and the triangular lattice. The word 'lattice' refers to their spatially periodic pattern, which repeats in two independent directions. The figures necessarily show a finite portion of the pattern, so you should ignore the edges. The same goes for Figures 17–20 below. Figure 16 (left) and (right) both show five rows of spheres, and in each row they touch their neighbours. However, the triangular lattice is slightly squashed: its rows are closer together. So the spheres in the triangular lattice are more closely packed than those in the square lattice.

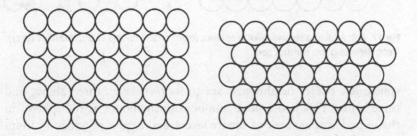

Fig 16 *Left*: Square lattice packing. *Right*: Triangular (also called hexagonal) lattice packing.

Next, Kepler asks how successive layers of this kind can be placed on top of each other, and he considers four cases. For the first two, all layers are square lattices. One way to stack the layers is to place the spheres in each layer directly above those below. Then every sphere will have six immediate neighbours: four within its layer, one above, and one below. This packing is like a three-dimensional chessboard made of cubes, and it would turn into that if you inflated the spheres until they could expand no further. But this, says Kepler, 'will not be the tightest pack'. It can be made tighter by sliding the second layer sideways, so that its spheres fit neatly into the dents between the spheres in the layer below, Figure 17 (left). Repeat this process, layer by layer, Figure 17 (right). Now each sphere has twelve neighbours: four in its own layer, four above, and four below. If you inflate them you fill the space with rhombic dodecahedrons.

In the other two cases, the layers are triangular lattices. If they are stacked so that the spheres in each layer lie directly above those below, then each sphere has eight neighbours: six in its own layer, one above, and one below. Alternatively, the spheres in the next layer can again be fitted into the dents between the spheres in the layer below. Now each

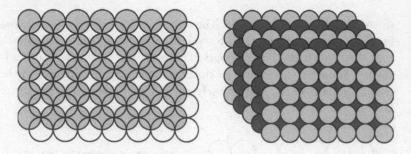

Fig 17 *Left*: Adding a second layer of spheres (open circles) on top of the first layer (grey). *Right*: Repeating this construction.

sphere has twelve neighbours: six in its own layer, three above, and three below. This is the same number of neighbours as the spheres in the second arrangement of square layers, and Kepler provides a careful analysis of the geometry to show that this fourth arrangement is actually the same as the second one. The only difference is that the square layers are no longer horizontal, but slanted at an angle. He writes: 'Thus in the closest pack in three dimensions, the triangular pattern, cannot exist without the square, and vice versa.' I'll come back to that: it's important.

Having sorted out the basic geometry of sphere packing, Kepler returns to the snowflake and its sixfold symmetry. He is reminded of the triangular lattice packing of spheres in a plane, in which each sphere is surrounded by six others, which form a perfect hexagon. This, he decides, must be why snowflakes are six-sided.

This chapter isn't primarily about snowflakes, but Kepler's explanation of their symmetry is very similar to the one we would offer today, so it would be a shame to stop here. Why are they – how *can* they be – so diverse, yet symmetric? When water crystallises to create ice, the atoms of hydrogen and oxygen that form the molecules of water pack together into a symmetric structure, the crystal lattice. This lattice is more complicated than any of Kepler's arrangements of spheres, but its dominant symmetry is sixfold. A snowflake grows from a tiny 'seed' with just a few atoms, arranged like a small piece of the lattice. This seed has the same sixfold symmetry, and it sets the scene for the growth of the ice crystal as the winds blow it this way and that inside a storm cloud.

The great variety of snowflake patterns is a consequence of changing conditions in the cloud. Depending on temperature and humidity, crystal growth may be uniform, with atoms being added at the same rate all along the boundary, leading to straight-sided hexagons, or it can be dendritic, with a growth rate that varies from place to place, resulting in treelike structures. As the growing flake is transported up and down through the cloud, these conditions keep changing, randomly. But the flake is so tiny that at any given moment, the conditions are essentially the same at all six corners. So they all do the same thing. Every snowflake carries traces of its history. In practice the sixfold symmetry is never exact, but it's often very close. Ice is strange stuff, and other shapes are possible too – spikes, flat plates, hexagonal prisms, prisms with plates on their ends. The full story is very complicated, but everything hinges on how the atoms in ice crystals are arranged.[38] In Kepler's day, atomic theory was at best a vague suggestion by a few ancient Greeks; it's amazing how far he got on the basis of folklore observations, thought experiments, and a feeling for pattern.

The Kepler conjecture is not about snowflakes as such. It is his offhand remark that stacking layers of close-packed spheres, so that successive layers fit into the gaps between those in the previous layer, leads to 'the closest pack in three dimensions'. The conjecture can be summarised informally: if you want to pack a large number of oranges into a big box, filling as much of the box as possible, then you should pack them together the way any greengrocer would.

The difficulty is not to find the *answer*. Kepler told us that. What's difficult is to prove he was right. Over the centuries, plenty of indirect evidence accumulated. No one could come up with a closer packing. The same arrangement of atoms is common in crystals, where efficient packing presumably corresponds to minimising energy, a standard principle that governs many natural forms. This kind of evidence was good enough to satisfy most physicists. On the other hand, no one could come up with a proof that there *wasn't* anything better. Simpler questions of the same kind, such as packing circles in the plane, turned out to have hidden depths. The whole area was difficult and full of surprises. All this worried mathematicians, even though most of them thought Kepler had the right answer. In 1958 C. Ambrose Rogers described the Kepler conjecture as something that 'many mathematicians believe, and all physicists know'.[39] This chapter describes how mathematicians turned belief into certainty.

To understand what they did, we need to take a close look at Kepler's arrangement of spheres, which is known as the face-centred cubic lattice. When we do, the subtleties of the problem start to emerge. The first question that springs to mind is why we use square layers. After all, the tightest pack in a single layer occurs for the *triangular* lattice. The answer is that we can also obtain the face-centred cubic lattice using triangular layers; this is the essence of Kepler's remark that 'the triangular pattern cannot exist without the square'. However, it is easier to describe the face-centred cubic lattice using square layers. As a bonus, we see that the Kepler conjecture is not quite as straightforward as greengrocers packing oranges.

Suppose that we start with a flat layer of spheres arranged in triangles, Figure 16 (right). In between the spheres are curved triangular dents, and a further layer of spheres can fit into these. When we began with a square layer, we were able to use all of the dents, so the position of the second layer, and those that followed, was uniquely determined. This is no longer the case if we begin with a triangular arrangement. We can't use all of the dents, because they are too close together. We can use only half of them. One choice is shown in Figure 18 (left), using small grey dots for clarity, and Figure 18 (right) shows how the next layer of spheres should be placed. The second way to fit a new layer into the dents of layer 1 is shown in Figure 19 (left) using darker dots. These dots coincide with dents in layer 2, so we add layer 3 in the corresponding positions: the result is Figure 19 (right).

The distinction between these choices doesn't actually make any difference when we have just these two layers. If we rotate the second arrangement through 60 degrees, we get the first. They are the same 'up to symmetry'. But after the first two layers have been positioned, there are two genuinely different choices for the third layer. Each new layer has two systems of dents, shown by the light and dark dots in Figure 19 (left). One system matches the centres of the layer immediately below, visible as small light grey triangles in Figure 19 (right). The other matches dents in the layer below that, visible as

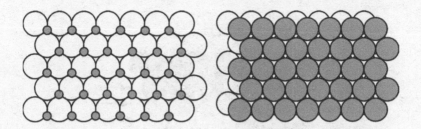

Fig 18 Fitting a triangular lattice into a set of gaps in the layer below.

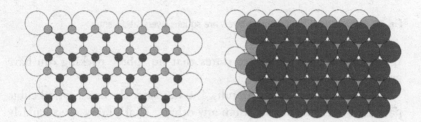

Fig 19 Stacking triangular lattices on top of each other.

triangles containing a tiny white hexagon in Figure 19 (right). To get the face-centred cubic lattice we must use the dark grey positions for the third layer, and then continue the same pattern indefinitely.

It's not entirely obvious that the result is the face-centred cubic lattice. Where are the squares? The answer is that they are present, but they slope at an angle. Figure 20 shows six successive triangular layers, with a number of spheres removed. The arrows indicate the rows and columns of a square lattice, hidden away inside. Layers parallel to this one arc also square lattices, and they fit together in exactly the way I constructed the face-centred cubic lattice.

How 'tight' is this packing? We measure the tightness (efficiency, closeness) of a packing by its density: the proportion of space occupied by spheres.[40] The bigger the density, the tighter the pack. Cubes pack together with density 1, filling the whole of space. Spheres obviously have to leave gaps, so the density is less than 1. For the face-centred cubic lattice the density is exactly $\pi/\sqrt{18}$, which is roughly 0.7405. So for this packing, the spheres fill slightly less than three quarters of the

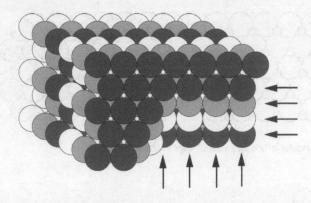

Fig 20 Hidden inside the triangular layers are square layers, at a slant.

space. The Kepler conjecture states that no sphere packing can have greater density than this.

I've stated that rather carefully. I've not said 'the face-centred cubic packing has greater density than any other'. That's false, spectacularly so. To see why, go back to the construction of the face-centred cubic lattice using triangular layers. I said that once the first two layers are determined, there are two choices for the third layer. The face-centred cubic lattice appears if we use the second one, the dark grey dots. What happens if we use the other one, the light grey dots? Now layer 3 sits exactly above layer 1. If we continue like that, placing each new layer immediately above the layer two stages below, we get a second lattice packing: the hexagonal lattice. It is genuinely different from the face-centred cubic packing, but it has the same density. This is intuitively clear because the two different ways to place the third layer are related by a rotational symmetry, so it fits the previous layer exactly as tightly, either way.

These are the only two *lattice* packings that can be obtained from successive triangular layers, but in 1883 the geologist and crystallographer William Barlow pointed out that we can choose the location of each successive layer at random from the two possibilities. Since either position makes the same contribution to the density, all of these packings have density $\pi/\sqrt{18}$. There are infinitely many random sequences, leading to infinitely many different packings, all having that density.

In short, there is no such thing as 'the' densest sphere packing. Instead, there are infinitely many of them, all equally dense. This lack of uniqueness is a warning: this is not a straightforward problem. The optimal *density* is unique, if Kepler was right, but infinitely many different arrangements have that density. So a proof that this density really is optimal is not just a matter of successively fitting each new sphere as tightly as possible. There are choices.

Impressive though the experience of greengrocers may be – and the face-centred cubic lattice was surely present in predynastic Egyptian markets – it's not remotely conclusive. In fact, it's a bit of an accident that the greengrocer's method yields a good answer. The problem facing greengrocers isn't that of packing oranges as closely as possible in space, where any arrangement is in principle possible. It is to pile up oranges stably, in a world where the ground is flat and gravity acts downwards. Greengrocers naturally start by making a layer; then they add another layer, and so on. They are likely to make the first layer into a square lattice if they're putting the oranges inside a rectangular box. If the oranges are unconfined then either a square or a triangular lattice is natural. As it happens, both give the same face-centred cubic lattice – at least, if the layers are appropriately placed in the triangular case. The square lattice actually looks like a poor choice, because it's not the densest way to pack a layer. By luck more than judgement, that turns out not to matter.

Physicists aren't interested in oranges. What they want to pack together are atoms. A crystal is a regular, spatially periodic arrangement of atoms. The Kepler conjecture explains the periodicity as a natural consequence of the atoms packing as closely as possible. As far as most physicists are concerned, the existence of crystals is evidence enough, so the conjecture is evidently true. However, we've just seen that there are infinitely many ways to pack spheres just as densely as the face-centred cubic and hexagonal lattices do, none of which are spatially periodic. So why does nature use periodic patterns for crystals? A possible answer is that we shouldn't model atoms as spheres.

Mathematicians aren't interested in oranges either. Like Kepler, they prefer to work with perfect, identical spheres. They don't find the physicists' argument convincing. If we shouldn't model atoms as spheres, the existence of crystals ceases to be evidence in favour of the

Kepler conjecture. We can't have it both ways. Even if you argue that the conjecture sort-of explains the crystal lattice, and the crystal lattice sort-of shows that the conjecture is correct ... there's a logical gap. Mathematicians want a proof.

Kepler didn't call his statement a conjecture: he just put it in his book. It's totally unclear whether he intended it to be interpreted in such a sweeping manner. Was he claiming that the face-centred cubic lattice was the 'closest pack in three dimensions' among all conceivable ways to pack spheres? Or did he just mean it was the closest pack of the three he had considered? We can't go back and ask. Whatever the historical reality, the interpretation of interest to mathematicians and physicists was the sweeping one, the ambitious one. The one that asked you to contemplate every possible way to pack infinitely many spheres together in infinite space – and show that none of them has greater density than the face-centred cubic lattice.

It is very easy to underestimate the difficulty of the Kepler conjecture. Surely the way to get the tightest pack is to add spheres one by one, making each one touch as many of the others as possible as you go? That leads inevitably to Kepler's pattern. And so it does if you add the spheres in the right order, placing them in the right positions when there are alternatives. However, there's no guarantee that this step-by-step process, adding spheres one at a time, can't be trumped by something more far-reaching. Anyone who has packed holiday gear into the boot of a car learns that fitting things in one by one can leave gaps into which nothing will fit, but starting again from the beginning and taking more care sometimes squeezes more in. Admittedly, part of the problem in packing holiday gear is the different shapes and sizes of the objects you're trying to fit in, but the logical point is clear enough: ensuring the tightest arrangement in a small region could have knock-on effects, and fail to lead to the tightest arrangement in a bigger region.

The arrangements that Kepler considers are very special. It is conceivable that some entirely different arrangement might be able to pack identical spheres together even more tightly. Maybe bumpy layers would be more efficient. Maybe 'layers' is the wrong idea. And even if you are absolutely certain that it's the right idea, you still have to prove that.

Not convinced? Still think it's obvious? So obvious that it doesn't *need* a proof? Let me try to destroy your confidence in your intuition for sphere packing. Here's a much simpler question, involving circles in the plane. Suppose I give you 49 identical circles, each of diameter 1 unit. What is the size of the *smallest* square that can contain them, if they are packed together without overlapping? Figure 21 (left) shows the obvious answer: pack them like milk bottles in a crate. The side of the crate is exactly 7 units. To prove this is the best, observe that each circle is held rigidly in place by all the others, so there's no way to create extra room. Figure 21 (right) shows that this answer is wrong. Pack them in the irregular manner shown, and they will fit into a square crate whose side is slightly less than 6.98 units.[41] So the proof is wrong as well. Being rigid is no guarantee that you can't do better.

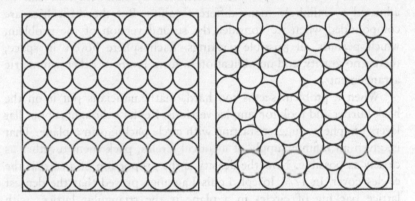

Fig 21 *Left*: 49 circles in a 7 × 7 square. *Right*: How to fit 49 circles into a slightly smaller square.

In fact, it's easy to see that the reasoning leading to the answer '7' can't possibly be right. Just consider larger squares. Using a square lattice, n^2 circles of diameter 1 pack into a square of side n. There is no way to improve the density by moving these circles in a continuous manner, because the packing is rigid. But there must be denser packings for large enough n, because a triangular lattice is more efficient than a square lattice. If we take a really big square, and fit as many circles into it as we can using a triangular lattice, the advantage that this has over the square lattice will eventually cause it to win, despite 'edge effects' at the boundary where you have to leave gaps. The size of the boundary is $4n$, which gets arbitrarily small compared to n^2. As it happens, the exact point at which the triangular lattice takes over is when $n = 7$. That's not obvious and it takes a lot of detailed work to establish it, but some n has to work. Rigidity is not enough.

There are really two versions of the Kepler conjecture. One considers only lattice packings, where the centres of the spheres form a spatially periodic pattern, repeating itself indefinitely in three independent directions like a kind of solid wallpaper. Even then the problem is hard, because there are many different lattices in space. Crystallographers recognise 14 types, classified by their symmetries, and some of those types are determined by numbers that can be adjusted to infinitely many different values. But the difficulties are compounded when we consider the second version of the problem, which permits all possible packings. Each sphere hovers in space, there's no gravity, and no obligation to form layers or other symmetric arrangements.

When a problem seems too hard, mathematicians put it on the back burner and look for simpler versions. Kepler's thoughts about flat layers of spheres suggest starting with circle packings in a plane. That is, given an infinite supply of identical circles, pack them together as closely as possible. Now the density is the proportion of *area* that the circles cover. In 1773 Joseph Louis Lagrange proved that the densest lattice packing of circles in a plane is the triangular lattice, with density $\pi/\sqrt{12} = 0.9069$. In 1831 Gauss was reviewing a book by Ludwig Seeber, who had generalised some of Gauss's number-theoretic results to equations in three variables. Gauss remarked that Seeber's results prove that the face-centred cubic and hexagonal lattices provide the densest lattice packing in three-dimensional space. An enormous amount is now known about lattice packings in spaces of higher dimension – 4, 5, 6, and so on. The 24-dimensional case is especially well understood. (The subject is like that.) Despite its air of impracticality, this area actually has implications for information theory and computer codes.

Non-lattice packings are another matter entirely. There are infinitely many, and they don't have any nice regular structure. So why not go to the other extreme and try random packings? In his *Vegetable Staticks* of 1727, Stephen Hales reported experiments in which he 'compressed several fresh parcels of pease [peas] in the same pot', finding that when they were all pressed together they formed 'pretty regular dodecahedrons'. He seems to have meant that the regular dodecahedrons were pretty, not that the dodecahedrons were pretty regular, but the second interpretation is better because regular dodecahedrons can't fill space. What he saw were probably rhombic dodecahedrons, which we've seen are associated with the face-centred cubic packing. G. David Scott put lots of ball bearings into a container, and shook it thoroughly, observing that the highest density was 0.6366. In 2008 Chaoming Song, Ping Wang, and Hernán Makse derived this figure analytically.[42] However, their result does not imply that Kepler was right – if only because, as stated, it would imply that the face-centred cubic lattice, with density 0.74, can't exist. The simplest way to explain this discrepancy is that their result ignores extremely rare exceptions. The face-centred cubic lattice, the hexagonal lattice, and all the arrangements of randomly chosen triangular layers are all exceptions of this kind. By the same token, there might exist some other arrangement with an even greater density. It can't be a lattice, but a random search will never find it because its probability is zero. So the study of random packings, while relevant to many questions in physics, doesn't tell us a lot about the Kepler conjecture.

The first real breakthrough came in 1892, when Axel Thue gave a lecture to the Scandinavian Natural Science Congress, sketching a proof that no circle packing in the plane can be denser than the triangular lattice. His lecture was published, but the details are too vague to reconstruct the proof he had in mind. He gave a new proof in 1910, which seemed convincing, save for a few technical points which he simply assumed could be sorted out. Instead of filling these gaps, László Fejes Tóth obtained a complete proof by other methods in 1940. Soon after, Beniamino Segre and Kurt Mahler found alternative proofs. In 2010 Hai-Chau Chang and Lih-Chung Wang put a simpler proof on the web.[43]

Finding the greatest density for circle or sphere packings, under
specified conditions, falls into a general class of mathematical
questions known as optimisation problems. Such a problem seeks
the maximum or minimum value of some function, which is a
mathematical rule for calculating a quantity that depends on some set
of variables in a specific manner. The rule is often specified by a
formula, but this is not essential. For instance, the milk crate problem,
with 49 circles, can be formulated in this manner. The variables are the
coordinates of the centres of the 49 circles; since each circle needs two
coordinates, there are 98 variables. The function is the size of the
smallest square, with sides parallel to the coordinate axes, that
contains a given set of non-overlapping circles. The milk crate problem
is equivalent to finding the minimum value that this function can attain
as the variables range over all packings.

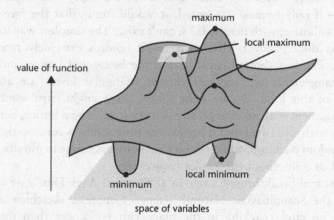

Fig 22 Peaks and valleys for a function.

A function can be thought of as a many-dimensional landscape.
Each point in the landscape corresponds to a choice of the variables,
and the height at that point is the corresponding value of the function.
The maximum of the function is the height of the highest peak, and
the minimum is the depth of the deepest valley. In principle,
optimisation problems can be solved by calculus: the function must
be horizontal at a peak or valley, Figure 22, and calculus expresses this

condition as an equation. To solve the milk-crate problem by that method we would have to solve a system of 98 equations in the 98 variables.

One snag with optimisation problems is that equations like these often have a large number of solutions. A landscape can have lots of local peaks, only one of which is the highest. Think of the Himalayas: hardly anything *but* peaks, yet only Everest holds the altitude record. Methods for finding peaks, of which the most obvious is 'head uphill if you can', often get trapped in a local peak. Another snag is that as the number of variables grows, so does the likely number of local peaks. Nevertheless, this method sometimes works. Even partial results can be useful: if you find a local peak, the maximum must be at least that high. This is how the improved arrangement of circles in the milk-crate problem was found.

For lattice packings, the function whose maximum is sought depends only on finitely many variables, the directions and lengths along which the lattice repeats. For non-lattice packings, the function depends on infinitely many variables: the centres of all the circles or spheres. In such cases the direct use of calculus or other optimisation techniques is hopeless. Tóth's proof used a clever idea to reformulate the non-lattice packing problem for circles as an optimisation problem in a *finite* set of variables. Later, in 1953, he realised that the same trick could in principle be applied to the Kepler conjecture. Unfortunately, the resulting function depends on about 150 variables, far too many for hand calculation. But Tóth presciently saw a possible way out: 'Mindful of the rapid development of our computers, it is imaginable that the minimum may be determined with great exactitude.'

At that time, computing was in its infancy, and no sufficiently powerful machine existed. So subsequent progress on the Kepler conjecture followed different lines. Various mathematicians placed bounds – upper limits – on how dense a sphere packing could be. For example in 1958 Rogers proved that it is at most 0.7797: no rare exceptions, this bound applied to all sphere packings. In 1986 J.H. Lindsey improved the bound to 0.77844, and Douglas Muder shaved off a tiny bit more in 1988 to obtain a bound of 0.77836.[44] These results show that you can't do *much* better than the face-centred cubic lattice's 0.7405. But there was still a gap, and little prospect of getting rid of it.

In 1990 Wu-Yi Hsiang, an American mathematician, announced a proof of the Kepler conjecture. When the details were made public, however, doubts quickly set in. When Tóth reviewed the paper in *Mathematical Reviews*, he wrote: 'If I am asked [whether the paper provides] a proof of the Kepler conjecture, my answer is: no. I hope that Hsiang will fill in the details, but I feel that the greater part of the work has yet to be done.'

Thomas Hales, who had been working on the conjecture for many years, also doubted that Hsiang's method could be repaired. Instead, he decided it was time to take Tóth's approach seriously. A new generation of mathematicians had grown up, for whom reaching for a computer was more natural than reaching for a table of logarithms. In 1996 Hales outlined a proof strategy based on Tóth's idea. It required identifying all possible ways to arrange several spheres in the immediate vicinity of a given one. A sphere packing is determined by the centres of the spheres; for unit spheres, these must be at least 2 units apart. Say that two spheres are *neighbours* if their centres are at most 2.51 units apart. This value is a matter of judgement: make it too small and there isn't enough room to rearrange neighbours to improve the density; make it too big and the number of ways to arrange the neighbours becomes gigantic. Hales found that 2.51 was an effective compromise. Now we can represent how neighbours are arranged by forming an infinite network in space. Its dots are the centres of the spheres, and two dots are joined by a line if they are neighbours. This network is a kind of skeleton of the packing, and it contains vital information about the neighbourhood of each sphere.

For any given sphere, we can look at its immediate neighbours in the network and consider only the lines between these neighbours, omitting the original sphere. The result is a sort of cage surrounding the dot at the centre of the original sphere. Figure 23 (left pair) shows the neighbours of a sphere in the face-centred cubic lattice and the associated cage. Figure 23 (right pair) does the same for a special arrangement of spheres, the pentagonal prism, which turned out to be a key player in the proof. Here there are two bands of pentagons parallel to the 'equator' of the central sphere, plus a single sphere at each pole.

The cages form a solid with flat faces, and the geometry of this solid controls the packing density near the central sphere.[45] The key

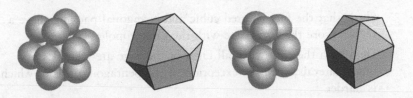

Fig 23 *From left to right*: Neighbourhood of a sphere in the face-centred cubic lattice; the cage formed by its neighbours; neighbourhood of a sphere of pentagonal prism type; the cage formed by its neighbours.

idea is to associate to each cage a number, known as its score, which can be thought of as a way to estimate the density with which the sphere's neighbours are packed. The score isn't the density itself, but a quantity that is better behaved and easier to calculate. In particular, you can find the score of the cage by adding up scores related to its faces, which doesn't work for density. In general many different notions of score satisfy that condition, but they all agree on one thing: for the face-centred cubic and hexagonal lattices the score is always 8 'points', no matter what choices are made in its definition. Here a point is a specific number:

$$4 \arctan \frac{\sqrt{2}}{5} - \frac{\pi}{3} = 0.0553736$$

So 8 points is actually 0.4429888. This curious number comes from the special geometry of the face-centred cubic lattice. Hales's key observation relates the Kepler conjecture to this number: if every cage has a score of 8 points or less, then the Kepler conjecture is true. So the focus shifts to cages and scores.

Cages can be classified by their topology: how many faces they have with a given number of sides, and how those faces adjoin. For a given topology, however, the edges can have many different lengths. These lengths affect the score, but the topology lumps lots of different cages together, and these can be dealt with in the same general way. In his eventual proof, Hales considered about 5000 types of cage, but the main calculations concentrated on a few hundred. In 1992, he proposed a five-stage programme:

1 Prove the desired result when all faces of the cage are triangles.

2 Show that the face-centred cubic and hexagonal packings have a higher score than any cage with the same topology.

3 Deal with the case when all faces of the cage are triangles and quadrilaterals, with the exception of the pentagonal prism, which is harder.

4 Deal with any cage having a face with more than four edges.

5 Sort out the only case remaining, when the cage is a pentagonal prism.

Part 1 was solved in 1994, and part 2 in 1995. As the programme developed, Hales modified the definition of a cage to simplify the argument (his term is 'decomposition star'). The new definition doesn't alter the two cages illustrated, and it didn't have any serious effects on those parts of the proof that had already been obtained. By 1998, using this new concept, all five stages had been completed. Hales's student Samuel Ferguson solved Part 5, the tricky case of a pentagonal prism.

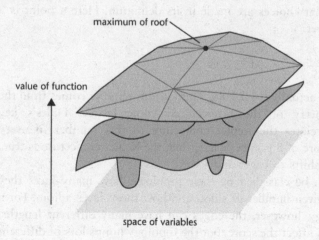

maximum of roof

value of function

space of variables

Fig 24 Fitting a roof over the top of a function.

The analysis involved heavy use of a computer at all stages. The trick is to pick, for each local network, a notion of score that makes the calculation relatively easy. Geometrically, replacing the density by the score puts a kind of roof over the top of the smooth landscape whose peak is being sought. The roof is made from numerous flat pieces, Figure 24. Shapes like this are easier to work with than smooth surfaces, because the maxima must occur at the corners, and these can be found by solving much simpler equations. There are efficient methods for doing this, known as linear programming. If the roof has cunningly been constructed so that its peak coincides with the peak of the smooth surface, then this simpler computation locates the peak of the smooth surface.

There is a price to pay for this approach: you have to solve about 100,000 linear programming problems. The calculations are lengthy, but well within the capabilities of today's computers. When Hales and Ferguson prepared their work for publication it ran to about 250 pages of mathematics, plus 3 gigabytes of computer files.

In 1999 Hales submitted the proof to the *Annals of Mathematics*, and the journal chose a panel of 12 expert referees. By 2003, the panel declared itself '99 per cent certain' that the proof was correct. The remaining uncertainty concerned the computer calculations; the panel had repeated many of them, and otherwise checked the way the proof was organised and programmed, but they were unable to verify some aspects. After a delay, the journal published the paper. Hales recognised that this approach to the proof would probably never be certified 100 per cent correct, so in 2003 he announced that he was starting a project to recast the proof in a form that could be verified by a computer using standard automated proof-checking software.

This may sound like stepping out of the frying-pan to fall into the fire, but it actually makes excellent sense. The proofs that mathematicians publish in journals are intended to convince humans. As I said in chapter 1, that sort of proof is a kind of story. Computers are poor at telling stories, but they excel at something we are hopeless at: performing long, tedious calculations without making a mistake. Computers are ideal for the formal concept of proof in the undergraduate textbooks: a series of logical steps, each following from previous ones.

Computer scientists have exploited this ability. To check a proof, get a computer to verify each logical step. It ought to be easy, but the proofs in the journals aren't written like that. They leave out anything routine or obvious. The traditional phrases are easy to spot: 'It is easy to verify that ...' 'Using the methods of Cheesberger and Fries, modified to take account of isolated singularities, we see that ...' 'A short calculation establishes ...' Computers can't (yet) handle that kind of thing. But humans can rewrite proofs with all these gaps filled in, and computers can then verify each step.

The reason that we're not hopping straight back into fire territory is straightforward: the software that does the verification needs to be checked only *once*. It is general-purpose software, applicable to all proofs written in the right format. All the worries about computer proofs are concentrated in that one piece of software. Verify that, and it can be used to verify everything else. You can even 'bootstrap' the process by writing the proof verification software in a language that can be checked by a much simpler piece of proof verification software.

In recent years, proofs of many key mathematical theorems have been verified in this manner. Often the proofs have to be presented in a style that is more suitable for computer manipulation. One of the current triumphs is a verified proof of the Jordan curve theorem: every closed curve in the plane that does not cross itself divides the plane into two distinct connected regions. This may sound obvious, but the pioneers of topology had trouble finding a rigorous proof. Camille Jordan finally succeeded in 1887 with a proof over 80 pages long, but he was later criticised for making unsubstantiated assumptions. Instead, the credit went to Oswald Veblen, who gave a more detailed proof in 1905, saying '[Jordan's] proof ... is unsatisfactory to many mathematicians. It assumes the theorem without proof in the important special case of a simple polygon, and of the argument from that point on, one must admit at least that all details are not given.' Later mathematicians accepted Veblen's criticism without demur, but recently Hales went over Jordan's proof and found 'nothing objectionable' in it. In fact, Veblen's remark about a polygon is bizarre: the theorem is straightforward for a polygon, and Jordan's proof doesn't rely on that version anyway.[46] Storytelling proofs have their own dangers. It's always worth checking whether the popular version of the story is the same as the original one.

As a warm-up to the Kepler conjecture, Hales gave a formal computer-verified proof of the Jordan curve theorem in 2007, using

60,000 lines of computer code. Soon after, a team of mathematicians produced another formal proof using different software. Computer verification isn't totally foolproof, but neither are traditional proofs. In fact, many mathematical research papers probably contain a technical error somewhere. These errors show up from time to time, and mostly they turn out to be harmless. Serious errors are usually spotted because they introduce inconsistencies, so that something visibly makes no sense. This is another downside of the storytelling approach: the price we pay for making a proof comprehensible by humans is that a ripping yarn can sometimes be very convincing even when it's wrong.

Hales calls his approach *Project FlysPecK* – the F, P, and K standing for 'formal proof of Kepler'. Initially, he estimated that it would take about 20 years to complete the task.[47] Nine years into the project, considerable progress has already been made. It may finish early.

6

New solutions for old
Mordell Conjecture

NOW WE'RE HEADING BACK INTO the realms of number
theory, aiming towards Fermat's last theorem. To prepare
the ground I'll start with a less familiar but arguably even more
important problem. In 2002 Andrew Granville and Thomas Tucker
introduced it like this:[48]

> In [1922] Mordell wrote one of the greatest papers in the history of
> mathematics... At the very end of the paper, Mordell asked five
> questions which were instrumental in motivating much of the
> important research in Diophantine arithmetic in the twentieth
> century. The most important and difficult of these questions was
> answered by Faltings in 1983 by inventing some of the deepest and
> most powerful ideas in the history of mathematics.

Mordell is the British number theorist Louis Mordell, who was
born in the United States to a Jewish family of Lithuanian origin, and
Faltings is the German mathematician Gerd Faltings. The question
referred to became known as the Mordell conjecture, and the quote
gives away its current status: proved, brilliantly, by Faltings.

The Mordell conjecture belongs to a major area of number theory:
Diophantine equations. They are named after Diophantus of
Alexandria, who wrote a famous book, the *Arithmetica*
('Arithmetic') around AD 250. It is thought that originally the
Arithmetica included 13 books, but only six have survived, all later
copies. This was not an arithmetic text in the sense of addition and
multiplication sums. It was the first algebra text, and it collected most

of what the Greeks knew about how to solve equations. It even had a rudimentary form of algebraic notation, which is believed to have used a variant ς of the Greek letter sigma for the unknown (our x), Δ^Y for its square (our x^2), and K^Y for its cube (our x^3). Addition was denoted by placing symbols next to each other, subtraction had its own special symbol, the reciprocal of the unknown (our $1/x$) was ς^x, and so on. The symbols have been reconstructed from later copies and translations, and may not be entirely accurate.

In the spirit of classical Greek mathematics, the solutions of equations sought in the *Arithmetica* were required to be rational numbers – that is, fractions like 22/7 formed from whole numbers. Often they were required to *be* whole numbers. All numbers involved were positive: negative numbers were introduced several centuries later in China and India. We now call such problems Diophantine equations. The book includes some remarkably deep results. In particular Diophantus appears to be aware that every whole number can be expressed as the sum of four perfect squares (including zero). Lagrange gave the first proof in 1770. The result that interests us here is a formula for all Pythagorean triples, in which two perfect squares add to form another perfect square. The name comes from Pythagoras's theorem: this relation holds for the sides of a right-angled triangle. The best known example is the celebrated 3-4-5 triangle: $3^2 + 4^2 = 5^2$. Another is $5^2 + 12^2 = 13^2$. There are infinitely many of these Pythagorean triples, and there is a recipe for finding them all in two lemmas (auxiliary propositions) preceding Propositions 29 and 30 in Book X of Euclid's *Elements*.

Euclid's procedure yields infinitely many Pythagorean triples. Mordell knew several other Diophantine equations for which there exists a formula that yields infinitely many solutions. He also knew another type of Diophantine equation with infinitely many solutions, *not* prescribed by a formula. These are the so-called elliptic curves – a rather silly name since they have virtually nothing to do with ellipses – and the infinitude of solutions arises because any two solutions can be combined to give another one. Mordell himself proved one of the basic properties of these equations: you need only a finite number of solutions to generate them all through this process.

Aside from these two types of equation, every other Diophantine equation that Mordell could think of fell into one of two categories.

Either it was known to have only finitely many solutions, including none at all, or no one knew whether the number of solutions was finite or infinite. This alone wasn't news, but Mordell thought he could spot a pattern that no one else had noticed. It wasn't a number-theoretic pattern at all; it came from topology. What mattered was how many holes the equation had. And to make sense of that, you had to think about its solutions in complex numbers, not rational numbers or integers. Which somehow seemed contrary to the entire spirit of Diophantine equations.

It's worth filling in a few details here. They will help a lot later. Don't be put off by the algebra; it's mainly there to give me something specific to refer to. Concentrate on the story behind the algebra.

Pythagorean triples are solutions, in integers, of the Pythagorean equation

$$x^2 + y^2 = z^2$$

Dividing by z^2 gives

$$(x/z)^2 + (y/z)^2 = 1$$

By chapter 3, this tells us that the pair of rational numbers $(x/z, y/z)$ lies on the unit circle in the plane. Now, the Pythagorean equation originated in geometry, and its interpretation is that the associated triangle has a right angle. The formula I've just derived provides a slightly different geometric interpretation, not just of one Pythagorean triple, but of all of them. The solutions of the Pythagorean equation correspond directly and naturally to all of the rational points on the unit circle. Here a point is said to be rational provided both of its coordinates are.

You can deduce a lot of interesting facts from this connection. With a bit of trigonometry, or by direct algebra, you can discover that for any number t the point

$$\left(\frac{2t}{t^2 + 1}, \frac{t^2 - 1}{t^2 + 1} \right)$$

lies on the unit circle. Moreover, if t is rational, so is this point. All rational points arise in this manner, so we have a complete formula for all solutions of the Pythagorean equation. It is equivalent to Euclid's formula, which is the same as Diophantus's. As an example, if $t = 22/7$ then the formula yields

$$\left(\frac{308}{533}, \frac{435}{533}\right)$$

and you can check that $308^2 + 435^2 = 533^2$. For us, the precise formula is not terribly important; what matters is that there is one.

This is not the only Diophantine equation for which a formula gives all solutions, but they are relatively rare. Others include the so-called Pell equations, such as $x^2 = 2y^2 + 1$. This has infinitely many solutions, such as $3^2 = 2 \times 2^2 + 1$, $17^2 = 2 \times 12^2 + 1$, and there is a general formula. However, Pythagorean triples have more structure than that, also derived from the geometry. Suppose you have two Pythagorean triples. Then there are two corresponding solutions of the Pythagorean equation – rational points on the circle. Geometry provides a natural way to 'add' those points together. Start from the point $(1,0)$ at which the circle cuts the horizontal axis, and find the angles between this point and the two solutions. Add the two angles together, Figure 25, and see what point results. It certainly lies on the circle. A short calculation shows that it is rational. So from any two solutions, we can derive a third. Mathematicians had already noticed many facts like this. Most of them make immediate sense if you think of rational points on a circle.

The 'short calculation' that I slid over makes use of trigonometry. The classical trigonometric functions such as the sine and cosine are intimately related to the geometry of a circle. The calculation alluded to uses standard, rather elegant formulas for the sine and cosine of the sum of two angles in terms of the sines and cosines of the angles themselves. There are many ways to set up sines and cosines, and a rather neat one comes from integral calculus. If you integrate the algebraic function $1/\sqrt{1-x^2}$, the result can be expressed in terms of the sine function. In fact, what we need is the inverse function of the sine: the angle whose sine is the number we're thinking of.[49]

The integral arises when we try to derive a formula for the arc length of a circle using calculus, and the geometry of the circle has a

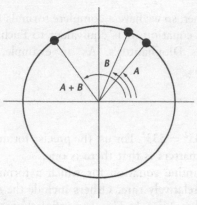

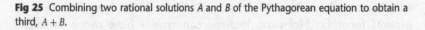

Fig 25 Combining two rational solutions A and B of the Pythagorean equation to obtain a third, $A + B$.

simple but very important implication for the result. The circumference of the unit circle is 2π, so going round the circle a distance 2π brings you back to the exact same point. The same goes for any integer multiple of 2π: by the standard mathematical convention, positive integers correspond to the anticlockwise direction, negative ones to the clockwise direction. It follows that the sine and cosine of a number remain unchanged if an integer multiple of 2π is added to that number. We say that the functions are periodic, with period 2π.

The analysts of the eighteenth and nineteenth centuries discovered a vast generalisation of this integral, along with a host of interesting new functions analogous to the familiar trigonometric ones. These new functions were intriguing; they were periodic, like the sine and cosine, but in a more elaborate way. Instead of having one period, like 2π (and its integer multiples), they had two independent periods. If you try to do that with real functions, all you get are constants, but for complex functions, the possibilities are much richer.

The area was initiated by the Italian mathematician Giulio di Fagnano and the prolific Euler. Fagnano was trying to find the arc length of an ellipse using calculus, but he couldn't find an explicit formula – no longer a surprise since we now know there isn't one.

However, he did notice a relationship between the lengths of various special arcs, which he published in 1750. Euler noticed the same relationship in the same context, and presented it as a formal relation between integrals. They are similar to the one associated with the sine function, but the quadratic expression $1 - x^2$ under the square root is replaced by a cubic or quartic polynomial, for example the quartic $(1 - x^2)(1 - 4x^2)$.

In 1811 Adrien-Marie Legendre published the first book in a massive three-volume treatise on these integrals, which are known as elliptic integrals because of their connection with the arc length of a segment of an ellipse. He managed to overlook the most significant feature of these integrals, however: the existence of new functions, analogous to the sine and cosine, whose *inverse* functions express the value of the integral in a simple way.[50] Gauss, Niels Henrik Abel, and Carl Jacobi quickly spotted the oversight. Gauss, rather typically, kept the discovery to himself. Abel submitted a paper to the French Academy in 1826, but Cauchy, the president, mislaid the manuscript, and it was not published until 1841, twelve years after Abel's tragic early death from lung disease. However, another paper by Abel on the same topic was published in 1827. Jacobi made these new 'elliptic functions' the basis of a huge tome, published in 1829, which propelled complex analysis along an entirely new trajectory.

What emerged was a beautiful package of interrelated properties, analogous to those of the trigonometric functions. The relationship noticed by Fagnano and Euler could be reinterpreted as a simple list of formulas relating elliptic functions of the sum of two numbers to elliptic functions of the numbers themselves. The most wonderful feature of elliptic functions outdoes trigonometric functions in a spectacular way. Not only are elliptic functions periodic: they are doubly periodic. A line is one-dimensional, so patterns can repeat in only one direction, along the line. The complex plane is two-dimensional, so patterns can repeat like wallpaper: down the strip of paper, and also sideways along the wall into adjacent strips of paper. Associated with each elliptic function are two independent complex numbers, its periods, and adding either of them to the variable does not change the value of the function.

Repeating this process, we conclude that the value of the function does not change if we add any integer combination of the two periods to the variable. These combinations have a geometric interpretation: they determine a lattice in the complex plane. The lattice specifies a tiling of the plane by parallelograms, and whatever happens in one parallelogram is copied in all the others, Figure 26. If we consider just one parallelogram, the way it joins up with adjacent copies means that we have to identify opposite edges, in the same way that a torus is defined by identifying opposite edges of a square, Figure 12. A parallelogram with opposite edges identified is also a topological torus. So, just as the sine and cosine are related to the circle, elliptic functions are related to a torus.

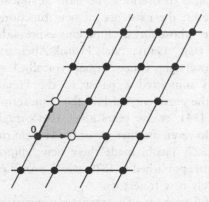

Fig 26 Lattice in the complex plane. The arrows point to the two periods, shown as white dots. The value of the function in the shaded parallelogram determines it in every other parallelogram.

There is also a link to number theory. I said that the inverse sine function is obtained by integrating a formula involving the square root of a quadratic polynomial. Elliptic functions are similar, but the quadratic is replaced by a cubic or quartic polynomial. The quartic case was mentioned briefly above, because historically it came first, but now let's focus on the cubic case. If we denote the square root by y, and the polynomial by $ax^3 + bx^2 + cx + d$ where a, b, c, d are numerical coefficients, then x and y satisfy the equation

$$y^2 = ax^3 + bx^2 + cx + d$$

This equation can be thought of in several different contexts, depending on what restrictions are placed on the variables and coefficients. If they are real, the equation defines a curve in the plane. If they are complex, algebraic geometers still call the set of solutions a curve, by analogy. But now it is a curve in the space of pairs of complex numbers, which is four-dimensional in real coordinates. And the curve is actually a surface, from this real-number viewpoint.

Figure 27 shows the real elliptic curves $y^2 = 4x^3 - 3x + 2$ and $y^2 = 4x^3 - 3x$, which are typical. Because y appears as its square, the curve is symmetric about the horizontal axis. Depending on the coefficients, it is either a single sinuous curve or it has a separate oval component. Over the complex numbers, the curve is always one connected piece.

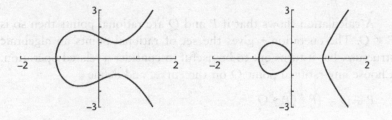

Fig 27 Typical real elliptic curves. *Left*: $y^2 = 4x^3 - 3x + 2$. *Right*: $y^2 = 4x^3 - 3x$.

Number theory comes into play when we require the variables and coefficients to be rational. Now we are looking at a Diophantine equation. It is rather confusingly called an elliptic curve, even though it looks nothing like an ellipse, because of the link to elliptic functions. It's like calling a circle a triangular curve because of the link to trigonometry. Unfortunately, the name is now inscribed in tablets of stone, so we have to live with it.

Because elliptic functions have a deep and rich theory, number theorists have discovered innumerable beautiful properties of elliptic curves. One is closely analogous to the way we can combine two solutions of the Pythagorean equation by adding the associated angles. Two points on an elliptic curve can be combined by drawing a straight line through them and seeing where it meets the curve for a third time, Figure 28. (There always is such a third point, because the equation is a cubic. However, it might be 'at infinity', or it might coincide with one of the first two points if the line cuts the curve at a tangent.) If the two points are P and Q, denote the third one by $P * Q$.

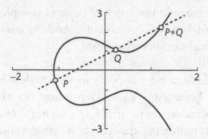

Fig 28 Combining points P, Q to get the point $P * Q$.

A calculation shows that if P and Q are rational points then so is $P * Q$. The operation $*$ gives the set of rational points an algebraic structure, but it turns out to be useful to consider a related operation. Choose any rational point O on the curve, and define

$$P + Q = (P * Q) * O$$

This new operation obeys some basic laws of ordinary algebra, with O behaving like zero, and it turns the set of all rational points into what algebraists call a group, chapter 10. The essential point is that like Pythagorean triples, you can 'add' any two solutions to get a third. The occurrence of this 'group law' on the rational points is striking, and in particular it means that once we have found two rational solutions of the Diophantine equation, we automatically get many more.

Around 1908 Poincaré asked whether there exists a finite number of solutions from which all other solutions can be obtained by applying the group operation over and over again. This result is important because it implies that *all* rational solutions can be characterised by writing down a finite list. In his spectacular 1922 paper Mordell proved that the answer to Poincaré's question is 'yes'. Now elliptic curves became of central importance in number theory, because it was unusual to have that kind of control over any Diophantine equation.

Both the Pythagorean equation and elliptic curves, then, have infinitely many rational solutions. In contrast, many Diophantine equations have only finitely many solutions, often none at all. I'm going to digress slightly to discuss an entire family of such equations, and the recent remarkable proof that the obvious solutions are the only ones that exist.

The Pythagoreans were interested in their equation because they believed that the universe is founded on numbers. In support of this philosophy, they discovered that simple numerical ratios govern musical harmony. They observed this experimentally using a stretched string. A string of the same tension that is half as long plays a note one octave higher. This is the most harmonious combination of two notes: so harmonious that it sounds a bit bland. In Western music the next most important harmonies are the fourth, where one string is $\frac{3}{4}$ as long as the other, and the fifth, where one string is $\frac{2}{3}$ as long as the other.[51]

Starting with 1 and repeatedly multiplying by 2 or 3 yields numbers 2, 3, 4, 6, 8, 9, 12, and so on – numbers of the form $2^a 3^b$. Because of the musical connection these became known as harmonic numbers. In the thirteenth century a Jewish writer who lived in France wrote *Sha'ar ha-Shamayim* ('Door of heaven'), an encyclopaedia based on Arab and Greek sources. He divided it into three parts: physics, astronomy, and metaphysics. His name was Gerson ben Solomon Catalan. In 1343 the Bishop of Meaux persuaded Gerson's son (well, historians think it was probably his son) Levi ben Gerson to write a mathematical book, *The Harmony of Numbers*. It included a problem raised by the composer and musical theorist Philippe de Vitry: when can two harmonic numbers differ by 1? It is easy to find such pairs: de Vitry knew of four, namely (1, 2), (2, 3), (3, 4), and (8, 9). Ben Gerson proved that these are the only possible solutions.

Among de Vitry's pairs of harmonic numbers, the most interesting is (8, 9). The first is a cube, 2^3; the second is a square, 3^2. Mathematicians started to wonder whether other squares and cubes might differ by 1, and Euler proved that they could not, aside from the trivial case (0, 1), and also (−1, 0) if negative integers are allowed. In 1844 the second Catalan in the story committed to print a more sweeping claim, which many mathematicians must have thought of but hadn't bothered to make explicit. He was the Belgian mathematician

Eugène Charles Catalan, and in 1844 he wrote to one of the leading mathematics journals of the day, the *Journal für die Reine und Angewandte Mathematik* ('Journal for pure and applied mathematics'):

> I beg you, sir, to please announce in your journal the following theorem that I believe true although I have not yet succeeded in completely proving it; perhaps others will be more successful. Two consecutive whole numbers, other than 8 and 9, cannot be consecutive powers; otherwise said, the equation $x^m - y^n = 1$ in which the unknowns are positive integers only admits a single solution.

This statement became known as the Catalan conjecture. The exponents m and n are integers greater than 1.

Despite partial progress, the Catalan conjecture stubbornly refused to surrender, until it was dramatically proved in 2002 by Preda Mihăilescu. Born in Romania in 1955, he had settled in Switzerland in 1973, and had only recently completed his PhD. The title of his thesis was 'Cyclotomy of rings and primality testing', and it applied number theory to primality testing, chapter 2. This problem had no particular bearing on the Catalan conjecture, but Mihăilescu came to realise that his methods most certainly did. They derived from ideas that I mentioned in chapter 3: Gauss's construction of the regular 17-gon and associated algebraic equations, whose solutions are called cyclotomic numbers. The proof was highly technical and came as a shock to the mathematical community. It tells us that whichever values we choose for the two powers, the number of solutions is finite – and aside from obvious solutions using 0 and ± 1, the only interesting one is $3^2 - 2^3 = 1$.

The above examples show that some Diophantine equations have infinitely many solutions, others don't. Big deal – those alternatives cover everything. If you start asking which equations are of which type, however, it gets more interesting. Mordell, an expert on Diophantine equations, was writing a seminal textbook. In his day the area looked like early biology: a lot of butterfly collecting and very little in the way of systematic classification. Here was a Painted Pythagorean, there a Large Blue Elliptic, and in the bushes were caterpillars of the Speckled

Pellian. The field was just as Diophantus had left it, only more so: a structureless list of separate tricks, one for each type of equation. This is poor textbook material, and it desperately needed organising, so Mordell set out to do just that.

At some point he must have noticed that all of the equations known to have infinitely many rational solutions – such as the Pythagorean equation and elliptic curves – had a common feature. He focused on one class of equations, those that (after being converted to rational number equations, as I did for Pythagoras) involve just two variables. There are two cases where we know how to find infinitely many solutions. One is exemplified by the Pythagorean equation in the equivalent form $x^2 + y^2 = 1$. Here there is a formula for the solutions. Plug any rational number into the formula and you get a rational solution, and all solutions arise. The other is exemplified by elliptic curves: there is a *process* that generates new solutions from old ones, and a guarantee that if you start with a suitable finite set of solutions, this process produces them all.

Mordell's conjecture states that whenever there are infinitely many rational solutions, one of these two features must apply. Either there is a general formula, or there is a process that generates all solutions from a suitable finite set of them. In all other cases, the number of rational solutions is finite, for example the equations $x^m - y^n = 1$ that feature in the Catalan conjecture. In a sense, the solutions are then just coincidences, with no underlying structure.

Mordell arrived at this observation in a slightly different way. He noticed that every equation with infinitely many rational solutions has a striking topological feature. It has genus 0 or 1. Recall from chapter 4 that genus is a concept from the topology of surfaces, and it counts how many holes the surface has. A sphere has genus 0, a torus genus 1, a 2-holed torus has genus 2, and so on. How do surfaces come into a problem in number theory? From coordinate geometry. We saw that the Pythagorean equation, interpreted in terms of rational numbers and extended to allow real numbers as solutions, determines a circle. Mordell went one step further, and allowed complex numbers as solutions. Any equation in two complex variables determines what algebraic geometers call a complex curve. However, from the point of view of real numbers and the human visual system, every complex number is two-dimensional: it has two real components, its real and imaginary parts. So a 'curve' to complex eyes is a surface to you and me. Being a surface, it has a genus: there you go.

Whenever an equation was known to have only finitely many solutions, its genus was at least 2. Important equations whose status was unknown also had genus at least 2. In a wild and courageous leap, based on what seemed at the time to be pretty flimsy evidence, Mordell conjectured that any Diophantine equation with genus 2 or more has only finitely many rational solutions. At a stroke, the Diophantine butterflies were neatly arranged into related families; appropriately, by genus.

There was only one tiny snag with Mordell's conjecture. It related two extremely different things: rational solutions and topology. At the time, any plausible link was tenuous in the extreme. If a connection existed, no one knew how to go about finding it. So the conjecture was wild, unsubstantiated speculation, but the potential payoff was huge.

In 1983 Faltings published a dramatic proof that Mordell's wild speculation had, in fact, been right. His proof used deep methods from algebraic geometry. A very different proof, based on approximating real numbers by rational ones, was soon found by Paul Vojta, and Enrico Bombieri published a simplified proof along the same lines in 1990. There is an application of Faltings's theorem to Fermat's last theorem, a problem that we treat at length in chapter 7. This states that for any integer n greater than or equal to 3, the equation $x^n + y^n = 1$ has only finitely many integer solutions. The genus of the associated curve is $(n-1)(n-2)/2$, and this is at least 3 if n is 4 or greater. Faltings's theorem immediately implies that for any $n \geqslant 4$, the Fermat equation has at most a finite number of rational solutions. Fermat claimed it had none except when x or y is zero, so this was a big advance. In the next chapter we take up the story of Fermat's last theorem, and see how Fermat's claim was vindicated in full.

7

Inadequate margins
Fermat's Last Theorem

WE FIRST ENCOUNTERED FERMAT in chapter 2, where his elegant theorem about powers of numbers provided a method for testing numbers to see whether they are prime. This chapter is about a far more difficult assertion: Fermat's last theorem. It sounds very mysterious. 'Theorem' seems clear, but who was Fermat, and why was it his *last* theorem? Is the name a cunning marketing ploy? As it happens, no: the name became attached to the problem in the eighteenth century, when only a few leading mathematicians had heard of it or cared about it. But Fermat's last theorem really is mysterious.

Pierre Fermat was born in France in 1601, according to some sources, and in 1607–8, according to others. The discrepancy may perhaps come from confusion with a sibling of the same name. His father was a successful leather merchant and held a high position in local government, and his mother came from a family of parliamentary lawyers. He went to the University of Toulouse, moved to Bordeaux in the late 1620s, and there he showed promising signs of mathematical talent. He was fluent in several languages, and he put together a restoration of a lost work of classical Greek mathematics by Apollonius. He shared his many discoveries with leading mathematicians of the day.

In 1631, having taken a law degree at the University of Orléans, he was appointed a councillor of the High Court of Judicature in Toulouse. This entitled him to change his surname to 'de Fermat', and he remained a councillor for the rest of his life. His passion, though, was mathematics. He published little, preferring to write letters outlining his discoveries, usually without proof. His work received due recognition from the professionals, with many of whom he was on first-name terms, while retaining his amateur status. But Fermat was so talented that he was really a professional; he just didn't hold an official position in mathematics.

Some of his proofs have survived in letters and papers, and it is clear that Fermat knew what a genuine proof was. After his death, many of his deepest theorems remained unproved, and the professionals set to work on them. Within a few decades, only one of Fermat's statements still lacked a proof, so naturally, this became known as his last theorem. Unlike the others, it failed to succumb, and quickly became notorious for the contrast between its ease of statement and the apparent difficulty of finding a proof.

Fermat seems to have conjectured his last theorem around 1630. The exact date is not known, but that was when Fermat started reading a recently published edition of Diophantus's *Arithmetica*. And that's where he got the idea. The last theorem first saw print in 1670, five years after Fermat's death, when his son Samuel published an edition of the *Arithmetica*. This edition had a novel feature. It incorporated marginal notes that Pierre had written in the margins of his personal copy of the 1621 Latin translation by Claude Gaspard Bachet de Méziriac. The last theorem is stated as a note attached to Diophantus's question VIII of Book II, Figure 29.

The problem solved there is to write a perfect square as the sum of two perfect squares. In chapter 6 we saw that there are infinitely many of these Pythagorean triples. Diophantus asks a related but harder question: how to find the two smaller sides of the triangle, given the largest. A specific square must be 'divided' into two squares, that is, expressed as their sum. He shows how to solve this problem when the largest side of the triangle is 4, obtaining the answer

$$4^2 = (16/5)^2 + (12/5)^2$$

in rational numbers. Multiplying through by 25 we obtain $20^2 = 16^2 + 12^2$, and dividing by 16 yields the familiar $3^2 + 4^2 = 5^2$. Diophantus typically illustrated general methods with specific

QVÆSTIO VIII.

PROPOSITVM quadratum diuidere in duos quadratos. Imperatum sit vt 16. diuidatur in duos quadratos. Ponatur primus 1 Q. Oportet igitur 16 − 1 Q. æquales esse quadrato. Fingo quadratum à numeris quotquot libuerit, cum defectu tot vnitatum quod continet latus ipsius 16. esto à 2 N. − 4. ipse igitur quadratus erit, 4 Q. + 16. − 16 N. hæc æquabuntur vnitatibus 16 − 1 Q. Communis adiiciatur vtrimque defectus, & à similibus auferantur similia, fient 5 Q. æquales 16 N. & fit 1 N. ⅕ Erit igitur alter quadratorum ¹⁴⁴⁄₂₅. alter verò ²⁵⁶⁄₂₅ & vtriusque summa est ⁴⁰⁰⁄₂₅ seu 16. & vterque quadratus est.

ΤΟΝ ἐπιταχθέντα τετράγωνον διελεῖν εἰς δύο τετραγώνους. ἐπιτετάχθω δὴ τ ι̅ς̅ διελεῖν εἰς δύο τετραγώνους. καὶ τετάχθω ὁ πρῶτος δυνάμεως μιᾶς. δεήσει ἄρα μονάδας ι̅ς̅ λείψει δυνάμεως μιᾶς ἴσας εἶ̄ τετραγώνῳ. πλάσσω τὸν τετράγωνον ἀπὸ ς̅ς̅. ὅσων δὴ ποτε λείψει τοσούτων ἢ ὅσων ἐστὶν ἥ τ ι̅ς̅. μ̄ πλήθω ἴσω ς̅ β̅ λείψει μ̄ δ̄. αὐτὸς ἄρα ὁ τετράγωνος ἔσται δυνάμεων δ̄ μ̄ ι̅ς̅ λείψει ς̅ ι̅ς̅. ταῦτα ἴσα μονάσι ι̅ς̅ λείψει δυνάμεως μιᾶς. κοινὴ προσκείσθω ἡ λεῖψις, κὴ ἀπὸ ὁμοίων ὅμοια. δυνάμεις ἄρα ε̄ ἴσαι ἀριθμοῖς ι̅ς̅. κὴ γίνεται ὁ ἀριθμὸς ι̅ς̅. πέμπτων. ἔσαι ὁ μὲ͂ν ὅτε εἰκοστόπεμπτον. ὁ δὲ ρμδ̄ εἰκοστόπεμπτον, ἓ οἱ δύο συντιθέντες ποιοῦσι

 υ εἰκοσόπεμπτα, ἥτοι μονάδας ι̅ς̅. καὶ ἔστιν ἑκάτερος τετράγωος.

OBSERVATIO DOMINI PETRI DE FERMAT.

CVbum autem in duos cubes, aut quadratoquadratum in duos quadratoquadratos & generaliter nullam in infinitum vltra quadratum potestatem in duos eiusdem nominis fas est diuidere cuius rei demonstrationem mirabilem sane detexi. Hanc marginis exiguitas non caperet.

Fig 29 Fermat's marginal note, published in his son's edition of the *Arithmetica* of Diophantus.

examples, a tradition that goes back to ancient Babylon, and he didn't provide proofs.

Fermat's personal copy of the *Arithmetica* has not survived, but he must have written his marginal note in it, because Samuel says so. Fermat is unlikely to have left such a treasure unopened for very long, and his conjecture is such a natural one that he probably thought of it as soon as he read question VIII of Book II. He evidently wondered whether anything similar could be achieved using cubes instead of squares, a natural question for a mathematician to ask. He found no examples – we can be sure of that since none exist – and he was equally unsuccessful when he tried higher powers, for example fourth powers. He decided that these questions had no solutions. His marginal note says as much; it translates into English as:

It is impossible to divide a cube into two cubes, or a fourth power into two fourth powers, or in general, any power higher than the second, into two like powers. I have discovered a truly marvellous proof of this, which this margin is too narrow to contain.

In algebraic language, Fermat was claiming to have proved that the Diophantine equation

$$x^n + y^n = z^n$$

has no whole number solutions if n is any integer greater than or equal to 3. Clearly he was ignoring trivial solutions in which x or y is zero. To avoid repeating the formula all over the place, I'll refer to it as the Fermat equation.

If Fermat really did have a proof, no one has ever found it. The theorem was finally proved true in 1995, more than three and a half centuries after he first stated it, but the methods go way beyond anything that was available in his day, or that he could have invented. The search for a proof had a huge influence on the development of mathematics. It pretty much caused the creation of algebraic number theory, which flourished in the nineteenth century because of a failed attempt to prove the theorem and a brilliant idea that partially rescued it. In the late twentieth and twenty-first centuries, it sparked a revolution.

Early workers on Fermat's last theorem tried to pick off powers one by one. Fermat's general proof, alluded to in his margin, may or may not have existed, but we do know how he proved the theorem for fourth powers. The main tool is Euclid's recipe for Pythagorean triples. The fourth power of any number is the square of the square of that number. So any solution of the Fermat equation for fourth powers is a Pythagorean triple, for which all three numbers are themselves squares. This extra condition can be plugged into Euclid's recipe, and after some cunning manoeuvres, what emerges is *another* solution to the Fermat equation for fourth powers.[52] This might not seem like progress; after a page of algebra, the problem is reduced to the same problem. However, it really is reduced: the numbers in the second solution are smaller than those in the first, hypothetical, solution. Crucially, if the first solution is not trivial – if x and y are nonzero – then the same is true of the second solution. Fermat pointed out that repeating this procedure would lead to a sequence of solutions in which the numbers become perpetually smaller. However, any decreasing sequence of whole numbers must eventually stop. This is a logical contradiction, so the hypothetical solution does not exist. He called this method 'infinite descent'. We now recognise it as a proof by mathematical induction, mentioned in chapter 4, and it can be rephrased in terms of minimal criminals. Or, in this case, minimal models of virtue. Suppose there exists a virtuous citizen, a nontrivial solution of the equation. Then there exists a minimal virtuous citizen. But Fermat's argument then implies the existence of an even smaller virtuous citizen – contradiction. Therefore no citizens can be virtuous. Different proofs for fourth powers have been appearing ever since, and around 30 are now known.

Fermat exploited the simple fact that a fourth power is a special kind of square. The same idea shows that in order to prove Fermat's last theorem, it can be assumed that the power n is either 4 or an odd prime. Any number n greater than two is divisible by 4 or by an odd prime p, so every nth power is either a fourth power or a pth power. Over the next two centuries, Fermat's last theorem was proved for exactly three odd primes: 3, 5, and 7. Euler dealt with cubes in 1770; although there is a gap in the published proof, it can be filled using a result that Euler published elsewhere. Legendre and Peter Lejeune-Dirichlet dealt with fifth powers around 1825. Gabriel Lamé proved Fermat's last theorem for seventh powers in 1839. Many different proofs were later found for these cases. Along the way, several mathematicians developed proofs when the power is 6, 10, and 14, but these were superseded by the proofs for 3, 5, and 7.

Each proof makes extensive use of algebraic features that are special to the power concerned. There was no hint of any general structure that might prove the theorem for all powers, or even for a significant number of different powers. As the powers got bigger, the proofs became ever more complicated. Fresh ideas were needed, and they had to break new ground. Sophie Germain, one of the great woman mathematicians, divided Fermat's last theorem for a prime power p into two subcases. In the first case, none of the numbers x, y, z is divisible by p. In the second case, one of them is. By considering special 'auxiliary' primes related to p she proved that the first case of Fermat's last theorem has no solutions for an odd prime power less than 100. However, it was difficult to prove much about auxiliary primes in general.

Germain corresponded with Gauss, at first using a masculine pseudonym, and he was very impressed by her originality. When she revealed she was a woman, he was even more impressed, and said so. Unlike many of his contemporaries, Gauss did not assume that women were incapable of high intellectual achievement, in particular mathematical research. Later, Germain made an unsuccessful attempt to prove the first case of Fermat's last theorem for all even powers, where again it is possible to exploit Euclid's characterisation of Pythagorean triples. Guy Terjanian finally disposed of the even powers in 1977. The second case seemed a much harder nut to crack, and no one got very far with it.

In 1847 Lamé, moving on from his proof for seventh powers, had a wonderful idea. It required the introduction of complex numbers, but by that time everyone was happy with those. The vital ingredient was the same one that Gauss had exploited to construct a regular 17-sided polygon, chapter 3. Every number theorist knew about it, but until Lamé, no one had seriously wondered whether it might be just the job for proving Fermat's last theorem.

In the system of real numbers, 1 has exactly one pth root (when p is odd), namely 1 itself. But in complex numbers, 1 has many pth roots; in fact, exactly p of them. This fact is a consequence of the fundamental theorem of algebra, because these roots satisfy the equation $x^p - 1 = 0$, which has degree p. There is a nice formula for these complex pth roots of unity, as they are called, and it shows that they are the powers $1, \zeta, \zeta^2, \zeta^3, \ldots, \zeta^{p-1}$ of a particular complex number ζ.[53] The defining property of these numbers implies that $x^p + y^p$ splits into p factors:

$$x^p + y^p = (x + y)(x + \zeta y)(x + \zeta^2 y) \cdots (x + \zeta^{p-1} y)$$

By the Fermat equation, this expression is also equal to z^p, which is the pth power of an integer. Now, it is easy to see that if a product of numbers, having no common factor, is a pth power, then each number is itself a pth power. So, give or take a few technicalities, Lamé could write each factor as a pth power. From this he deduced a contradiction.

Lamé announced the resulting proof of Fermat's last theorem to the Paris Academy in March 1847, giving credit for the basic idea to Joseph Liouville. Liouville thanked Lamé, but pointed out a potential issue. The crucial statement implying that each factor is a pth power is not a done deal. It depends on uniqueness of prime factorisation – not just for ordinary integers, where that property is true, but for the new kinds of number that Lamé had introduced. These combinations of powers of ζ are called cyclotomic integers; the word means 'circle-cutting', and refers to the connection that Gauss had exploited. Not only was the property of unique prime factorisation not proved for cyclotomic integers, said Liouville: it might be false.

Others already had doubts. Three years earlier, in a letter, Gotthold Eisenstein wrote:

> If one had the theorem which states that the product of two complex numbers can be divisible by a prime number only when one of the factors is – which seems completely obvious – then one would have the whole theory [of algebraic numbers] at a single blow; but this theorem is totally false.

The theorem to which he alludes is the main step needed for a proof of uniqueness of prime factorisation. Eisenstein was referring not just to the numbers that Lamé needed, but to similar numbers arising from other equations. They are called algebraic numbers. An algebraic number is a complex number that satisfies a polynomial equation with rational coefficients. An algebraic integer is a complex number that satisfies a polynomial equation with integer coefficients, provided the coefficient of the highest power of x is 1. For each such polynomial, we obtain an associated algebraic number field (meaning that you can add, subtract, multiply, and divide such numbers to get numbers of the same kind) and its ring (similar but omit 'divide') of algebraic integers. These are the basic objects that are studied in algebraic number theory.

If, for instance, the polynomial is $x^2 - 2$, then it has a solution $\sqrt{2}$. The field consists of all numbers $a + b\sqrt{2}$ with a, b rational; the ring of integers consists of the numbers of this form with a, b integers. Again prime factors can be defined, and are unique. There are some surprises: the polynomial $x^2 + x - 1$ has a solution $(\sqrt{5} - 1)/2$, so despite the fraction, this is an algebraic *integer*.

In algebraic number theory, the difficulty is not to define factors. For example, a cyclotomic integer is a factor of (that is, divides) another if the second is equal to the first multiplied by some cyclotomic integer. The difficulty is not to define primes: a cyclotomic integer is prime if it has no factors, other than trivial 'units', which are the cyclotomic integers that divide 1. There is no problem about resolving a cyclotomic integer, or any other algebraic number, into prime factors. Just keep factorising it until you run out of factors. There is a simple way to prove that the procedure stops, and when it does, every factor must be prime. So what's the difficulty? Uniqueness. If you run the procedure again, making different choices along the way, it might stop with a different list of prime factors.

At first sight, it's hard to see how this can happen. The prime factors are the smallest possible pieces into which the number can be split. It's like taking a Lego toy and pulling it apart into its component bricks. If there were another way to do that, it would end up pulling one of those bricks apart into two or more pieces. But then it wouldn't be a brick. Unfortunately, the analogy with Lego is misleading. Algebraic numbers aren't like that. They are more like bricks with movable links, able to lock together in different ways. Break up a brick in one way, and the resulting pieces lock together and can't be pulled apart any further. Break it up in a different way, and again the resulting pieces lock together. But now they are different.

I'll give you two examples. The first uses only ordinary integers; it's easy to grasp but it has some unrepresentative features. Then I'll show you a genuine example.

Suppose we lived in a universe where the only numbers that existed were 1, 5, 9, 13, 17, 21, 25, and so on – numbers which in our actual universe have the form $4k + 1$. If you multiply two such numbers together you get another number of the same kind. Define such a number to be 'prime' if it is not the product of two smaller numbers *of that kind*. For example, 25 is not prime because it is 5×5, and 5 is a number in the list. But 21 *is* prime, in this new sense, because its usual factors 3 and 7 are not in the list. They are of the form $4k + 3$, not $4k + 1$. It is easy to see that every number of the specified kind is a product of primes in the new sense. The reason is that factors, if they exist, must get smaller. Eventually, the process of factorisation has to stop. When it does, the factors concerned are primes.

However, this type of prime factorisation is not unique. Consider the number 4389, which is $4 \times 1097 + 1$, so is of the required form. Here are three distinct factorisations into numbers of the required form:

$$4389 = 21 \times 209 = 33 \times 133 = 57 \times 77$$

I claim that, with our current definition, all of these factors are primes. For example, 57 is prime, because its usual factors 3 and 19 are not of the required form. The same goes for 21, 33, 77, 133, and 209. Now we can explain the lack of uniqueness. In ordinary integers

$$4389 = 3 \times 7 \times 11 \times 19$$

and all of these factors have the 'wrong' form, $4k + 3$. The three different prime factorisations, in the new sense, arise by grouping these numbers in pairs:

$$(3 \times 7) \times (11 \times 19) \quad (3 \times 11) \times (7 \times 19) \quad (3 \times 19) \times (7 \times 11)$$

We need to use pairs because two numbers of the form $4k + 3$, multiplied together, yield a number of the form $4k + 1$.

This example shows that the argument 'the factors must be unique because they are the smallest pieces' doesn't work. It's true that there are *smaller* pieces hanging around ($21 = 3 \times 7$, for instance) but those pieces aren't in the system concerned. The main reason why this example isn't entirely representative is that although multiplying together numbers of the form $4k + 1$ produces numbers of the same form, that's not true for addition. For instance, $5 + 5 = 10$ is not of the required form. So, in the jargon of abstract algebra, we're not working in a ring.

The second example doesn't have that defect, but in compensation, it's a bit harder to analyse. It is the ring of algebraic integers for the polynomial $x^2 - 15$. This ring consists of all numbers $a + b\sqrt{15}$ where a and b are integers. In it, the number 10 has two distinct factorisations:

$$10 = 2 \times 5 = (5 + \sqrt{15}) \times (5 - \sqrt{15})$$

All four factors 2, 5, $5 + \sqrt{15}$, $5 - \sqrt{15}$ can be proved prime.[54]

All this is much clearer now than it was in 1847, but it didn't take long to show that Liouville's doubts were justified. A fortnight after he expressed them, Wantzel informed the academy that uniqueness was true for some small values of p, but his method of proof failed for 23rd powers. Shortly afterwards, Liouville told the academy that unique prime factorisation is *false* for cyclotomic integers corresponding to $p = 23$. Ernst Kummer had discovered this three years earlier, but hadn't told anyone because he was working on a method for getting round the obstacle. Lamé's proof worked for smaller values of p, including some new ones: 11, 13, 17, 19. But for the general case, the proof was in tatters. It was an object lesson in not assuming that plausible mathematical statements are obvious. They may not even be true.

Kummer had been thinking about Fermat's last theorem, along similar lines to Lamé's. He noticed the potential obstacle, took it seriously, investigated it, and discovered that it wrecked that approach. He found an explicit example of non-unique prime factorisation for cyclotomic integers based on 23rd roots of unity. But Kummer was not one to give up easily, and he found a way to get round the obstacle – or, at least, to mitigate its worst effects. His idea is especially transparent in the case of those $4k + 1$ numbers. The way to restore unique factorisation is to throw in some *new* numbers, outside the system we're interested in. For that example, what we need are the missing $4k + 3$ numbers. Or we can go the whole hog and throw in the even integers as well; then we get the integers, which are closed under addition and multiplication. That is, if you add or multiply two integers, the result is an integer.

Kummer came up with a version of the same idea. For instance, we can restore unique prime factorisation in the ring of all numbers $a + b\sqrt{15}$ by throwing in a new number, namely $\sqrt{5}$. To obtain a ring, it turns out that we must also throw in $\sqrt{3}$. Now

$$2 = (\sqrt{5} + \sqrt{3}) \times (\sqrt{5} - \sqrt{3}) \qquad 5 = \sqrt{5} \times \sqrt{5}$$

and

$$5 + \sqrt{15} = \sqrt{5} \times (\sqrt{5} + \sqrt{3}) \qquad 5 - \sqrt{15} = \sqrt{5} \times (\sqrt{5} - \sqrt{3})$$

So the two factorisations arise by grouping the four numbers $\sqrt{5}$, $\sqrt{5}$, $\sqrt{5} + \sqrt{3}$, $\sqrt{5} - \sqrt{3}$ in two different ways.

Kummer called these new factors ideal numbers, because in his general formulation they weren't exactly numbers at all. They were symbols that behaved a lot like numbers. He proved that every cyclotomic integer can be factorised uniquely into prime ideal numbers. The set-up was subtle: neither the cyclotomic integers, nor the ideal numbers, had unique prime factorisation. But if you used the ideal numbers as the ingredients for prime factorisation of cyclotomic integers, the result was unique.

Later Richard Dedekind found a more civilised reinterpretation of Kummer's procedure, and this is the one we now use. To each ideal number outside the ring concerned, he associated a *set* of numbers inside the ring. He called such a set an ideal. Every number in the ring defines an ideal: it consists of all multiples of that number. If prime factorisation is unique, every ideal is like that. When it's not, there are extra ideals. We can define product and sums of ideals, and prime ideals, and Dedekind proved that prime factorisation *of ideals* is unique for all rings of algebraic integers. This suggests that for most problems you should work with ideals, not the algebraic numbers themselves. Of course, that introduces new complexities, but the alternative is usually to get stuck.

Kummer was able to work with his ideal numbers – well enough to prove a version of Fermat's last theorem with some extra hypotheses. But other mortals found ideal numbers rather difficult, if not a bit mystical. Once viewed in Dedekind's way, however, ideal numbers made perfect sense, and algebraic number theory took off. One important idea that emerged was a way to measure how badly unique factorisation fails in a ring of algebraic integers. To each such ring there corresponds a whole number called its *class number*. If the class number is 1, prime factorisation is unique; otherwise it's not. The bigger the class number, the 'less unique' prime factorisation is, in a meaningful sense.

Being able to quantify the lack of uniqueness was a big step forward, and with extra effort it rescued Lamé's strategy – sometimes. In 1850 Kummer announced that he could prove Fermat's last theorem for a great many primes, those that he called regular. Among the primes up to 100, only 37, 59, and 67 are irregular. For all other primes up to that limit, and many beyond it, his methods proved Fermat's last theorem. The definition of a regular prime requires the class number: a prime is regular if it does not divide the class number of the corresponding ring of cyclotomic integers. So for a regular prime, although prime factorisation is not unique, the way it fails to be unique does not involve the prime concerned in an essential manner.

Kummer claimed that there exist infinitely many regular primes, but this assertion remains unproved. Ironically, in 1915 K.L. Jensen proved that there are infinitely many irregular primes. A bizarre criterion for a prime to be regular emerged from connections with analysis. It involves a sequence of numbers discovered independently by the Japanese mathematician Seki Takazu (or Kōwa) and the Swiss mathematician Jacob Bernoulli, called Bernoulli numbers. This criterion shows that the first ten irregular primes are 37, 59, 67, 101, 103, 131, 149, 157, 233, and 257. By digging more deeply into the structure of cyclotomic integers, Dmitri Mirimanoff disposed of the first irregular prime, 37, in 1893. By 1905 he had proved Fermat's last theorem up to $p = 257$. Harry Vandiver developed computer algorithms that extended this limit. Using these methods, John Selfridge and Bary Pollack proved the theorem up to the 25,000th power in 1967, and S. Wagstaff increased that to 100,000 in 1976.

The evidence for the truth of Fermat's last theorem was piling up, but the main implication was that if the theorem were false, then a counterexample – an example exhibiting its falsity – would be so gigantic that no one would ever be able to find it. The other implication was that methods like Kummer's were running into the same problems that afflicted the work of the earlier pioneers: bigger powers required special, complicated treatment. So this line of attack slowly ground to a halt.

When you get stuck on a mathematical problem, Poincaré's advice is spot on: go away and do something else. With luck and a following wind, a new idea will eventually turn up. Number theorists didn't consciously follow his advice, but nevertheless, they did what he had recommended. As Poincaré had insisted it would, the tactic worked. Some number theorists turned their attention to elliptic curves, chapter 6. Ironically, this area eventually turned out to have a startling and unexpected link to Fermat's last theorem, leading to Wiles's proof. To describe this link, one further concept is required: that of a modular function. The discussion is going to get a bit technical, but there is a sensible story behind the ideas and all we'll need is a broad outline. Bear with me.

In chapter 6 we saw that the theory of elliptic functions had a profound effect on complex analysis. In the 1830s Joseph Liouville discovered that the variety of elliptic functions is fairly limited. Given the two periods, there is a special elliptic function, the Weierstrass function, and every other elliptic function with those two periods is a simple variant. This implies that the only doubly periodic functions you needed to understand are the Weierstrass functions – one for each pair of periods.

Geometrically, the doubly periodic structure of an elliptic function can be phrased in terms of a lattice in the complex plane: all integer combinations $mu + nv$ of the two periods u and v, for integers m and n, Figure 30. If we take a complex number z and add one of these lattice points to it, the elliptic function at this new point has the same value that it had at the original point. In other words, the elliptic function has the same symmetry as the lattice.

Analysts had discovered a much richer source of symmetries of the complex plane, known as Möbius transformations. These change z to $(az + b)/(cz + d)$, for complex constants a, b, c, d. Lattice symmetries are special kinds of Möbius transformation, but there are others. Sets of points analogous to the lattice still exist in this more general setting. A lattice defines a tiling pattern in the Euclidean plane: use a parallelogram for the tiles and place its corners at the lattice points, Figures 26 and 30. Using Möbius transformations, we can construct tiling patterns in a suitable non-Euclidean geometry, the hyperbolic plane. We can identify this geometry with a region of the complex plane, in which straight lines are replaced by arcs of circles.

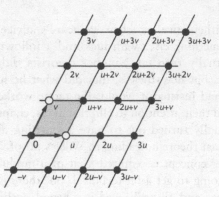

Fig 30 The lattice is formed from all integer combinations of the two periods.

There are highly symmetric tiling patterns in hyperbolic geometry. For each of them, we can construct complex functions that repeat the same values on every tile. These are known as modular functions, and they are natural generalisations of elliptic functions. Hyperbolic geometry is a very rich subject, and the range of tiling patterns is much more extensive than it is for the Euclidean plane. So complex analysts started thinking seriously about non-Euclidean geometry. A profound link between analysis and number theory then appeared. Modular functions do for elliptic curves what trigonometric functions do for the circle.

Recall that the unit circle consists of points (x,y) such that $x^2 + y^2 = 1$. Suppose A is a real number, and set

$$x = \cos A \qquad y = \sin A$$

Then the definition of sine and cosine tells us that this point lies on the unit circle. Moreover, every point on the unit circle is of this form. In the jargon, the trigonometric functions *parametrise* the circle. Something very similar happens for modular functions. If we define x and y using suitable modular functions of a parameter A, the corresponding point lies on an elliptic curve – the same elliptic curve, whatever value A takes. There are more abstract ways to make this statement precise, and workers in the area use those because they are more convenient, but this version brings out the analogy with trigonometry and the circle. This connection produces an elliptic curve for each modular function, and the variety of modular functions is huge – all symmetric tilings of the hyperbolic plane. So an awful lot of elliptic curves can be related to modular functions. Which elliptic curves can be obtained in this way? That turned out to be the heart of the matter.

This 'missing link' first came to prominence in 1975, when Yves Hellegouarch noticed a curious connection between Fermat's last theorem and elliptic curves. Gerhard Frey developed the idea further in two papers published in 1982 and 1986. Suppose, as always, that p is an odd prime. Assume – hoping to derive a contradiction – that there exist nonzero integers a, b, and c that satisfy the Fermat equation, so $a^p + b^p = c^p$. Now extract the rabbit from the hat with a theatrical flourish: consider the elliptic curve

$$y^2 = x(x - a^p)(x - b^p)$$

This is called the Frey elliptic curve. Frey applied the machinery of elliptic curves to this one, and what emerged was a string of ever more bizarre coincidences. His hypothetical elliptic curve is very strange indeed. It doesn't seem to make sense. Frey proved that it makes so little sense that it can't exist. And that, of course, proves Fermat's last theorem, by providing the required contradiction.

However, there was a gap, one that Frey was well aware of. In order to prove that his hypothetical elliptic curve does not exist, you have to show that if it did exist, it would be modular – that is, one of the curves that arise from modular functions. We've just seen that such curves are common, and at that time no one had ever found an elliptic curve that was *not* modular. It seemed likely that the Frey curve should be modular – but it was a hypothetical curve, the numbers a, b, c were not known, and if the curve *were* modular, then it wouldn't exist at all. However, there was one way to deal with all those issues: prove that *every* elliptic curve is modular. Then the Frey curve, hypothetical or not, would have to be modular if it existed. And if it didn't exist, the proof was complete anyway.

The statement that every elliptic curve is modular is called the Taniyama-Shimura conjecture. It is named for two Japanese mathematicians, Yutaka Taniyama and Goro Shimura. They met by accident, both wanting to borrow the same book from the library at the same time for the same reason. This triggered a long collaboration. In 1955 Taniyama was at a mathematics conference in Tokyo, and the younger participants were invited to collect together a list of open questions. Taniyama contributed four, all hinting at a relationship between modular functions and elliptic curves. He had calculated some numbers associated with a particular modular function, and noticed that the exact same numbers turned up in connection with a particular elliptic curve. This kind of coincidence is often a sign that it's not actually a coincidence at all; that there must be some sensible explanation. The equality of these numbers is now known to be equivalent to the elliptic curve being modular; in fact, that's the preferred definition in the research literature. Anyway, Taniyama was sufficiently intrigued that he calculated the numbers for a few other modular functions, finding that these, too, correspond to specific elliptic curves.

He began to wonder whether something similar would work for every elliptic curve. Most workers in the field considered this too good to be true, a pipedream for which there was very little evidence. Shimura was one of the few who felt that the conjecture had merit. But in 1957–8 Shimura visited Princeton for a year, and while he was away, Taniyama committed suicide. He left a note that read, in part, 'As to the cause of my suicide, I don't quite understand it myself, but it is not the result of a particular incident, nor of a specific matter. Merely may I say, I am in the frame of mind that I lost confidence in my future.' At the time he had been planning to get married, and the prospective bride, Misako Suzuki, killed herself about a month later. Her suicide note included 'Now that he is gone, I must go too in order to join him.'

Shimura continued to work on the conjecture, and as evidence in favour accumulated, he started to think it might actually be true. Most other workers in the area disagreed. Simon Singh[55] reports an interview with Shimura, in which he recalled trying to explain this to one of his colleagues:

> The professor enquired, 'I hear that you propose that some elliptic equations can be linked to modular forms.'

'No, you don't understand,' replied Shimura. 'It's not just *some* elliptic equations, it's *every* elliptic equation!'

Despite this kind of scepticism, Shimura persevered, and after many years the proposal became sufficiently respectable to be referred to as the Taniyama-Shimura conjecture. Then André Weil, one of the great number theorists of the twentieth century, found a lot more evidence in the conjecture's favour, publicised it, and expressed the belief that it might well be true. It became known as the Taniyama-Shimura-Weil conjecture. The name never quite settled down, and many permutations of subsets of the three mathematicians have been associated with it. I will stick to 'Taniyama-Shimura conjecture'.

In the 1960s another heavyweight, Robert Langlands, realised that the Taniyama-Shimura conjecture could be viewed as just one element in a much broader and more ambitious programme, which would unify algebraic and analytic number theory. He formulated a whole raft of conjectures related to this idea, now known as the Langlands programme. It was even more speculative than the Taniyama-Shimura conjecture, but it had a compelling elegance, the sort of mathematics that ought to be true because it was so beautiful. Throughout the 1970s the mathematical world became accustomed to the wild beauty of the Langlands programme, and it started to become accepted as one of the core aims of algebraic number theory. The Langlands programme seemed to be the right way forward, if only someone could take the first step.

At this point Frey noticed that applying the Taniyama-Shimura conjecture to his elliptic curve would prove Fermat's last theorem. However, by then another problem with Frey's idea had emerged. When he gave a lecture about it in 1984, the audience spotted a gap in his key argument: the curve is so bizarre that it can't be modular. Jean-Pierre Serre, one of the leading figures in the area, quickly filled the gap, but he had to invoke another result that also lacked a proof, the special level reduction conjecture. By 1986, however, Ken Ribet had proved the special level reduction conjecture. Now the only obstacle to a proof of Fermat's last theorem was the Taniyama-Shimura conjecture, and the consensus began to shift. Serre predicted that Fermat's last theorem would probably be proved within a decade or so. Exactly how was another matter, but there was a general feeling in the air: the techniques related to modular functions were becoming so powerful that someone would soon make Frey's approach work.

That someone was Andrew Wiles. In the television programme about his proof, he said:

> I was a ten-year-old and ... I found a book on math and it told a bit about the history of this problem [Fermat's last theorem] – that someone had [posed] this problem 300 years ago, but no one had ever seen a proof, no one knew if there was a proof, and people ever since have looked for the proof. And here was a problem that I, a ten-year-old, could understand, but none of the great mathematicians of the past had been able to resolve. And from that moment of course I just tried to solve it myself. It was such a challenge, such a beautiful problem.

In 1971 Wiles took a mathematics degree at Oxford, moving to Cambridge for his PhD. His supervisor, John Coates, advised him (correctly) that Fermat's last theorem was too difficult for a PhD. So instead, Wiles set to work on elliptic curves, then considered to be a far more promising area of research. By 1985–6 he was in Paris at the Institut des Hautes Études Scientifiques (Institute of Advanced Scientific Studies), one of the world's leading mathematical research institutes. Most top researchers pass through it at some point; if you are a mathematician, it's a great place to hang out. Among the visitors was Ribet, and his proof of the special level reduction conjecture electrified Wiles. Now he could pursue entirely respectable research into elliptic curves, by trying to prove the Taniyama-Shimura conjecture, and at the same time he could try to fulfil his childhood dream of proving Fermat's last theorem.

Because everyone in the area now knew of the connection, there was a worry. Suppose Wiles managed to put together an almost complete proof, with a few small gaps that needed extra work. Suppose someone else learned of this, and filled the gap. Then technically, this person would be the one who had proved Fermat's last theorem. Mathematicians generally manage not to behave like that, but when the prize is so great, it is wise to take precautions. So Wiles carried out his research in secret, something that mathematicians seldom do. It wasn't that he didn't trust his colleagues. It was just that he couldn't take the tiniest risk of being pipped at the post.

He laboured for seven years, tucked away in the roof of his house where there was an office. Only his wife and his head of department knew what he was working on. In peace and seclusion, he attacked the problem with every technique he could learn, until the castle walls began to shake under the onslaught. In 1991 Coates put him on to some new results proved by Mattheus Flach. The crack in the wall began to widen ever more rapidly as the siege took its toll.

By 1993 the proof was complete. Now it had to be revealed to the world. Still cautious, Wiles didn't want to risk proclaiming his solution only for some mistake to surface – something that happened to Yoichi Miyaoka in 1988, whose claim of a proof hit the media, only for a fatal error to be found. So Wiles decided to deliver a series of three lectures at the Isaac Newton Institute in Cambridge, a newly founded international research centre for mathematics. The title was innocuous and technical: 'Modular forms, elliptic curves, and Galois theory.' Few were fooled: they knew Wiles was on to something big.

In the third lecture, Wiles outlined a proof of a special case of the Taniyama-Shimura conjecture. He had discovered that something a little less ambitious would also work. Prove that the Frey curve, if it exists, must belong to a special class of elliptic curves, the 'semistable' ones, and prove that all curves *in that class* must be modular. Wiles then proved both results. At the end of the lecture, he wrote a corollary – a supplementary theorem that follows directly from whatever has just been proved – on the blackboard. The corollary was Fermat's last theorem.

When Shimura heard about Wiles's announcement, his comment was brief and to the point. 'I told you so.'

If only it had been that straightforward. But fate had a twist in store. The proof still had to be refereed by experts, and as usual that process turned up a few points that needed further explanation. Wiles dealt with most of these comments, but one of them forced a rethink. Late in 1993 he issued a statement saying that he was withdrawing his claim until he could patch up a logical gap that had emerged. But now he was forced to operate in the full glare of publicity, exactly what he had hoped to avoid.

By March 1994, no repaired proof had appeared, and Faltings expressed a widespread view in the mathematical community: 'If [repairing the proof] were easy, he'd have solved it by now. Strictly speaking, it was not a proof when it was announced.' Weil remarked 'I believe he has some good ideas ... but the proof is not there ... proving Fermat's last theorem is like climbing Everest. If a man wants to climb Everest and falls short of it by 100 yards, he has not climbed Everest.' Everyone could guess how it was going to end. They'd seen it all before. The proof had collapsed, it would have to be retracted completely, and Fermat's last theorem would live to fight another day.

Wiles refused to concede defeat, and his former student Richard Taylor joined the quest. The root of the difficulty was now clear: Flach's results weren't quite suited to the task. They tried to modify Flach's methods, but nothing seemed to work. Then, in a flash of inspiration, Wiles suddenly understood what the obstacle was. 'I saw that the thing that had stopped [Flach's method] working was something that would make another method I had tried previously work.' It was as though the soldiers besieging the castle had realised that their battering-ram would never work because the defenders kept dropping rocks on it, but those self-same rocks could be stuffed into a trebuchet and used to break down the door instead.

By April 1995 the new proof was finished, and this time there were no gaps or errors. Publication quickly followed, two papers in the ultra-prestigious *Annals of Mathematics*. Wiles became an international celebrity, was awarded several major prizes and a knighthood ... and went back to his research, carrying on pretty much as before.

The really important feature of Wiles's solution is not Fermat's last theorem at all. As I said, nothing vital hinges on the answer. If someone had found three 100-digit numbers and a 250-digit prime that provided a counterexample to Fermat's claim, then the theorem would have been false, but no crucial area of mathematics would have been in any way diminished. Of course a direct attack by computer would not be able to search through numbers that large, so you would have to be amazingly clever to establish any such thing, but a negative result would not have caused any heartaches.

The real importance of the solution lies in the proof of the semistable case of the Taniyama-Shimura conjecture. Within six years Christophe Breuil, Brian Conrad, Fred Diamond, and Taylor extended Wiles's methods to handle not just the semistable case, but all elliptic curves. They proved the full Taniyama-Shimura conjecture, and number theory would never be the same. From that point on, whenever anyone encountered an elliptic curve, it was guaranteed to be modular, so a host of analytic methods would open up. Already these methods have been used to solve other problems in number theory, and more will turn up in future.

8

Orbital chaos
Three-Body Problem

ACCORDING TO A TIME-HONOURED joke, you can tell how advanced a physical theory is by the number of interacting bodies that it can't handle. Newton's law of gravity runs into problems with three bodies. General relativity has difficulty dealing with two bodies. Quantum theory is over-extended for one body, and quantum field theory runs into trouble with *no* bodies – the vacuum. Like many jokes, this one contains a grain of truth.[56] In particular, the gravitational interaction of a mere three bodies, assumed to obey Newton's inverse square law of gravity, stumped the mathematical world for centuries. It still does, if what you want is a nice formula for the orbits of those bodies. In fact, we now know that three-body dynamics is chaotic – so irregular that it has elements of randomness.

All this is a huge contrast to the stunning success of Newtonian gravitational theory, which explained, among many other things, the orbit of a planet round the Sun. The answer is what Kepler had already deduced empirically from astronomical observations of Mars: an ellipse. Here only two bodies occur: Sun and planet. The obvious next step is to use Newton's law of gravity to write down the equation for three-body orbits, and solve it. But there is no tidy geometric characterisation of three-body orbits, not even a formula in coordinate geometry. Until the late nineteenth century, very little was known about the motion of three celestial bodies, even if one of them were so tiny that its mass could be ignored.

Our understanding of the dynamics of three (or more) bodies has grown dramatically since then. A large part of that progress has been an ever-increasing realisation of how difficult the question is, and why. That may seem a retrograde step, but sometimes the best way to advance is to make a strategic retreat and try something else. For the three-body problem, this plan of campaign has scored some real successes, when a head-on attack would have become hopelessly bogged down.

Early humans cannot have failed to notice that the Moon moves gradually across the night sky, relative to the background of stars. The stars also appear to move, but they do so as a whole, like tiny pinpricks of light on a vast spinning bowl. The Moon is clearly special in another way: it is a great shining disc, which changes shape from new moon to full moon and back again. It's not a pinprick of light like a star.

A few of those pinpricks of light also disobey the rules. They go walkabout. They don't change their position relative to the stars as quickly as the Moon, but even so, you don't have to watch the sky for many nights to see that some of them are moving. Five of these wandering stars are visible to the naked eye; the Greeks called them *planetes* – wanderers. They are, of course, the planets, and the five that have been recognised since ancient times are what we now call Mercury, Venus, Mars, Jupiter, and Saturn – all named after Roman gods. With the aid of telescopes we now know of two more: Uranus and Neptune. Plus our own Earth, of course. No longer does Pluto count as a planet, thanks to a controversial decision on terminology made by the International Astronomical Union in 2006.

As ancient philosophers, astronomers, and mathematicians studied the heavens, they realised that the planets do not just wander around at random. They follow convoluted but fairly predictable paths, and they return to much the same position in the night sky at fairly regular intervals of time. We now explain these patterns as periodic motion round a closed orbit, with a small contribution from the Earth's own orbital motion. We also recognise that the periodicity is not exact – but it comes close. Mercury takes nearly 88 days to circle the Sun, while Jupiter takes nearly 12 years. The further from the Sun the planet is, the longer it takes to complete one orbit.

The first quantitatively accurate model of the motion of the planets was the Ptolemaic system, named for Claudius Ptolemy, who described it in his *Almagest* ('The greatest [treatise]') of about AD 150. It is a geocentric – Earth-centred – model, in which all celestial bodies orbit the Earth. They move as if supported by a series of gigantic spheres, each rotating at a fixed rate about an axis that may itself be supported by another sphere. Combinations of many rotating spheres were required to represent the complex motion of the planets in terms of the cosmic ideal of uniform rotation in a circle – the equator of the sphere. With enough spheres and the right choices for their axes and speeds, the model corresponds very closely to reality.

Nicolaus Copernicus modified Ptolemy's scheme in several ways. The most radical was to make all bodies, other than the Moon, revolve round the Sun instead, which simplified the description considerably. It was a heliocentric model. This proposal fared ill with the Catholic church, but eventually the scientific view prevailed, and educated people accepted that the Earth revolves round the Sun. In 1596 Kepler defended the Copernican system in his *Mysterium Cosmographicum* ('The cosmographic mystery'), whose high point was his discovery of a mathematical relationship between a planet's distance from the Sun and its orbital period. Moving outwards from the Sun, the ratio of the increase in period from one planet to the next is twice the increase in distance. Later, he decided that this relationship was too inaccurate to be correct, but it sowed the seeds of a more accurate relationship in his future work. Kepler also explained the spacing of the planets in terms of the five regular solids, nested neatly inside each other, separated by the spheres that held them. Five solids explained why there were five planets, but we now recognise eight, so this feature is no longer an advantage. There are 120 different ways to arrange five solids in order, and one of these is likely to come close to the celestial proportions given by planetary orbits. So it's just an accidental approximation, shoehorning nature into a meaningless pattern.

In 1600 the astronomer Tycho Brahe hired Kepler to help him analyse his observations, but political problems intervened. After Brahe's death, Kepler was appointed imperial mathematician to Rudolph II. In his spare time, he worked on Brahe's observations of Mars. One outcome was *Astronomia Nova* ('A new astronomy') of 1609, which presented two more laws of planetary motion. Kepler's first law states that planets move in ellipses – he had established this for Mars, and it seemed likely that the same would be true of the other planets. Initially he assumed that an egg shape would fit the data, but that didn't work out, so he tried an ellipse. This, too, was rejected, and he found a different mathematical description of the orbit's shape. Finally he realised that this was actually just another way to define an ellipse:[57]

> I laid [the new definition] aside, and fell back on ellipses, believing that this was quite a different hypothesis, whereas the two, as I shall prove in the next chapter, are one in the same... Ah, what a foolish bird I have been!

Kepler's second law states that the planet sweeps out equal areas in equal times. In 1619, in his *Harmonices Mundi* ('Harmonies of the world'), Kepler completed his three laws with a far more accurate relationship between distances and periods: the cube of the distance (half the length of the major axis of the ellipse) is proportional to the square of the period.

The stage was now set for Isaac Newton. In his *Philosophiae Naturalis Principia Mathematica* ('Mathematical principles of natural philosophy') of 1687, Newton proved that Kepler's three laws are equivalent to a single law of gravitation: two bodies attract each other with a force that is proportional to their masses and inversely proportional to the square of the distance between them. Newton's law had a huge advantage: it applied to any system of bodies, however many there might be. The price to be paid was the way the law prescribed the orbits: not as geometric shapes, but as solutions of a differential equation, which involved the planets' accelerations. It is not at all clear how to find the shapes of planetary orbits, or the planets' positions at a given time, from this equation. It is not terribly clear how to find their accelerations, to be blunt. Nevertheless, the equation *implicitly* provided that information. The problem was to make it explicit. Kepler had already done that for two bodies, and the answer was elliptical orbits pursued with speeds that sweep out areas at a constant rate.

What about three bodies?

It was a good question. According to Newton's law, all bodies in the solar system influence each other gravitationally. In fact, all bodies in the entire universe influence each other gravitationally. But no one in their right mind would try to write down differential equations for every body in the universe. As always, the way forward was to simplify the problem – but not too much. The stars are so far away that their gravitational effect on the solar system is negligible, unless you want to describe how the Sun moves as the galaxy rotates. The motion of the Moon is mainly influenced by two other bodies: the Earth and the Sun, except for some subtle effects involving other planets. In the early 1700s, this question escaped the realms of astronomy and acquired practical significance, when it was realised that the motion of the Moon could be useful for navigation. (No GPS in those days; not even chronometers to measure longitude.) But this method required more accurate predictions than existing theories could provide. The obvious place to start was to write down the implications of Newton's law for three bodies, which for this purpose could be treated as point masses, because the planets are exceedingly small compared to the distances between them. Then you solved the resulting differential equations. However, the tricks that led from two bodies to ellipses failed when an extra body entered the mix. A few preliminary steps worked, but then the calculation hit an obstruction. In 1747 Jean d'Alembert and Alexis Clairaut, bitter rivals, both competed for a prize from the Paris Academy of Sciences on the 'problème des trois corps', which they both approached through numerical approximations. The three-body problem had acquired its name, and it soon became one of the great enigmas of mathematics.

Some special cases could be solved. In 1767 Euler discovered solutions in which all three bodies lie on a rotating straight line. In 1772 Lagrange found similar solutions where the bodies form a rotating equilateral triangle, which expands and contracts. Both solutions were periodic: the bodies repeated the same sequence of movements indefinitely. However, even drastic simplifications failed to produce anything more general. You could assume that one of the bodies had negligible mass, you could assume that the other two moved in perfect circles about their mutual centre of mass, a version known as the 'restricted' three-body problem ... and *still* you couldn't solve the equations exactly.

In 1860 and 1867 the astronomer and mathematician Charles-Eugène Delaunay attacked the specific case of the Sun-Earth-Moon system using perturbation theory, which views the effect of the Sun's gravity on the Moon as a small change imposed on the effect of the Earth, and derived approximate formulas in the form of series: many successive terms added together. He published his results in 1860 and 1867; each volume was 900 pages long, and consisted largely of the formulas. In the late 1970s his calculations were checked using computer algebra, and only two small and unimportant errors were found.

It was a heroic calculation, but the series approached its limiting value too slowly to have much practical use. However, it spurred others to seek series solutions that converged more rapidly. It also uncovered a big technical obstacle to all such approaches, known as small denominators. Some terms in the series are fractions, and the denominator (the part on the bottom) becomes very small if the bodies are near a resonance: a periodic state in which their periods are rational multiples of each other. For example, Jupiter's three innermost moons Io, Europa, and Ganymede have periods of revolution around the planet of 1.77 days, 3.55 days, and 7.15 days, in almost exact 1 : 2 : 4 ratios. Secular resonances, rational relations between the rates at which the axes of two nearly elliptical orbits turn, are an especial nuisance, because the probable error in evaluating a fraction gets very large when its denominator is small.

If the three-body problem was difficult, the n-body problem – any number of point masses moving under Newtonian gravity – was surely harder. Yet nature presents us with an important example: the entire solar system. This contains eight planets, several dwarf planets such as Pluto, and thousands of asteroids, many quite large. Not to mention satellites, some of which – Titan, for instance – are larger than the planet Mercury. So the solar system is a 10-body problem, or a 20-body problem, or a 1000-body problem, depending on how much detail you want to include.

For short-term predictions, numerical approximations are effective, and in astronomy, a thousand years is short. Understanding how the solar system will evolve over hundreds of millions of years is quite another matter. And one big question depends on that kind of long-term view: the stability of the solar system. The planets seem to be moving in relatively stable, almost-elliptical orbits. The orbits change a little bit when other planets perturb them, so the period might change by a fraction of a second, or the size of the ellipse might not be exactly constant. Can we be sure that this gentle jostling is all that will happen in future? Is it typical of what happened in the past, especially in the early stages of the solar system? Will the solar system remain stable, or will two planets collide? Could a planet be flung out into the distant reaches of the universe?

The year 1889 was the 60th birthday of Oscar II, king of Norway and Sweden. As part of the celebrations, the Norwegian mathematician Gösta Mittag-Leffler persuaded the king to announce a prize for the solution of the n-body problem. This was to be achieved not by an exact formula – by then it was clear that this was asking too much – but by some kind of convergent series. Poincaré became interested, and he decided to start with a very simple version: the restricted three-body problem, where one body has negligible mass, like a tiny particle of dust. If you apply Newton's law naively to such a particle, the force exerted on it is the product of the masses divided by the square of the distance, and one of the masses is zero, so the product is zero. That's not very helpful, because the dust particle simply goes its own way, decoupled from the other two bodies. Instead, you set up the model so that the dust particle feels the effect of the other two bodies, but they ignore it completely. So the orbits of the two massive bodies are circular, and they move at a fixed speed. All of the complexity of the motion is invested in the dust particle.

Poincaré didn't solve the problem that King Oscar posed. That was just too ambitious. But his methods were so innovative, and made so much progress, that he was awarded the prize anyway. His prizewinning research was published in 1890, and it suggested that even the restricted three-body problem might not possess the kind of answer that had been stipulated. Poincaré divided his analysis into several distinct cases, depending on general features of the motion. In most of them, series solutions might well be obtainable. But there was one case in which the dust particle's orbit became extraordinarily messy.

Poincaré deduced this inescapable messiness from some other ideas he had been developing, which made it possible to describe solutions to differential equations without actually solving them. This 'qualitative theory of differential equations' was the seed from which modern nonlinear dynamics has grown. The basic idea was to explore the geometry of the solutions; more precisely their topology, a topic in which Poincaré was also deeply interested, chapter 10. In this interpretation, the positions and speeds of the bodies are coordinates in a multidimensional space. As time passes, any initial state follows a curved path through this space. The topology of this path, or the entire system of all possible paths, tells us many useful things about the solutions.

A periodic solution, for example, is a path that closes up on itself to form a loop. As time passes, the state goes round and round the loop, repeating the same behaviour indefinitely. The system is then periodic. Poincaré suggested that a good way to detect such loops is to place a multidimensional surface so that it cuts through the loop. We now call it a Poincaré section. Solutions that start on this surface may eventually return to the surface; the loop itself returns to exactly the same point, and solutions through nearby points always return to the section after roughly one period. So a periodic solution can be interpreted as a fixed point of the 'first return map', which tells us what happens to points on the surface when they first come back to it, if they do. This may not seem much of an advance, but it reduces the dimension of the space – the number of variables in the problem. This is almost always a good thing.

Poincaré's great idea starts to come into its own when we pass to the next most complex kind of solution, combinations of several periodic motions. As a simple example, the Earth goes round the Sun roughly every 365 days and the Moon goes round the Earth roughly every 28 days. So the Moon's motion combines these two different periods. Of course the whole point of the three-body problem is that this description is not entirely accurate, but 'quasiperiodic' solutions of this kind are quite common in many-body problems. The Poincaré section detects quasiperiodic solutions: when they return to the surface they don't hit exactly the same point, but the point where they hit moves round and round a closed curve *on the surface*, in small steps.

Poincaré realised that if every solution were like this, he would be able to set up suitable series to model them quantitatively. But when he analysed the topology of the first return map, he noticed that it could be more complicated. Two particular curves, related by the dynamics, might cross each other. That wasn't too bad in itself, but when you pushed the curves along until they hit the surface again, the resulting curves still had to cross – but at a different place. Push them round again, and they crossed yet again. Not only that: these new curves that arose by pushing the original ones along were not actually new. They were parts of the original curves. Sorting out the topology took some clear-headed thinking, because no one had really played this kind of game before. What emerges is a very complex picture like a crazy net, in which the curves repeatedly zigzag to and fro, crossing each other, and the zigzags themselves zigzag to and fro, and so on to any level of complexity. Poincaré in effect declared himself to be stumped:

> When one tries to depict the figure formed by these two curves and their infinity of intersections, each of which corresponds to a doubly asymptotic solution, these intersections form a kind of net, web or infinitely tight mesh... One is struck by the complexity of this figure that I am not even attempting to draw.

Fig 31 Part of a homoclinic tangle. A complete picture would be infinitely complicated.

We now call his picture a homoclinic ('self-connected') tangle, Figure 31. Thanks to new topological ideas introduced in the 1960s by Stephen Smale, we now recognise this structure as an old friend. Its most important implication is that the dynamics is *chaotic*. Although the equations have no explicit element of randomness, their solutions are very complicated and irregular, sharing certain features of genuinely random processes. For example, there are orbits – most of them, in fact – for which the motion exactly mimics the repeated random tossing of a coin. The discovery that a deterministic system – one whose entire future is uniquely determined by its present state – can nevertheless have random features is remarkable, and it has changed many areas of science. No longer do we automatically assume that simple rules cause simple behaviour. This is what is colloquially known as chaos theory, and it all goes back to Poincaré and his Oscar award.

Well, almost. For many years, historians of mathematics told the story that way. But around 1990 June Barrow-Green found a printed copy of Poincaré's memoir in the depths of the Mittag-Leffler Institute in Stockholm, thumbed through it, and realised that it was different from the version that could be found in innumerable mathematics libraries around the globe. It was, in fact, the official printing of Poincaré's prizewinning memoir, and there was a mistake in it. When Poincaré submitted his work for the prize, he had overlooked the chaotic solutions. He spotted the error before the memoir was published, worked out what he should have deduced – namely chaos – and paid (more than the prize was worth) to have the original version destroyed and a corrected version to be printed. For some reason the Mittag-Leffler Institute archives retained a copy of the original faulty version, but this became forgotten until Barrow-Green unearthed it, publishing her discovery in 1994.

Poincaré seems to have thought that these chaotic solutions were incompatible with series expansions, but that turns out to be wrong too. It was an easy assumption to make: series seem too regular to represent chaos; only topology can do that. Chaos is complicated behaviour caused by simple rules, so the inference isn't watertight, but the structure of the three-body problem definitely precludes simple solutions of the kind that Newton derived for two bodies. The two-body problem is 'integrable', which means that the equations have enough conserved quantities, such as energy, momentum, and angular momentum, to determine the orbits. 'Conserved' means that these quantities do not change as the bodies pursue their orbits. The three-body problem is known not to be integrable.

Even so, series solutions do exist, but they are not universally valid. They fail for initial states with zero angular momentum – a measure of the total spin – which are infinitely rare because zero is a single number among the infinity of all real numbers. Moreover, they are not series in the time variable as such: they are series in its cube root. The Finnish mathematician Karl Fritiof Sundman discovered all this in 1912. Something similar even holds for the n-body problem, again with rare exceptions, a result obtained in 1991 by Qiudong Wang. But for four or more bodies, we don't have any classification of the precise circumstances in which the series fail to converge. We know that such a classification must be very complicated, because there exist solutions where all bodies escape to infinity, or oscillate with infinite rapidity, after a finite time, chapter 12. Physically, these solutions are artefacts of the assumption that the bodies are single (massive) points. Mathematically, they tell us where to look for wild behaviour.

Dramatic progress has been made on the n-body problem when all of the bodies have the same mass. This is seldom a realistic assumption in celestial mechanics, but it is sensible for some non-quantum models of elementary particles. The main interest is mathematical. In 1993 Cristopher Moore found a solution to the three-body problem in which all three bodies play a game of follow-my-leader along the same orbit. Even more surprising is the shape of the orbit: a figure-eight, shown in Figure 32. Although the orbit crosses itself, the bodies never collide.

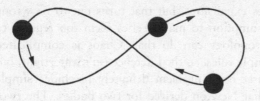

Fig 32 The figure-eight choreography.

Moore's calculation was numerical, on a computer. His solution was rediscovered independently in 2001 by Alain Chenciner and Richard Montgomery, who combined a long-standing principle of classical mechanics, known as 'least action', with some distinctly sophisticated topology to give a rigorous proof that such a solution exists. The orbits are time-periodic: after a fixed interval of time the bodies all return to their initial positions and velocities, and thereafter repeat the same movements indefinitely. For a given common mass, there is at least one such solution for any period.

In 2000 Carles Simó performed a numerical analysis indicating that the figure-eight is stable, except perhaps for a very slow long-term drift known as Arnold diffusion, related to the detailed geometry of Poincaré's return map. For this kind of stability, *almost* all perturbations lead to an orbit very close to the one concerned, and as the perturbation becomes smaller, the proportion of such perturbations approaches 100 per cent. For the small proportion of perturbations that do not behave in this stable manner, the orbit drifts away from its original location extremely slowly. Simó's result was a surprise, because stable orbits are rare in the equal-mass three-body problem. Numerical calculations show that the stability persists, even when the three masses are slightly different. So it is possible that somewhere in the universe, three stars with almost identical masses are chasing each other in a figure-eight. In 2000 Douglas Heggie estimated that the number of such triple stars lies somewhere between one per galaxy and one per universe.

The figure-eight has an interesting symmetry. Start with three bodies A, B, and C. Follow them for one third of the orbital period. You will then find three bodies with the same positions and velocities that they had to begin with – but now the corresponding bodies are B, C, and A. After two thirds of the period the same happens for C, A, and B. A full period restores the original labels of the bodies. This kind of solution is known as a choreography: a planetary dance in which everyone swaps position every so often. Numerical evidence reveals the existence of choreographies for more than three bodies: Figure 33 shows some examples. Simó in particular has found a huge number of choreographies.[58]

Even here, many questions remain unanswered. We lack rigorous proofs of the existence of these choreographies. For more than three bodies they all appear to be unstable; this is most likely correct, but

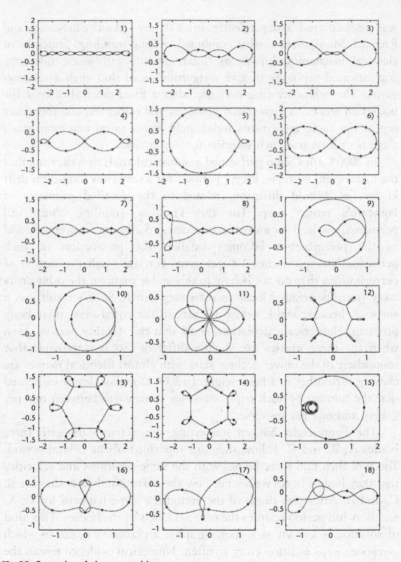

Fig 33 Examples of choreographies.

remains to be proved. The figure-eight orbit for three bodies of given mass and given period appears to be unique, but again no proof is known, although in 2003 Tomasz Kapela and Piotr Zgliczynski provided a computer-aided proof that it is locally unique – no nearby orbit works. Choreographies could be another great problem in the making.

Is the solar system stable, then?

Maybe, maybe not.

By following up Poincaré's great insight, the possibility of chaos, we now understand much more clearly the theoretical issues involved in establishing stability. They turn out to be subtle, complex, and – ironically – not related to the existence of series solutions in any very useful way. Work of Jürgen Moser and Vladimir Arnold has led to proofs that various simplified models of the solar system are stable for almost all initial states, except perhaps for the effect of Arnold diffusion, which prevents stronger kinds of stability in almost all problems of this kind. In 1961 Arnold proved that an idealised model solar system is stable in this sense, but only under the assumption that the planets have very small masses compared to the central star, and the orbits are very close to circular and very close to a common plane. As far as rigorous proof goes, 'very close' here means 'differing by a factor of at most 10^{-43}, and even then, the full statement is that the probability of being unstable is zero. In this kind of perturbation argument, the results are often valid for much larger discrepancies than anything that can be rigorously proved, so the inference is that planetary systems reasonably close to this ideal are probably stable. However, in our solar system the relevant numbers are about 10^{-3} for masses and 10^{-2} for circularity and inclination. Those comfortably exceed 10^{-43}. So the applicability of Arnold's result was moot. It was nevertheless encouraging that *anything* could be said with certainty.

The practical issues in such problems have also become clearer, thanks to the development of powerful numerical methods to solve equations approximately by computer. This is a delicate matter because chaos has an important consequence: small errors can grow very rapidly and ruin the answers. Our theoretical understanding of chaos, and of equations like those for the solar system where there is no friction, has led to the development of numerical methods that are immune to many of the most annoying features of chaos. They are called symplectic integrators. Using them, it turns out that Pluto's orbit is chaotic. However, that doesn't imply that Pluto rushes all over the solar system causing havoc. It means that in 200 million year's time, Pluto will still be somewhere close to its present orbit, but we don't have a clue whereabouts in that orbit it will be.

In 1982 Archie Roy's Project Longstop modelled the outer planets (Jupiter outwards) on a supercomputer, which found no large-scale instability, although some of the planets acquired energy at the expense of others in strange ways. Since then two research groups, in particular, have developed these computational methods and applied them to many different problems about our solar system. They are run by Jack Wisdom and Jacques Laskar. In 1984 Wisdom's group predicted that Saturn's satellite Hyperion, rather than spinning regularly, should tumble chaotically, and subsequent observations confirmed this. In 1988, in collaboration with Gerry Sussman, the group built its own computer, tailored to the equations of celestial mechanics: the digital orrery. An orrery is a mechanical device with cogs and gears, simulating the movement of the planets, which are little metal balls on sticks.[59] The original computation followed the next 845 million years of the solar system, and revealed the chaotic nature of Pluto. With its successors, Wisdom's group has explored the dynamics of the solar system for the next few billion years.

Laskar's group published its first results on the long-term behaviour of the solar system in 1989, using an averaged form of the equations that goes back to Lagrange. Here some of the fine detail is fuzzed out and ignored. The group's calculations showed that the Earth's position in its orbit is chaotic, much like Pluto's: if we measure where the Earth is today, and are wrong by 15 metres, then its position in orbit 100 million years from now cannot be predicted with any certainty.

One way to mitigate the effects of chaos is to perform many simulations, with slightly different initial data, and get a picture of the range of possible futures and how likely each of them is. In 2009 Laskar and Mickaël Gastineau applied this technique to the solar system, following 2500 different scenarios. The differences are extraordinarily small – move Mercury by 1 metre, for example. In about 1 per cent of these futures, Mercury becomes unstable: it collides with Venus, plunges into the Sun, or gets flung out into space.

In 1999 Norman Murray and Matthew Holman investigated the

inconsistency between results like Arnold's, indicating stability, and simulations, indicating instability. 'Are the numerical results incorrect, or are the classical calculations simply inapplicable?' they asked. Using analytic methods, not numerical ones, they demonstrated that the classical calculations don't apply. The perturbations needed to reflect reality are too big. The main source of chaos in the solar system is a near-resonance among Jupiter, Saturn, and Uranus, plus a less important one involving Saturn, Uranus, and Neptune. They also used numerical methods to check this contention, showing that the prediction horizon – a measure of the time it takes small errors to become large enough to have a significant effect – is about 10 million years.[60] Their simulations show that Uranus undergoes occasional near encounters with Saturn, as the eccentricity of its orbit changes chaotically, and there is a chance that it would eventually be ejected from the solar system altogether. However, the likely time is around 10^{18} years. The Sun will blow up into a red giant much sooner, about 5 billion years from now, and this will affect all of the planets, not least because the Sun will lose 30 per cent of its mass. The Earth will move outwards, and might just escape being engulfed by the greatly expanded Sun. However, it is now thought that tidal interactions will eventually pull the Earth into the Sun. The Earth's oceans will have boiled away long before. But since the typical lifetime of a species, in evolutionary terms, is no more than 5 million years, we really don't need to worry about any of these potential catastrophes. Something else will get us first.

The same methods can be used to investigate the solar system's past: use the same equations and just run time backwards, a simple mathematical trick. Until recently astronomers tended to assume the planets have always been close to their present orbits, ever since they condensed out of a cloud of gas and dust surrounding the nascent Sun. In fact, their orbits and composition have been used to infer the size and composition of that primal dust cloud. It now looks as though the planets did not start out in their present orbits. As the dust cloud coalesced under its own gravitational forces, Jupiter – the most massive planet – began to organise the positions of the other bodies, and these in turn influenced each other. This possibility was proposed in 1984 by Julio Fernandez and Wing-Huen Ip, but for a time their work was viewed as a minor curiosity. In 1993 Renu Malhotra started thinking seriously about the way changes in Neptune's orbit might influence the other giant planets, others took up the tale, and a picture of a very dynamic early solar system emerged.

As the planets continued to aggregate, there came a time when Jupiter, Saturn, Uranus, and Neptune were nearly complete, but among them circulated huge numbers of rocky and icy planetesimals, small bodies about 10 kilometres across. From that point on, the solar system evolved through the migration and collision of planetesimals. Many were ejected, which reduced the energy and angular momentum of the four giant planets. Since these worlds had different masses and were at different distances from the Sun, they reacted in different ways. Neptune was one of the winners in the orbital energy stakes, and migrated outwards. So did Uranus and Saturn, to a lesser extent. Jupiter was the big loser, energywise, and moved inwards. But it was so massive that it didn't move very far.

The other, smaller bodies of the solar system were also affected by these changes. Our solar system's current, apparently stable, plan arose through an intricate dance of the giants, in which they threw the smallest bodies at each other in a riot of chaos. So is the solar system stable? Probably not, but we won't be around to find out.

9

Patterns in primes
Riemann Hypothesis

I
N CHAPTER 2 WE LOOKED AT the properties of prime numbers as individuals, and I compared them to the often erratic and unpredictable behaviour of human beings. Humans have free will; they can make their own choices for their own reasons. Primes have to do whatever the logic of arithmetic imposes upon them, but they often seem to have a will of their own as well. Their behaviour is governed by strange coincidences and often lacks any sensible structure.

Nevertheless, the world of primes is not ruled by anarchy. In 1835 Adolphe Quetelet astounded his contemporaries by finding genuine mathematical regularities in social events that depended on conscious human choices or the intervention of fate: births, marriages, deaths, suicides. The patterns were statistical: they referred not to individuals, but to the average behaviour of large numbers of people. This is how statisticians extract order from individual free will. At much the same time, mathematicians began to realise that the same trick works for primes. Although each is a rugged individualist, collectively they conform to the rule of law. There are hidden patterns.

Statistical patterns appear when we think about entire ranges of primes. For instance: how many primes are there up to some specified limit? That's a very difficult question to answer exactly, but there are excellent approximations, and the larger the limit, the better those approximations become. Sometimes the difference between the approximation and the exact answer can be made very small, but usually that's asking too much. Most approximations in this area are asymptotic, meaning that the ratio of the approximation to the exact answer can be made very close to 1. The absolute error in the approximation can grow to any size, even though the percentage error is shrinking towards zero.

If you're wondering how this can be possible, suppose the approximate sequence of numbers, for some abstruse property of primes, is the powers of 100:

$$100 \quad 10,000 \quad 1,000,000 \quad 100,000,000$$

but the exact numbers are

$$101 \quad 10,010 \quad 1,000,100 \quad 100,001,000$$

where the extra 1 moves one place to the left at each stage. Then the ratios of the corresponding numbers get closer and closer to 1, but the differences are

$$1 \quad 10 \quad 100 \quad 1000$$

which become as large as we please. This kind of behaviour happens if the errors – the differences between the approximation and the exact answer – grow without limit, but increase more slowly than the numbers themselves.

The quest for asymptotic formulas related to primes inspired new methods in number theory, based not on whole numbers but on complex analysis. Analysis is the rigorous formulation of calculus, which has two key aspects. One, differential calculus, is about the rate at which some quantity, called a function, changes with respect to another quantity. For example, a body's position depends on – is a function of – time, and the rate at which that position changes as time passes is the body's instantaneous velocity. The other aspect, integral calculus, is about calculating areas, volumes, and the like by adding together large numbers of very small pieces, a process called integration. Remarkably, integration turns out to be the reverse of differentiation. The original formulation of calculus by Newton and Gottfried Leibniz required some manoeuvres with infinitely small quantities, raising questions about the theory's logical validity. Eventually these conceptual issues were sorted out by defining the notion of a limit, a value that can be approximated as closely as required, but need not actually be attained. When presented in this more rigorous way, the subject is called analysis.

In Newton's and Leibniz's day, the quantities concerned were real numbers, and the subject that emerged was real analysis. As complex numbers became widely accepted among mathematicians, it was natural to extend analysis to complex quantities. This subject is complex analysis. It turned out to be extraordinarily beautiful and powerful. When it comes to analysis, complex functions are much better behaved than real ones. They have their peculiarities, mind you, but the advantages of working with complex functions greatly outweigh the disadvantages.

It came as a great surprise when mathematicians discovered that arithmetical features of whole numbers can profitably be reformulated in terms of complex functions. Previously, these two number systems had asked very different questions and used very different methods. But now, complex analysis, an extraordinarily powerful body of techniques, could be used to discover special features of number-theoretic functions; from these, asymptotic formulas and much else could be extracted.

In 1859 a German mathematician, Bernhard Riemann, picked up an old idea of Euler's and developed it in a dramatic new way, defining the so-called zeta function. One of the consequences was an *exact* formula for the number of primes up to some limit. It was an infinite sum, but analysts were accustomed to those. It wasn't just a clever but useless trick; it provided genuine new insights into the primes. There was just one tiny snag. Although Riemann could prove that his formula was exact, its most important potential consequences depended on a simple statement about the zeta function, and Riemann couldn't prove this statement. A century and a half later, we still can't. It's called the Riemann hypothesis, and it is the holy grail of pure mathematics.

In chapter 2 we saw that primes tend to thin out as they get bigger. Since exact results on their distribution seemed out of the question, why not look for statistical patterns instead? In 1797–8 Legendre counted how many primes occur up to various limits, using tables of primes that had recently been provided by Jurij Vega and Anton Felkel.

Vega must have liked lengthy calculations; he constructed tables of logarithms and in 1789 he held the world record for calculating π, to 140 decimal places (126 correct). Felkel just liked calculating primes. His main work is the 1776 *Tafel aller Einfachen Factoren der durch 2, 3, 5 nicht theilbaren Zahlen von 1 bis 10 000 000* ('Table of all prime factors of numbers up to 10 million, except for those divisible by 2, 3, or 5'). There are easy ways to test for factors 2, 3, and 5, mentioned in chapter 2, so he saved a lot of space by omitting those numbers. Legendre discovered an empirical approximation to the number of primes less than a given number x, which is denoted $\pi(x)$. If you've seen π only as a symbol for the number 3.14159, this takes a little getting used to, but it's not hard to work out which is intended, even if you don't notice that they use different fonts. Legendre's 1808 text on number theory stated that $\pi(x)$ seems to be very close to $x/(\log x - 1.08366)$.

In an 1849 letter to the astronomer Johann Encke, Gauss stated that when he was about 15 he wrote a note in his logarithm tables, stating that the number of primes less than or equal to x is $x/\log x$ for large x. As with many of his discoveries, Gauss did not publish this approximation, perhaps because he had no proof. In 1838 Dirichlet informed Gauss of a similar approximation that he had discovered, which boils down to the logarithmic integral function[61]

$$\mathrm{Li}(x) = \int_0^x \frac{\mathrm{d}t}{\log t}$$

The ratio of $\mathrm{Li}(x)$ to $x/\log x$ tends to 1 as x becomes large, which implies that if one is asymptotic to $\pi(x)$ then so is the other, but Figure 34 suggests (correctly) that $\mathrm{Li}(x)$ is a better approximation than $x/\log x$. The accuracy of $\mathrm{Li}(x)$ is quite impressive; for example,

$$\pi(1,000,000,000) = 50,847,534$$
$$\mathrm{Li}(1,000,000,000) = 50,849,234.9$$

That of $x/\log x$ is poorer: here it is 48,254,942.4.

The approximation formula – either using $\mathrm{Li}(x)$ or $x/\log x$ – became known as the prime number theorem, where 'theorem' was used in the sense of 'conjecture'. The quest for a proof that these formulas are asymptotic to $\pi(x)$ became one of the key open problems

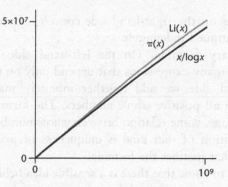

Fig 34 On this scale, $\pi(x)$ and Li(x) (grey) are indistinguishable. However, $x/\log x$ (black) is visibly smaller. Here x runs horizontally and the value of the function is plotted on the vertical axis.

in number theory. Many mathematicians attacked it using traditional methods of that area, and a few came close; however, there always seemed to be some tricky assumption that evaded proof. New methods were needed. They came from a curious reformulation of two of Euclid's ancient theorems about primes.

The prime number theorem was a response to Euclid's theorem that the primes go on for ever. Another basic Euclidean theorem is the uniqueness of prime factorisation: every positive integer is a product of primes *in exactly one way.* In 1737 Euler realised that the first theorem can be restated as a rather startling formula in real analysis, and the second statement becomes a simple consequence of that formula. I'll start by presenting the formula, and then try to make sense of it. Here it is:

$$\frac{1}{1-2^{-s}} \times \frac{1}{1-3^{-s}} \times \cdots \times \frac{1}{1-p^{-s}} \times \cdots$$

$$= \frac{1}{1^s} + \frac{1}{2^s} + \frac{1}{3^s} + \frac{1}{4^s} + \frac{1}{5^s} + \frac{1}{6^s} + \frac{1}{7^s} + \cdots$$

Here p runs through all the primes and s is constant. Euler was mainly interested in the case when s is a whole number, but his formula works for real numbers as well, provided s is greater than 1. This condition is required to make the series on the right-hand side converge: have a meaningful value when continued indefinitely.

This is an extraordinary formula. On the left-hand side we multiply together infinitely many expressions that depend only on the primes. On the right-hand side we add together infinitely many expressions that depend on all positive whole numbers. The formula expresses, in analytic language, some relation between whole numbers and primes. The main relation of that kind is uniqueness of prime factorisation, and this is what justifies the formula.

I'll sketch the main step to show that there is a sensible idea behind all this. Using school algebra we can expand the expression in p into a series, rather like the right-hand side of the formula but involving only powers of p. Specifically,

$$\frac{1}{1-p^{-s}} = \frac{1}{1^s} + \frac{1}{p^s} + \frac{1}{p^{2s}} + \frac{1}{p^{3s}} + \cdots$$

When we multiply all of these series together, over all primes p, and 'expand' to obtain a sum of simple terms, we get every combination of prime powers – that is, every whole number. Each occurs as the reciprocal of (1 divided by) its sth power, and each occurs exactly once by uniqueness of prime factorisation. So we get the series on the right.

No one has ever found a simple algebraic formula for this series, although there are many using integrals. So we give it a special symbol, the Greek letter zeta (ζ), and define a new function

$$\zeta(s) = \frac{1}{1^s} + \frac{1}{2^s} + \frac{1}{3^s} + \frac{1}{4^s} + \frac{1}{5^s} + \frac{1}{6^s} + \frac{1}{7^s} + \cdots$$

Euler didn't actually use the symbol ζ, and he considered only positive integer values of s, but I will call the above series the Euler zeta function. Using his formula, Euler deduced that there exist infinitely many primes by allowing s to get very close to 1. If there are finitely many primes, the left-hand side of the formula has a finite value, but the right-hand side becomes infinite. This is a contradiction, so there must be infinitely many primes. Euler's main aim was to obtain formulas like $\zeta(2) = \pi^2/6$, giving the sum of the series for even integers s. He didn't take his revolutionary idea much further.

Other mathematicians spotted what Euler had missed, and considered values of s that are not integers. In two papers of 1848 and 1850 the Russian mathematician Pafnuty Chebyshev had a bright idea: try to prove the prime number theorem using analysis.[62] He started from the link between prime numbers and analysis provided by the Euler zeta function. He didn't quite succeed, because he assumed s to be real, and the analytic techniques available in real analysis were too limited. But he managed to prove that when x is large, the ratio of $\pi(x)$ to $x/\log x$ lies between two constants: one slightly bigger than 1, and one slightly smaller. There was genuine payoff, even with this weaker result, because it allowed him to prove Bertrand's postulate, conjectured in 1845: if you take any integer and double it, there exists a prime between the two.

The stage was now set for Riemann. He also recognised that the zeta function holds the key to the mystery of the prime number theorem, but to make this approach work he had to propose an ambitious extension: define the zeta function not just for a real variable, but for a complex one. Euler's series is a good place to start. It converges for all real s greater than 1, and if exactly the same formula is used for complex s then the series converges whenever the real part of s is greater than 1. However, Riemann discovered he could do much better than that. Using a procedure called analytic continuation he extended the definition of $\zeta(s)$ to *all* complex numbers other than 1. That value of s is excluded because the zeta function becomes infinite when $s = 1$.[63]

In 1859 Riemann put his ideas about the zeta function together into a paper, whose title translates as 'On the number of primes less than a given magnitude'.[64] In it he gave an explicit, exact formula for $\pi(x)$.[65] I'll describe a simpler formula, equivalent to Riemann's, to show how the zeros of the zeta function appear. The idea is to count how many primes or prime powers there are, up to any chosen limit. However, instead of counting each of them once, which is what $\pi(x)$ does for the primes, larger primes are given extra weight. In fact, any power of a prime is counted according to the logarithm of that prime. For example, up to a limit of 12 the prime powers are

$$2, 3, 4 = 2^2, 5, 7, 8 = 2^3, 9 = 3^2, 11$$

so the weighted count is

$$\log 2 + \log 3 + \log 2 + \log 5 + \log 7 + \log 2 + \log 3 + \log 11$$

which is about 10.23.

Using analysis, information about this more sophisticated way to count primes can be turned into information about the usual way. However, this way leads to simpler formulas, a small price to pay for the use of the logarithm. In these terms, Riemann's exact formula states that this weighted count up to a limit x is equal to

$$-\sum_\rho \frac{x^\rho}{\rho} + x - \tfrac{1}{2}\log(1 - x^{-2}) - \log 2\pi$$

where Σ indicates a sum over all numbers ρ for which $\zeta(\rho)$ is zero, excluding negative even integers. These are called the nontrivial zeros of the zeta function. The trivial zeros are the negative even integers -2, -4, -6, ... The zeta function is zero at these values because of the formula used in the definition of the analytic continuation, but these zeros turn out not to be important for Riemann's formula, or indeed for much else.

In case the formula looks a bit daunting, let me pick out the main point: a fancy way to count primes up to a limit x, which can be turned into the usual way with a bit of analytic trickery, is *exactly* equal to a sum over all nontrivial zeros of the zeta function of the simple expression x^ρ/ρ, *plus a straightforward function of x*. If you are a complex analyst, you will see immediately that the prime number theorem is equivalent to proving that the weighted count up to the limit x is asymptotic to x. Using complex analysis, this will be true if all nontrivial zeros of the zeta function have real parts between 0 and 1. Chebyshev couldn't prove that, but he got close enough to obtain useful information.

Why are the zeros of the zeta function so important? A basic theorem in complex analysis states that subject to some technical conditions, a function of a complex variable is completely determined by the values of the variable for which that function is zero or infinity, together with some further information about its behaviour at those points. These special places are known as the function's zeros and poles. This theorem doesn't work in real analysis – one of many reasons why complex analysis became the preferred setting, despite requiring the square root of minus one. The zeta function has one pole, at $s = 1$, so everything about it is determined by its zeros provided we bear this single pole in mind.

For convenience, Riemann mostly worked with a related function, the xi function $\xi(x)$, which is closely related to the zeta function, and emerges from the method of analytic continuation. He remarked that:

> It is very probable that all [zeros of the xi function] are real. One would, however, wish for a strict proof of this; I have, though, after some fleeting futile attempts, provisionally put aside the search for such, as it appears unnecessary for the next objective of my investigation.

This statement about the xi function is equivalent to one about the related zeta function. Namely, all nontrivial zeros of the zeta function are complex numbers of the form $\frac{1}{2} + it$: they lie on the *critical line* 'real part equals $\frac{1}{2}$,' Figure 35. This version of his remark is the famous Riemann hypothesis.

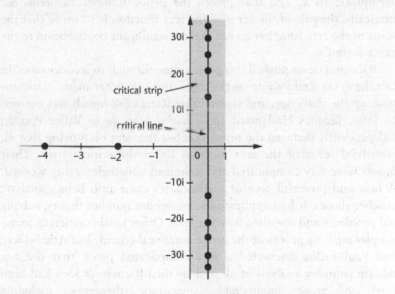

Fig 35 Zeros of the zeta function, the critical line, and the critical strip.

Riemann's remark is rather casual, as if the Riemann hypothesis isn't terribly important. For his programme to prove the prime number theorem, it wasn't. But for many other questions, the reverse is true. In fact, the Riemann hypothesis is widely considered to be the most important unanswered question in mathematics.

To understand why, we must pursue Riemann's thinking a little further. He had his sights on the prime number theorem. His exact formula suggested a way to achieve this: understand the zeros of the zeta function, or equivalently the xi function. The full Riemann hypothesis is not required; you just have to prove that all nontrivial zeros of the zeta function have real parts between 0 and 1. That is, they lie within distance $\frac{1}{2}$ of Riemann's critical line, in the so-called critical strip. This property of the zeros implies that the sum over the zeros of the zeta function, in the exact formula above, is a finite constant. Asymptotically, for large x, it might as well not be there at all. Among the terms in the formula, the only one that remains important as x becomes very large is x. All of the complicated stuff disappears asymptotically by comparison with x. Therefore the weighted count is asymptotic to x, and that proves the prime number theorem. So, ironically, the role of the zeros of the zeta function is to prove that the zeros of the zeta function do not make a significant contribution to the exact formula.

Riemann never pushed this programme through to a conclusion. In fact, he never again wrote on the topic. But two other mathematicians took up the challenge, and showed that Riemann's hunch was correct. In 1896, Jacques Hadamard and Charles Jean de la Vallée Poussin independently deduced the prime number theorem by proving that all nontrivial zeros of the zeta function lie in the critical strip. Their proofs were very complicated and technical; nonetheless, they worked. A new and powerful area of mathematics came into being: analytic number theory. It had applications throughout number theory, solving old problems and revealing new patterns. Other mathematicians found simpler analytic proofs of the prime number theorem, and Atle Selberg and Paul Erdős discovered a very complicated proof that did not require complex analysis at all. But by then Riemann's idea had been used to prove innumerable important theorems, including approximations to many number-theoretic functions. So their new proof added an ironic footnote, but otherwise had little effect. In 1980

Donald Newman found a much simpler proof, using only one of the most basic results in complex analysis, known as Cauchy's theorem.

Although Riemann declared his hypothesis to be unnecessary for his immediate objectives, it turned out to be vital to many other questions in number theory. Before discussing the Riemann hypothesis, it is worth taking a look at some of the theorems that would follow if the hypothesis could be proved true.

One of the most important implications is the size of the error in the prime number theorem. The theorem states that for large x, the ratio of $\pi(x)$ to $\mathrm{Li}(x)$ gets closer and closer to 1. That is, the size of the difference between those two functions shrinks to zero, *relative to the size of x*. However, the actual difference can (and does) become larger and larger. It just does so at a slower rate than x itself does.[66] Computer experiments suggest that the size of the error is roughly proportional to $\sqrt{x}\log x$. If the Riemann hypothesis is true, this statement can be proved. In 1901 Helge von Koch proved that the Riemann hypothesis is logically equivalent to the estimate

$$|\pi(x) - \mathrm{Li}(x)| \leqslant \frac{1}{8\pi}\sqrt{x}\log x$$

for all $x \geqslant 2657$. Here the vertical bars $|\;|$ indicate the absolute value: the difference multiplied by ± 1 to make it positive. This formula provides the best possible bound on the difference between $\pi(x)$ and $\mathrm{Li}(x)$.

The Riemann hypothesis implies many estimates for other number-theoretic functions. For example, it is equivalent to the sum of the divisors of n being less than

$$e^{\gamma} n \log\log n$$

for all $n \geqslant 5040$, where $\gamma = 0.57721 \ldots$ is Euler's constant.[67] These facts may seem like oddities, but good estimates for important functions are vital to many applications, and most number theorists would give their right arm to prove any of them.

The Riemann hypothesis also tells us how big the gaps between consecutive primes can be. We can deduce the typical size of this gap from the prime number theorem: on average, the gap between a prime

p and the next one is comparable to $\log p$. Some gaps are smaller, some bigger, and it would make mathematicians' lives easier if they knew how large the biggest gaps can become. Harald Cramér proved in 1936 that if the Riemann hypothesis is true, the gap at prime p is no bigger than a constant times $\sqrt{p} \log p$.

The true significance of the Riemann hypothesis lies much deeper. There are far-reaching generalisations, plus a strong hunch that anyone who can prove the Riemann hypothesis can probably prove the corresponding generalised Riemann hypothesis. Which in turn would give mathematicians a lot of control over wide-ranging areas of number theory.

The generalised Riemann hypothesis arises from a finer description of prime numbers. All primes other than 2 are odd, and we saw in chapter 2 that the odd ones can be classified into two kinds: those that are 1 greater than a multiple of 4, and those that are 3 greater than a multiple of 4. They are said to be of the form $4k + 1$ or $4k + 3$, where k is what you multiply by 4 to get them. Here's a short list of the first few primes of each type, together with the corresponding multiples of 4:

multiple of 4	0	4	8	12	16	20	24	28	32	36
plus 1	0	5	•	13	17	•	•	29	•	37
plus 3	3	7	11	•	19	23	•	•	•	•

The dot indicates that the number concerned is not prime.

How many primes of each kind are there? How are they distributed among the primes, or among all integers? Euclid's proof that there are infinitely many primes can be modified, without much effort, to prove that there are infinitely many primes of the form $4k + 3$. It is much harder to prove that there are infinitely many primes of the form $4k + 1$; it can be done, but only by using some fairly difficult theorems. The difference arises because any number of the form $4k + 3$ has some factor of that form; the same is not always true of numbers $4k + 1$.

There's nothing sacred about the numbers here. Apart from 2 and 3, all primes are either of the form $6k + 1$ or $6k + 5$, and we can ask similar questions. For that matter, all primes except 5 take one of the forms $5k + 1$, $5k + 2$, $5k + 3$, $5k + 4$. We leave out $5k$ because these are the multiples of 5, so all of them except 5 are not prime.

It's not hard to come up with a sensible guess for all questions of this kind – primes in an arithmetic sequence. The $5k$ case is typical. Experiment quickly suggests that numbers of the four types listed above have much the same chance of being prime. Here's a similar table:

multiple of 5	5	10	15	20	25	30	35	40
plus 1	•	11	•	•	•	31	•	41
plus 2	7	•	17	•	•	•	37	•
plus 3	•	13	•	23	•	•	•	43
plus 4	•	•	19	•	29	•	•	•

So there should be infinitely many of each individual type, and on average about one quarter of the primes, up to some given limit, should be of any specific form.

Simple proofs show that some forms lead to infinitely many primes, more sophisticated proofs work for other forms, but until the mid-1800s no one could prove there are infinitely many primes of each possible form, let alone that the proportions are roughly equal. Lagrange assumed this without proof in his work on the law of quadratic reciprocity – a deep property of squares to a prime modulus – in 1785. The results clearly had useful consequences, and it was high time someone proved them. In 1837 Dirichlet discovered how to adapt Riemann's ideas about the prime number theorem to prove both of those statements. The first step was to define analogues of the zeta function for these types of prime. The resulting functions are called Dirichlet L-functions. An example, arising in the $4k + 1/4k + 3$ case, is

$$L(s, \chi) = 1 - 3^{-s} + 5^{-s} - 7^{-s} + 9^{-s} - \cdots$$

where the coefficients are $+1$ for numbers of the form $4k + 1$, -1 for numbers $4k + 3$, and 0 for the rest. The Greek letter χ is called a Dirichlet character, and it reminds us to use these signs.

For Riemann's zeta function what matters is not just the series but its analytic continuation, which gives the function a meaning for all complex numbers. The same goes for the L-function, and Dirichlet defined a suitable analytic continuation. By adapting the ideas used to prove the prime number theorem, he was then able to prove an analogous theorem for primes of specific forms. For example, the number of primes of the form $5k + 1$ less than or equal to x is asymptotic to $\text{Li}(x)/4$, and the same goes for the other three cases $5k + 2$, $5k + 3$, $5k + 4$. In particular there are infinitely many primes of each form.

The Riemann zeta function is a special case of a Dirichlet L-function for primes of the form $1k + 0$, that is, all primes. The generalised Riemann hypothesis is the obvious generalisation of the original Riemann hypothesis: the zeros of any Dirichlet L-function either have real part $\frac{1}{2}$, or they are 'trivial zeros' with real part either negative or greater than 1.

If the generalised Riemann hypothesis is true, then so is the Riemann hypothesis. Many of the consequences of the generalised Riemann hypothesis are analogues of those for the Riemann hypothesis. For example, similar error bounds can be proved for the analogous versions of the prime number theorem, applied to primes of any specific form. However, the generalised Riemann hypothesis implies many things that are quite different from anything that we can derive using the ordinary Riemann hypothesis. Thus in 1917 Godfrey Harold Hardy and John Edensor Littlewood proved that the generalised Riemann hypothesis implies a conjecture of Chebyshev, to the effect that (in a precise sense) primes of the form $4k + 3$ are more common than those of the form $4k + 1$. Both types are equally likely, in the long run, by Dirichlet's theorem, but that doesn't stop the $4k + 3$ primes outcompeting the $4k + 1$ primes if you set up the right game.

The generalised Riemann hypothesis also has important implications for primality tests, such as Miller's 1976 test mentioned in chapter 2. If the generalised Riemann hypothesis is true, then Miller's test provides an efficient algorithm. Estimates of the efficiency of more recent tests also depend on the generalised Riemann hypothesis. There are significant applications to algebraic number theory, too. Recall from chapter 7 that Dedekind's reformulation of Kummer's ideal numbers led to a new and fundamental concept, ideals. Prime factorisation in rings of algebraic integers exists, but may not be unique. Prime factorisation of ideals is much tidier: both existence and uniqueness are valid. So it makes sense to reinterpret all questions about factors in terms of ideals. In particular, there is a notion of a 'prime ideal', a reasonable and tractable analogue of a prime number.

Knowing this, it is natural to ask whether Euler's link between ordinary primes and the zeta function has an analogue for prime ideals. If so, the whole powerful machinery of analytic number theory becomes available for algebraic numbers. It turns out that this can be done, with deep and vital implications. The result is the Dedekind zeta function: one such function for each system of algebraic numbers. There is a deep link between the complex analytic properties of the Dedekind zeta function and the arithmetic of prime numbers for the corresponding algebraic integers. And, of course, there is an analogue of the Riemann hypothesis: all nontrivial zeros of the Dedekind zeta function lie on the critical line. The phrase 'generalised Riemann hypothesis' now includes this conjecture as well.

Even this generalisation is not the end of the story of the zeta function. It has inspired the definition of analogous functions in several other areas of mathematics, from abstract algebra to dynamical systems theory. In all of these areas there are even more far-reaching analogues of the Riemann hypothesis. A few of them have even been proved to be true. In 1974 Pierre Deligne proved such an analogue for varieties over finite fields. Generalisations known as Selberg zeta functions satisfy an analogue of the Riemann hypothesis. The same goes for the Goss zeta function. However, there exist other generalisations, Epstein zeta functions, for which the appropriate analogue of the Riemann hypothesis is false. Here infinitely many nontrivial zeros lie on the critical line, but some do not, as Edward Titchmarsh demonstrated in 1986. On the other hand, these zeta functions do not have an Euler product formula, so they fail to resemble the Riemann zeta functions in what may well be a crucial respect.

The circumstantial evidence in favour of the truth of the Riemann hypothesis – either the original, or its generalisations – is extensive. Many beautiful things would follow from the truth of the hypothesis. None of these things has ever been disproved: to do so would be to disprove the Riemann hypothesis, but neither proof nor disproof is known. There is a widespread feeling that a proof of the original Riemann hypothesis would open the way to a proof of its generalisations as well. In fact, it might be better to attack the generalised Riemann hypothesis in all its glory, exploiting the wealth of methods now available, and then to deduce the original Riemann hypothesis as a special case.

There is also a vast amount of experimental evidence for the truth of the Riemann hypothesis – or what certainly looks like a vast amount until someone throws cold water on that claim. According to Carl Ludwig Siegel, Riemann calculated the first few zeros of his zeta function numerically but did not publish the results: they are located at

$$\tfrac{1}{2} \pm 14.135i \qquad \tfrac{1}{2} \pm 21.022i \qquad \tfrac{1}{2} \pm 25.011i$$

The nontrivial zeros always come in $\pm$ pairs like this. I've written $\frac{1}{2}$ here rather than 0.5 because the real part is known *exactly* in these cases, by exploiting general results in complex analysis and known properties of the zeta function. The same goes for the computer calculations reported below. They don't just show that zeros are very close to the critical line; they are actually on it.

In 1903 Jorgen Gram showed numerically that the first ten ($\pm$ pairs of) zeros lie on the critical line. By 1935 Titchmarsh had increased the number to 195. In 1936 Titchmarsh and Leslie Comrie proved that the first 1041 pairs of zeros are on the critical line – the last time anyone did such computations by hand. Alan Turing is best known for his wartime efforts at Bletchley Park, where he helped to break the German Enigma code, and for his work on the foundations of computing and artificial intelligence. But he also took an interest in analytic number theory. In 1953 he discovered a more efficient method for calculating zeros of the zeta function, and used a computer to deduce that the first 1104 pairs of zeros are on the critical line. Evidence for all zeros up to some limit being on the critical line piled up; the current record, obtained by Yannick Saouter and Patrick Demichel in 2004, is 10 trillion (10^{13}). Various mathematicians and computer scientists have also checked other ranges of zeros. To date, every nontrivial zero that has been computed lies on the critical line.

This might seem conclusive, but mathematicians are ambivalent about this kind of evidence, with good reason. Numbers like 10 trillion may sound large, but in number theory what often matters is the logarithm of the number, which is proportional to the number of digits. The logarithm of 10 trillion is just under 30. In fact, many problems hinge on the logarithm of the logarithm, or even the logarithm of the logarithm of the logarithm. In those terms, 10 trillion is *tiny*, so numerical evidence up to 10 trillion carries hardly any weight.

There is also some general analytic evidence, which is not subject to this objection. Hardy and Littlewood proved that infinitely many zeros lie on the critical line. Other mathematicians have shown, in a precise sense, that almost all zeros lie very close to the critical line. Selberg proved that a nonzero proportion of zeros lie on the critical line. Norman Levinson proved that this proportion is at least one third, a figure now improved to at least 40 per cent. All of these results suggest that if the Riemann hypothesis is false, zeros that do not lie on the critical line are very large, and very rare. Unfortunately, the main implication is that if such exceptions exist, finding them will be extraordinarily hard.

Why bother? Surely the numerical evidence ought to satisfy any sensible person? Unfortunately not. It doesn't satisfy mathematicians, and in this instance they are not just being pedantic: they are indeed acting as sensible people. In mathematics in general, and especially in number theory, apparently extensive 'experimental' evidence often carries much less weight than you might imagine.

An object lesson is provided by the Pólya conjecture, stated in 1919 by the Hungarian mathematician George Pólya. He suggested that at least half of all whole numbers up to any specific value have an odd number of prime factors. Here repeated factors are counted separately, and we start from 2. For example, up to 20 the number of prime factors looks like Table 2, where the column 'percentage' gives the percentage of numbers up to this size with an odd number of prime factors.

All the percentages in the final column are bigger than 50 per cent, and more extensive calculations make it reasonable to conjecture that this is always true. In 1919, with no computers available, experiments could not find any numbers that disproved the conjecture. But in 1958

number	factorisation	how many primes?	percentage
2	2	1	100
3	3	1	100
4	2^2	2	66
5	5	1	75
6	2×3	2	60
7	7	1	66
8	2^3	3	71
9	3^2	2	62
10	2×5	2	55
11	11	1	60
12	$2^2 \times 3$	3	63
13	13	1	66
14	2×7	2	61
15	3×5	2	57
16	2^4	4	60
17	17	1	62
18	2×3^2	3	64
19	19	1	66
20	$2^2 \times 5$	3	65

Table 2 Percentages of numbers, up to a given size, that have an odd number of prime factors.

Brian Haselgrove used analytic number theory to prove that the conjecture is false for some number – less than 1.845×10^{361}, to be precise. Once computers arrived on the scene, Sherman Lehman showed that the conjecture is false for 906,180,359. By 1980 Minoru Tanaka has proved that the smallest such example is 906,150,257. So you could have amassed experimental evidence in the conjecture's favour for nearly all numbers up to a billion, even though it is false.

Still, it's nice to know that the number 906,150,257 is unusually interesting.

Of course, today's computers would disprove the conjecture in a few seconds, if suitably programmed. But sometimes even computers don't help. A classic example is Skewes's number, where apparently huge amounts of numerical evidence initially suggested that a famous conjecture should be true, but in fact it is false. This gigantic number appeared in a problem closely related to the Riemann hypothesis: the approximation of $\pi(x)$ by $\mathrm{Li}(x)$. As we've just seen, the prime number theorem states that the ratio of these two quantities tends to 1 as x becomes large. Numerical calculations seem to indicate something stronger: the ratio is always less than 1; that is, $\pi(x)$ is less than $\mathrm{Li}(x)$. In 2008 Tadej Kotnik's numerical computations showed that this is true whenever x is less than 10^{14}. By 2012 Douglas Stoll and Demichel had improved this bound to 10^{18}, a figure obtained independently by Andry Kulsha. Results of Tomás Oliveira e Silva suggest it can be increased to 10^{20}.

This might sound definitive. It's stronger than the best numerical results we have for the Riemann hypothesis. But in 1914 Littlewood proved that this conjecture is false – spectacularly so. As x runs through the positive real numbers, the difference $\pi(x)-\mathrm{Li}(x)$ changes sign (from negative to positive or the reverse) *infinitely often*. In particular, $\pi(x)$ is *bigger* than $\mathrm{Li}(x)$ for some sufficiently large values of x. However, Littlewood's proof gave no indication of the size of such a value.

In 1933 his student, the South African mathematician Stanley Skewes, estimated how big x must be: no more than 10^10^10^34, where ^ indicates 'raised to the power'. That number is so gigantic that if all of its digits were printed in a book – rather a boring book, consisting of 1 followed by endless 0s – the universe would not be big enough to contain it, even if every digit were the size of a subatomic particle. Moreover, Skewes had to assume the truth of the Riemann hypothesis to make his proof work. By 1955 he had found a way to avoid the Riemann hypothesis, but at a price: his estimate increased to 10^10^10^963.

These numbers are too big even for the adjective 'astronomical', but further research reduced them to something that might be termed cosmological. In 1966 Lehman replaced Skewes's numbers by 10^{1165}. Te Riele reduced this to 7×10^{370} in 1987, and in 2000 Carter Bays and Richard Hudson reduced it to 1.39822×10^{316}. Kuok Fai Chow and Roger Plymen chipped a bit off, and got the number down to 1.39801×10^{316}. This may seem a negligible improvement, but it's about 2×10^{313} smaller. Saouter and Demichel made a further improvement to $1.3971667 \times 10^{316}$. Meanwhile in 1941 Aurel Wintner had proved that a small but nonzero proportion of integers satisfy $\pi(x) > \text{Li}(x)$. In 2011 Stoll and Demichel computed the first 200 billion zeros of the zeta function, which gives control of $\pi(x)$ when x is anything up to $10^{10,000,000,000,000}$, and found evidence that if x is less than 3.17×10^{114} then $\pi(x)$ is smaller than $\text{Li}(x)$.[68] So for this particular problem, the evidence up to at least 10^{18}, and very possibly up to 10^{114} or more, is completely misleading. The fickle gods of number theory are having an amusing joke at human expense.

Over the years, many attempts have been made to prove or disprove the Riemann hypothesis. Matthew Watkins's website 'Proposed proofs of the Riemann hypothesis' lists around 50 of them since 2000.[69] Mistakes have been found in many of these attempts, and none has been accepted as correct by qualified experts.

One of the most widely publicised efforts, in recent years, was that of Louis de Branges in 2002. He circulated a lengthy manuscript claiming to deduce the Riemann hypothesis by applying a branch of analysis that dealt with operators on infinite-dimensional spaces, known as functional analysis. There were reasons to take de Branges seriously. He had previously circulated a proof of the Bieberbach conjecture, about series expansions of complex functions. Although his original proof had errors, it was eventually established that the underlying idea worked. However, there now seem to be good reasons to think that de Branges's proposed method for proving the Riemann hypothesis has no chance of succeeding. Some apparently fatal obstacles have been pointed out by Brian Conrey and Xian-Jin Li.[70]

Perhaps the greatest hope for a proof comes from new or radically different ways to think about the problem. As we've seen repeatedly, breakthroughs on great problems often arise when someone links them to some totally different area of mathematics. Fermat's last theorem is a clear example: once it was reinterpreted as a question about elliptic curves, progress was rapid.

De Branges's tactics now seem questionable, but his approach is strategically sound. It has its roots in a verbal suggestion made around 1912 by David Hilbert, and independently by George Pólya. The physicist Edmund Landau asked Pólya for a physical reason why the

Riemann hypothesis should be true. Pólya related in 1982 that he had come up with an answer: the zeros of the zeta function should be related to the eigenvalues of a so-called self-adjoint operator. These are characteristic numbers associated with special kinds of transformation. In quantum physics, one of the important applications, these numbers determine the energy levels of the system concerned, and a standard and easy theorem states that the eigenvalues of this special kind of operator are always real. As we have seen, the Riemann hypothesis can be rephrased as the statement that all zeros of the xi function are real. If some self-adjoint operator had eigenvalues that were the same as the zeros of the xi function, the Riemann hypothesis would be an easy consequence. Pólya didn't publish this idea – he couldn't write down such an operator, and until someone could, it was pie in the sky. But in 1950 Selberg proved his 'trace formula', which relates the geometry of a surface to the eigenvalues of an associated operator. This made the idea a little more plausible.

In 1972 Hugh Montgomery was visiting the Institute for Advanced Study in Princeton. He had noticed some surprising statistical features of the nontrivial zeros of the zeta function. He mentioned them to the physicist Freeman Dyson, who immediately spotted a similarity to statistical features of random Hermitian matrices, another special type of operator used to describe quantum systems such as atomic nuclei. In 1999 Alain Connes came up with a trace formula, similar to Selberg's, whose validity would imply the truth of the generalised Riemann hypothesis. And in 1999 physicists Michael Berry and Jon Keating suggested that the required operator might arise by quantising a well-known concept from classical physics, related to momentum. The resulting Berry conjecture can be viewed as a more specific version of the Hilbert-Pólya conjecture.

These ideas, relating the Riemann hypothesis to core areas of mathematical physics, are remarkable. They show that progress may eventually come from apparently unrelated areas of mathematics, and raise hopes that the Riemann hypothesis may one day be settled. However, they have not yet led to any definitive breakthrough that encourages us to think that a solution is just around the corner. The Riemann hypothesis remains one of the most baffling and irritating enigmas in the whole of mathematics.

Today there is a new reason to try to prove the Riemann hypothesis: a substantial prize.

There is no Nobel prize in mathematics. The most distinguished prize in mathematics is the Fields medal, more properly the International Medal for Outstanding Discoveries in Mathematics. It is named after the Canadian mathematician John Fields, who endowed the award in his will. Every four years, at the International Congress of Mathematicians, up to four of the world's leading young (under 40) mathematicians receive a gold medal and a cash award, currently $15,000. As far as mathematicians are concerned, the Fields medal is equivalent in prestige to a Nobel prize.

Many mathematicians consider the lack of a Nobel in their subject to be a good thing. A Nobel prize is currently worth just over a million dollars, an amount that could easily distort research objectives and lead to arguments about priority. However, the absence of a major mathematical prize may also have distorted the public perception of the value and utility of mathematics. It's easy to imagine that if no one is willing to pay for it, it can't be worth much.

Recently two new high-prestige mathematical awards have come into being. One is the Abel prize, awarded annually by the Norwegian Academy of Science and Letters, and named after the great Norwegian mathematician Niels Henrik Abel. The other new award consists of seven Clay Mathematics Institute millennium prizes. The Clay Institute was founded by Landon Clay and his wife Lavinia. Landon Clay is an American businessman active in mutual funds, with a love of, and respect for, mathematics. In 1999 he established a new foundation for mathematics in Cambridge, Massachusetts, which runs meetings, awards research grants, organises public lectures, and administers an annual research award.

In 2000 Sir Michael Atiyah and John Tate, leading mathematicians in Britain and the United States, announced that the Clay Mathematics Institute had set up a new award, intended to encourage the solution of seven of the most important open problems in mathematics. They would be known as the millennium problems, and a properly published and refereed solution of any one of them would be worth 1 million dollars. Collectively, these problems draw attention to some of the central unanswered questions in mathematics, carefully selected by some of the world's top mathematicians. The substantial prize makes a very clear point to the public: mathematics is valuable. Everyone involved is aware that its intellectual value may be deeper than mere money, but a cash prize does help to concentrate minds. The best-known millennium problem, and the one that goes back furthest historically, is the Riemann hypothesis. It is the only question to feature both in Hilbert's list of 1900 and the list of millennium problems. The other six millennium problems are discussed in chapters 10–15. Mathematicians are not especially obsessed with prizes, and they would work on the Riemann hypothesis without one. A new and promising idea would be all the motivation they would need.

It's worth remembering that conjectures, however time-honoured, may not be true. Today, most mathematicians seem to think that a proof of the Riemann hypothesis will eventually be found. A few, however, think it may be false: somewhere out in the wilderness of very big numbers there may lurk a zero that does not sit on the critical line. If such a 'counterexample' exists it is likely to be very, very big.

However, opinions count for little at the frontiers of mathematics. Expert intuition is often very good indeed, but there have been plenty of occasions when it was wrong. The conventional wisdom can be both conventional and wise, without being true. Littlewood, one of the great experts in complex analysis, was unequivocal: in 1962 he said that he was sure the Riemann hypothesis was false, adding that there was no imaginable reason for it to be true. Who is right? We can only wait and see.

10

What shape is a sphere?
Poincaré Conjecture

HENRI POINCARÉ WAS ONE OF the greatest mathematicians of the late nineteenth century, a bit of an eccentric, but an astute operator. He became a member of France's Bureau des Longitudes, whose job was to improve navigation, time-keeping, and measurement of the Earth and planets. This appointment led him to propose setting up international time zones; it also inspired him to think about the physics of time, anticipating some of Einstein's discoveries in special relativity. Poincaré left his mark all over the mathematical landscape, from number theory to mathematical physics.

In particular, he was one of the founders of topology, the mathematics of continuous transformations. Here, in 1904, he ran into an apparently simple question, having belatedly realised that he had tacitly assumed the answer in earlier work, but couldn't find a proof. 'This question would lead us too far astray,' he wrote, which rather slid round the real issue: it wasn't leading him *anywhere*. Although he phrased the problem as a question, it became known as the Poincaré conjecture because everyone expected the answer to be 'yes'. It is another of the seven Clay millennium prize problems, and rightly so, because it turned out to be one of the most baffling problems in the whole of topology. Poincaré's question was finally answered in 2002 by a young Russian, Grigori Perelman. The solution introduced a whole raft of new ideas and methods, so much so that it took the mathematical community a few years to digest the proof and accept that it was correct.

For his success, Perelman was awarded a Fields medal, the most prestigious mathematical prize, but he declined it. He didn't want publicity. He was offered the million-dollar Clay prize for proving the Poincaré conjecture, and turned it down. He didn't want money, either. What he had wanted was for his work to be accepted by the mathematical community. Eventually it was, but unfortunately, for sensible reasons, that took a while. And it was always unrealistic to expect acceptance without publicity or the offer of prizes. But these unavoidable consequences of success didn't suit Perelman's sometimes reclusive nature.

We encountered topology in connection with the four colour theorem, and I resorted to the cliché 'rubber-sheet geometry'. Euclid's geometry deals with straight lines, circles, lengths, and angles. It takes place in a plane, or in a space of 3 dimensions when it becomes more advanced. A plane is like an infinite sheet of paper, and it shares one basic feature of paper: it doesn't stretch, shrink, or bend. You can roll paper into a tube, and it can shrink or stretch a tiny bit, especially if you spill coffee on it. But you can't wrap a sheet of paper round a sphere without creating creases. Mathematically, the Euclidean plane is rigid. In Euclid's geometry, two objects – triangles, squares, circles – are the same if you can transform one of them into the other by a rigid motion. And 'rigid' means that distances don't change.

What if you use an elastic sheet instead of paper? That does stretch, it does bend, and with a bit of effort it can be compressed. Lengths and angles have no fixed meaning on a sheet of elastic. Indeed, if it is elastic *enough*, neither do triangles, squares, or circles. You can deform a triangle on a sheet of rubber to give it an extra corner. You can even turn it into a circle, Figure 36. Whatever the concepts of rubber-sheet geometry are, they don't include the traditional Euclidean ones.

It might seem that geometry on a sheet of rubber would be so flexible that nothing would have a fixed meaning, in which case little of substance could be proved. Not so. Draw a triangle and place a point inside it. If you stretch and deform the sheet until the triangle becomes

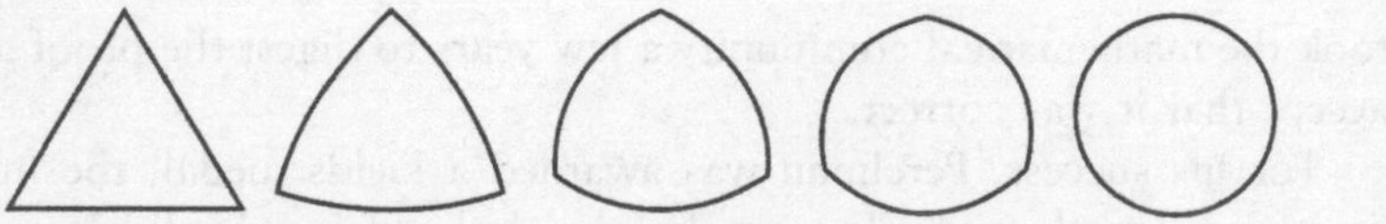

Fig 36 Topological deformation of a triangle to a circle.

a circle, one feature of your diagram doesn't change: the point remains on the inside. Agreed, it's now inside a circle, not a triangle, but it's not *outside*. In order to move the point to the outside, you have to tear the sheet. That breaks the rules of this particular game.

There's another feature that survives distortion, too. A triangle is a simple closed curve. It is a line that joins up with itself so that there are no free ends, and does not cross itself. A figure-eight is a closed curve, but it's not simple – it crosses itself. When you deform the rubber sheet, the triangle may change shape, but it always remains a simple closed curve. There is no way to turn it into a figure-eight, for example, without tearing the sheet.

In three-dimensional topology, the whole of space becomes elastic. Not like a block of rubber, which twangs back into its original shape if you let go, but like a gel that can be deformed without any resistance. A topological space is infinitely deformable; you can take a region the size of a grain of rice and blow it up to the size of the Sun. You can pull out tentacles until the region is shaped like an octopus. The one thing you are not allowed to do is to introduce any kind of discontinuity. You mustn't tear space, or perform any kind of distortion that rips nearby points apart.

What features of a shape in space survive all continuous deformations? Not length, area, or volume. But being knotted does. If you tie a knot in a curve and join the ends to make a loop, then the knot can never escape. However you deform the space, the curve remains knotted. So we are working with a new kind of geometry in which the important and meaningful concepts seem rather fuzzy: 'inside', 'closed', 'simple', 'knotted'. This new geometry has a respectable name: topology. It may seem rather esoteric, perhaps even absurd, but it has turned out to be one of the major areas of twentieth-century mathematics, and it remains equally vital in the twenty-first. And one of the main people we have to thank for that is Poincaré.

The story of topology began to take off almost a century before Poincaré, in 1813. Simon Antoine Jean Lhuilier, a Swiss mathematician, didn't exactly set the world of mathematics alight during his lifetime, even though he turned down a large sum of money that a relative had promised to pay him if he entered the Church. Lhuilier preferred a career in mathematics. He specialised in a mathematical backwater: Euler's theorem for polyhedrons. In chapter 4 we came across this curious, apparently isolated result: if a polyhedron has F faces, V vertexes, and E edges, then $F - E + V = 2$. Lhuilier spent much of his career investigating variants of this formula, and in retrospect he took a vital step in the direction of topology when he discovered that Euler's formula is sometimes wrong. Its validity depends on the qualitative shape of the polyhedron.

The formula is correct for polyhedrons without any holes, which can be drawn on the surface of a sphere or continuously deformed into a shape of that kind. But when the polyhedron has holes, the formula fails. A picture-frame made from wood with a rectangular cross-section, for example, has 16 faces, 32 edges, and 16 vertexes; here $F - E + V = 0$. Lhuilier modified Euler's formula to cover these more exotic polyhedrons: if there are g holes, then $F - E + V = 2 - 2g$. This was the first discovery of a significant topological invariant: a quantity associated with a space, which does not change when the space is deformed continuously. Lhuilier's invariant provides a rigorous way to count how many holes a surface has, without the need to define 'hole'. This is useful, because the concept of a hole is tricky. A hole is not part of the surface, nor is it the region outside the surface. It appears to be a feature of how the surface sits in its surrounding space. But Lhuilier's discovery shows that what we interpret as the *number* of holes is an intrinsic feature, independent of any surrounding space. It is not necessary to define holes and then count them; in fact, it's better not to.

After Lhuilier, the next key figure in the prehistory of topology is Gauss. He encountered several other topological invariants when working on various core areas of mathematics. His work in complex analysis, especially the proof that every polynomial equation has at least one solution in complex numbers, led him to consider the winding number of a curve in the plane: how many times it winds round a given point. Problems in electricity and magnetism led to the linking number of two closed curves: how many times one of them winds through the other. These and other examples led Gauss to wonder whether there might exist some as-yet undiscovered branch of mathematics that would provide a systematic way to understand qualitative features of geometrical figures. He published nothing on the topic, but he mentioned it in letters and manuscripts.

He also passed the idea on to his student Johann Listing and his assistant August Möbius. I've mentioned the Möbius band, a surface with only one side and also one edge, which he published in 1865, and it can be found in Figure 9 of chapter 4. Möbius pointed out that 'having only one side', while intuitively clear, is difficult to make precise, and proposed a related property that could be defined in complete rigour. This property was orientability. A surface is orientable if you can cover it with a network of triangles, with arrows circulating round each triangle, so that whenever two triangles have a common edge the arrows point in opposite directions. If you draw a network on a plane and make all arrows run clockwise, for example, this is what happens. On a Möbius band, no such network exists.

Listing's first publication in topology came earlier, in 1847. Its title was *Vorstudien zur Topologie* ('Lectures on topology'), and it was the first text to employ that word. He had been using the term informally for about a decade. Another term used at that time is the Latin phrase *analysis situs*, 'analysis of position', but this eventually fell out of favour. Listing's book contains little of great significance, but it does set up a basic notion: covering a surface with a network of triangles. In 1861, four years ahead of Möbius, he described the Möbius band and studied connectivity: whether a space can be split into two or more disconnected parts. Building on Listing's work, a number of mathematicians, among them Walther von Dyck, put together a complete topological classification of surfaces, assuming them to be closed (no edge) and compact (of finite extent). The answer is that every orientable surface is topologically equivalent to a sphere, to which a finite number g of handles has been attached, Figure 11 (middle and right) in chapter 4. The number g is called the genus of the surface, and it is what Lhuillier's invariant determines. If $g = 0$ we have the sphere, and if $g > 0$ we obtain a torus with g holes. A similar sequence of surfaces, starting with the simplest non-orientable surface, the projective plane, classifies all non-orientable surfaces. The method was extended to allow surfaces with edges as well. Each edge is a closed loop, and the only extra information needed is how many of these loops occur.

The Poincaré conjecture will make more sense if we first take a look at one of the basic techniques employed in classifying surfaces. Earlier, I described topology in terms of deforming a shape made of rubber or gel, and emphasised the need to use *continuous* transformations. Ironically, one of the central techniques in topology involves what at first sight seems to be a discontinuous transformation: cut the shape into pieces. However, continuity is restored by a series of rules, describing which piece is joined to which, and in what manner. An example is the way we defined a torus by identifying opposite edges of a square, Figure 12 of chapter 4.

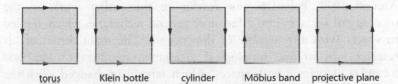

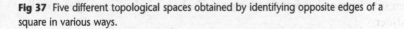

| torus | Klein bottle | cylinder | Möbius band | projective plane |

Fig 37 Five different topological spaces obtained by identifying opposite edges of a square in various ways.

Identifying points that appear to be distinct allows us to represent complicated topological spaces using simpler ingredients. A square is a square is a square, but a square with identification rules can be a torus, a Klein bottle, a cylinder, a Möbius band, or a projective plane, depending on the choice of rules, Figure 37. So when I explained a continuous transformation in terms of stretching and bending a rubber sheet, I asked for more than is strictly necessary. We are also permitted to cut the sheet, at an intermediate stage, provided that eventually we either join the edges together again exactly as they were to begin with, or we specify rules that have the same effect. As far as a topologist is concerned, *stating* a rule for gluing edges together is the same as actually implementing the rule. Provided you don't forget the rule in whatever else you do later.

The classic method for classifying surfaces begins by drawing a network of triangles on the surface. Then we cut enough of the edges to fold the triangles out flat to make a polygon. Gluing rules, derived from how we made the cuts, then specify how to identify various edges of the polygon, reconstructing the original surface. At that point, all of the interesting topology is implicit in the gluing rules. The classification is proved by manipulating the rules algebraically, and transforming them into rules that define a g-holed torus or one of the analogous non-orientable surfaces. Modern topology has other ways to achieve the same result, but it often uses this kind of 'cut and paste' construction. The method generalises without difficulty to spaces of any dimension, but it is too restricted to lead to a classification of higher-dimensional topological spaces without further assistance.

Around 1900, Poincaré was developing the earlier work on the topology of surfaces into a far more general technique, which applied to spaces with any number of dimensions. The main thrust of his research was to find topological invariants: numbers or algebraic formulas associated with spaces, which remain unchanged when the space is continuously deformed. If two spaces have different invariants, then one cannot be deformed into the other, so they are topologically distinct.

He started from the Italian mathematician Enrico Betti's generalisation of Lhuilier's topological invariant $F - E + V$, which is now rather unfairly known as the Euler characteristic, to higher-dimensional spaces, achieved in 1870. Betti had noticed that the largest number of closed curves that can be drawn on a surface of genus g, without dividing it into disconnected pieces, is $g - 1$. This is another way to characterise the surface topologically. He generalised this idea to 'connectivity numbers' of any dimension, which Poincaré called

Betti numbers, a term still in use today. The k-dimensional Betti number counts the number of k-dimensional holes in the space.

Poincaré developed Betti's connectivity numbers into a more sensitive invariant called homology, which has a lot more algebraic structure. We will discuss homology in more detail in chapter 15. Suffice it to say that it looks at collections of multidimensional 'faces' in this kind of network, and asks which of them form the boundary of a topological disc. A disc has no holes, unlike a torus, so we can be sure that within any collection of faces that constitutes a boundary, there are no holes. Conversely, we can detect holes by playing off collections of faces that do not form boundaries against collections that do. In this manner we can construct a series of invariants of a space, known as its homology groups. 'Group' here is a term from abstract algebra; it means that any two objects in the group can be combined to give something else in the same group, in a way that is subject to several nice algebraic rules. I'll say a bit more later when we need this idea. There is one such group for each dimension from 0 to n, and for each space we get a series of topological invariants, with all sorts of fascinating algebraic properties.

Listing had classified all topological surfaces – spaces of dimension 2. The obvious next step was to look at dimension 3. And the simplest space to start with was a sphere. In everyday language the word 'sphere' has two different meanings: it can be a solid round ball, or just the surface of the ball. When working on the topology of surfaces, the word 'sphere' is always interpreted in the second sense: the infinitely thin surface of a ball. Moreover, the inside of the sphere isn't considered to be part of it: it is just a consequence of the usual way we embed a spherical surface in space. Intrinsically, all we have is a surface, topologically equivalent to the surface of a ball. You can think of the sphere as a hollow ball with an infinitely thin skin.

The 'correct' 3-dimensional analogue of a sphere, called a 3-sphere, is *not* a solid ball. A solid ball is 3-dimensional, but it has a boundary: its surface, the sphere. A sphere doesn't have a boundary, and neither should its 3-dimensional analogue. The simplest way to define a 3-sphere is to mimic the coordinate geometry of an ordinary sphere. This leads to a space that is a little tricky to visualise: I can't show you a model in 3 dimensions because the 3-sphere – even though it has only

3 dimensions – doesn't embed in ordinary 3-dimensional space. Instead, it embeds in 4-dimensional space.

The usual unit sphere, in 3-dimensional space, consists of all points that are distance 1 from a specified point: the centre. Analogously, the unit 3-sphere in 4-dimensional space consists of all points that are unit distance from the centre. In coordinates we can write down a formula for this set using a generalisation of Pythagoras's theorem to define distance.[71] More generally, a 3-sphere is *any* space that is topologically equivalent to the unit 3-sphere, just as all sorts of lumpy versions of a unit 2-sphere are topological 2-spheres, and of course the same goes in higher dimensions.

If you're not satisfied by that, and want a more geometric image, try this one. A 3-sphere can be represented as a solid ball whose entire surface is identified with a single point. This is another example of a gluing rule, and in this case it is analogous to a way to turn a circular disc into a 2-sphere. If you run a cord round the edge of a cloth disc and draw it tight, like closing a bag, the result is topologically the same as a 2-sphere. Now perform the analogous operation on a solid ball, but as usual, do not try to visualise the result: just think of a solid ball and implement the gluing rules conceptually.

Anyway, Poincaré was very interested in the 3-sphere, because it was presumably the simplest 3-dimensional topological space that had no boundary and was of finite extent. In 1900 he published a paper in which he claimed that homology groups were a sufficiently powerful invariant to characterise the 3-sphere topologically. Specifically, if a 3-dimensional topological space has the same homology groups as a 3-sphere, then it is topologically equivalent to (can be continuously deformed into) a 3-sphere. By 1904, however, he had discovered that this claim is wrong. There is at least one 3-dimensional space that is not a 3-sphere, but has the same homology groups as a 3-sphere. The space was a triumph for the gluing rules philosophy, and the proof that it was not a 3-sphere involved the creation of a new invariant, necessarily more powerful than homology.

First, the space. It is called Poincaré's dodecahedral space, because a modern construction uses a solid dodecahedron. Poincaré was unaware of its relation to a dodecahedron: he glued two solid toruses together in a very obscure manner. The dodecahedron interpretation was published in 1933, some 21 years after Poincaré's death, by

Herbert Seifert and Constantin Weber, and it is much easier to comprehend. The analogy to bear in mind is the construction of a torus by gluing together opposite edges of a square. As always, you don't try to *do* the gluing; you just remember that corresponding points are considered to be the same. Now we do the same thing, but using opposite faces of a dodecahedron, Figure 38.

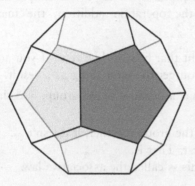

Fig 38 To make the Poincaré dodecahedral space, take a dodecahedron and glue all pairs of opposite faces (such as the shaded pair) together, with a twist to make them fit.

The Pythagoreans knew about dodecahedrons, 2500 years ago. The boundary of a dodecahedron consists of 12 regular pentagons, joined to make a roughly spherical cage, with three pentagons meeting at each corner. Now glue each face of the dodecahedron to the opposite face ... except that there's a twist. Literally. Each face has to be rotated through a suitable angle before it is glued to the opposite one. The angle is the smallest one that aligns the corresponding faces, which is 36 degrees. You can think of this rule as an elaborate version of the Möbius band rule: twist an edge through 180 degrees and then glue it to the opposite edge.

That's the space. Now let's look at the invariant. I'm not just wool-gathering: we need all this to understand the Poincaré conjecture.

Poincaré called his new invariant the fundamental group. We still use this name today, but we also refer to it as the (first) homotopy group. Homotopy is a geometrical construction that can be carried out entirely inside the space, and it provides information on the topological type of that space. It does that using an abstract algebraic structure known as a group. A group is a collection of mathematical objects, any two of which can be combined to give another object in the group. This law of combination – often called multiplication or addition, even when it's not the usual arithmetical operation with that name – is required to satisfy a few simple and natural conditions. If we call the operation addition, the main conditions are:

- The group contains an element that behaves like zero: if you add it to anything in the group, you get the same thing as a result.

- Every member of the group has a negative in the group: add the two and you get zero.

- If you add three members of the group together, it doesn't matter which two you add first. That is, $(a + b) + c = a + (b + c)$. This is called the associative law.

The one algebraic law that is *not* imposed (although sometimes it is true as well) is the commutative law $a + b = b + a$.[72]

Poincaré's fundamental group is a sort of simplified skeleton of the space. It is a topological invariant: topologically equivalent spaces have the same fundamental group. To gain useful insight, and very possibly to reconstruct part of Poincaré's motivation, let's see how it works for a circle, by stealing an image that goes back to Gauss. Imagine an ant whose entire universe is the circle. How can it find out what shape its universe is? Can it distinguish the circle from, say, a line? Bear in mind throughout that the ant is not permitted to step outside its universe, look at it, and see that it is circular. All it can do is wander around inside its universe, whatever that may be. In particular, the ant won't realise that its universe is bent, because its version of a light ray is also confined to the circle. Please ignore practicalities such as objects having to pass through each other – it's going to be a very loose analogy.

The ant can discover the shape of its universe in several ways. I'll focus on a method that generalises to any topological space. For the purposes of this discussion, the ant is a point. It lives at a bus stop, which is also a point. Every day it starts from home, takes the bus (which, of course, is a point) and ends up back home again. The most straightforward trip is the number 0 bus, which just sits at the stop and goes nowhere. For a more interesting excursion, the ant catches the number 1 bus, which goes round the universe exactly once in an anticlockwise direction and stops when it returns home. The number 2 bus goes round twice, the number 3 bus goes round three times, and so on: one anticlockwise bus for each positive integer. There are also negative buses, which go the other way. The number −1 bus goes round once clockwise, the number −2 bus goes round twice clockwise, and so on.

The ant quickly notices that two successive trips on the number 1 bus are essentially the same as a single trip on the number 2 bus, and three trips on the number 1 bus are essentially the same as a single trip on the number 3 bus. Similarly, a trip on the number 5 bus followed by a trip on the number 8 bus is essentially the same as a trip on the number 13 bus. In fact, given any two positive numbers, a trip on the bus with the first number, followed by a trip on the bus with the second number, boils down to a trip on the bus whose number is their sum.

The next step is more subtle. The same relationship *nearly* holds for buses whose numbers are negative or zero. A trip on the number 0 bus, followed by a trip on the number 1 bus, is very similar to a trip on the number 1 bus. However, there is a slight difference. On the $0 + 1$ trip, bus 0 waits around for a time at the start, which doesn't happen for a single trip on bus 1. So we introduce a notion with the forbidding name of homotopy ('same place' in Greek). Two loops are homotopic if one can be continuously deformed into the other. If we allow bus itineraries to be changed by homotopies, we can gradually shrink the time that the ant spends sitting at the bus stop on the number 0 bus, until the stationary period vanishes. Now the difference between the $0 + 1$ trip and the 1 trip has disappeared, so 'up to homotopy' the result is just a trip on the number 1 bus. That is, the bus number equation $0 + 1 = 1$ remains valid − not for trips, but for homotopy classes of trips.

What about a trip on the number 1 bus followed by a trip on the number −1 bus? We'd like this to be a trip on the number 0 bus, but it's not. It goes all the way round anticlockwise, and then comes back again all the way round clockwise. This is clearly different from spending the whole trip sitting at the bus stop on the number 0 bus. So $1 + (−1)$, that is, $1 − 1$, is not equal to 0. But again homotopy comes to the rescue: the combination of buses 1 and −1 is homotopic to the same overall trip as bus 0. To see why, suppose the ant follows the combined route for buses 1 and −1 by car, but just before it gets all the way round to the bus stop it reverses direction and goes home again. This trip is very close to the double bus trip: it just misses out one tiny part of the journey. So the original double bus trip has 'shrunk', continuously, to a slightly shorter car journey. Now the ant can shrink the journey again, by turning back slightly earlier. It can keep shrinking the journey, gradually turning back earlier and earlier, until eventually all it does is to sit in a parked car at the bus stop, going nowhere. This shrinking process is also a homotopy, and it shows that a trip on the number 1 bus followed by a trip on the number −1 bus is homotopic to a trip on the number 0 bus. That is, $1 + (−1) = 0$ for homotopy classes of trips.

It is now straightforward, for an algebraist, to prove that a trip on any bus, followed by a trip on a second bus, is homotopic to a trip on the bus you get by adding the two bus numbers. This is true for positive buses, negative buses, and the zero bus. So if we add bus trips together – well, homotopy classes of bus trips – we obtain a group. In fact, it's a very familiar group. Its elements are the integers (bus numbers) and its operation is addition. Its conventional symbol is $\mathbb{Z}$, from the German Zahl ('integer').

A lot more hard work proves that in a circular universe, *any* round trip by car – even if it involves lots of backtracking, reversing, going to and fro over the same stretch of road – is homotopic to one of the standard bus trips. Moreover, bus trips with different numbers are not homotopic. The proof requires some technique; the basic idea is Gauss's winding number. This counts the total number of times that the trip winds round the circle in the anticlockwise direction.[73] It tells you which bus route your journey is homotopic to.

Once the details are filled in, this description proves that the fundamental group of a circle is the same as the group $\mathbb{Z}$ of integers under addition. To add trips, just add their winding numbers. The ant could use this topological invariant to distinguish a circular universe from, say, an infinite line. On a line, any trip, however much it jiggles around, must at some stage reach a maximum distance from home. Now we can shrink the trip continuously by shrinking all distances from home by the same amount – first to 99 per cent, then 98 per cent, and so on. So on a line, *any* journey is homotopic to zero: staying at home. The fundamental group of the line has only one element: 0. Its algebraic properties are trivial: $0 + 0 = 0$. So it is called the trivial group, and since it differs from the group of all integers, the ant can tell the difference between living on a line and living on a circle.

As I said, there are other methods, but this is how the ant can do it using Poincaré's fundamental group.

Now we up the ante (pun intentional). Suppose that the ant lives on a surface. Again, that is its entire universe; it can't step outside and take a look to see what kind of surface it inhabits. Can it work out the topology of its universe? In particular, can it tell the difference between a sphere and a torus? Again the answer is 'yes', and the method is the same as for a circular universe: get on a bus and make round trips that start and finish from home. To add trips together, perform them in turn. The zero trip is 'stay at home', the inverse of a trip is the same trip in the opposite direction, and we get a group provided we work with homotopy classes of trips. This is the fundamental group of the surface. Compared to a circular universe there is more freedom to create trips and to deform them continuously into other trips, but the same basic idea works.

The fundamental group is again a topological invariant, and the ant can use it to find out whether it lives on a sphere or a torus. If its universe is a sphere, then no matter which trip the ant takes, it can gradually be deformed into the zero trip: stay at home. This is not the case if its universe is a torus. Some trips can be deformed to zero, but a trip that winds once through the central hole, as in Figure 39 (left), cannot be. That statement needs proof, but this can be provided. There are standard bus trips on the torus, but now the bus numbers are pairs of integers (m, n). The first number m specifies how many times the trip winds through the central hole; the second number n specifies how many times the trip winds round the torus. Figure 39 (right) shows the trip $(5,2)$, which winds five times through the hole and twice round the torus. To add trips, add the corresponding numbers; for example, $(3,6) + (2,4) = (5,10)$. The fundamental group of the torus is the group $\mathbb{Z}^2$ of *pairs* of integers.

Any topological space has a fundamental group, defined in exactly

Fig 39 *Left*: Bus trips (1,0) and (0,1) on the torus. *Right*: Bus trip (5,2). Grey lines are at the back.

the same way using trips – more properly known as loops – that start and finish at the same point. Poincaré invented the fundamental group to prove that his dodecahedral space is not a 3-sphere, despite having exactly the same homology invariants. His original recipe is nicely adapted to the calculation of its fundamental group. The more modern 'twist and glue' recipe is even better adapted. The answer turns out to be a group with 120 elements related to the dodecahedron. In contrast, the fundamental group of a 3-sphere has only one element: the zero loop. So the dodecahedral space is *not* topologically equivalent to a sphere, despite having the same homology, and Poincaré had proved that his 1900 claim was wrong.

He went on to speculate about his new invariant: was that the missing ingredient in a topological characterisation of the 3-sphere? Perhaps any 3-dimensional space with the same fundamental group as a 3-sphere – that is, the trivial group – must actually *be* a 3-sphere? He phrased this suggestion in a negative way as a question: 'Consider a compact 3-dimensional manifold [topological space] V without boundary. Is it possible that the fundamental group of V could be trivial, even though V is not [topologically equivalent to] the 3-dimensional sphere?' He left the question open, but the very plausible belief that the answer is the obvious one – 'no', when the question is stated that way – quickly became known as the Poincaré conjecture. And it equally quickly became one of the most notorious open questions in topology.

'Trivial fundamental group' is another way to say 'every loop can be continuously deformed to a point'. Not only does a 3-sphere possess that property; so does an analogous n-sphere for any dimension n. So we can make the same conjecture for a sphere of any dimension. This statement is the n-dimensional Poincaré conjecture. It is true when $n = 2$, by the classification theorem for surfaces. And for over 50 years, that was as far as anyone could get.

In 1961 Stephen Smale borrowed a trick from the classification of surfaces, and applied it in higher dimensions. One way to think of a g-holed torus is to start with a sphere and add g handles – just like the handle on a teacup or a jug. Smale generalised this construction to any number of dimensions, calling the process handle decomposition. He analysed how handles could be modified without changing the topology of the space, and deduced the Poincaré conjecture in all dimensions greater than or equal to 7. His proof didn't work for lower dimensions, but other mathematicians found ways to repair it: John Stallings for dimension 6 and Christopher Zeeman for dimension 5. However, one vital step, known as the Whitney trick, failed for dimensions 3 and 4 because in these spaces there isn't enough room to perform the required manoeuvres, and no one could find effective substitutes. A general feeling emerged that topology, for spaces of these two dimensions, might be unusual.

This conventional wisdom was shaken in 1982 when Michael Freedman discovered a proof of the 4-dimensional Poincaré conjecture that did not require the Whitney trick. It was extremely complicated, but it worked. So, after 50 years with little progress and 20 years of frenzied activity, topologists had polished off the Poincaré conjecture in every dimension except the one Poincaré had originally asked about. The successes were impressive, but the methods used to obtain them provided very little insight into the 3-dimensional case. A different way of thinking was needed.

What finally broke the deadlock was rather like the traditional list of wedding gifts: something old, something new, something borrowed ... and, stretching a point, something blue. The old idea was to revisit an area of topology which, after all of the flurry of activity on spaces of higher dimension, was generally thought to have been mined out: the topology of surfaces. The new idea was to rethink the classification of surfaces from a point of view that at first seemed completely alien:

classical geometry. The borrowed idea was the Ricci flow, which took its motivation from the mathematical formalism of Einstein's theory of general relativity. And the blue idea was 'blue sky' speculation: some far-reaching suggestions based on a dash of intuition and a lot of hope.

Recall that orientable surfaces without boundary can be listed: each is topologically equivalent to a torus with some number of holes. This number is the genus of the surface, and when it is zero, the surface is a sphere with no handles – that is, a sphere. The very word reminds us that among all topological spheres, there is one surface that stands out from all others as the archetype. Namely, the unit sphere in Euclidean space. Forget, for a second, all that rubber-sheet stuff. We'll put that back in a moment. Concentrate on the good old Euclidean sphere. It has all sorts of extra mathematical properties, coming from the rigidity of Euclid's geometry. Paramount among those properties is curvature. Curvature can be quantified; at each point on a geometric surface there is a number that measures how curved the surface is near that point. The sphere is the only closed surface in Euclidean space whose curvature is the same at every point, and is positive.

This is weird, because constant curvature is not a topological property. Even weirder: the sphere is not alone. There is also one standard geometric surface that stands out as the archetypal torus. Namely, start with a square in the plane, and identify opposite edges, Figure 12 of chapter 4. When we draw the result in 3-dimensional space, rolling up the square to make its edges meet, the result looks curved. But from an intrinsic point of view, we can work entirely with the square plus the gluing rules. A square has a natural geometric structure: it is a region in the Euclidean plane. The plane also has constant curvature, but now the constant is *zero*. A torus with this particular geometry also has zero curvature, so it is called the *flat torus*. The name may sound like an oxymoron, but for an ant living on a flat torus, carting a ruler and protractor around to measure lengths and angles, the local geometry would be identical to that of the plane.

The geometers of the eighteenth century, trying to understand Euclid's axiom about the existence of parallel lines, set out to deduce that axiom from the rest of Euclid's basic assumptions, failed repeatedly, and ended up realising that no such deduction is possible. There are three different kinds of geometry, each of which obeys every condition that Euclid requires, save for the parallel axiom. Those geometries are called Euclidean (the plane, where the parallel axiom is valid), elliptic (geometry on the surface of a sphere, with a few bells and whistles, where any two lines meet and parallels do not exist), and hyperbolic (where some lines fail to meet, and parallels fail to be unique). Moreover, the classical mathematicians interpreted these geometries as the geometry of curved spaces. Euclidean geometry corresponds to zero curvature, elliptic/spherical geometry corresponds to constant positive curvature, and hyperbolic geometry corresponds to constant negative curvature.

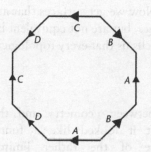

Fig 40 Making a 2-holed torus from an octagon by identifying edges in pairs (*AA, BB, CC, DD*).

We have just seen how to obtain the first two of these geometries: they occur on the sphere and the flat torus. In terms of the classification theorem, these are *g*-holed toruses for *g* = 0 and 1. The only thing missing is hyperbolic geometry. Does every *g*-holed torus have a natural geometric structure, based on taking some polygon in hyperbolic space and identifying some of its edges? The answer is striking: it is 'yes' for *any* value of *g* greater than or equal to 2. Figure 40 shows an example for *g* = 2 based on an octagon. I'll skip the hyperbolic geometry and the identification of this surface as a 2-torus, but they can be sorted out. Different values of *g* arise if we take different polygons, but every *g* occurs. In the jargon, a torus with two or more holes has a natural hyperbolic structure. So now we can reinterpret the list of standard surfaces:

- Sphere, $g = 0$: elliptic geometry.
- Torus, $g = 1$: Euclidean geometry.
- g-holed torus, $g = 2, 3, 4, \ldots$: hyperbolic geometry.

It may seem that we have thrown the baby out with the bathwater, because topology is supposed to be about rubber-sheet geometry, not rigid geometry. But now we can easily put the rubber back. Rigid geometry is used here only to *define* the standard surfaces. It provides simple descriptions, which happen to have extra rigid structure. Now relax the rigidity – in effect, allow the space to *become* like rubber. Let it deform in ways that rigidity prohibits. Now we get surfaces that are topologically equivalent to the standard ones, but are not equivalent by rigid motions. The classification theorem tells us that every topological surface can be obtained in this manner.

Topologists were aware of this link between geometry and the classification theorem for surfaces, but it looked like a funny coincidence, no doubt a consequence of the rather limited possibilities in two dimensions. Everyone knew that the 3-dimensional case was much richer, and in particular spaces of constant curvature did not exhaust the possibilities. It took one of the world's top geometers, William Thurston, to realise that rigid geometry might still be relevant to 3-dimensional topology. There were already a few hints: Poincaré's 3-sphere has a natural elliptic/ spherical geometry, coming from its definition. Although a standard dodecahedron lives in Euclidean space, the angle between adjacent faces is less than 120 degrees, so three such angles do not make a full circle. To remedy that, we have to inflate the dodecahedron so that its faces bulge slightly: this makes the natural geometry spherical, not Euclidean. Analogously, triangles on a sphere also bulge. The 3-torus, obtained by identifying opposite faces of a cube, has a flat – that is, Euclidean – geometry, just like its 2-dimensional analogue. Max Dehn and others had discovered a few 3-dimensional topological spaces with natural hyperbolic geometries.

Thurston began to see hints of a general theory, but two innovations were needed to make it even remotely plausible. First, the range of 3-dimensional geometries had to be extended. Thurston wrote down reasonable conditions, and proved that exactly eight geometries satisfy them. Three of them are the classics: spherical, Euclidean, and hyperbolic geometry. Two more are like cylinders: flat in one direction, curved in two other directions. The curved part is either positively curved, the 2-sphere, or negatively curved, the hyperbolic plane. Finally, there are three other, rather technical, geometries.

Second: some 3-dimensional spaces did not support any of the eight geometries. The answer was to cut the space into pieces. One piece might have a spherical geometric structure, another a hyperbolic one, and so on. In order to be useful, the cutting had to be done in a very tightly constrained manner, so that reassembling the pieces conveyed useful information. The good news was that in many examples, this turned out to be possible. In 1982, in a great leap of imagination, Thurston stated his geometrisation conjecture: *every* 3-dimensional space can be cut up, in an essentially unique manner, into pieces, each of which has a natural geometric structure corresponding to one of the eight possible geometries. He also proved that if his geometrisation conjecture were true, then the Poincaré conjecture would be a simple consequence.

Meanwhile, a second line of attack was emerging, also geometric, also based on curvature, but coming from a very different area: mathematical physics. Gauss, Riemann, and a school of Italian geometers had developed a general theory of curved spaces, called manifolds, with a concept of distance that extended Euclidean and classical non-Euclidean geometry enormously. Curvature need no longer be constant: it could vary smoothly from one point to another. A shape like a dog's bone, for instance, is positively curved at each end, but negatively curved in between, and the amount of curvature varies smoothly from one region to the next. Curvature is quantified using mathematical gadgets known as tensors. Around 1915 Albert Einstein realised that curvature tensors were exactly what he needed to extend his theory of special relativity, which was about space and time, to general relativity, which also included gravity. In this theory, the gravitational field is represented as the curvature of space, and

Einstein's field equations describe how the associated measure of curvature, the curvature tensor, changes in response to the distribution of matter. In effect, the curvature of space *flows* as time passes; the universe or some part of it spontaneously changes its shape.

Richard Hamilton, a specialist in Riemannian geometry, realised that the same trick might apply more generally, and that it might lead to a proof of the Poincaré conjecture. The idea was to work with one of the simplest measures of curvature, called Ricci curvature after the Italian geometer Gregorio Ricci-Curbastro. Hamilton wrote down an equation that specified how Ricci curvature should change over time: the Ricci flow. The equation was set up so that the curvature should gradually redistribute itself in as even a manner as possible. This is a bit like the cat under a carpet in chapter 4, but now, even though the cat can't escape, it can smear itself out in an even layer. (A topological cat is essential here.)

For example, in the 2-dimensional case, start with a pear-shaped surface, Figure 41. This has a region at one end that is strongly and positively curved; a region at the other, fatter, end that is also positively curved, but not so strongly; and a band in between where the curvature is negative. The Ricci flow in effect transports curvature from the strongly curved end (and to a lesser extent from the other end) into the negatively curved band, until all of the negative curvature has been gobbled up. At that stage the result is a bumpy surface with positive curvature everywhere. The Ricci flow continues to redistribute curvature, taking it away from the highly curved regions and moving it to the less curved regions. As time becomes ever larger, the surface gets closer and closer to one that has constant positive curvature – that is, a Euclidean sphere. The topology remains the same, even though the detailed shape changes, so we can prove that the original pear-shaped surface is topologically equivalent to a sphere by following the Ricci flow.

positive ——
negative ——
positive ——

Fig 41 How the Ricci flow turns a pear into a sphere.

In this example the topological type of the surface was obvious to start with, but the same general strategy works for any manifold. Start with a complicated shape and follow the Ricci flow. As time passes, the curvature redistributes itself more evenly, and the shape becomes simpler. Ultimately, you should end up with the simplest shape having the same topology as the original manifold, whatever that may be. In 1981 Hamilton proved that this strategy works in 2 dimensions, providing a new proof of the classification theorem for surfaces.

He also made significant progress on the analogous strategy for 3-dimensional manifolds, but now there was a serious obstacle. In 2 dimensions, every surface automatically simplifies itself by following the Ricci flow. The same is true in 3 dimensions if the initial manifold has strictly positive curvature at every point: never zero or negative. Unfortunately, if there are points at which the curvature is zero, and there often are, the space can get tangled up in itself as it flows. This creates singularities: places where the manifold ceases to be smooth. At such points, the equation for the Ricci flow breaks down, and the redistribution of curvature has to stop. The natural way to get round this obstacle is to understand what the singularities look like and redesign the manifold – perhaps by cutting it into pieces – so that the Ricci flow can be given a jump-start. Provided you have enough control over how the topology of the remodelled manifold relates to that of the original one, this modified strategy can be successful. Unfortunately, Hamilton also realised that for 3-dimensional spaces, the singularities in the Ricci flow can be very complicated indeed – apparently too complicated to use that kind of trick. The Ricci flow quickly became a standard technique in geometry, but it fell short of proving the Poincaré conjecture.

By 2000 mathematicians had still not cracked the conjecture, and its importance was recognised more widely when it was made one of the seven millennium problems. By then it had also become clear that if Hamilton's idea could somehow be made to work in sufficient generality, it wouldn't just imply the Poincaré conjecture. It would prove Thurston's geometrisation conjecture as well. The prize was glittering, but it remained tantalisingly out of reach.

Mathematics is like the other branches of science: in order for research to be accepted as correct, it has to be published, and for that to happen, it has to survive peer review. Experts in the field have to read the paper carefully, check the logic, and make sure the calculations are correct. This process can take a long time for a complicated and important piece of mathematics. As mentioned in chapter 1, the remedy used to be a preprint, but nowadays there is a standard website, the arXiv ('archive'), where electronic preprints can be posted, subject to a partial refereeing process and an endorsement procedure to weed out rubbish. Today, most researchers first encounter new results on the arXiv or the author's own website.

In 2002 Grigori Perelman put a preprint on the arXiv about the Ricci flow. It made a remarkable claim: the flow is gradient-like. That is, there is a well-defined 'downhill' direction, a single numerical quantity associated with the shape of the manifold, and the manifold flows downhill in the sense that this quantity always decreases as time passes. It's analogous to height in a landscape, and it provides a quantitative measure of what 'simplifying' a manifold means. Gradient-like flows are fairly restricted: they can't go round and round in circles or behave chaotically. No one seems to have suspected that the Ricci flow would be so tame. But Perelman didn't just make the claim: he proved it. He ended by outlining an argument that would prove the Thurston geometrisation conjecture – which implies the Poincaré conjecture but goes much further – promising further details in subsequent postings to the arXiv. Over the next eight months he posted two follow-up papers containing many of the promised details.

The first posting caused quite a stir. Perelman was claiming to have carried out the entire Hamilton programme, by using the Ricci flow to simplify a 3-dimensional manifold and proving that the result was exactly what Thurston had predicted. The other two postings added further weight to the feeling that Perelman knew what he was talking about, and that his ideas went well beyond outlining a plausible strategy with the odd logical gap or unproved assumption. The usual scepticism of the mathematical community towards claims of a solution to a great problem was muted; there was a general feeling that he might well have succeeded.

However, the devil is in the detail, and in mathematics the detail can be devilish indeed. The work had to be checked, at length and in depth, by people who understood the areas involved and were aware of the potential pitfalls. And that wasn't straightforward, because Perelman had combined at least four very different areas of mathematics and mathematical physics, and few people understood more than one or two of them. Deciding whether his proof was correct would need a lot of teamwork and a lot of effort. Moreover, the preprints on the arXiv did not include full details at the normal level for a published paper. They were pretty clearly written, for preprints, but they didn't always dot the i's and cross the t's. So the experts had to reconstruct a certain amount of Perelman's thinking – and he had been deeply immersed in the work for years.

It all took time. Perelman lectured on his proof and answered e-mails questioning various steps. Whenever anyone found what seemed to be a gap, he responded quickly with further explanation, filling it. The signs were encouraging. But no one was going to risk their reputation by stating publicly that Perelman had proved the Poincaré conjecture, let alone the more difficult geometrisation conjecture, until they were very confident that there were no mistakes in the proof. So despite the generally favourable view of Perelman's work, public acceptance was initially withheld. This was unavoidable, but also unfortunate, because as the waiting dragged on, Perelman became increasingly irritated by what seemed to be fence-sitting. He *knew* his proof was correct. He understood it so well that he couldn't see why others were having trouble. He declined to write up the work in more detail or to submit it to a journal. As far as he was concerned, it was a done deal, and the arXiv preprints contained everything that was required. He stopped answering questions about allegedly missing details. To him, they weren't missing. Come on guys, you can figure this out without further help from me. It's not that difficult.

Some reports have suggested that in this respect the mathematical community was unfair to Perelman. But this misunderstands how the mathematical community functions when great problems are allegedly solved. It would have been irresponsible to just slap him on the back, say 'well done!', and ignore the missing steps in his preprints. It was entirely proper, indeed unavoidable, to ask him to prepare more extensive treatments, suitable for publication. On a problem of this importance, a rush job is dangerous and unacceptable. The experts went out of their way to spend a lot of time on Perelman's proof, and they kept their natural scepticism at bay to an unusual extent. His treatment was if anything *more* favourable than usual. And eventually, when this process was complete, his work was accepted as correct.

By that time, however, Perelman had lost patience. It perhaps didn't help that he had solved such an important problem that nothing else was likely to match it. He was like a mountaineer who has climbed Everest solo without oxygen. There were no comparable challenges left. Media publicity repelled him: he wanted acceptance by his peers, not by television presenters. So it is not terribly surprising that when his peers finally agreed he was right, and offered him a Fields medal and the Clay prize, he didn't want to know.

Perelman's proof is deep and elegant, and opens up a new world of topology. It implements Hamilton's Ricci flow programme by finding clever ways to get round the occurrence of singularities. One is to change the scales of space and time to get rid of the singularity. When this approach fails, the singularity is said to collapse. In such cases, he analyses the geometry of the Ricci flow in some detail, classifying how a collapse can occur. In effect, the space puts out ever-thinner tentacles, perhaps in profusion, like the branches of a tree. Whenever a tentacle is close enough to collapsing, it can be cut, and the spiky, sharply curved end can be cut off and replaced by a smooth cap. For some of these tentacles, the Ricci flow grinds to a halt: if so, leave them alone. If not, the Ricci flow can be restarted. So some tentacles end in smooth caps, and others are temporarily interrupted, but continue to flow.

This cut-and-paste capping procedure chops the space up in much the same way as Thurston's dissection into pieces, each with one of his eight geometries, and the two procedures turn out to give more or less identical results. One technical point is vital: the capping-off operations do not pile up ever faster, so that infinitely many of them occur in a finite time. This is one of the most complicated parts of the proof.

Some commentators have criticised the mathematical community for treating Perelman unfairly. No one should be immune to criticism, and there were a few incidents that might well be classed as unfair or in other ways ill-considered, but the mathematical community reacted rapidly and positively to Perelman's work. It also reacted cautiously, which is absolutely standard in mathematics and science, for excellent reasons. The inevitable glare of publicity, heightened by the million-dollar prize, had an impact on everyone, Perelman included.

From Perelman's first posting on the arXiv in November 2002 to the announcement in March 2010 that he had been awarded the Clay prize took eight years. That sounds like a lengthy delay, perhaps an unreasonable one. However, that first posting tackled only part of the problem. Most of the rest was posted in March 2003. By September 2004, eighteen months after this second posting, the Ricci flow and topological communities had already worked through the proof – a process that had started mere days after the first posting – and the main experts announced that they 'understood the proof'. They had found mistakes, they had found gaps, but they were convinced that those could all be put right. Eighteen months is actually remarkably quick when something so important is at stake.

Late in 2005 the International Mathematical Union approached Perelman and offered him a Fields medal, the subject's highest honour, to be awarded at the International Congress of Mathematicians in 2006. The ICM is held every four years, so this was the first opportunity to recognise his work in this manner. Because a few doubts remained about the complete proof of the Poincaré conjecture – mistakes were still turning up – the medal was officially awarded for advances in understanding the Ricci flow, the part of Perelman's preprints that were now considered error-free.

The conditions for the award of the prize are stated on the Clay Institute's website. In particular, a proposed solution has to be published in a refereed journal and still be accepted by the mathematical community two years later. After that, a special advisory committee looks into the matter and recommends whether or not to award the prize. Perelman has not complied with the first condition, and it doesn't seem likely that he ever will. In his view, the arXiv preprints suffice. Nevertheless, the Clay Institute waived that requirement and started the statutory two-year wait to see whether any mistakes or other issues emerged. That ended in 2008, after which the Institute's procedures, carefully structured to avoid awarding the prize prematurely, had to be followed.

It is true that some experts were slow to express their belief that the proof was correct. The reason is straightforward: they were genuinely uncertain. It's no great exaggeration to say that the only person capable of grasping Perelman's proof quickly was another Perelman. You can't read a mathematical proof like a musician sight-reads a score. You have to convince yourself it all makes sense. Whenever the argument gets very complicated, you know there is a serious chance of a mistake. The same applies when the ideas get too simple; many a prospective proof has fallen foul of an assertion so evident that no proof seemed necessary. Until the experts were genuinely sure that proof was basically correct – at which point they gave Perelman full credit *despite* the remaining gaps and errors – it was sensible to suspend judgement. Think about all the fuss over the eventually discredited work on cold fusion. Caution is the correct professional response, and the cliché applies: extraordinary claims require extraordinary evidence.

Why did Perelman reject the Fields medal and decline the Clay prize? He alone knows, but he wasn't interested in that kind of recognition and he repeatedly said so. He had already refused smaller prizes. He made it clear at the start that he did not want premature publicity; ironically, this is the same reason why the experts were understandably reluctant to take the plunge too soon. To be realistic, there wasn't the slightest chance that the media would *not* notice his work. For years the mathematical community has been making a big effort to get newspapers, radio, and television interested in the subject. It doesn't make a lot of sense to complain when this effort succeeds, or to expect the media to ignore the hottest mathematics story since Fermat's last theorem. But Perelman didn't see it that way, and he retreated into his shell. There is an offer on the table to make the prize money available for educational or other purposes, if he agrees. So far, he has failed to respond.

11

They can't all be easy
P/NP Problem

NOWADAYS, MATHEMATICIANS ROUTINELY use computers to solve problems, even great problems. Computers are good at arithmetic, but mathematics goes far beyond mere 'sums', so putting a problem on a computer is seldom straightforward. Often the hardest part of the work is to convert the problem into one that a computer calculation can solve, and even then the computer may struggle. Many of the great problems that have been solved recently involve little or no work with a computer. Fermat's last theorem and the Poincaré conjecture are examples.

When computers have been used to solve great problems, like the four colour theorem or the Kepler conjecture, the computer effectively plays the role of servant. But sometimes the roles are reversed, with mathematics as the servant of computer science. Most of the early work on computer design made good use of mathematical insights, for example the connection between Boolean algebra – an algebraic formulation of logic – and switching circuits, developed in particular by the engineer Claude Shannon, the inventor of information theory. Today, both practical and theoretical aspects of computers rely on the extensive use of mathematics, from many different areas.

One of the Clay millennium problems lies in the borderland of mathematics and computer science. It can be viewed both ways: computer science as a servant of mathematics, and mathematics as a servant of computer science. What it requires, and is helping to bring about, is more balanced: a partnership. The problem is about computer algorithms, the mathematical skeletons from which computer programs are made. The crucial concept here is how efficient the algorithm is: how many computational steps it takes to get an answer for a given amount of input data. In practical terms, this tells us how long the computer will take to solve a problem of given size.

The word algorithm goes back to the Middle Ages, when Muhammad ibn Mūsā al-Khwārizmī wrote one of the earliest books on algebra. Earlier, Diophantus had introduced one element we associate with algebra: symbols. However, he used the symbols as abbreviations, and his methods for solving equations were presented through specific – though typical – examples. Where we would now write something like '$x + a = y$, therefore $x = y - a$', Diophantus would write 'suppose $x + 3 = 10$, then $x = 10 - 3 = 7$' and expect his readers to understand that the same idea would work if 3 and 10 were replaced by any other numbers. He would explain his illustrative example using symbols, but he wouldn't manipulate the symbols as such. Al-Khwārizmī made the general recipe explicit. He did this using words, not symbols, but he had the basic idea, and is generally considered to be the father of algebra. In fact, that name comes from the title of his book: *al-Kitāb al-Mukhtasar fī Hisāb al-Jabr wa'l-Muqābala* ('The compendious book on calculation by completion and balancing'). Al-jabr became algebra. The word 'algorithm' comes from a medieval version of his name, Algorismus, and it is now used to mean a specific mathematical process for solving a problem, one that is guaranteed to find the solution provided you wait long enough.

Traditionally, mathematicians considered a problem to be solved if, in principle, they could write down an algorithm leading to an answer. They seldom used that word, preferring to present, say, a formula for the solution, which is a particular kind of algorithm in symbolic language. Whether it was possible to apply the formula in practice wasn't terribly important: the formula *was* the solution. But the use of computers changed that view, because formulas that had been too complicated for hand calculation might become practical with the aid of a computer. However, it was a bit disappointing to find, as sometimes happened, that the formula was still too complicated: although the computer could try to run through the algorithm, it was too slow to reach the answer. So attention shifted to finding efficient algorithms. Both mathematicians and computer scientists had a vested interest in developing algorithms that really did give answers in a reasonable period of time.

Given an algorithm, it is relatively straightforward to estimate how long it will take (measured by the number of computational steps required) to solve a problem with a given size of input. This may require a certain amount of technique, but you know what process is involved and you know a lot about what it is doing. It is much more difficult to devise a more efficient algorithm if the one you start from turns out to be inefficient. And it is even harder to decide how good or bad the most efficient algorithm for a given problem can be, because that involves contemplating all possible algorithms, and you don't know what these are.

Early work on such questions led to a coarse but convenient dichotomy between algorithms that were efficient, in a simple but rough and ready sense, and those that were not. If the length of the computation grows relatively slowly as the size of the input increases, the algorithm is efficient and the problem is easy. If the length of the computation grows ever faster as the size of the input increases, the algorithm is inefficient and the problem is hard. Experience tells us that although some problems are easy, in this sense, most seem to be hard. Indeed, if all mathematical problems were easy, mathematicians would be out of a job. The millennium prize problem asks for a rigorous proof that at least one hard problem exists – or that contrary to experience, all problems are easy. It is known as the P/NP problem, and no one has a clue how to solve it.

We've already encountered a rough-and-ready measure of efficiency in chapter 2. An algorithm is class P if it has polynomial running time. In other words, the number of steps it takes to get the answer is proportional to some fixed power, such as the square or cube, of the size of the input data. Such algorithms are efficient, speaking very broadly. If the input is a number, that size is how many digits it has, not the number itself. The reason is that the quantity of information needed to specify the number is the space it occupies in the computer's memory, which is (proportional to) the number of digits. A problem is class P if there exists a class P algorithm that solves it.

Any other algorithm or problem belongs to the class not-P, and most of these are inefficient. Among them are those whose running time is exponential in the input data: approximately equal to some fixed number raised to the power of the size of the input. These are class E, and they are definitely inefficient.

Some algorithms are so efficient that they run much faster than polynomial time. For example, to determine whether a number is even or odd, look at its last digit. If (in decimal notation) this is 0, 2, 4, 6, or 8, the number is even; otherwise it is odd. The algorithm has at most six steps:

Is the last digit 0? If yes, then STOP. The number is even.
Is the last digit 2? If yes, then STOP. The number is even.
Is the last digit 4? If yes, then STOP. The number is even.
Is the last digit 6? If yes, then STOP. The number is even.
Is the last digit 8? If yes, then STOP. The number is even.
STOP. The number is odd.

So the running time is at most 6, independently of the input size. It belongs to the class 'constant time'.

Sorting a list of words into alphabetical order is a class P problem. A straightforward way to perform this task is the bubble sort, so named because in effect words move up the list like bubbles in a glass of fizzy drink if they are further down the list than words that should be below them in alphabetical order. The algorithm repeatedly works through the list, compares adjacent words, and swaps them if they are in the wrong order. For example, suppose the list starts out as

PIG DOG CAT APE

On the first run through this becomes

DOG PIG CAT APE
DOG **CAT PIG** APE
DOG CAT **APE PIG**

where the bold words are the ones that have just been compared. On the second run this becomes

CAT DOG APE PIG

CAT **APE DOG** PIG
CAT APE **DOG** PIG

The third run goes

APE CAT DOG PIG
APE **CAT DOG** PIG
APE CAT **DOG** PIG

On the fourth run, nothing moves so we know we've finished. Notice how APE bubbles up step by step to the top (that is, front).

With four words, the algorithm runs through three comparisons at each stage, and there are four stages. With n words, there are $n-1$ comparisons per stage and n stages, a total of $n(n-1)$ steps. This is a little less than n^2, so the running time is polynomial, indeed quadratic. The algorithm may terminate sooner, but in the worst case, when the words are in exactly the reverse order, it takes $n(n-1)$ steps. The bubble sort is obvious and class P, but it is nowhere near the most efficient sorting algorithm. The fastest comparison sort, which is set up in a more clever way, runs in $n \log n$ steps.

A simple algorithm with exponential running time, class E, is 'print a list of all binary numbers with n digits.' There are 2^n numbers in the list, and printing each (and calculating it) takes roughly n steps, so the running time is roughly $2^n n$, which is bigger than 2^n but less than 3^n when n is sufficiently large. However, this example is a bit silly because what makes it so slow is the size of the output, not the complexity of the calculation, and this observation will turn out to be crucial later on.

A more typical class E algorithm solves the travelling salesman problem. A salesman has to visit a number of cities. He can do so in any order. Which route visits them all in the shortest total distance? The naive way to solve this is to list all possible routes, calculate the total distance for each, and find the shortest one. With n cities there are

$$n! = n \times (n-1) \times (n-2) \times \cdots \times 3 \times 2 \times 1$$

routes (read 'factorial n'). This grows faster than any exponential.[74] A more efficient method, called dynamic programming, solves the travelling salesman problem in exponential time. The first such method, the Held-Karp algorithm, finds the shortest tour in $2^n n^2$ steps, which again lies between 2^n and 3^n when n is sufficiently large.

Even though these algorithms are 'inefficient', special tricks can be used to shorten the computation when the number of cities is large by human standards, but not too large for the tricks to cease to be effective. In 2006 D.L. Applegate, R.M. Bixby, V. Chvátal, and W.J. Cook solved the travelling salesman problem for 85,900 cities, and this was still the record in mid-2012.[75]

These examples of algorithms don't just illustrate the concept of efficiency. They also drive home my point about the difficulty of finding improved algorithms, and the even greater difficulty of finding one that is as efficient as possible. All of the known algorithms for the travelling salesman problem are class E, exponential time – but that does not imply that no efficient algorithm exists. It just shows that we haven't found one yet. There are two possibilities: we haven't found a better algorithm because we're not clever enough, or we haven't found a better algorithm because no such thing exists.

Chapter 2 is a case in point. Until Agrawal's team found their class P algorithm for primality testing, the best known algorithm was not-P. It was still pretty good, with running time $n \log n$ for n-digit numbers, which is actually better than the Agrawal-Kayal-Saxena algorithm until we reach numbers with 10^{1000} digits. Before their algorithm was discovered, opinion on the status of primality testing was divided. Some experts suspected it was class P and a suitable algorithm would be found; some thought it wasn't. The new algorithm came from nowhere, one of thousands of ideas that someone could have tried; this one happened to work. The precedent here is sobering: we don't know, we can't tell, and the experts' best guess may or may not be a good one.

The great problem that concerns us here asks for the answer to a more fundamental question. Are there any hard problems? Might all problems be easy, if only we were clever enough? The actual statement is subtler, because we've already seen one instance of a problem that is indubitably hard: print a list of all binary numbers with n digits. As I remarked, this is a bit silly: the difficulty lies not in the calculation, but the sheer drudgery of printing out a very long answer. We know there's no short cut, because the answer is that long *by definition*. If it were shorter, it wouldn't be the answer.

In order to pose a sensible question, trivial examples like this one have to be eliminated. The way to do that is to introduce another class of algorithm, class NP. This is not the class not-P; it is the class of algorithms that run in nondeterministic polynomial time. The jargon means that however long the algorithm takes to come up with its answer, we can *check that the answer is right* in polynomial time. Finding the answer may be hard, but once found, there's an easy test of its validity.

The word 'nondeterministic' is used here because it's possible to solve an NP problem by making an inspired guess. Having done so, you can confirm that it really is correct (or not). For instance, if the problem is to factorise the number 11,111,111,111, you might guess that one factor is the prime number 21,649. As it stands, that's just a wild guess. But it's easy to check: just divide by it and see what you get. The result turns out to be 513,239, exactly, no remainder. So the guess was correct. If I'd guessed 21,647 – which is also prime – instead, then division would lead to the result 513,286 plus a remainder of 9069. So that guess would have been wrong.

Making a correct guess is basically a miracle here, or else there's a trick (I worked out the factors of 11,111,111,111 before 'guessing'). But that's actually what we want. If it weren't miraculous, you could turn a class NP algorithm into a class P algorithm by just making lots and lots of guesses until one turns out to be right. My example suggests why this won't work: you need too many guesses. Indeed, all we are doing here is 'trial division' by all possible primes, until one works. We know from chapter 2 that this is a hopeless way to find factors.

Class NP rules out silly examples like my very long list. If someone guesses a list of all binary digits of length n, then it doesn't just take exponential time to print the list out. It also takes exponential time to read it, so it takes even longer to check it's correct. It would be a truly horrible proofreading task. Class P is definitely contained in class NP. If you can find the answer in polynomial time, with a guarantee that it's correct, then you've already checked it. So the check automatically requires nothing worse than polynomial time. If someone presented you with the supposed answer, you could just run the entire algorithm again. That's the check.

Now we can state the millennium problem. Is NP bigger than P, or are they the same? More briefly: is P equal to NP?

If the answer is 'yes' then it would be possible to find fast, efficient algorithms for scheduling airline flights, optimising factory output, or performing a million other important practical tasks. If the answer is 'no', we will have a cast-iron guarantee that all of the apparently hard problems really are hard, so we will be able to stop wasting time trying to find fast algorithms for them. We win either way. What's a nuisance is not knowing which way it goes.

It would make life much simpler for mathematicians if the answer were 'yes', so the pessimist in every human being immediately suspects that life isn't going to be that simple, and the likely answer is 'no'. Otherwise we're all getting a free lunch, which we don't deserve and haven't earned. I suspect most mathematicians would actually prefer the answer to be 'no', because that should keep them in business until the end of civilisation. Mathematicians prove themselves by solving hard problems. For whatever reason, most mathematicians and computer scientists expect the answer to the question 'does P equal NP?' to be 'no'. Hardly anyone expects it to be 'yes'.

There are two other possibilities. It might be possible to prove that P is equal to NP without actually finding a polynomial-time algorithm for any specific NP problem. Mathematics has a habit of providing existence proofs that are not constructive; they show that something exists but don't tell us what it is. Examples include primality tests, which cheerfully inform us that a number is not prime without producing any specific factor, or theorems in number theory asserting that the solutions to some Diophantine equation are bounded – less than some limit – without providing any specific bound. A polynomial-time algorithm might be so complicated that it's impossible to write it down. Then the natural pessimism about free lunches would be justified, even if the answer turned out to be affirmative.

More drastically, some researchers speculate that the question may be undecidable within the current formal logical framework for mathematics. If so, neither 'yes' nor 'no' can be proved. Not because we're too stupid to find the proof: because there isn't one. This option became apparent in 1931 when Kurt Gödel set the cat of undecidability loose among the philosophical pigeons that infested the foundations of mathematics, by proving that some statements in arithmetic are undecidable. In 1936 Alan Turing found a simpler undecidable problem, the halting problem for Turing machines. Given an algorithm, is there always a proof that it either stops, or a proof that it must go on for ever? Turing's surprising answer was 'no'. For some algorithms no proof either way exists. The P/NP problem could, perhaps, be like that. It would explain why no one can either prove it or disprove it. But no one can prove or disprove that the P/NP problem is undecidable, either. Maybe its undecidability is undecidable ...

The most direct way to approach the P/NP problem would be to select some question known to be in class NP, assume that there exists a polynomial-time algorithm to solve it, and somehow derive a contradiction. For a time, people tried this technique on a variety of problems, but in 1971 Stephen Cook realised that the choice of problem often makes no difference. There is a sense in which all such problems – give or take some technicalities – stand or fall together. Cook introduced the notion of an NP-complete problem. This is a specific NP problem, with the property that if there exists a class P algorithm to solve it, then *any* NP problem can be solved using a class P algorithm.

Cook found several NP-complete problems, including SAT, the Boolean satisfiability problem. This asks whether a given logical expression can be made true by choosing the truth or falsity of its variables in a suitable manner. He also obtained a deeper result: a more restrictive problem, 3-SAT, is also NP-complete. Here the logical formula is one that can be written in the form 'A or B or C or ... or Z', where each of A, B, C, ..., Z is a logical formula involving only three variables. Not necessarily the same three variables each time, I hasten to add. Most proofs that a given problem is NP-complete trace back to Cook's theorem about 3-SAT.

Cook's definition implies that all NP-complete problems are on the same footing. Proving that one of them is class P would prove that all of them are class P. This result leaves open a tactical possibility: some NP-complete problems might be easier to work with than others. But strategically, it suggests that you may as well pick one specific NP-complete problem and work with that one. NP-complete problems all stand or fall together because an NP-complete problem can simulate any NP problem whatsoever. Any NP problem can be converted into a special case of the NP-complete one by 'encoding' it, using a code that can be implemented in polynomial time.

For a flavour of this procedure, consider a typical NP-complete problem: find a Hamiltonian cycle in a network. That is, specify a closed path along the edges of the network that visits every vertex (dot) exactly once. Closed means that the path returns to its starting point. The size of input data here is the number of edges, which is less than or equal to the square of the number of dots since each edge joins two dots. (We assume at most one edge joins a given pair.) No class P algorithm to solve this problem is known, but suppose, hypothetically, that there were one. Now choose some other problem, and call it problem X. Suppose that problem X can be rephrased in terms of finding such a path in some network associated to problem X. If the method for translating the data of problem X into data about the network, and conversely, can be carried out in polynomial time, then we automatically obtain a class P algorithm for problem X, like this:

1 Translate problem X into the search for a Hamiltonian cycle on the related network, which can be done in polynomial time.
2 Find such a cycle in polynomial time using the hypothetical algorithm for the network problem.
3 Translate the resulting Hamiltonian cycle back into a solution of problem X, which again can be done in polynomial time.

Since three polynomial-time steps combined run in polynomial time, this algorithm is class P.

To show how this works, I'll consider a less ambitious version of the Hamiltonian cycle problem in which the path is not required to be closed. This is called the Hamiltonian path problem. A network may possess a Hamiltonian path without possessing a cycle: Figure 42 (left) is an example. So a solution to the Hamiltonian cycle problem may not solve the Hamiltonian path problem. However, we can convert the Hamiltonian path problem into a Hamiltonian cycle problem on a related, but different, network. This is obtained by adding one extra dot, joined to every dot in the original network as in Figure 42 (right). Any Hamiltonian cycle in the new network can be converted to a

Hamiltonian path in the original one: just omit the new vertex and the two edges of the cycle that meet it. Conversely, any Hamiltonian path in the original network yields a Hamiltonian cycle in the new network: just join the two ends of the Hamiltonian path to the new dot. This 'encoding' of the path problem as a cycle problem introduces only one new dot and one new edge per dot in the original. So this procedure, and its inverse, run in polynomial time.

Fig 42 *Left*: Network with a Hamiltonian path (solid line) but no Hamiltonian cycle. *Right*: Add an extra dot (grey) and four more lines to convert the Hamiltonian path into a Hamiltonian cycle (solid line). The two grey edges are not in the cycle but are needed for the construction of the larger network.

Of course all I've done here is to encode one specific problem as a Hamiltonian cycle problem. To prove that the Hamiltonian cycle problem is NP-complete, we have to do the same for any NP problem. This can be done: the first proof was found by Richard Karp in 1972, in a famous paper that proved 21 different problems are NP-complete.[76]

The travelling salesman problem is 'almost' NP-complete, but there is a technical issue: it is not known to be NP. Over 300 specific NP-complete problems are known, in areas of mathematics that include logic, networks, combinatorics, and optimisation. Proving that any one of them can or cannot be solved in polynomial time would prove the same for every one of them. Despite this embarrassment of riches, the P/NP problem remains wide open. It wouldn't surprise me if it still is, a hundred years from now.

12

Fluid thinking
Navier-Stokes Equation

FIVE OF THE MILLENNIUM PROBLEMS, including the three discussed so far, come from pure mathematics, although the P/NP problem is also fundamental to computer science. The other two come from classical applied mathematics and modern mathematical physics. The applied mathematics problem arises from a standard equation for fluid flow, the Navier-Stokes equation, named after the French engineer and physicist Claude-Louis Navier and the Irish mathematician and physicist George Stokes. Their equation is a partial differential equation, which means that it involves the rate of change of the flow pattern in both space and time. Most of the great equations of classical applied mathematics and physics are also partial differential equations – we just met one, Laplace's – and those that are not are ordinary differential equations, involving only the rate of change with respect to time.

In chapter 8 we saw how the motion of the solar system is determined by Newton's laws of gravity and motion. These relate the accelerations of the Sun, Moon, and planets to the gravitational forces that are acting. Acceleration is the rate of change of velocity with respect to time, and velocity is the rate of change of position with respect to time. So this is an ordinary differential equation. As we saw, solving such equations can be very difficult. Solving partial differential equations is generally a lot harder.

For practical purposes, the equations for the solar system can be solved numerically using computers. That's still hard, but good methods now exist. The same is true for practical applications of the Navier-Stokes equations. The techniques employed are known as computational fluid dynamics, and they have a vast range of important applications: aircraft design, car aerodynamics, even medical problems like the flow of blood in the human body.

The millennium prize problem does not ask mathematicians to find explicit solutions to the Navier-Stokes equation, because this is essentially impossible. Neither is it about numerical methods for solving the equations, important though these are. Instead, it asks for a proof of a basic theoretical property: the *existence* of solutions. Given the state of a fluid at some instant of time – the pattern in which it is moving – does there exist a solution of the Navier-Stokes equation, valid for all future time, starting from the state concerned? Physical intuition suggests that the answer must surely be 'yes', because the equation is a very accurate model of the physics of real fluids. However, the mathematical issue of existence is not so clear-cut, and this basic property of the equation has never been proved. It might not even be true.

The Navier-Stokes equation describes how the pattern of fluid velocities changes with time, in given circumstances. The equation is often referred to using the plural, Navier-Stokes equations, but it's the same either way. The plural reflects the classical view: in three-dimensional space, velocity has three components, and classically each component contributed one equation, making three in all. In the modern view, there is one equation for the velocity *vector* (a quantity with both size and direction), but this equation can be applied to each of three components of the velocity. The Clay Institute website uses the classical terminology, but here I will follow the modern practice. I mention this to avoid possible confusion.

The equation dates from 1822, when Navier wrote down a partial differential equation for the flow of a viscous – sticky – fluid. Stokes's contributions occurred in 1842 and 1843. Euler had written down a partial differential equation for a fluid with zero viscosity – no stickinesss – in 1757. Although this equation remains useful, most real fluids including water and air are viscous, so Navier and Stokes modified the Euler equation to take account of viscosity. Both scientists derived essentially the same equation independently, so it is named after them both. Navier made some mathematical errors but ended up with the right answer; Stokes got the mathematics right, which is how we know Navier's answer is correct despite his mistake. In its most general form, the equation applies to compressible fluids like air. However, there is an important special case in which the fluid is assumed to be incompressible. Such a model applies to fluids like water, which do compress under huge forces, but only very slightly.

There are two ways to set up the mathematical description of fluid flow: you can either describe the path that each particle of fluid takes as time passes, or you can describe the velocity of the flow at each point in space and each instant of time. The two descriptions are related: given one, you can – with effort – deduce the other. Euler, Navier, and Stokes all used the second point of view, because it leads to an equation that is much more tractable mathematically. So their equations refer to the fluid's velocity field. At each fixed instant of time, the velocity field specifies the speed and direction of every particle of fluid. As time varies, this description may change. This is why rates of change both in space and in time occur in the equation.

The Navier-Stokes equation has an excellent physical pedigree. It is based on Newton's laws of motion, applied to each tiny particle (small region) of fluid, and expresses, in that context, the law of conservation of momentum. Each particle moves because forces act on it, and Newton's law of motion states that the particle's acceleration is proportional to the force. The main forces are friction, caused by viscosity, and pressure. There are also forces generated by the particle's acceleration. The equation follows classical practice, and treats the fluid as an infinitely divisible continuum. In particular it ignores the fluid's discrete atomic structure at very small scales.

Equations of themselves are of little value: you have to be able to solve them. For the Navier-Stokes equation, this means calculating the velocity field: the speed and direction of the fluid at each point in space and each instant of time. The equation provides constraints on these quantities, but it doesn't prescribe them directly. Instead, we have to apply the equation to relate future velocities to current ones. Partial differential equations like Navier-Stokes have many different solutions; indeed, infinitely many. This is no surprise: fluids can flow in many different ways; the flow over the surface of a car differs from that over the wings of an aircraft. There are two main ways to select a particular flow from this multitude of possibilities: initial conditions and boundary conditions.

Initial conditions specify the velocity field at some particular reference time, usually taken to be time zero. The physical idea is that once you know the velocity field at this instant, the Navier-Stokes equation determines the field a very short time later, in a unique way. If you start by giving the fluid a push, it keeps going while obeying the laws of physics. Boundary conditions are more useful in most applications, because it is difficult to set up initial conditions in a real fluid, and in any case these are not entirely appropriate to applications like car design. What matters there is the shape of the car. Viscous fluid sticks to surfaces. Mathematically, this feature is modelled by specifying the velocity on these surfaces, which form the boundary of the region occupied by the fluid, which is where the equation is valid. For example, we might require the velocity to be zero on the boundary, or whichever other condition best models reality.

Even when initial or boundary conditions are specified, it is highly unusual to be able to write down an explicit formula for the velocity field, because the Navier-Stokes equation is nonlinear. The sum of two solutions is not normally a solution. This is one reason why the three-body problem of chapter 8 is so hard – though not the sole reason, because the two-body problem is also nonlinear yet has an explicit solution.

For practical purposes, we can solve the Navier-Stokes equation on a computer, representing the velocity field as a list of numbers. This list can be turned into elegant graphics, and used to calculate quantities of interest to engineers, such as the stresses on aircraft wings. Since computers can't handle infinite lists of numbers, and they can't handle numbers to infinite accuracy, we have to replace the actual flow by a discrete approximation, that is, a list of numbers that samples the flow at finitely many locations and times. The big issue is to ensure that the approximation is good enough.

The usual approach is to divide space into a large number of small regions to form a computational grid. The velocity is calculated only for the points at the corners of the grid. The grid might be just an array of squares (or cubes in three dimensions), like a chessboard, but for cars and aircraft it has to be more complicated, with smaller regions near the boundary, to capture finer detail of the flow. The grid may be dynamic, changing shape as time passes. Time is generally assumed to progress in steps, which may all be the same size, or may change size according to the prevailing state of the calculation.

The basis of most numerical methods is the way 'rate of change' is defined in calculus. Suppose that an object moves from one location to another in a very short period of time. Then the rate of change of position – the velocity – is the change in position divided by the time taken, with a small error that vanishes as the time period gets smaller. So we can approximate the rate of change, which is what enters into the Navier-Stokes equation, by this ratio of the spatial change to the temporal change. In effect, the equation now tells us how to push a known initial state – a specified list of velocities – one time step into the future. Then we have to repeat the calculation many times, to see what happens further into the future. There is a similar way to approximate solutions when the one we want is determined by boundary conditions. There are also many sophisticated ways to achieve the same result more accurately.

The more finely the computational grid is divided, and the shorter the time intervals are, the more accurate the approximation becomes. However, the computation also takes longer. So there is a compromise between accuracy and speed. Broadly speaking, an approximate answer obtained by computer is likely to be acceptable provided the flow does not have significant features that are smaller than the grid size. There are two main types of fluid flow, laminar and turbulent. In laminar flow, the pattern of movement is smooth, and layers of fluid glide neatly past each other. Here a small enough grid ought to be suitable. Turbulent flow is more violent and frothy, and the fluid gets mixed up in extremely complex ways. In such circumstances, a discrete grid, however fine, could easily cause trouble.

One of the characteristics of turbulence is the occurrence of vortexes, like small whirlpools, and these can be very tiny indeed. A standard image of turbulence consists of a cascade of ever-smaller vortexes. Most of the fine detail is smaller than any practical grid. To get round this difficulty, engineers often resort to statistical models in questions about turbulent flow. Another worry is that the physical model of a continuum might be inappropriate for turbulent flow, because the vortexes may shrink to atomic sizes. However, comparisons between numerical calculations and experiments show that the Navier-Stokes equation is a very realistic and accurate model – so good that many engineering applications now rely solely on computational fluid dynamics, which is cheap, instead of performing experiments with scale models in wind tunnels, which are expensive. However, experimental checks like these are still used when human safety is vital, for example when designing aircraft.

In fact, the Navier-Stokes equation is so accurate that it even seems to apply when physics suggests that it should stand a reasonable chance of failing: turbulent flow. At least, that is the case if it can be solved accurately enough. The main problem is a practical one: numerical methods for solving the equation take enormous quantities of computer time when the flow becomes turbulent. And they always miss some small-scale structure.

Mathematicians are always uneasy when the main information they have about a problem is based on some kind of approximation. The millennium prize for the Navier-Stokes equation tackles one of the key theoretical issues. Its solution would reinforce the gut feeling that the numerical methods usually work very well. There is a subtle distinction between the approximations used by the computer, which make small changes to the equation, and the accuracy of the answer, which is about small changes to the solution. Is an exact answer to an approximate question the same as an approximate answer to the exact question? Sometimes the answer is 'no'. The exact flow for a fluid with very small viscosity, for instance, often differs from an approximate flow for a fluid with zero viscosity.

One step towards understanding these issues is so simple that it can easily be overlooked: proving that an exact solution exists. There has to be something for the computer calculations to be approximations *to*. This observation motivates the millennium prize for the Navier-Stokes equation. Its official description on the Clay Institute website consists of four problems. Solving any one of them is enough to win the prize. In all four, the fluid is assumed incompressible. They are:

1 *Existence and smoothness of solutions in three dimensions.* Here the fluid is assumed to fill the whole of infinite space. Given any

initial smooth velocity field, prove that a smooth solution to the equation exists for all positive times, coinciding with the specified initial field.

2 *Existence and smoothness of solutions in the three-dimensional flat torus.* The same question, but now assuming that space is a flat torus – a rectangular box with opposite faces identified. This version avoids potential problems caused by the infinite domain assumed in the first version, which does not match reality and might cause bad behaviour for silly reasons.

3 *Breakdown of solutions in three dimensions.* Prove that (1) is wrong. That is, find an initial field for which a smooth solution does not exist for all positive times, and prove that statement.

4 *Breakdown of solutions in the three-dimensional flat torus.* Prove that (2) is wrong.

The same problems remain open for the Euler equation, which is the same as the Navier-Stokes equation but assumes no viscosity, but no prize is on offer for the Euler equation.

The big difficulty here is that the flow under consideration is three-dimensional. There is an analogous equation for fluid flowing in a plane. Physically, this represents either a thin layer of fluid between two flat plates, assumed not to cause friction, or a pattern of flow in three dimensions in which the fluid moves in exactly the same manner along a system of parallel planes. In 1969 the Russian mathematician Olga Alexandrovna Ladyzhenskaya proved that (1) and (2) are true, while (3) and (4) are false, for the two-dimensional Navier-Stokes equation and the two-dimensional Euler equation.

Perhaps surprisingly, the proof is harder for the Euler equation, even though that equation is simpler than the Navier-Stokes equation, omitting terms involving viscosity. The reason is instructive. Viscosity 'damps down' bad behaviour in the solution, which potentially might lead to some kind of singularity that prevents the solution existing for all time. If the viscosity term is missing, no such damping occurs, and this shows up as mathematical issues in the existence proof.

Ladyzhenskaya made other vital contributions to our understanding of the Navier-Stokes equation, proving not only that solutions do exist but also that certain computational fluid dynamics schemes approximate them as accurately as we wish.

The millennium prize problems refer to incompressible flow because compressible flows are well known to be badly behaved. The equations for an aircraft, for example, run into all sorts of trouble if the aircraft travels faster than sound. This is the famous 'sound barrier', which worried engineers trying to design supersonic jet fighters, and the problem is related to the compressibility of air. If a body moves through an incompressible fluid, it pushes the fluid particles out of the way, like tunnelling through a box filled with ball bearings. If the particles pile up, they slow the body down. But in a compressible fluid, where there is a limit to the speed with which waves can travel – the speed of sound – that doesn't happen. At supersonic speeds, instead of being pushed out of the way, the air piles up ahead of the aircraft, and its density there increases without limit. The result is a shockwave. Mathematically, this is a discontinuity in the pressure of the air, which suddenly jumps from one value to a different one across the shockwave. Physically, the result is a sonic boom: a loud bang. If not understood and taken into account, a shockwave can damage the aircraft, so the engineers were right to worry. But the speed of sound isn't really a barrier, just an obstacle. The presence of shockwaves implies that the compressible Navier-Stokes equations need not have smooth solutions for all time, even in two dimensions. So the answer is already known in that case, and it's negative.

The mathematics of shockwaves is a substantial area within partial differential equations, despite this breakdown of solutions. Although the Navier-Stokes equation alone is not a good physical model for compressible fluids, it is possible to modify the mathematical model by adding extra conditions to the equations, which take discontinuities of shockwaves into account. But shockwaves do not occur in the flow of an incompressible fluid, so it is at least conceivable that in that context solutions should exist for all time, no matter how complicated the initial flow might be, provided it is smooth.

Some positive results are known for the three-dimensional Navier-Stokes equation. If the initial flow pattern involves sufficiently small velocities, so that the flow is very sluggish, then (1) and (2) are both true. Even if the velocities are large, (1) and (2) are true for some nonzero interval of time. There may not exist a solution valid for all future times, but there is a definite amount of time over which a solution does exist. It might seem that we can repeat this process, pushing a solution forward in time by small amounts, and then using the end result as a new initial condition. The problem with this line of reasoning is that the time intervals may shrink so rapidly that infinitely many such steps take a finite time. For instance, if each successive step takes half the time of the previous one, and the first step takes, say, 1 minute, then the whole process is over in a time $1 + \frac{1}{2} + \frac{1}{4} + \frac{1}{8} + \cdots$, which equals 2. If the solution ceases to exist – at the present time a purely hypothetical assumption, but one we can still contemplate – then the solution concerned is said to *blow up*. The time it takes for this to happen is the blowup time.

So the four questions ask whether solutions can, in fact, blow up. If they can't, (1) and (2) are true; if they can, (3) and (4) are. Perhaps solutions can blow up in an infinite domain, but not in a finite one. By the way, if the answer to (1) is 'yes' then so is the answer to (2), because we can interpret any flow pattern in a flat torus as a spatially periodic flow pattern in the whole of infinite space. The idea is to fill space with copies of the rectangular box involved, and copy the same flow pattern in each. The gluing rules for a torus ensure that the flow remains smooth when it crosses these flat interfaces. Similarly if the answer to (4) is 'yes' then so is the answer to (3), for the same reason. We just make the initial state spatially periodic. But for all we currently know, the answer to (2) might be 'yes' but the answer to (1) could be 'no'.

We do know one striking fact about blowups. If there is a solution with a finite blowup time, then the maximum velocity of the fluid, at all points in space, must become arbitrarily large. This could occur, for instance, if a jet of fluid forms, and the speed of the jet increases so rapidly that it diverges to infinity after a finite amount of time has passed.

These objections are not purely hypothetical. There are precedents for this kind of singular behaviour in other equations of classical mathematical physics. A remarkable example occurs in celestial mechanics. In 1988 Zhihong Xia proved that there exists an initial configuration of five point masses in three-dimensional space, obeying Newton's law of gravity, for which four particles disappear to infinity after a finite period of time – a form of blowup – and the fifth undergoes ever wilder oscillations. Earlier, Joseph Gerver had indicated that five bodies in a plane might all disappear to infinity in finite time, but he was unable to complete the proof for the scenario he envisaged. In 1989 he proved that this kind of escape definitely can occur in a plane if the number of bodies is large enough.

It is remarkable that this behaviour is possible, given that such systems obey the law of conservation of energy. Surely, if all bodies are moving arbitrarily fast, the total kinetic energy must increase? The answer is that there is also a decrease in potential energy, and for a point particle, the total gravitational potential energy is infinite. The bodies must also conserve angular momentum, but they can do that provided some of them move faster and faster in ever-decreasing circles.

The physical point involved is the famous slingshot effect, which is used routinely to dispatch space probes to distant worlds in the solar system. A good example is NASA's Galileo probe, whose mission was to travel to Jupiter, to study that giant planet and its many satellites. It launched in 1989 and arrived at Jupiter in 1995. One of the reasons it took so long was that its route was distinctly indirect. Although Jupiter's orbit lies outside that of the Earth, Galileo began by heading inwards, towards Venus. It passed close to Venus, returned to fly past the Earth, and headed out into space to look at the asteroid 951 Gaspra. Then it came back towards Earth, passed round our home planet *again*, and finally headed off towards Jupiter. Along the way it approached another asteroid, Ida, discovering that it had its own tiny moon, a new asteroid named Dactyl.

Why such a convoluted trajectory? Galileo gained energy, hence speed, from each close encounter. Imagine a space probe heading towards an approaching planet, not on a collision course, but getting pretty close to the surface, swinging round the back of the planet, and being flung off into space. As the probe passes behind the planet, each attracts the other. In fact, they've been attracting each other all along, but at this stage the force of attraction is at its greatest and so has the greatest effect. The planet's gravity gives the probe a speed boost. Energy must be conserved, so in compensation the probe slows the planet down very slightly in its orbit round the Sun. Since the probe has a very small mass and the planet has a very large mass, the effect on the planet is negligible. The effect on the probe is not: it can speed up dramatically.

Galileo got within 16,000 kilometres of Venus's surface, and gained 2.23 kilometres per second in speed. It then passed within 960 kilometres of Earth, and again within 300 kilometres, adding a further 3.7 kilometres per second. These manoeuvres were essential to get it to Jupiter, because its rockets were not powerful enough to take it there directly. The original plan had been to do just that, using the Centaur-G liquid-hydrogen-fuelled booster. But the disaster in which the space shuttle *Challenger* exploded shortly after takeoff caused this plan to be abandoned, because the Centaur-G was prohibited. So Galileo had to use a weaker solid-fuel booster instead. The mission was a huge success, and the scientific payoff included observing the collision between comet Shoemaker-Levy 9 and Jupiter in 1994, while the probe was still *en route* to Jupiter.

Xia's scenario also makes use of the slingshot effect. Four planets of equal mass form two close pairs, revolving round their common centres of mass in two parallel planes.[77] These two-body racquets play celestial tennis with a fifth, lighter body that shuttles to and fro between them at right angles to those planes. The system is set up so that every time this 'tennis ball' passes by a pair of planets, slingshot effects speed up the ball, and push the pair of planets outwards along the line joining the two pairs, so the tennis court gets longer and the players get further apart. Energy and momentum are kept in balance because the two planets concerned get slightly closer together, and revolve ever faster around their centre of mass. With the right initial setup, the pairs of planets move apart ever faster, and their speed increases so rapidly that they get to infinity after a finite amount of time. Meanwhile the tennis ball oscillates between them ever faster. Gerver's escape scenarios also use the slingshot effect.

Is this disappearing act relevant to real celestial bodies? Not if taken literally. It relies on the bodies being point masses. For many problems in celestial mechanics, that's a sensible approximation, but not if the bodies get arbitrarily close to each other. If bodies of finite size did that, they would eventually collide. Relativistic effects would stop the bodies moving faster than light, and change the law of gravity.

Anyway, the initial conditions, and assumptions that some masses are identical, would be too rare to happen in practice. Nonetheless, these curious examples show that even though the equations of celestial mechanics model reality very well in most circumstances, they can have complicated singularities that prevent solutions existing for all time. It has also recently been realised that slingshot effects in triple-star systems, where three stars orbit each other in complicated paths, can expel one of the stars at high velocity. So innumerable orphan stars, flung out of their systems by their siblings, may be roaming the galaxy – or even intergalactic space – cold, lonely, unwanted, and unnoticed.

When a differential equation behaves so strangely that its solutions cease to make sense after some finite period of time, we say that there is a singularity. The above work on the many-body problem is really about various types of singularity. The millennium prize problem about the Navier-Stokes equation asks whether singularities can occur in initial-value problems, for a fluid occupying either the whole of space or a flat torus. If a singularity can form in finite time, the result is likely to be blowup, unless the singularity somehow unravels itself subsequently, which seems unlikely.

There are two main ways to approach these questions. We can try to prove that singularities never occur, or we can try to find one by choosing suitable initial conditions. Numerical solutions can help, either way: they can suggest useful general features of flows, and they can provide strong hints about the possible nature of potential singularities. However, the potential lack of accuracy in numerical solutions means that any such hints must be treated with caution and justified more rigorously.

Attempts to prove regularity – the absence of singularities – employ a variety of methods to gain control over the flow. These include complicated estimates of how big or small certain key variables can become, or more abstract techniques. A popular approach is by way of so-called weak solutions, which are not exactly flows at all, but more general mathematical structures with some of the properties of flows. It is known, for instance, that the set of singularities of a weak solution of the three-dimensional Navier-Stokes equations is always small, in a specific technical sense.

Many different scenarios that might lead to singularities have been investigated. The standard model of turbulence as a cascade of ever-decreasing vortexes goes back to Andrei Kolmogorov in 1941, and he suggested that on very small scales, all forms of turbulence look very similar. The proportions of vortexes of given size, for example, follow a universal law. It is now known that as the vortexes become smaller, they change shape, and get longer and thinner, forming filaments. The law of conservation of angular momentum implies that the vorticity – how much the vortex is spinning – must increase. This is called vortex-stretching, and it is the kind of behaviour that might cause a singularity – for example, if the very small vortexes could become infinitely long in finite time, and the vorticity could become infinite at some points.

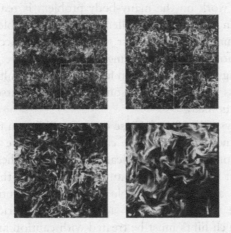

Fig 43 Zooms into a turbulent flow, simulated with the computer system VAPOR.

Figure 43 shows a zoom into very small scales of a turbulent flow, simulated by Pablo Mininni and colleagues using VAPOR, the Visualization and Analysis Platform for Ocean, Atmosphere, and Solar Research. The images show the vorticity intensity: how rapidly the fluid is spinning. They illustrate the formation of vortex filaments, the long thin structures in the figures, and show that they can cluster together to form larger-scale patterns. Their setup can perform simulations on cubic grids with more than 3 billion grid points.

In his article about this problem on the Clay Institute website,[78] Charles Fefferman writes:

> There are many fascinating problems and conjectures about the behavior of solutions of the Euler and Navier–Stokes equations... Since we don't even know whether these solutions exist, our understanding is at a very primitive level. Standard methods from [partial differential equations] appear inadequate to settle the problem. Instead, we probably need some deep, new ideas.

The complexity of the flow in images like Figure 43 drives home the difficulties that are likely to be encountered when seeking these ideas. Undaunted, mathematicians are soldiering on, seeking simple principles within the apparent complexities.

13

Quantum conundrum
Mass Gap Hypothesis

A FEW KILOMETRES NORTH OF GENEVA there is a sharp kink in the border between Switzerland and France. On the surface, all you see are minor roads and small villages. But between 50 and 175 metres underground is the largest scientific instrument on the planet. It is a gigantic circular tunnel, over 8 kilometres in diameter, joined to a second circular tunnel about a quarter as big. Most of it is under France but two sections are in Switzerland. Within the tunnels run pairs of pipes, which meet at four points.

It is the Large Hadron Collider, it cost €7.5 billion (about $9 billion) and it is probing the frontiers of particle physics. The key aim of the 10,000 scientists from over 100 nations who collaborated on it was to find the Higgs boson – or not to find it, if that's the way the continuum crumbled. They were looking for it to complete the Standard Model of particle physics, in which everything in the universe is made from 17 different fundamental particles. According to theory, the Higgs boson is what gives all particles mass.

In December 2011 ATLAS and CMS, two experimental divisions of the Large Hadron Collider, independently found tentative evidence for a Higgs boson with a mass of about 125 GeV (gigaelectronvolts, units used interchangeably for mass and energy in particle physics, since both are equivalent). On 4 July 2012 CERN, the European particle physics laboratory that operates the Large Hadron Collider, announced, to a packed audience of scientists and science journalists, that the continuum had crumbled in favour of the Higgs. Both groups had collected large amounts of additional data, and the chance that their data showed a random fluctuation, rather than a new particle with Higgs-like properties, had dropped below 1 in 2 million. This is the degree of confidence traditionally required in particle physics before breaking out the champagne.

Further experiments will be needed to make sure that the new particle has all of the features that a theoretical Higgs boson should possess. For example, theory predicts that the Higgs boson should have spin 0; at the time of the announcement, the observations showed it to be either 0 or 2. There is also a possibility that 'the' Higgs boson may turn out to be composed of other, smaller, particles, or that it is just the first in a new family of Higgs-like particles. So either the current model of fundamental particles will be cemented in place, or we will have new information that will eventually lead to a better theory.

The last of the seven millennium prize problems is closely related to the Standard Model and the Higgs boson. It is a central question in quantum field theory, the mathematical framework in which particle physics is studied. It is called the mass gap hypothesis, and it places a specific lower limit on the possible mass of a fundamental particle. It is one representative problem chosen from a series of big unsolved questions in this deep and very new area of mathematical physics. It has connections that range from the frontiers of pure mathematics to the long-sought unification of the two main physical theories, general relativity and quantum field theory.

In classical Newtonian mechanics, the basic physical quantities are space, time, and mass. Space is assumed to be three-dimensional Euclidean, time is a one-dimensional quantity independent of space, and mass signifies the presence of matter. Masses change their location in space under the influence of forces, and the rate at which their position changes is measured with respect to time. Newton's law of motion describes how a body's acceleration (the rate of change of velocity, which is itself the rate of change of position) relates to the mass of the body and the force applied.

The classical theories of space, time, and matter were brought to their peak in James Clerk Maxwell's equations for electromagnetism.[79] This elegant system of equations unified two of nature's forces, previously thought to be distinct. In place of electricity and magnetism, there was a single electromagnetic field. A field pervades the whole of space, as if the universe were filled with some kind of invisible fluid. At each point of space we can measure the strength and direction of the field, as if that fluid were flowing in mathematical patterns. For some purposes the electromagnetic field can be split into two components, the electric field and the magnetic field. But a moving magnetic field creates an electric one, and conversely, so when it comes to dynamics, both fields must be combined into a single more complex one.

This cosy picture of the physical world, in which the fundamental scientific concepts bear a close resemblance to things that our senses perceive, changed dramatically in the early years of the twentieth century. At that point, physicists began to realise that on very small scales, much too small to be observed in any microscope then available, matter is very different from what everyone had imagined. Physicists and chemists started to take seriously a rather wild theory that went back more than two millennia to the philosophical musings of Democritus in ancient Greece and other scholars in India. This was the idea that although the world seems to be made of countless different materials, all matter is built from tiny particles: atoms. The word comes from the Greek for 'indivisible'.

The nineteenth-century chemists found indirect evidence for atoms: the elements that combine together to form more complex molecules do so in very specific proportions, often close to whole numbers. John Dalton formulated these observations as his law of multiple proportions, and put forward atoms as an explanation. If each chemical compound consisted of fixed numbers of atoms of various kinds, this kind of ratio would automatically appear. For example, we now know that each molecule of carbon dioxide consists of two oxygen atoms and one carbon atom, so the numbers of atoms will be in the ratio two to one. However, there are complications: different atoms have different masses, and many elements occur as molecules formed from several atoms – for example, the oxygen molecule is composed of two oxygen atoms. If you don't realise what's going on, you will think that an atom of oxygen is twice as massive as it actually is. And some apparent elements are actually mixtures of different 'isotopes' – atomic structures. For instance, chlorine occurs in nature as a mixture of two stable forms, now called chlorine-35 and chlorine-

37, in proportions of about 76 per cent and 24 per cent respectively. So the observed 'atomic weight' is 35.45, which in the fledgling stages of atomic theory was interpreted as 'the chlorine atom is composed of thirty-five and a half hydrogen atoms'. And that means that an atom is not indivisible. As the twentieth century opened, most scientists still felt that the leap to atomic theory was too great, and the numerical evidence was too weak to justify taking it.

Some scientists, notably Maxwell and Ludwig Boltzmann, were ahead of the curve, convinced that gases are thinly distributed collections of molecules and that molecules are made by assembling atoms. What convinced most of their fellows seems to have been Albert Einstein's explanation of Brownian motion, erratic movements of tiny particles suspended in a fluid, that were visible under a microscope. Einstein decided that these movements must be caused by collisions with randomly moving molecules of the fluid, and he carried out some quantitative calculations to support that view. Jean Perrin confirmed these predictions experimentally in 1908. Being able to see the effect of the alleged indivisible particles of matter, and make quantitative predictions, carried more conviction than philosophical musings and curious numerology. In 1911 Amedeo Avogadro sorted out the problem with isotopes, and the existence of atoms became the scientific consensus.

While this was going on, a few scientists started to realise that atoms are not indivisible. They have some kind of structure, and it is possible to knock small pieces off them. In 1897 Joseph John Thomson was experimenting with so-called cathode rays, and he discovered that atoms could be made to emit even tinier particles, electrons. Not only that: the atoms of different elements emitted the same particles. By applying a magnetic field, Thomson showed that electrons carry a negative electric charge. Since an atom is electrically neutral, there must also be some part of atoms with a positive charge, leading Thomson to propose the plum pudding model: an atom is like a positively charged pudding riddled with negatively charged plums. But in 1909 one of Thomson's ex-students, Ernest Rutherford, performed experiments showing that most of the mass of an atom is concentrated near its centre. Puddings aren't like that.

How can experiments probe such tiny regions of space? Imagine a plot of land, which may or may not have buildings or other structures on it. You're not permitted to enter the area, and it's pitch dark so you can't see what's there. However, you do have a rifle and many boxes of ammunition. You can shoot bullets at random into the plot, and observe the direction in which they exit. If the plot is like a plum pudding, most bullets will go straight through. If you occasionally have to duck as a bullet ricochets straight back at you, there's something pretty solid somewhere. By observing how frequently the bullet exits at a given angle, you can estimate the size of the solid object.

Rutherford's bullets were alpha particles, nuclei of helium atoms, and his plot of land was a thin sheet of gold foil. Thomson's work had shown that the electron plums have very little mass, so almost all of the mass of an atom should be found in the pudding. If the pudding didn't have lumps, most alpha particles ought to go straight through, with very few being deflected, and then not by much. Instead, a small but significant proportion experienced large deflections. So the plum pudding picture didn't work. Rutherford suggested a different metaphor, one we still use informally today despite it being superseded by more modern images: the planetary model. An atom is like the solar system; it has a huge central nucleus, its sun, around which electrons orbit like planets. So, like the solar system, the interior of an atom is mostly empty space.

Rutherford went on to find evidence that the nucleus is composed of two distinct types of particle: protons, with positive charge; and neutrons, with zero charge. The two have very similar masses, and both are about 1800 times as massive as an electron. Atoms, far from being indivisible, are made from even smaller subatomic particles. This theory explains the integer numerology of chemical elements: what are being counted are the numbers of protons and neutrons. It also explains isotopes: adding or subtracting a few neutrons changes the mass, but keeps the charge zero and leaves the number of electrons – equal to the number of protons – unchanged. An atom's chemical properties are mostly controlled by its electrons. For instance, chlorine-35 has 17 protons, 17 electrons, and 18 neutrons; chlorine-37 has 17 protons, 17 electrons, and 20 neutrons. The 35.45 figure arises because natural chlorine is a mixture of these two isotopes.

By the early twentieth century there was a new theory in town, applicable to matter on the scales of subatomic particles. It was quantum mechanics, and once it became available, physics would never be the same. Quantum mechanics predicted a host of new phenomena, many of which were quickly observed in the laboratory. It explained a great many strange and previously baffling observations. It predicted the existence of new fundamental particles. And it told us that the classical image of the universe we live in, despite its previously excellent agreement with observations, is wrong. Our human-scale perceptions are poor models of reality at its most fundamental level.

In classical physics, matter is made from particles and light is a wave. In quantum mechanics, light is also a particle, the photon; conversely, matter, for instance electrons, can sometimes behave like a wave. The previously sharp dividing line between waves and particles did not so much become blurred as vanish altogether, replaced by wave/particle duality. The planetary model of the atom didn't work very well if you took it literally, so a new image arose. Instead of orbiting the nucleus like planets, electrons form a fuzzy cloud centred on the nucleus, a cloud not of stuff but of probabilities. The density of the cloud corresponds to the likelihood of finding an electron in that location.

As well as protons, neutrons, and electrons, physicists knew one further subatomic particle, the photon. Soon others appeared. An apparent failure of the law of conservation of energy led Wolfgang Pauli to propose patching it up by postulating the existence of the neutrino, an invisible and virtually undetectable new particle that would provide the missing energy. It was just detectable enough for its existence to be confirmed in 1956. And that opened the floodgates. Soon there were pions, muons, and kaons, the latter discovered by observing cosmic rays. Particle physics was born, and it continued to use Rutherford's method to probe the incredibly tiny spatial scales involved: to find out what's inside something, throw a lot of stuff at it and observe what bounces off. Ever larger particle accelerators – in effect, the guns that shot the bullets – were built and operated. The Stanford linear accelerator was 3 kilometres long. To avoid having to build accelerators whose lengths would span continents, they were bent into a circle, so particles could go round and round huge numbers of times at colossal speeds. That complicated the technology, because particles moving in circles radiate energy, but there were fixes.

The first fruit of these labours was an ever-growing catalogue of allegedly fundamental particles. Enrico Fermi expressed his frustration: 'If I could remember the names of all these particles, I'd be a botanist.' But every so often, new ideas from quantum theory collapsed the list again, as new kinds of ever-smaller particles were proposed to unify the structures already observed.

Early quantum mechanics applied to individual wave-like or particle-like things. But initially, no one could describe a good quantum-mechanical analogue of a field. It was impossible to ignore this gap because particles (describable by quantum mechanics) could and did interact with fields (not describable by quantum mechanics). It was like wanting to find out how the planets of the solar system move, when you knew Newton's laws of motion (how masses move when forces are applied), but didn't know his law of gravity (what those forces are).

There was another reason to want to model fields, rather than just particles. Thanks to wave/particle duality, they are intimately related. A particle is essentially a bunched-up chunk of field. A field is a sea of tightly packed particles. The two concepts are inseparable. Unfortunately, the methods developed to that date relied on particles being like tiny points, and they didn't extend in any sensible way to fields. You couldn't just stick lots of particles together and call the result a field, because particles *interact* with each other.

Imagine a crowd of people in ... well, a field. Perhaps they are at a rock concert. Viewed from a passing helicopter, the crowd resembles a fluid, sloshing around the field – often literally, for example at the Glastonbury Festival, renowned for becoming a sea of mud. Down on the ground, it becomes clear that the fluid is really a seething mass of individual particles: people. Or perhaps dense clusters of people, as a few friends walk close together, forming an indivisible unit, or as a group of strangers comes together with a common purpose, such as getting to the bar. But you can't model the crowd accurately by adding together what the people would do if they were on their own. As one group heads for the bar, it blocks the path of another group. The two groups collide and jostle. Setting up an effective quantum field theory is like doing this when the people are localised quantum wavefunctions.

By the end of the 1920s this kind of reasoning had convinced physicists that however hard the task might be, quantum mechanics had to be extended to take care of fields as well as particles. The natural place to start was the electromagnetic field. Somehow the electrical and magnetic components of this field had to be quantised: rewritten in the formalism of quantum mechanics. Mathematically, that formalism was unfamiliar and not terribly physical. Observables – things you could measure – were no longer represented using good old numbers. Instead, they corresponded to operators on a Hilbert space: mathematical rules for manipulating waves. These operators violated the usual assumptions of classical mechanics. If you multiply two numbers together, the result is the same whichever comes first. For instance, 2×3 and 3×2 are the same. This property, called commutativity, fails for many pairs of operators, much as putting on your socks and then your shoes does not have the same effect as putting on your shoes and then your socks. Numbers are passive creatures, operators are active. Which action you take first sets the scene for the other.

Commutativity is a very pleasant mathematical property. Its absence is a bit of a nuisance, and this is just one of the reasons why quantising a field turns out to be tricky. Nonetheless, it can sometimes be done. The electromagnetic field was quantised in a series of stages, starting with Dirac's theory of the electron in 1928, and completed by Sin-Itiro Tomonaga, Julian Schwinger, Richard Feynman, and Freeman Dyson in the late 1940s and early 1950s. The resulting theory is known as quantum electrodynamics.

The point of view that was used there hinted at a method that might work more generally. The underlying idea went right back to Newton. When mathematicians attempted to solve the equations supplied by Newton's law, they discovered some useful general tricks, known as conservation laws. When a system of masses moves, some quantities remain unchanged. The most familiar is energy, which comes in two flavours, kinetic and potential. Kinetic energy is related to how fast the body moves, and potential energy is the work done by forces. When a rock is pushed off the edge of a cliff it trades potential energy, due to gravity, for kinetic energy; in ordinary language, it falls and speeds up. Other conserved quantities are momentum, which is mass times velocity, and angular momentum, which is related to the body's rate of spin. These conserved quantities relate the different variables used to describe the system, and therefore reduce their number. That helps when solving the equations, as we saw for the two-body problem in chapter 8.

By the 1900s the source of these conservation laws had been understood. Emmy Noether proved that every conserved quantity corresponds to a continuous group of symmetries of the equations. A symmetry is a mathematical transformation that leaves the equations unchanged, and all symmetries form a group, with the operation being 'do one transformation, then the other'. A continuous group is a group of symmetries defined by a single real number. For example, rotation about a given axis is a symmetry, and the angle of rotation can be any real number, so the rotations – through any angle – about a given axis form a continuous family. Here the associated conserved quantity is angular momentum. Similarly, momentum is the conserved quantity associated with the family of translations in a given direction. What about energy? That is the conserved quantity corresponding to time symmetries – the equations are the same at all instants of time.

When physicists tried to unify the basic forces of nature, they became convinced that symmetries were the key. The first such unification was Maxwell's, combining electricity and magnetism into a single electromagnetic field. Maxwell achieved this unification without considering symmetry, but it soon became clear that his equations possess a remarkable kind of symmetry that had not previously been noticed: gauge symmetry. And that looked like a strategic lever that might open up more general quantum field theories.

Rotations and translations are global symmetries: they apply uniformly across the whole of space and time. A rotation about some axis rotates every point in space through the same angle. Gauge symmetries are different: they are local symmetries, which can vary from point to point in space. In the case of electromagnetism, these local symmetries are changes of phase. A local oscillation of the electromagnetic field has both an amplitude (how big it is) and a phase (the time at which it reaches its peak). If you take a solution of Maxwell's field equations and change the phase at each point, you get another solution, provided you make a compensating change to the description of the field, incorporating a local electromagnetic charge.

Gauge symmetries were introduced by Hermann Weyl in an abortive attempt to bring about a further unification, of electromagnetism and general relativity. That is, the electromagnetic and gravitational forces. The name came about because of a misapprehension: he thought that the correct local symmetries should be changes of spatial scale, or 'gauge'. This idea didn't work out, but the formalism of quantum mechanics led Vladimir Fock and Fritz London to introduce a different type of local symmetry. Quantum mechanics is formulated using complex numbers, not just real numbers, and every quantum wavefunction has a complex phase. The relevant local symmetries rotate the phase through any angle in the complex plane. Abstractly this group of symmetries consists of all rotations, but in complex coordinates these are 'unitary transformations' (U) in a space with one complex dimension (1), so the group formed by these symmetries is denoted by U(1). The formalism here is not just an abstract mathematical game: it allowed physicists to write down, and then solve, the equations for charged quantum particles moving in an electromagnetic field. In the hands of Tomonaga, Schwinger, Feynman, and Dyson, this point of view led to the first relativistic quantum field theory of electromagnetism: quantum electrodynamics. Symmetry under the gauge group U(1) was fundamental to their work.

The next step, unifying quantum electrodynamics with the weak nuclear force, was achieved by Abdus Salam, Sheldon Glashow, Steven Weinberg, and others in the 1960s. Alongside the electromagnetic field with its U(1) gauge symmetry, they introduced fields associated with four fundamental particles, the so-called bosons W^+, W^0, W^-, and B^0. The gauge symmetries of this field, which in effect rotate combinations of these particles to produce other combinations, form another group, called SU(2) – unitary (U) transformations in a two-dimensional complex space (2) that are also special (S), a simple technical condition. The combined gauge group is therefore $U(1) \times SU(2)$, where the × indicates that the two groups act independently on the two fields. The result, called the electroweak theory, required a difficult mathematical innovation. The group U(1) for quantum electro-dynamics is commutative: applying two symmetry transformations in turn gives the same result, whichever one is applied first. This pleasant property makes the mathematics much simpler, but it fails for SU(2). This was the first application of a non-commutative gauge theory.

The strong nuclear force comes into play when we consider the internal structure of particles like protons and neutrons. The big breakthrough in this area was motivated by a curious mathematical pattern in one particular class of particles, called hadrons. The pattern was known as the eightfold way. It inspired the theory of quantum chromodynamics, which postulated the existence of hidden particles called quarks, and used them as basic components for the large zoo of hadrons.

In the standard model, everything in the universe is made from sixteen genuinely fundamental particles, whose existence has been confirmed by accelerator experiments. Plus a seventeenth, for which the Large Hadron Collider is currently searching. Of the particles known to Rutherford, only two remain fundamental: the electron and the photon. The proton and the neutron, in contrast, are made from quarks. The name was coined by Murray Gell-Mann, who intended it to rhyme with 'cork'. He came across a passage in James Joyce's *Finnegans Wake*:

Three quarks for Muster Mark!
Sure he has not got much of a bark
And sure any he has it's all beside the mark.

This would seem to hint at a pronunciation rhyming with 'mark', but Gell-Mann found a way to justify his intention. Both pronunciations are now common.

The Standard Model envisages six quarks, arranged in pairs. They have curious names: up/down, charmed/strange, and top/bottom. There are six leptons, also in pairs: the electron, muon, and tauon (today usually just called tau) and their associated neutrinos. These twelve particles collectively are called fermions, after Fermi. Particles are held together by forces, which are of four kinds: gravity, electromagnetism, the strong nuclear force, and the weak nuclear force. Leaving out gravity, which has not yet been fully reconciled with the quantum picture, this gives three forces. In particle physics, forces are produced by an exchange of particles, which 'carry' or 'mediate' the force. The usual analogy is two tennis players being held together by their mutual attention to the ball. The photon mediates the electromagnetic force, the Z- and W-bosons mediate the weak nuclear force, and the gluon mediates the strong nuclear force. Well, technically it mediates the colour force, which holds quarks together, and the strong force is what we observe as a result. The proton consists of two up quarks plus one down quark; the neutron consists of one up quark plus two down quarks. In each of these particles, the quarks are held together by gluons. These four force carriers are known collectively as bosons, after Chandra Bose. The distinction between fermions and bosons is important: they have different statistical properties. Figure 44 (left) shows the resulting catalogue of conjecturally fundamental particles. Figure 44 (right) shows how to make a proton and a neutron from quarks.

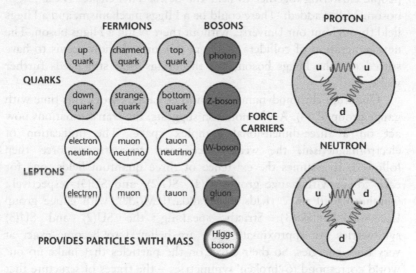

Fig 44 *Left*: The 17 particles of the Standard Model. *Right*: How to make a proton and a neutron from quarks. *Right top*: Proton = two up quarks + one down quark. *Right bottom*: Neutron = one up quark + two down quarks.

The Higgs boson completes this picture by explaining why the other 16 particles of the Standard Model have nonzero masses. It is named for Peter Higgs, one of the physicists who suggested the idea.

Others involved include Philip Anderson, François Englert, Robert Brout, Gerald Guralnik, Carl Hagen, and Thomas Kibble. The Higgs boson is the particle incarnation of a hypothetical quantum field, the Higgs field, with an unusual but vital feature: in a vacuum, the field is nonzero. The other 16 particles are influenced by the Higgs field, which makes them behave as if they have mass.

In 1993 David Miller, responding to a challenge from the British science minister William Waldegrave, presented a striking analogy: a cocktail party. People are spread uniformly around the room when the guest of honour (an ex–prime minister) walks in. Immediately everyone bunches up around her. As she moves across the room, different people join and leave the bunch, and the moving bunch gives her extra mass, making her harder to stop. This is the Higgs mechanism. Now imagine a rumour passing through the room, with people clustering together to hear the news. This cluster is the Higgs boson. Miller added: 'There could be a Higgs mechanism, and a Higgs field throughout our Universe, without there being a Higgs boson. The next generation of colliders will sort this out.' It now seems to have sorted out the Higgs boson, but the Higgs field still needs further work.

Quantum chromodynamics is another gauge theory, this time with gauge group SU(3). As the notation suggests, the transformations now act on a three-dimensional complex space. The unification of electromagnetism, the weak force, and the strong force then followed. It assumes the existence of three quantum fields, one for each force, with gauge groups U(1), SU(2), and SU(3) respectively. Combining all three yields the Standard Model, with gauge group U(1) × SU(2) × SU(3). Strictly speaking, the SU(2) and SU(3) symmetries are approximate; they are believed to become exact at very high energies. So their effect on the particles that make up our world correspond to 'broken' symmetries – the traces of structure that remain when the ideal perfectly symmetric system is subjected to small perturbations.

All three groups contain continuous families of symmetries: one such family of U(1), three for SU(2), and eight for SU(3). Associated with these are various conserved quantities. The symmetries of Newtonian mechanics again provide energy, momentum, and angular momentum. The conserved quantities for the U(1) × SU(2) × SU(3) gauge symmetries are various 'quantum numbers', which characterise particles. These are analogous to such quantities as spin and charge, but apply to quarks; they have names like colour charge, isospin, and hypercharge. Finally, there are some additional conserved quantities for U(1): quantum numbers for the six leptons, such as electron number, muon number, and tau number. The upshot is that the symmetries of the equations of the Standard Model, via Noether's theorem, explain all of the core physical variables of fundamental particles.

The important message for our story is the overall strategy and outcome. To unify physical theories, find their symmetries and unify those. Then devise a suitable theory with that combined group of symmetries. I'm not suggesting that the process is straightforward; it is actually technically very complex. But so far, this is how quantum field theory has developed, and only one of the four forces of nature currently falls outside its scope: gravity.

Not only does Noether's theorem explain the main physical variables associated with fundamental particles: that was how many of the underlying symmetries were found. Physicists worked backwards from observed and inferred quantum numbers to deduce what symmetries the model ought to have. Then they wrote down suitable equations with those symmetries, and confirmed that these equations fitted reality very closely. At the moment, this final step requires choosing the values of 19 parameters – numbers that must be plugged into the equations to provide quantitative results. Nine of these are masses of specific particles: all six quarks and the electron, muon, and tau. The rest are more technical, things like mixing angles and phase couplings. Seventeen of these parameters are known from experiments, but two are not; they describe the still-hypothetical Higgs *field*. But now there is a good prospect of measuring them, because physicists know where to look.

The equations employed in these theories belong to a general class of gauge field theories, known as Yang-Mills theories. In 1954 Chen-Ning Yang and Robert Mills attempted to develop gauge theories to explain the strong force and the particles associated with it. Their first attempts ran into trouble when the field was quantised, because this required the particles to have zero mass. In 1960 Jeffrey Goldstone,

Yoichiro Nambu, and Giovanni Jona-Lasinio found a way round this problem: start with a theory that predicted massless particles, but then modify it by breaking some of the symmetries. That is, change the equations a little by introducing new asymmetric terms. When this idea was used to modify Yang-Mills theory, the resulting equations performed very well both in the electroweak theory and in quantum chromodynamics.

Yang and Mills assumed the gauge group was a special unitary group. For the particle applications this was either SU(2) or SU(3), the special unitary group for two or three complex dimensions, but the formalism worked for any number of dimensions. Their theory tackles head-on a difficult but unavoidable mathematical difficulty. The electromagnetic field is in one respect misleadingly simple: its gauge symmetries commute. Unlike most quantum operators, the order in which you change phases doesn't affect the equations. What physicists had their eyes on was a quantum field theory for subatomic particles. There, the gauge group was non-commutative, which made quantising the equations very difficult.

Yang and Mills succeeded by using a diagrammatic representation of particle interactions introduced by Richard Feynman. Any quantum state can be thought of as a superposition of innumerable particle interactions. For example, even a vacuum involves pairs of particles and antiparticles momentarily winking into existence and then winking out again. A simple collision between two particles splits up into a bewildering dance of temporary appearances and disappearances of intermediary particles, shuttling back and forth, splitting and combining. What saves the day is a combination of two things. The field equations can be quantised for each specific Feynman diagram, and all of these contributions can be added together to represent the effect of the full interaction. Moreover, the most complicated diagrams are rare, so they don't contribute a lot to the sum. Even so, there is a serious problem. The sum, interpreted straightforwardly, is infinite. Yang and Mills found a way to 'renormalise' the calculation so that an infinity of terms that shouldn't really matter were removed. What was left was a finite sum, and its value matched reality very closely. This technique was totally mysterious when first invented, but it now makes sense.

In the 1970s, mathematicians got in on the act and Michael Atiyah generalised Yang-Mills theory to a broad class of gauge groups. The mathematics and physics began to feed off each other, and Edward Witten and Nathan Seiberg's work on topological quantum field theories led to the concept of supersymmetry, in which all known particles have new 'supersymmetric' counterparts: electrons and selectrons, quarks and squarks. This simplified the mathematics and led to physical predictions. However, these new particles have not yet been observed, and some probably should have shown up by now in the experiments carried out using the Large Hadron Collider. The mathematical value of these ideas is well established, but their direct relevance to physics is not. However, they shed useful light on Yang-Mills theory.

Quantum field theory is one of the fastest-moving frontiers of mathematical physics, so the Clay Institute wanted to include something on the topic as one of its millennium prizes. The mass gap hypothesis sits squarely in this rich area, and it addresses an important mathematical issue linked to particle physics. The application of Yang-Mills fields to describe fundamental particles in terms of the strong nuclear force depends on a specific quantum-theoretic feature known as a mass gap. In relativity, a particle that travels at the speed of light acquires infinite mass, unless its mass is zero. The mass gap allows quantum particles to have finite nonzero masses, even though the associated classical waves travel with the speed of light. When a mass gap exists, any state that is not the vacuum has an energy that exceeds that of the vacuum by at least some fixed amount. That is, there is a nonzero lower limit to the mass of a particle.

Experiments confirm the existence of a mass gap, and computer simulations of the equations support the mass gap hypothesis. However, we can't assume that a model matches reality and then use reality to verify mathematical features of the model, because the logic becomes circular. So theoretical understanding is needed. One key step would be a rigorous proof that quantum versions of Yang-Mills theory exist. The classical (non-quantum) version is fairly well understood nowadays, but the quantum analogue is bedevilled by the problem of renormalisation – those pesky infinities that have to be spirited away by mathematical trickery.

One attractive approach begins by turning continuous space into a discrete lattice and writing down a lattice analogue of the Yang-Mills equation. The main issue is then to show that as the lattice becomes increasingly fine, approximating a continuum, this analogue converges to a well-defined mathematical object. Some necessary features of the mathematics can be inferred from physical intuition, and it would be possible to prove that a suitable quantum Yang-Mills theory exists if these features could be established rigorously. The mass gap hypothesis involves a more detailed understanding of how the lattice theories approximate this hypothetical Yang-Mills theory. So existence of the theory, and the mass gap hypothesis, are closely intertwined.

That's where everyone is stuck. In 2004 Michael Douglas wrote a report on the status of the problem, and said: 'So far as I know, no breakthroughs have been made on this problem in the last few years. In particular, while progress has been made in lower dimensional field theories, I know of no significant progress towards a mathematically rigorous construction of the quantum Yang-Mills theory.' That assessment still seems to be correct.

Progress has been more impressive on some related problems, however, which may shed useful light. Special quantum field theories, known as two-dimensional sigma models, are more tractable, and the mass gap hypothesis has been established for one such model. Supersymmetric quantum field theories, involving hypothetical super-partners of the usual fundamental particles, have pleasant mathematical features which in effect remove the need for renormalisation. Physicists such as Edward Witten have been making progress towards solving related questions in the supersymmetric case. The hope is that some of the methods that emerge from this work might suggest new ways to tackle the original problem. But whatever the physical implications may be, and however the mass gap hypothesis eventually pans out, many of these developments have already enriched mathematics with important new concepts and tools.

14

Diophantine dreams
Birch–Swinnerton-Dyer Conjecture

IN CHAPTER 7 WE ENCOUNTERED the *Arithmetica* of
Diophantus, and I remarked that six of its 13 books survive
as Greek copies. Around AD 400, when ancient Greek civilisation went
into decline, Arabia, China, and India took up the torch of
mathematical innovation from Europe. Arabic scholars translated
many of the classical Greek works, and often these translations are our
main historical source for their contents. The Arab world knew about
the *Arithmetica*, and built upon it. Four Arabic manuscripts discovered
in 1968 may be translations of other 'missing' books from the
Arithmetica.

Some time near the end of the tenth century AD, the Persian
mathematician al-Karaji asked a question that could easily have
occurred to Diophantus. Which integers can occur as the common
difference between three rational squares that form an arithmetic
sequence? For example the integer squares 1, 25, and 49 have common
difference 24. That is, $1 + 24 = 25$ and $25 + 24 = 49$. Al-Karaji lived
between about AD 953 and 1029, so he may have had access to a copy
of the *Arithmetica*, but the earliest known translation was made by
Abu'l-Wafā in 998. Leonard Dickson, who wrote a three-volume
synopsis of the history of number theory, suggested that the problem
might have originated some time before 972 in an anonymous Arab
manuscript.

In algebraic language the problem becomes: for which integers d
does there exist a rational number x such that $x - d$, x, and $x + d$ are

all perfect squares? It can be restated in a form that is equivalent, though not obviously so: which whole numbers can be the area of a right-angled triangle with rational sides? That is: if a, b, and c are rational and $a^2 + b^2 = c^2$, what are the possible integer values for $ab/2$? Integers that satisfy these equivalent conditions are called congruent numbers. The term does not relate to other uses of the word 'congruent' in mathematics, and that makes it slightly confusing to a modern reader. Its origins are explained below.

Some numbers are not congruent: for example, it can be proved that 1, 2, 3, and 4 are not congruent. Others, such as 5, 6, and 7, are congruent. Indeed, the 3-4-5 triangle has area $3 \times 4/2 = 6$, proving that 6 is congruent. To prove 7 is congruent, observe that $(24/5)^2$, $(35/12)^2$, and $(337/60)^2$ have common difference 7. I'll come back to 5 in a moment. Proceeding case by case in this manner provides a lengthy list of congruent numbers, but sheds little light on their nature. No amount of case by case construction of examples can prove that a particular whole number is *not* congruent. For centuries no one knew whether 1 is congruent.

We now know that the problem goes far beyond anything that Diophantus could have solved. In fact, this deceptively simple question has still not been fully answered. The closest we've got is a characterisation of congruent numbers, discovered by Jerrold Tunnell in 1983. Tunnell's idea provides an algorithm for deciding whether a given integer can or cannot occur by counting its representations as two different combinations of squares. With a little ingenuity this calculation is feasible for quite large integers. The characterisation has only one serious disadvantage: it has never been proved correct. Its validity depends on solving one of the millennium problems, the Birch–Swinnerton-Dyer conjecture. This conjecture provides a criterion for an elliptic curve to have only finitely many rational points. We encountered these Diophantine equations in chapters 6 on the Mordell conjecture and 7 on Fermat's last theorem. Here we see further evidence for their prominent role at the frontiers of number theory.

The earliest European work referring to these questions was written by Leonardo of Pisa. Leonardo is best known for a sequence of strange numbers that he seems to have invented, which arose in an arithmetic problem about the progeny of some very unrealistic rabbits. These are the Fibonacci numbers

0 1 1 2 3 5 8 13 21 34 55 89 . . .

in which each, after the first two, is the sum of the previous two. Leonardo's father was a customs official named Bonaccio, and the famous nickname means 'son of Bonaccio'. There is no evidence that it was used during Leonardo's lifetime, and it is thought to have been an invention of the French mathematician Guillaume Libri in the nineteenth century.[80] Be that as it may, Fibonacci numbers have many fascinating properties, and they are widely known. They even appear in Dan Brown's crypto-conspiracy thriller *The Da Vinci Code*.

Leonardo introduced the Fibonacci numbers in a textbook on arithmetic, the *Liber Abbaci* ('Book of calculation') of 1202, whose main aims were to draw European attention to the new arithmetical notation of the Arabs, based on the ten digits 0–9, and to demonstrate its utility. The idea had already reached Europe through al-Khwārizmī's text of 825 in its Latin translation *Algoritmi de Numero Indorum* ('On the calculation with Hindu numerals'), but Leonardo's book was the first to be written with the specific intention of promoting the uptake of decimal notation in Europe. Much of the book is devoted to practical arithmetic, especially currency exchange. But Leonardo wrote another book, not as well known, which in many ways was a European successor to Diophantus's *Arithmetica*: his *Liber Quadratorum* ('Book of squares').

Like Diophantus, he presented general techniques using special examples. One arose from al-Karaji's question. In 1225 Emperor Frederick II visited Pisa. He was aware of Leonardo's mathematical reputation, and seems to have decided that it would be fun to put it to the test in a mathematical tournament. Such public contests were common at the time. Contestants set each other questions. The emperor's team consisted of John of Palermo and Master Theodore. Leonardo's team consisted of Leonardo. The emperor's team challenged Leonardo to find a square which remains a square when 5 is added or subtracted. As usual, the numbers should be rational. In other words, they wanted a proof that 5 is a congruent number, by finding a specific rational x for which $x - 5$, x, and $x + 5$ are square.

This is by no means trivial – the smallest solution is

$$x = \frac{1681}{144} = \left(\frac{41}{12}\right)^2$$

in which case

$$x - 5 = \frac{961}{144} = \left(\frac{31}{12}\right)^2 \quad and \quad x + 5 = \frac{2401}{144} = \left(\frac{49}{12}\right)^2$$

Leonardo found a solution, and included it in the *Liber Quadratorum*. He got the answer using a general formula related to the Euclid/Diophantus formula for Pythagorean triples. From it, he obtained three integer squares with common difference 720, namely 31^2, 41^2, and 49^2. Then he divided by $12^2 = 144$ to get three squares with common difference 720/144, which is 5.[81] In terms of Pythagorean triples, take the 9, 40, 41 triangle with area 180 and divide by 36 to get a triangle with sides 20/3, 3/2, 41/6. Then its area is 5.

It is in Leonardo that we find the Latin word *congruum* for a set of three squares in arithmetic sequence. Later Euler used the word *congruere*, 'come together'. The first ten congruent numbers, and the corresponding simplest Pythagorean triples, are listed in Table 3. No simple patterns are evident.

d	Pythagorean triple
5	3/2, 20/3, 41/6
6	3, 4, 5
7	24/5, 35/12, 337/60
13	780/323, 323/30, 106921/9690
14	8/3, 63/6, 65/6
15	15/2, 4, 17/2
20	3, 40/3, 41/3
21	7/2, 12, 25/2
22	33/35, 140/3, 4901/105
23	80155/20748, 41496/3485, 905141617/72306780

Table 3 The first ten congruent numbers and corresponding Pythagorean triples.

Most of the early progress on this question was made by Islamic mathematicians, who showed that the numbers 5, 6, 14, 15, 21, 30, 34, 65, 70, 110, 154, and 190 are congruent, along with 18 larger numbers. To these, Leonardo, Angelo Genocchi (1855) and André Gérardin (1915) added 7, 22, 41, 69, 77, and 43 other numbers less than 1000. Leonardo stated in 1225 that 1 is not congruent, but gave no proof. In 1569 Fermat supplied one. By 1915 all congruent numbers less than 100 had been determined, but the problem yielded ground slowly, and by 1980 the status of many numbers less than 1000 remained unresolved. The difficulty can be judged by L. Bastien's discovery that 101 is congruent. The sides of the corresponding right-angled triangle are

$$\frac{711024064578955010000}{118171431852779451900}$$

$$\frac{3967272806033495003922}{118171431852779451900}$$

$$\frac{4030484925899520003922}{118171431852779451900}$$

He found these numbers in 1914, by hand. By 1986, with computers now on the scene, G. Kramarz had found all congruent numbers less than 2000.

At some point it was noticed that a different but related equation

$$y^2 = x^3 - d^2 x$$

has solutions x, y in whole numbers if and only if d is congruent.[82] This observation is obvious in one direction: the right-hand side is the product of x, $x - d$, and $x + d$, and if these are all squares, so is their product. The converse is fairly straightforward, too. This reformulation of the problem places it squarely within a rich and flourishing area of number theory. For any given d this equation sets y^2 equal to a cubic polynomial in x, and therefore defines an elliptic curve. So the problem of congruent numbers is a special case of a question that number theorists would dearly love to answer: when does an elliptic curve have at least one rational point? This question is far from straightforward, even for the special type of elliptic curve just mentioned. For instance, 157 is a congruent number, but the *simplest* right triangle with that area has hypotenuse

$$\frac{2244035177043369699245575130906674863160948472041}{8912332268928859588025535178967163570016480830}$$

Before proceeding further, we borrow Leonardo's trick, the one that led him from 720 to 5, and apply it in full generality. If we multiply any congruent number d by the square n^2 of an integer n, we also get a congruent number. Just take a rational Pythagorean triple corresponding to a triangle with area d, and multiply the numbers by n. The area of the triangle multiplies by n^2. The same is true if we divide the numbers by n; now the area divides by n^2. This process gives an integer only when the area has a square factor, so when seeking congruent numbers, it is enough to work with numbers that are squarefree – have no square factor. The first few squarefree numbers are

1 2 3 5 6 7 10 11 13 14 15 17 19

Now we can state Tunnell's criterion. An odd squarefree number d is congruent if and only if the number of (positive or negative) integer solutions x, y, z to the equation

$$2x^2 + y^2 + 8z^2 = d$$

is precisely twice the number of solutions to the equation

$$2x^2 + y^2 + 32z^2 = d$$

An even squarefree number d is congruent if and only if the number of integer solutions x, y, z to the equation

$$8x^2 + 2y^2 + 16z^2 = d$$

is precisely twice the number of solutions to the equation

$$8x^2 + 2y^2 + 64z^2 = d$$

These results are more useful than they might first seem. Because all the coefficients are positive, the sizes of x, y, z cannot exceed certain multiples of the square root of d. So the number of solutions is finite, and they can be found by a systematic search, with some useful short cuts. Here are the complete calculations for a few examples with small d:

- If $d = 1$ then the only solutions of the first equation are $x = 0$, $y = \pm 1$, $z = 0$. The same goes for the second equation. So both equations have two solutions, and the criterion fails to hold.

- If $d = 2$ then the only solutions of the first equation are $x = \pm 1$, $y = 0$, $z = 0$. The same goes for the second equation. So both equations have two solutions, and the criterion fails to hold.

- If $d = 3$ then the only solutions of the first equation are $x = \pm 1$, $y = \pm 1$, $z = 0$. The same goes for the second equation. So both equations have four solutions, and the criterion fails to hold.

- If $d = 5$ or 7 then the first equation has no solutions. The same goes for the second equation. Since twice zero is zero, the criterion is satisfied.

- If $d = 6$ we have to use the criterion for even numbers. Again both equations have no solutions, and the criterion is satisfied.

These simple calculations show that $1, 2, 3, 4(= 2^2 \times 1)$ are not congruent, but $5, 6$, and 7 are. The analysis can easily be extended, and in 2009 a team of mathematicians applied Tunnell's test to the first trillion numbers, finding exactly 3,148,379,694 congruent numbers. The researchers verified their results by performing the calculations twice, on different computers using different algorithms written by two independent groups. Bill Hart and Gonzalo Tornaria used the computer Selmer at the University of Warwick. Mark Watkins, David Harvey, and Robert Bradshaw used the computer Sage at the University of Washington.

However, there's a gap in all such calculations. Tunnell proved that if a number d is congruent, then it must satisfy his criterion. Therefore, if the criterion fails, the number is not congruent. This implies, for instance, that $1, 2, 3$, and 4 are not congruent. However, he was unable to prove the converse: if a number satisfies his criterion, then it must be congruent. This is what we need to conclude that $5, 6$, and 7 are congruent. In these particular cases we can find suitable Pythagorean triples, but that won't help with the general case. Tunnell did show that this converse follows from the Birch–Swinnerton-Dyer conjecture – but that remains unproved.

Like several of the millennium problems, the Birch–Swinnerton-Dyer conjecture is difficult even to state. (You think you can win a million dollars by doing something easy? I can sell you a lovely bridge, dead cheap ...) However, it rewards perseverance, because along the way we start to appreciate the depths, and long historical traditions, of number theory. If you look carefully at the name of the conjecture, one hyphen is longer than the other. It's not something conjectured by mathematicians named Birch, Swinnerton, and Dyer, but by Brian Birch and Peter Swinnerton-Dyer. Its full statement is technical, but it's about a basic issue in Diophantine equations – algebraic equations for which we seek solutions in whole or rational numbers. The question is simple: when do they have solutions?

In chapter 6 on Mordell's conjecture and chapter 7 on Fermat's last theorem we encountered some of the most wonderful gadgets in the whole of mathematics, elliptic curves. Mordell made what at the time was basically a wild guess, and conjectured that the number of rational solutions to an algebraic equation in two variables depends on the topology of the associated complex curve. If the genus is 0 – the curve is topologically a sphere – then the solutions are given by a formula. If the genus is 1 – the curve is topologically a torus, which is equivalent to it being an elliptic curve – then all rational solutions can be constructed from a suitable finite list by applying a natural group structure. If the genus is 2 or more – the curve is topologically a g-holed torus with $g \geqslant 2$ – then the number of solutions is finite. As we saw, Faltings proved this remarkable theorem in 1983.

The most striking feature of rational solutions to elliptic curve equations is that these solutions form a group, thanks to the geometric construction in Figure 28 of chapter 6. The resulting structure is called the Mordell-Weil group of the curve, and number theorists would dearly like to be able to calculate it. That involves finding a system of generators: rational solutions from which all others can be deduced by repeatedly using the group operation. Failing that, we would like at the very least to work out some of the basic features of the group, such as how big it is. Here, much detail is still not understood. Sometimes the group is infinite, so it leads to infinitely many rational solutions, sometimes it isn't, and the number of rational solutions is finite. It would be useful to be able to tell which is which. Indeed, what we would really like to know is the abstract structure of the group.

Mordell's proof that a finite list generates all solutions tells us that the group must be built from a finite group and a lattice group. A lattice group consists of all lists of integers of some fixed finite length. If the length is three, for instance, then the group consists of all lists (m_1, m_2, m_3) of integers, and lists are added in the obvious way:

$$(m_1, m_2, m_3) + (n_1, n_2, n_3) = (m_1 + n_1, m_2 + n_2, m_3 + n_3)$$

The length of the list is called the rank of the group (and geometrically it is the dimension of the lattice). If the rank is 0, the group is finite. If the rank is nonzero, the group is infinite. So to decide how many solutions there are, we don't need the full structure of the group. All we need is its rank. And that's what the Birch–Swinnerton-Dyer conjecture is about.

In the 1960s, when computers were just coming into being, the University of Cambridge had one of the earlier ones, called EDSAC. Which stands for electronic delay storage automatic calculator, and shows how proud its inventors were of its memory system, which sent sound waves along tubes of mercury and redirected them back to the beginning again. It was the size of a large truck, and I vividly remember being shown round it in 1963. Its circuits were based on thousands of valves – vacuum tubes. There were vast racks of the things along the walls, replacements to be inserted when a tube in the machine itself blew up. Which was fairly often.

Peter Swinnerton-Dyer was interested in the Diophantine side of elliptic curves, and in particular he wanted to understand how many solutions there would be if you replaced the curve by its analogue in a finite field with a prime number p of elements. That is, he wanted to study Gauss's trick of working 'modulo p'. He used the computer to calculate these numbers for lots of primes, and looked for interesting patterns.

Here's what he began to suspect. His supervisor John William Scott ('Ian') Cassels was highly sceptical at first, but as more and more data came in, he started to believe there might be something to the idea. What Swinnerton-Dyer's computer experiments suggested was this. Number theorists have a standard method which reinterprets any equation in ordinary integers in terms of integers to some modulus – recall 'clock arithmetic' to the modulus 12 in chapter 2. Because the rules of algebra apply in this version of arithmetic, any solution of the original equation becomes a solution of the 'reduced' equation to that modulus. Because the numbers involved form a finite list – only 12 numbers for clock arithmetic, for example – you can find all solutions by trial and error. In particular, you can count how many solutions there are, for any given modulus. Solutions to any modulus also impose conditions on the original integer solutions, and can sometimes even prove that such solutions exist. So it is a reflex among number theorists to reduce equations using various moduli, and primes are an especially useful choice.

So, to find out something about an elliptic curve, you can consider all primes up to some specific limit. For each prime, you can find how many points lie on the curve, modulo that prime. Birch noticed that Swinnerton-Dyer's computer experiments produce an interesting pattern if you divide the number of such points by the prime concerned. Then multiply all of these fractions together, for all primes less than or equal to a given one, and plot the results against successive primes on logarithmic graph paper. The data all seem to lie close to a straight line, whose slope is the rank of the elliptic curve. This led to a conjectured formula for the number of solutions associated with any prime modulus.[83]

The formula doesn't come from number theory, however: it involves complex analysis, the darling of the 1800s, which by some miracle is far more elegant than old-fashioned real analysis. In chapter 9 on the Riemann hypothesis we saw how analysis puts out tentacles in all directions, in particular having surprising and powerful connections with number theory. Swinnerton-Dyer's formula led to a more detailed conjecture about a type of complex function that I mentioned in chapter 9, called a Dirichlet L-function. This function is an analogue, for elliptic curves, of Riemann's notorious zeta function. The two mathematicians were definitely pushing the boat out, because at that time it wasn't known for sure that all elliptic curves *had* Dirichlet L-functions. It was a wild guess supported by the most tenuous of evidence. But as knowledge of the area grew, it came to appear ever more inspired. It wasn't a wild leap into the unknown: it was a wonderfully accurate, far-sighted stroke of refined mathematical intuition. Instead of standing on the shoulders of giants, Birch and Swinnerton-Dyer had stood on their own shoulders – giants who could hover in mid-air.

A basic tool in complex analysis is to express a function using a power series, like a polynomial but containing infinitely many terms, using bigger and bigger powers of the variable, which in this area is traditionally called s. To find out what a function does near some specific point, say 1, you use powers of $(s - 1)$. The Birch–Swinnerton-Dyer conjecture states that if the power series expansion near 1 of a Dirichlet L-function looks like

$$L(C, s) = c(s - 1)^r + \text{higher-order terms}$$

where c is a nonzero constant, then the rank of the curve is r, and conversely. In the language of complex analysis, this statement takes the form '$L(C, s)$ has a zero of order r at $s = 1$.'

The crucial point here is not the precise expression required: it is that given any elliptic curve, there exists an analytic calculation, using a related complex function, that tells us precisely how many independent rational solutions we have to find to specify them all.

Perhaps the simplest way to demonstrate that the Birch–Swinnerton-Dyer conjecture has genuine content is to observe that the largest known rank is 28. That is, there exists an elliptic curve having a set of 28 rational solutions, from which all rational solutions can be deduced. Moreover, no smaller set of rational solutions does that. Although curves of this rank are known to exist, no explicit examples have been found. The largest rank known for an explicit example is 18. The curve, found by Noam Elkies in 2006, is:

$$y^2 + xy = x^3 - 26175960092705884096311701787701203903556438 \\ 969515x + 51069381476131486489742177100373772089779 \\ 7910325389056784326$$

As it stands this is not in the standard 'y^2 = cubic in x' form, but it can be transformed into that form at the expense of making the numbers even bigger. It is thought that the rank can be arbitrarily large, but this has not been proved. For all we know, the rank can never exceed some fixed size.

Most of what we can prove concerns curves of rank 0 or 1. When the rank is 0, there are finitely many rational solutions. When it is 1, then one specific solution leads to almost all of the rest, with perhaps a finite number of exceptions. These two cases include all elliptic curves of the form $y^2 = x^3 + px$ when p is a prime of the form $8k + 5$ (such as 13, 29, 37, and so on). It is conjectured that in these cases the rank is always 1, which implies that there are infinitely many rational solutions. Andrew Bremner and Cassels have proved this to be true for all such primes up to 1000. It can be tricky to find solutions that lead to almost all others, even when the rank is known, and small. They found that when $p = 877$ the *simplest* solution of this kind is the rational number

$$\frac{375494528127162193105504069942092792346201}{62159877768715054254632207806972238044100}$$

A great many theorems related to the Birch–Swinnerton-Dyer conjective have been proved, usually with very technical assumptions, but progress towards a solution has been relatively slight. In 1976 Coates and Wiles found the first hint that the conjecture might be true. They proved that a special kind of elliptic curve has rank 0 if the Dirichlet L-function does not vanish at 1. For such an elliptic curve, the number of rational solutions to the Diophantine equation is finite, perhaps zero – and you can deduce that from the corresponding L-function. Since then there have been a number of technical advances, still mostly limited to ranks 0 and 1. In 1990 Victor Kolyvagin proved that the Birch–Swinnerton-Dyer conjecture is true for ranks 0 and 1.

More detailed conjectures, with plenty of computer support, relate the constant c in the Birch–Swinnerton-Dyer conjecture to various number-theoretic concepts. There are analogues – equally enigmatic – for algebraic number fields. It is also known, in a precise sense, that most elliptic curves have rank 0 or 1. In 2010 Manjul Bhargava and Arul Shankar announced they had proved that the average rank of an elliptic curve is at most 7/6. If this and a few other recently announced theorems hold up under scrutiny, the Birch–Swinnerton-Dyer conjecture is true for a nonzero proportion of all elliptic curves. However, these are the simplest ones and they don't really represent the curves with a richer structure: rank 2 or more. These are an almost total mystery.

15

Complex cycles
Hodge Conjecture

SOME AREAS OF MATHEMATICS can be related, fairly directly, to everyday events and concerns. We don't encounter the Navier-Stokes equation in our kitchens, but we all understand what fluids are and have a feel for how they flow. Some areas can be related to esoteric questions in frontier science: you may need a PhD in mathematical physics to understand quantum field theory, but analogies with electricity and magnetism, or semi-meaningful images like 'probability wave', go a long way. Some can be explained using pictures: the Poincaré conjecture is a good example. But some defy all of these methods for making difficult abstract concepts accessible.

The Hodge conjecture, stated by the Scottish geometer William Hodge in 1950, is one of them. It's not the proof that causes problems, because there isn't one. It's the statement. Here it is from the Clay Institute's website, in a slightly edited form:

> On any non-singular projective complex algebraic variety, any Hodge class is a rational linear combination of classes of algebraic cycles.

Clearly we have work to do. The only words that make immediate sense are 'on, any, is, a,' and 'of'. Others are familiar as words: 'variety, class, rational, cycle'. But the images these conjure up – choice in the supermarket, a roomful of school children, unemotional thinking, a device with two wheels and handlebars – are obviously not the meanings that the Clay Institute has in mind. The rest are even more evidently jargon. But not jargon for jargon's sake – complicated names for simple things. These are simple names for complicated things. There are no ready-made names for such concepts in ordinary language, so we borrow some and invent others.

Looking on the positive side, we have a real opportunity here (as in 'boy, do we have opportunities'). The Hodge conjecture is arguably more representative of real mathematics, as done by mathematicians of the twentieth and twenty-first centuries, than any other topic in this book. By approaching it in the right way, we gain valuable insight into just how conceptually advanced frontier mathematics really is. Compared to school mathematics, it's like Mount Everest beside a molehill.

Is it all just airy-fairy pretentious nonsense carried out in ivory towers, then? If no ordinary person can understand what it's about, why should anyone hand over good tax money to employ the people who think about such matters? Let me turn that round. Suppose any ordinary person *could* understand everything that mathematicians think about. Would you be happy handing over the tax money then? Aren't they being paid for their expertise? If everything were so easy and comprehensible that it made immediate sense to anyone pulled in at random from the street, what would be the point of having mathematicians? If everybody knew how to drain a central heating system and solder a joint, what would be the point of having plumbers?

I can't show you any killer app that relies on the Hodge conjecture. But I can explain its importance within mathematics. Modern mathematics is a unified whole, so any significant advance, in any core area, will eventually prove its worth in dollars and cents terms. We may not find it in our kitchen today, but tomorrow, who knows? Closely related mathematical concepts are already proving their worth in several areas of science, ranging from quantum physics and string theory to robots.

Sometimes practical applications of new mathematics appear almost instantly. Sometimes it takes centuries. In the latter case, it might seem more cost-effective to wait until the need for such results arises and then run a crash programme to develop them. All mathematical problems that don't have immediate, obvious uses should be put on the back burner until they do. However, if we did that we'd always be behind the game, as mathematics spent a few hundred years playing catch-up with the needs of applied science. And it might not be at all clear which idea we need. Would you be happy if no one started thinking about how to make bricks until you'd hired a builder to begin work on a house? The more original a mathematical concept is, the less likely it would be to emerge from a crash programme.

A better strategy is to let some parts of mathematics develop along their own lines, and stop expecting immediate payoff. Don't try to cherry-pick; allow the mathematical edifice to grow organically. Mathematicians are cheap: they don't need expensive equipment like particle physicists do (Large Hadron Collider: €7.5 billion and counting). They pay their way by teaching students. Allowing a few of them to work part time on the Hodge conjecture, if that's what grabs them, is hardly unreasonable.

I'm going to unpick the statement of the Hodge conjecture, word by word. The easiest concept is 'algebraic variety'. It is a natural consequence of Descartes's use of coordinates to link geometry to algebra, chapter 3. With its aid, the tiny toolkit of curves introduced by Euclid and his successors – line, circle, ellipse, parabola, hyperbola – became a bottomless cornucopia. A straight line, the basis of Euclidean geometry, is the set of points that satisfy a suitable algebraic equation, for example $y = 3x + 1$. Change 3 and 1 to other numbers, and you get other lines. Circles need quadratic equations, as do ellipses, parabolas, and hyperbolas. In principle, anything you can state geometrically can be reinterpreted algebraically, and conversely. Do coordinates make geometry obsolete, then? Do they make algebra obsolete? Why use two tools when each does the same job as the other?

In my toolbox in the garage I have a hammer and a large pair of pincers. The hammer's job is to knock nails into wood, the pincers' job is to pull them out again. In principle, though, I could bash the nails in using the pincers, and the hammer has a claw specifically intended for extracting nails. So why do I need both? Because the hammer is better at some jobs, and the pincers are better at others. It's the same with geometry and algebra: some ways of thinking are more natural using geometry, some are more natural using algebra. It's the link between them that matters. If geometric thinking gets stuck, switch to algebra. If algebraic thinking gets stuck, switch to geometry.

Coordinate geometry provides a new freedom to invent curves. Just write down an equation and look at its solutions. Unless you've chosen a silly equation like $x = x$, you should get a curve. (The equation $x = x$ has the entire plane as its solutions.) For instance, I might write down $x^3 + y^3 = 3xy$, whose solutions are drawn in Figure 45. This curve is the folium of Descartes, and you won't find it in Euclid. The range of new curves that anyone can invent is literally infinite.

Fig 45 The folium of Descartes.

An automatic reflex among mathematicians is generalisation. Once someone has found an interesting idea, we can ask whether anything similar happens in a more general context. Descartes's idea has at least three major generalisations or modifications, all of which are needed to make sense of the Hodge conjecture.

First, what happens if we work with spaces other than the plane? Three-dimensional Euclidean space has three coordinates (x, y, z) instead of two. In space, one equation typically defines a surface. Two equations define a curve, where the corresponding surfaces meet. Three equations typically determine a point. (By 'typically' I mean that sometimes there might be exceptions, but these are very unusual and satisfy special conditions. We saw something similar in the plane with the silly equation $x = x$.)

Again, we can define new surfaces or curves, not found in Euclid, by writing down new equations. In the nineteenth century there was a vogue for doing this. You could publish a new surface if you said something genuinely interesting about it. A typical example is a surface introduced by Kummer in 1864, with the equation

$$x^4 + y^4 + z^4 - y^2z^2 - z^2x^2 - x^2y^2 - x^2 - y^2 - z^2 + 1 = 0$$

Figure 46 shows a picture. The main features of interest are the 16 'double points', where the shape is like two cones joined tip to tip. This is the maximum number possible for a quartic surface, one whose equation has degree 4, and that was interesting enough to merit publication.

Fig 46 Kummer's quartic surface with its 16 double points.

By the nineteenth century, mathematicians had experienced the heady delights of higher-dimensional spaces. There is no need to stop with three coordinates; why not try four, five, six, ... a million? This is not idle speculation. It is the algebra of lots of equations in lots of variables, and those turn up all over the mathematical landscape – for example in chapter 5 on the Kepler conjecture and chapter 8 on celestial mechanics. Nor was it idle generalisation: being able to think about such things geometrically, as well as algebraically, is a powerful tool that should not be restricted to spaces of two or three dimensions, just because that's where you can draw pictures and make models.

The word 'dimension' may sound impressive and mystical, but in this context it has a straightforward meaning: how many coordinates you need. For instance, 4-dimensional space has four coordinates (x, y, z, w), and as far as mathematics is concerned, that defines it. In four dimensions, a single equation typically defines a three-dimensional 'hypersurface', two equations define a surface (two dimensions), three equations define a curve (one dimension), and four equations define a point (zero dimensions). Each new equation gets rid of one dimension – one variable. So we can predict that in space of 17 dimensions, 11 equations define a 6-dimensional object, except for rare (and detectable) cases where some of the equations are superfluous.

An object defined in this way is called an algebraic variety. The word 'variety' arose in languages like French and Spanish, and has a similar meaning to 'manifold' in English: basically, the word 'many'. For reasons lost in the mists of history, 'manifold' became associated with topology and differential geometry – topology combined with calculus – while 'variety' became associated with algebraic geometry.[84] Using different names avoids confusion, so they both stuck. An algebraic variety could have been called a 'multidimensional space defined by a system of algebraic equations', but you can see why no one did that.

A second attractive way to generalise the notions of coordinate geometry is to allow the coordinates to be complex numbers. Recall that the complex number system involves a new kind of number, i, whose square is −1. Why complicate everything in that way? Because algebraic equations are much better behaved over the complex numbers. Over the real numbers, a quadratic equation may have two solutions or none. (It can also have just one, but there is a meaningful sense in which that is the same solution occurring twice.) Over the complex numbers, a quadratic equation *always* has two solutions (again counting multiplicities correctly). For some purposes, this is a far more pleasant property. You can say 'solve the equation for the seventh variable' and be confident that such a solution does actually exist.

Pleasant though it may be in this respect, complex algebraic geometry has features that take a little getting used to. With real variables, a line may cut a circle, or be tangent to it, or miss it altogether. With complex variables, the third option disappears. Once you get used to these changes, however, complex algebraic varieties are much better behaved than real ones. Sometimes real variables are essential, but for most purposes the complex context is a better choice. At any rate, we now know what a complex algebraic variety is.

How about 'projective'? This is the third generalisation, and it requires a slightly different notion of space. Projective geometry arose from Renaissance painters' interest in perspective, and it eliminates the exceptional behaviour of parallel lines. In Euclid's geometry, two straight lines either meet or they are parallel: they don't meet, no matter how far they extend. Now, imagine yourself standing on an infinite plane, paintbrush in hand, easel set up, paintbox at the ready, with a pair of parallel lines heading off towards the distant sunset like infinitely long railway lines. What do you see, and what would you draw? Not two lines that fail to meet. Instead, the lines appear to converge, meeting on the horizon.

What part of the plane does the horizon correspond to? The part where parallel lines meet. But there's no such thing. The horizon is the boundary, on your picture, of the image of the plane. If all's right with the world, that surely ought to be the image of the boundary of the plane. But a plane has no boundary. It goes on for ever. This is all a bit puzzling. It's as if part of the Euclidean plane is missing. If you 'project' a plane (the one with the railway lines) on to another plane (the canvas on the easel) you get a line in the image, the horizon, that isn't the projection of any line in the plane.

There is a way to get rid of this puzzling anomaly: add a so-called line at infinity to the Euclidean plane, representing the missing horizon. Now everything becomes much simpler. Two lines always meet at a point; the old notion of parallel lines corresponds to the case where two lines meet each other at infinity. This idea, suitably interpreted, can be turned into perfectly sensible mathematics. The result is called projective geometry. It's a very elegant subject, and the mathematicians of the eighteenth and nineteenth centuries loved it. Eventually they ran out of things to say, until the mathematicians of the twentieth century decided to generalise algebraic geometry to multidimensional spaces and to use complex numbers. At that point it became clear that we might as well go the whole hog and study complex solutions of systems of algebraic equations in projective space rather than real solutions in Euclidean space.

Let me sum up. A projective complex algebraic variety is like a curve, defined by an algebraic equation, but:

- The number of equations and variables can be whatever we wish (algebraic variety).
- The variables can be complex rather than real (complex).
- The variables can take on infinite values in a sensible way (projective).

While we're at it, there's another term that can easily be dealt with: non-singular. It means that the variety is smooth, with no sharp ridges or places where the shape is more complicated than just a smooth piece of space. Kummer's surface is singular at those 16 double points. Of course we have to explain what 'smooth' means when the variables are complex and some can be infinite, but that's routine technique.

We're almost halfway along the statement of the Hodge conjecture. We know what we're talking about, but not how Hodge thought it ought to behave. Now we have to tackle the deepest and most technical aspects: algebraic cycles, classes, and (especially) Hodge classes. However, I can reveal the general gist straight away. They are technical gadgets that provide a partial answer to a very basic question about our generalised surface: *what shape is it?* The only remaining terms, 'rational linear combination', provide what everyone hopes is the right answer to that question.

See how far we've come. Already we understand what sort of statement the Hodge conjecture is. It tells us that given any generalised surface defined by some equations, you can work out what shape it is by doing some algebra with things called cycles. I could have told you that on the first page of this chapter, but at that stage it wouldn't have made any more sense than the formal statement did. Now that we know what a variety is, everything starts to hang together.

It also starts to sound like topology. 'Finding the shape by doing algebraic calculations' is strikingly reminiscent of Poincaré's ideas about algebraic invariants for topological spaces. So the next step requires a discussion of algebraic topology. Among Poincaré's discoveries were three important types of invariant, defined in terms of three concepts: homotopy, homology, and cohomology. The one we want is cohomology – and of course, wouldn't you just know it, that's the most difficult to explain.

I think we just have to jump in.

In three-dimensional space with real coordinates, a sphere and a plane meet (if they meet at all) in a circle. The sphere is a variety (I'll omit the adjective 'algebraic' when we speak of varieties), the circle is a variety, and the circle is contained in the sphere. We call it a *subvariety*. More generally, if we take the equations (many variables, complex, projective) that define some variety, and add some more equations, then we typically lose some of the solutions: those that fail to satisfy the new equations. The more equations we have, the smaller the variety becomes. The extended system of equations defines some part of the original variety, and this part is a variety in its own right – a subvariety.

When we count the number of solutions of a polynomial equation, it can be convenient to count the same point more than once. From this point of view, the set of solutions consists of a number of points, to each of which we 'attach' a number, its multiplicity. We might, for instance, have solutions 0, 1, and 2, with multiplicities 3, 7, and 4 respectively. The polynomial would then be $x^3(x-1)^7(x-2)^4$, if you want to know. Each of the three points $x = 0$, 1, or 2 is a (rather trivial) subvariety of the complex numbers. So the solutions of this polynomial can be described as a list of three subvarieties, with a whole number attached to each like a label.

An algebraic cycle is similar. Instead of single points, we use any finite list of subvarieties. To each of them we can attach a numerical label, which need not be a whole number. It could be a negative integer, it could be a rational number, it could be a real or even a complex number. For various reasons, the Hodge conjecture uses rational numbers as labels. This is what 'rational linear combination' refers to. So, for example, our original variety might be the unit sphere in 11-dimensional space, and this list might look like this:

A 7-dimensional hypersphere (with equations such and such) with label 22/7

A torus (with equations such and such) with label −4/5

A curve (with equations such and such) with label 413/6

Don't try to picture this, or if you do, think like a cartoonist: three squiggly blobs with little labels. Each such cartoon, each list, constitutes one algebraic cycle.

Why go to such fuss and bother to invent something so abstract? Because it captures essential aspects of the original algebraic variety. Algebraic geometers are borrowing a trick from topologists.

In chapter 10 on the Poincaré conjecture we thought about an ant whose universe is a surface. How can the ant work out what shape its universe is, when it can't pop outside and take a look? In particular, how can it distinguish a sphere from a torus? The solution presented there involved closed loops – topological bus trips. The ant pushes these loops around, finds out what happens when they are joined end to end, and computes an algebraic invariant of the space called its fundamental group. 'Invariant' means that topologically equivalent spaces have the same fundamental group. If the groups are different, so are the spaces. This is the invariant that led Poincaré to his conjecture. However, it's not easy for the poor ant to examine all possible loops in his universe, and this remark reflects genuine mathematical subtleties in the calculation of the fundamental group. A more practical invariant exists, and Poincaré investigated this as well. Pushing loops around is called homotopy. This alternative has a similar name: homology.

I'll show you the simplest, most concrete version of homology. Topologists quickly improved on this version, streamlined it, generalised it, and turned it into a huge mathematical machine called homological algebra. This simple version gives the barest flavour of how the topic goes, but that's all we need.

The ant starts by surveying its universe to make a map. Like a human surveyor, it covers its universe with a network of triangles. The crucial condition is that no triangle should surround a hole in the surface, and the way to ensure that is to create the triangles by slapping rubber patches on to the surface, like someone mending a bicycle tyre. Then each triangle has a well-defined interior that is topologically the same as the interior of an ordinary triangle in the plane. Topologists call such a patch a topological disc, because it is also equivalent to a circle and its interior. To see why, look at Figure 36 in chapter 10, where a triangle is deformed continuously into a circle. It's not possible to fit a patch of this kind to a triangle surrounding a hole, because the hole creates a tunnel that links the inside of the triangle to its outside. The patch would have to leave the surface, and the ant isn't allowed to do that.

The ant has now created a *triangulation* of its universe. The condition about patches ensures that the topology of the surface – its shape, in the sense of topological equivalence – can be reconstructed if all you know is the list of triangles, together with which triangles are adjacent to which. If you went to Ikea and bought an Ant Universe flatpack with suitably labelled triangles, and then glued edge A to edge AA, edge B to edge BB, and so on, you would be able to build the surface. The ant is confined to the surface, so it can't make a model, but it can be sure that in principle its map contains the information it needs. To extract that information, it has to perform a calculation. When doing so, the ant no longer has to contemplate the infinitude of all possible loops, but it does have to contemplate quite a lot of them: all closed loops that run along edges of its chosen network.

In homotopy, we ask whether a given loop can be shrunk continuously to a point. In homology, we ask a different question: does the loop form the boundary of a topological disc? That is, can you fit one or more triangular patches together so that the result is a region without any holes, and the boundary of this region is the loop concerned?

Figure 47 (left) shows part of a triangulation of a sphere, a closed loop, and the topological disc whose boundary it is. By setting up the right techniques, it can be proved that *any* loop in a triangulation of a sphere is a boundary: triangular patches, and more generally topological discs, are hole-detectors, and intuitively a sphere has no holes. However, a torus does have a hole, and indeed some loops on a torus are not boundaries. Figure 47 (right) shows such a loop, winding through the central hole. In other words: by running through a list of loops and finding out which of them are boundaries, the ant can distinguish a spherical universe from a toroidal one.

If the ant is as clever as Poincaré and the other topologists of his day, it can turn this idea into an elegant topological invariant, the homology group of its surface. The basic idea is to 'add' two loops by drawing both of them. However, that's not a loop, so we have to go back to the beginning and start again. Right back to the beginning, in fact; back to the days when we were first introduced to algebra. My mathematics

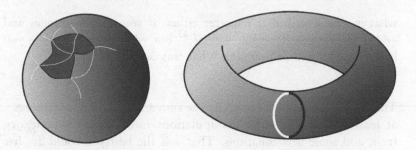

Fig 47 *Left*: Part of a triangulation of a sphere, a closed loop (black lines), and the disc whose boundary it is (dark shading). *Right*: Loop on a torus that is not the boundary of a disc (lighter part is at the back).

teacher started by pointing out that you can add a number of apples to a number of apples and get a total number of apples. But you can't add apples to oranges unless you count everything as fruit.

That's true in arithmetic, though even there you have to be careful not to use the same apple twice, but it's not true in algebra. There, you can add apples to oranges, while keeping them distinct. In fact, in advanced mathematics it is commonplace to add together things that you might imagine no one in their right mind would have invented at all, let alone want to add together. The freedom to do this kind of thing turns out to be amazingly useful and important, and the mathematicians who did it weren't mad after all – at least, not in that respect.

To understand some of the ideas that the Hodge conjecture brings together, we have to be able to add apples and oranges without lumping them all together as plain fruit. The way to add them is not actually very difficult. What's difficult is to accept that there's any point in doing so. Many of us have met a version of this potential conceptual block already. My teacher told the class that the letters stood for unknown numbers, with different letters for different unknowns. If you had a apples and another a apples, the total number of apples was $a + a = 2a$. And it worked whatever the number of apples might be. If you had $3a$ apples and added $2a$ apples, the result was $5a$, whatever the number of apples might be. The symbol, and what it represented, didn't matter either: if you had $3b$ oranges and added $2b$ oranges, the result was $5b$.[85] But what happened when you had $3a$ apples and $2b$ oranges? What was $3a$ plus $2b$?

$$3a + 2b.$$

That was it. You couldn't simplify the sum and make it 5 somethings – at least not without some manipulations involving a new category, fruit, and some new equations. That was the best you could do: live with it. However, once you took that step, you could do sums like

$$(3a + 2b) + (5a - b) = 8a + b$$

without introducing any new ideas. Or new kinds of fruit.

There were some caveats. I've already noted that if you add one apple to one apple, you only get two apples if the second apple is different from the first. The same goes for more complicated combinations of apples and oranges. Algebra assumes that for the purpose of doing the sums, all apples involved are different. In fact, it is often sensible to make this assumption, even in cases where two apples – or whatever else we are adding – might actually be the same. One apple plus the same apple is an apple with multiplicity two.

Once you get used to this idea, you can use it for anything. A pig plus the same pig is that pig with multiplicity two: pig + pig = 2 pig whatever pig is. A pig plus a cow is pig + cow. A triangle plus three circles is triangle + 3 circle. A superdupersphere plus three hyperelliptic quasiheaps is

superdupersphere + 3 hyperellipticquasiheap

whatever the jargon means (which, here, is nothing).

You can even allow negative numbers, and talk of three pigs minus eleven cows: 3 pig −11 cow. I have no idea what minus eleven cows look like, but I can be confident that if I add six cows to that, I've got minus five cows.[86] It's a formal game played with symbols, and no more realistic interpretation is needed, useful, or – often – possible. You could allow real numbers: π pigs minus $\sqrt{2}$ cows. Complex numbers. Any kind of fancy number that any mathematician has ever invented or will invent in future. The idea can be made a little more respectable if you think of the numbers as *labels*, attached to the pigs and cows. Now π pigs minus $\sqrt{2}$ cows can be thought of as a pig labelled π together with a cow labelled $-\sqrt{2}$. The arithmetic applies to the labels, not to the animals.

The Hodge conjecture involves this kind of construction, with extra bells and whistles. In place of animals, it uses curves, surfaces, and their higher-dimensional analogues. Strange as it may seem, the result is not just abstract nonsense, but a profound connection between topology, algebra, geometry, and analysis.

To set up the formalism of homology we want to add loops together, but not the way we did for the fundamental group. Instead, we do it the way my teacher told me. Just write the loops down and put a + sign in between. To make sense of that, we work not with single loops, but with finite sets of them. We label each loop with an integer that counts how often it occurs. Call such a labelled set a *cycle*. Now the ant can add any two cycles together by lumping them together and adding the corresponding labels, and the result is another cycle. Perhaps I should have used bicycles, not buses, as my image for the ant's travels in chapter 10.

When we were constructing the fundamental group, where 'addition' joins loops end to end, there was a technical snag. Adding the trivial loop to a loop didn't *quite* give the same loop, so the zero loop misbehaved. Adding a loop to its reversal didn't quite give the trivial loop, so inverses didn't behave correctly. The way out was to consider loops to be the same if one could be deformed into the other.

For homology, that's not the problem. There is a zero cycle (all labels zero), and every cycle does have an inverse (turn every label into its negative), so we do get a group. The trouble is, it's the wrong group. It tells us nothing about the topology of the space. To sort that out we use a similar trick, and take a more relaxed view of which cycles should count as zero. The ant cuts the space into triangular patches, and the boundary of each patch is topologically rather trivial: you can shrink it down to a point by pushing it all to the middle of its patch. So we require these boundary cycles to be equivalent to the zero cycle. It's a bit like turning ordinary numbers into clock arithmetic by pretending that the number 12 is irrelevant, so it can be set to zero. Here we turn cycles into homology by pretending that any boundary cycle is irrelevant.

The consequences of this pretence are dramatic. Now the algebra of cycles is affected by the topology of the space. The group of cycles modulo boundaries is a useful topological invariant, the homology group of the surface. At first sight it depends on which triangulation the ant chose, but as for the Euler characteristic, different triangulations of the same surface lead to the same homology group. So the ant has invented an algebraic invariant that can distinguish different surfaces. It's a bit fiddly, but you never get good invariants without doing some hard work somewhere along the line. This one is so effective that it can distinguish not just sphere from torus, but a 2-holed torus from a 5-holed torus, and similarly for any other numbers of holes.

Homology may seem a bit of a mouthful, but it started a rich vein of topological invariants, and it is based on simple geometric ideas: loops, boundaries, lumping sets together, doing arithmetic with labels. Considering that the poor ant is confined to its surface, it's astonishing that the creature can find out anything significant about the shape of its universe just by slapping down triangular patches, making a map, and doing some algebra.

There is a natural way to extend homology to higher dimensions. The 3-dimensional analogue of a triangle is a tetrahedron; it has 4 vertexes, 6 edges, 4 triangular faces, and a single 3-dimensional 'face', its interior. More generally, in n dimensions we can define an n-simplex with $n + 1$ vertexes, joined in pairs by all possible edges, which in turn form triangles, which assemble to create tetrahedrons, and so on. It is now easy to define cycles, boundaries, and homology, and again we can concoct a group by adding (homology classes of) cycles. In fact, we now get a whole series of groups: one for 0-dimensional cycles (points), one for 1-dimensional cycles (lines), one for 2-dimensional cycles (triangles), and so on, all the way up to the dimension of the space itself. These are the 0th, 1st, 2nd, and so on, homology groups of the space. Roughly speaking, they make precise the notion of holes, of various dimensions, in the space: do they exist, how many are there, and how do they relate to each other?

That, then, is homology, and it's almost what we need to

understand what the Hodge conjecture *says*. However, what we actually need is a closely related concept called *cohomology*. In 1893 Poincaré noticed a curious coincidence in the homology of any manifold: the list of homology groups reads the same in reverse. For a manifold of dimension 5, say, the 0th homology group is the same as the 5th, the 1st homology group is the same as the 4th, and the 2nd homology group is the same as the 3rd. He realised that this couldn't just be coincidence, and he explained it in terms of the dual of a triangulation, which we met in chapter 4 in connection with maps. This is a second triangulation in which each triangle is replaced by a vertex, each edge between two triangles by an edge that links the corresponding new vertexes, and each point by a triangle, as in Figure 9 of chapter 4. Notice how the dimensions appear in reverse order: 2-dimensional triangles become 0-dimensional points, and conversely; 1-dimensional edges remain 1-dimensional because 1 is in the middle.

It turns out to be useful to distinguish the two lists, even though they yield the same invariants. When the whole setup is generalised and formulated in abstract terms, triangulations disappear, and the dual triangulation no longer makes sense. What survives are two series of topological invariants, called homology groups and cohomology groups. Every concept in homology has a dual, usually named by adding 'co' at the front. So in place of cycle we have cocycles, and in place of two cycles being homologous we have two cocycles being cohomologous. The classes referred to in the Hodge conjecture are cohomology classes, and these are collections of cocycles that are cohomologous to each other.

Homology and cohomology don't tell us everything we would like to know about the shape of a topological space – distinct spaces can have the same homology and cohomology – but they do provide a lot of useful information, and a systematic framework in which to calculate it and use it.

An algebraic variety, be it real, complex, projective, or not, is a topological space. Therefore it has a shape. To find out useful things about the shape, we think like topologists and calculate the homology and cohomology groups. But the natural ingredients in algebraic geometry aren't geometric objects like triangulations and cycles. They are the things we can most easily describe by algebraic equations. Go back and look at the equation for Kummer's surface. How would that relate to a triangulation? There's nothing in the formula that hints at triangles.

Maybe we need to start again. Instead of triangles, we should use the natural building blocks for varieties, which are subvarieties, defined by imposing extra equations. Now we have to redefine cycles: instead of sets of triangles with integer labels, we use sets of subvarieties with whatever labels do the best job. For various reasons – mostly that the Hodge conjecture is false if we use integer labels – rational numbers are the sensible choice. Hodge's question boils down to this: does this new definition of homology and cohomology capture everything that the topological definition does? If his conjecture is true, then the algebraic cycle tool is sharp enough to match the cohomological chisel of topology. If it's false, then the algebraic cycle is a blunt instrument.

Except ... sorry, I've over-egged the pudding. The conjecture says that it is enough to use a particular *kind* of algebraic cycle, one that lives in a Hodge class. To explain that, we need yet another ingredient in an already rich mixture: analysis. One of the most important concepts in analysis is that of a differential equation, which is a condition about the rates at which variables change, chapter 8. Nearly all of the mathematical physics of the eighteenth, nineteenth, and twentieth centuries models nature using differential equations, and even in the twenty-first, most does. In the 1930s this idea led Hodge to a new body of technique, now called Hodge theory. It ties in naturally with a lot of other powerful methods in the general area of analysis and topology.

Hodge's idea was to use a differential equation to organise the cohomology classes into distinctive types. Each piece has extra structure, which can be used to advantage in topological problems. The pieces are defined using a differential equation that appeared in the late 1700s, notably in the work of Pierre-Simon de Laplace. Accordingly, it is called the Laplace equation. Laplace's main research was in celestial mechanics, the motion and form of planets, moons, comets, and stars. In 1783 he was working on the detailed shape of the Earth. By then it was known that the Earth is not a sphere, but it is flattened at the poles to form an oblate spheroid – like a beachball that someone is sitting on. But even that description misses some of the fine detail. Laplace found a method to calculate the shape to any required accuracy, based on a physical quantity that represents the Earth's gravitational field: not the field itself, but its gravitational potential. This is a measure of the energy contained in gravitation, a numerical quantity defined at each point in space. The force of gravity acts in whichever direction makes the potential decrease at the fastest rate, and the magnitude of the force is the rate of decrease.

The potential satisfies Laplace's equation: roughly speaking, this says that in the absence of matter – that is, in a vacuum – the average value of the potential over a very small sphere is equal to its value at the centre of the sphere. It's a kind of democracy: your value is the average of the values of your neighbours. Any solution of Laplace's equation is called a harmonic function. Hodge's special types of cohomology class are those that bear a particular relationship to harmonic functions. Hodge theory, the study of these types, opened up a deep and wonderful area of mathematics: relations between the topology of a space and a special differential equation on that space.

So now you have it. The Hodge conjecture postulates a deep and powerful connection between three of the pillars of modern mathematics: algebra, topology, and analysis. Take any variety. To understand its shape (topology, leading to cohomology classes) pick out special instances of these (analysis, leading to Hodge classes by way of differential equations). These special types of cohomology class can be realised using subvarieties (algebra: throw in some extra equations and look at algebraic cycles). That is, to solve the topology problem 'what shape is this thing?' for a variety, turn the question into analysis and then solve that using algebra.

Why is that important? The Hodge conjecture is a proposal to add two new tools to the algebraic geometer's toolbox: topological invariants and Laplace's equation. It's not really a conjecture about a mathematical theorem; it's a conjecture about new kinds of tools. If the conjecture is true, those tools immediately acquire new significance, and can potentially be used to answer an endless stream of questions. Of course, it might turn out to be false. That would be disappointing, but it's better to understand a tool's limitations than to keep hitting your thumb with it.

Now that we appreciate the nature of the Hodge conjecture, we can look at the evidence for it. What do we know? Precious little.

In 1924, before Hodge made his conjecture, Solomon Lefschetz proved a theorem which boils down to the Hodge conjecture for the dimension-2 cohomology of any variety. With a bit of routine algebraic topology, this implies the Hodge conjecture for varieties of dimension 1, 2, and 3. For higher-dimensional varieties, only a few special cases of the Hodge conjecture are known.

Hodge originally stated his conjecture in terms of integer labels. In 1961 Michael Atiyah and Friedrich Hirzebruch proved that in higher dimensions, this version of his conjecture is false. So today we interpret Hodge's conjecture using rational labels. For this version, there is a certain amount of encouraging evidence. The strongest evidence in its favour is that one of its deeper consequences, an even more technical theorem known as 'algebraicity of Hodge loci', has been proved – *without* assuming the Hodge conjecture. Eduardo Cattani, Pierre Deligne, and Aroldo Kaplan found such a proof in 1995.

Finally, there is an attractive conjecture in number theory that is analogous to the Hodge conjecture. It is called the Tate conjecture, after John Tate, and it links algebraic geometry to Galois theory, the circle of ideas that proved that there is no algebraic formula to solve polynomial equations of degree 5. Its formulation is technical, and it involves yet another version of cohomology. There are independent reasons to hope that the Tate conjecture might be true, but its status is currently open. But at least there is a sensible relative of the Hodge conjecture, even if it currently seems equally intractable.

The Hodge conjecture is one of those annoying mathematical assertions for which the evidence either for or against it is not very extensive, and not particularly convincing. There is a definite danger that the conjecture could be wrong. Perhaps there is a variety with a million dimensions that disproves the Hodge conjecture, for reasons that boil down to series of structureless calculations, so complicated that no one could ever carry them out. If so, the Hodge conjecture could be false for essentially silly reasons – it just happens not to be true – but virtually impossible to disprove. I know some algebraic geometers who suspect just that. If so, those million dollars will be safe for the foreseeable future.

16
Where next?

PREDICTION IS VERY DIFFICULT, especially about the future,[87] as Nobel-winning physicist Niels Bohr and baseball player and team manager Yogi Berra are supposed to have said.[88] Mind you, Berra also said: 'I never said most of the things I said.' Allegedly. Arthur C. Clarke, famous for his science fiction and the movie *2001: A Space Odyssey* and its sequels, was also a futurologist: he wrote books predicting the future of technology and society. Among the many predictions in his 1962 *Profiles of the Future* are:

Understanding the languages of whales and dolphins by 1970
Fusion power by 1990
Detection of gravity waves by 1990
Colonising planets by 2000

None of these has yet happened. On the other hand, he had some successes:

Planetary landings by 1980 (though he may have meant human landings)
Translating machines by 1970 (a bit premature, but they now exist on Google)
Personal radio by 1990 (mobile phones work like that)

He also predicted that we would have a global library by 2000, and this may be nearer the mark than we thought a few years ago, because this is one of the many functions of the Internet. With the advent of cloud computing, we may all end up using the same giant computer.

He missed some of the most important trends, such as the rise of the computer and genetic engineering, though he did predict this for 2030. With Clarke's uneven record as a warning, it would be foolhardy to predict the future of great mathematical problems in any detail. However, I can make some educated guesses, safe in the knowledge that most of them will turn out to be wrong.

In the introduction I mentioned Hilbert's 1900 list of 23 big problems. Most are now solved, and his brave war-cry 'We must know, we shall know' may seem vindicated. However, he also said 'in mathematics there is no *ignorabimus* [we shall be ignorant]' and Kurt Gödel knocked that idea firmly on the head with his incompleteness theorem: some mathematical problems may not have a solution within the usual logical framework of mathematics. Not just be impossible, like squaring the circle: they can be undecidable, meaning that no proof exists and no disproof exists. Possibly this could be the fate of some of the currently unsolved great problems. I'd be surprised if the Riemann hypothesis were like that, and amazed if anyone could prove it to be undecidable even if it were. On the other hand, the P/NP problem could well turn out to be undecidable, or to satisfy some other technical variation on the theme of 'it can't be done.' It has that kind of – well, *smell*.

I suspect that by the end of the twenty-first century we will have proofs of the Riemann hypothesis, the Birch–Swinnerton-Dyer conjecture, and the mass gap hypothesis, along with disproofs of the Hodge conjecture and the regularity of solutions of the Navier-Stokes equation in three dimensions. I expect P/NP still to be unsolved in 2100, but to succumb some time in the twenty-second century. So of course someone will disprove the Riemann hypothesis tomorrow and prove P is different from NP next week.

I'm on safer ground with general observations, because we can learn from history. So I'm reasonably confident that by the time the seven millennium problems have been solved, many of them will be seen as minor historical curiosities. 'Oh, they used to think *that* was important, did they?' This is what happened to some of the problems on Hilbert's hit-list. I can also be confident that within 50 years several major areas of mathematics that don't exist today will have come into being. It will then transpire that a few basic examples and some rudimentary theorems in these areas existed long before, but no one realised that these isolated snippets were clues to deep and important new areas. This is what happened with group theory, matrix algebra, fractals, and chaos. I don't doubt that it will happen again, because it's one of the standard ways in which mathematics develops.

These new areas will arise through two main factors. They will emerge from the internal structure of mathematics itself, or they will be responses to new questions about the outside world – often both in tandem. Like Poincaré's three-step process to problem-solving – preparation, incubation, and illumination – the relation between mathematics and its applications is not a single transition: science poses a problem, mathematics solves it, done. Instead, we find an intricate network of trade in questions and ideas, as new mathematics sparks further experiments or observations or theories, which in turn motivate new mathematics. And each node of this network turns out, on closer examination, to be a smaller network of the same kind.

There is more outside world than there used to be. Until recently, the main external source of inspiration for mathematics was the physical sciences. A few other areas played their part: biology and sociology influenced the development of probability and statistics, and philosophy had a big effect on mathematical logic. In the future we will see growing contributions from biology, medicine, computing, finance, economics, sociology, and very possibly politics, the movie industry, and sport. I suspect that some of the first new great problems will arise from biology, because that link is now firmly in place. One trend is an explosion in our ability to gather biological and biochemical data; small genomes can now be sequenced using a device the size of a memory stick, based on nanopore technology, for instance. Big genomes will rapidly follow using this or different technology, most of it already in existence.

These developments are potential game-changers, but we need to have better methods for understanding what the data imply. Biology isn't really about data as such. It is about processes. Evolution is a process, and so are the division of a cell, the growth of an embryo, the onset of cancer, the movement of a crowd, the workings of the brain, and the dynamics of the global ecosystem. The best way we currently know to take the basic ingredients of a process, and deduce what it does, is mathematics. So there will be great problems of new kinds – how dynamics unfolds in the presence of complex but specific organising information (DNA sequences); how genetic changes conspire with environment to constrain evolution; how rules for cell growth, division, mobility, stickiness, and death give developing organisms their shape; how the flow of electrons and chemicals in a network of nerve cells determines what it can perceive or how it will act.

Computing is another source of new mathematics that already has a track record. It is usually thought of as a tool for doing mathematics, but mathematics is equally a tool for understanding and structuring computations. This two-way trade is becoming increasingly important to the health and development of both areas, and they may even merge at some point in the future. Some mathematicians feel they should never have been allowed to split. Among the many trends visible here, the question of very large data sets again springs to mind. It relates not just to the DNA example mentioned earlier, but to earthquake prediction, evolution, the global climate, the stock market, international finance, and new technologies. The problem is to use large quantities of data to test and refine mathematical models of the real world, so that they give us genuine control over very complex systems.

The prediction about which I am most confident is in some ways negative, but it is also an affirmation of the continuing creativity of the mathematical community. All research mathematicians feel, from time to time, that their subject has a mind of its own. Problems work out the way mathematics wants them to, not how mathematicians want them to. We can choose what questions to ask, but we can't choose what answers we get. This feeling relates to two major schools of thought about the nature of mathematics. Platonists think that the 'ideal forms' of mathematics have some kind of independent existence, 'out there' in some realm distinct from the physical world. (There are subtler ways to say that which probably sound more sensible, but that's the gist.) Others see mathematics as a shared human construct. But unlike most such things – the legal system, money, ethics, morality – mathematics is a construct with a strong logical skeleton. There are severe constraints on what assertions you can or cannot share with everyone else. It is these constraints that give the impression that mathematics has its own agenda, and create the feeling in the minds of mathematicians that mathematics itself *exists* outside the domain of human activity. Platonism, I think, is not a description of what mathematics is. It is a description of what mathematics *feels like* when you are doing it. It is like the vivid sensation of 'red' that we experience when we see a rose, blood, or a traffic-light. Philosophers call these sensations qualia (singular: quale), and some of them think that our sensation of free will is actually a quale of the brain's way of making decisions. When we decide between alternatives, we feel that we have a genuine choice – whether or not the dynamic of the brain is actually deterministic in some sense. Similarly, Platonism is a quale of taking part in a shared human construct within a rigid framework of logical deduction.

So mathematics can seem to have a mind of its own, even though it is created by a collective of human minds. History tells us that the mathematical mind – in this sense – is more innovative and surprising than any single human mind can predict. All of which is a complicated way of getting to my main point: one thing we can safely predict about the future of mathematics is that it will be unpredictable. The most important mathematical questions of the next century will emerge as natural, perhaps even inevitable, consequences of our growing understanding of what we currently believe to be the great problems of mathematics. However, they will almost certainly be questions that we cannot currently conceive of. That is only right and proper, and we should celebrate it.

17

Twelve for the future

I don't want to leave you with the impression that most mathematical problems have been solved, aside from the odd really difficult one. Mathematical research is like exploring a new continent. As the area that we know about expands, the frontier that borders the unknown gets longer. I'm not suggesting that the more mathematics we discover, the less we know; I'm saying that the more mathematics we discover, the more we realise that we don't know. But what we don't know changes as time passes, with some old problems disappearing while new ones are added. In contrast, what we know just gets bigger – barring the occasional lost document.

To give you a tiny indication of what we currently *don't* know, aside from the great problems already discussed, here are twelve unsolved problems that have been baffling the world's mathematicians for quite a while. I've chosen them so that it is easy to understand the questions. As has amply been demonstrated, that carries no implications about how easy it may be to find the answers. Some of these problems may turn out to be great: that will mainly depend on the methods invented to solve them and what they lead to, not on the answer as such.

Brocard's Problem

For any whole number n, its factorial $n!$ is the product

$$n \times (n-1) \times (n-2) \times \cdots \times 3 \times 2 \times 1$$

This is the number of different ways to arrange n objects in order. For instance, the English alphabet with 26 letters can be arranged in

$$26! = 403,291,461,126,605,635,584,000,000$$

different orders. In articles written in 1876 and 1885, Henri Brocard noted that

$$4! + 1 = 24 + 1 = 25 = 5^2$$
$$5! + 1 = 120 + 1 = 121 = 11^2$$
$$7! + 1 = 5040 + 1 = 5041 = 71^2$$

are all perfect squares. He found no other factorials which became square when increased by 1, and asked whether any existed. The self-taught Indian genius Srinivasa Ramanujan independently asked the same question in 1913. Bruce Berndt and William Galway used a computer in 2000 to show that no further solutions exist for factorials of numbers up to 1 billion.

Odd Perfect Numbers

A number is perfect if it is equal to the sum of all of its proper divisors (that is, numbers that divide it exactly, excluding the number itself). Examples include

$$6 = 1 + 2 + 3$$
$$28 = 1 + 2 + 4 + 7 + 14$$

Euclid proved that if $2^n - 1$ is prime, then $2^{n-1}(2^n - 1)$ is perfect. The above examples correspond to $n = 2, 3$. Primes of this form are called Mersenne primes, and 47 of them are known, the largest to date being $2^{43,112,609} - 1$, also the largest known prime.[89] Euler proved that all even perfect numbers must be of this form, but no one has ever found an odd perfect number, or proved that they cannot exist. Pomerance has devised a non-rigorous argument which indicates that they don't. Any odd perfect number must satisfy several stringent conditions. It must be at least 10^{300}, it must have a prime factor greater than 10^8, its second largest prime factor must be at least 10^4, and it must have at least 75 prime factors and at least 12 distinct prime factors.

Collatz Conjecture

Take a whole number. If it is even, divide by 2. If it is odd, multiply by 3 and add 1. Repeat indefinitely. What happens?

For example, start with 12. Successive numbers are

$$12 \rightarrow 6 \rightarrow 3 \rightarrow 10 \rightarrow 5 \rightarrow 16 \rightarrow 8 \rightarrow 4 \rightarrow 2 \rightarrow 1$$

after which the sequence $4 \rightarrow 2 \rightarrow 1 \rightarrow 4 \rightarrow 2 \rightarrow 1$ repeats for ever. The Collatz conjecture states that the same end result occurs no matter which number you start with. It is named after Lothar Collatz who came up with it in 1937, but it has many other names: $3n + 1$ conjecture, hailstone problem, Ulam conjecture, Kakutani's problem, Thwaites conjecture, Hasse's algorithm, and Syracuse problem.

What makes the problem difficult is that the numbers can often explode. For instance, if we start with 27 then the sequence rises to 9232; even so, it finally gets down to 1 after 111 steps. Computer simulations verify the conjecture for all initial numbers up to 5.764×10^{18}. It has been proved that no cycles other than $4 \rightarrow 2 \rightarrow 1$ exist that involve fewer than 35,400 numbers. The possibility that some initial number leads to a sequence that contains ever-larger numbers, separated by smaller ones, has not been ruled out. Ilia Krasikov and Jeffrey Lagarias have proved that for initial values up to n, at least a contant times $n^{0.84}$ of them eventually get to 1. So exceptions, if they exist, are rare.[90]

Existence of Perfect Cuboids

This takes as its starting point the existence of, and formula for, Pythagorean triples, and it moves the problem into the third dimension. An Euler brick is a cuboid – a brick-shaped block – with integer sides, all of whose faces have integer diagonals. The smallest Euler brick was discovered in 1719 by Paul Halcke. Its edges are 240, 117, and 4; the face diagonals are 267, 244, and 125. Euler found formulas for such bricks, analogous to the formula for Pythagorean triples, but these do not give all solutions.

It is not known whether a perfect cuboid exists: that is, whether there is an Euler brick whose main diagonal, cutting through the interior of the brick from one corner to an opposite one, also has integer length. (There are four such diagonals but they all have the same length.) It is known that Euler's formulas cannot provide an example. Such a brick, if it exists, must satisfy several conditions – for instance, at least one edge must be a multiple of 5, one must be a multiple of 7, one must be a multiple of 11, and one must be a multiple of 19. Computer searches have shown that one of the sides must be at least one trillion.

There are some near-misses. The brick with sides 672, 153, and 104 has an integer main diagonal and two of the three lengths for face diagonals are also integers. In 2004 Jorge Sawyer and Clifford Reiter proved that perfect parallelepipeds exist.[91] A parallelepiped (the word comes from parallel-epi-ped, but is often misspelt 'parallelopiped') is like a cuboid but its faces are parallelograms. So it's tilted. The edges have lengths 271, 106, and 103; the minor face diagonals have lengths 101, 266, and 255; the major face diagonals have lengths 183, 312, and 323; and the body diagonals have lengths 374, 300, 278, and 272.

Lonely Runner Conjecture

This comes from an abstruse area of mathematics known as Diophantine approximation theory, and was formulated by Jörg Wills in 1967. Luis Goddyn coined the name in 1998. Suppose that n runners run round a circular track of unit length at uniform speed, with each runner's speed being different. Will every runner be lonely – that is, be more than a distance $1/n$ from all other runners – at some instant of time? Different times for different runners, of course. The conjecture is that the answer is always 'yes', and it has been proved when $n = 4, 5, 6,$ and 7.

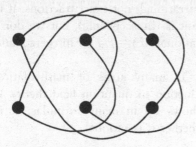

Fig 48 Example of a thrackle.

Conway's Thrackle Conjecture

A thrackle is a network drawn in the plane so that every two edges meet exactly once, Figure 48. They may either meet at a common dot (node, vertex) or they may cross at interior points, but not both. If they cross, they must do so transversely; that is, neither can remain on the same side of the other (which could happen if, for example, they are tangent to each other). In unpublished work, John Horton Conway conjectured that in any thrackle the number of lines is less than or equal to the number of dots. In 2011 Radoslav Fulek and János Pach proved that every thrackle with n dots has at most $1.428n$ lines.[92]

Irrationality of Euler's Constant

There is no known 'closed form' formula for the sum of the harmonic series

$$H_n = 1 + \frac{1}{2} + \frac{1}{3} + \frac{1}{4} + \frac{1}{5} + \cdots + \frac{1}{n}$$

and probably no such formula exists. However, there is an excellent approximation: as n increases, H_n gets ever closer to $\log n + \gamma$. Here γ is Euler's constant, with a numerical value of roughly 0.5772156649. Euler established this formula in 1734, and Lorenzo Mascheroni studied the constant in 1790. Neither used the symbol γ.

Euler's constant is one of those strange numbers that sometimes arise in mathematics, like π and e, which appear all over the place, but seem to be creatures of their own, not expressible in any nice manner in terms of simpler numbers. We saw in chapter 3 that both π and e are transcendental: they do not solve any algebraic equations with integer coefficients. In particular, they are irrational: not exact fractions. It is widely thought that Euler's constant is transcendental, but we don't even know for sure whether it is irrational. If $\gamma = p/q$ for integers p and q, then q is at least $10^{242,080}$.

Euler's constant is important in many areas of mathematics, ranging from the Riemann zeta function to quantum field theory. It appears in many contexts and shows up in many formulas. It is outrageous that we can't decide whether it is rational.

Real Quadratic Number Fields

In chapter 7 we saw that some algebraic number fields have unique prime factorisation and some do not. The best understood algebraic number fields are the quadratic ones, obtained by taking the square root of some number d that is not a perfect square; indeed, has no square prime factors. The corresponding ring of algebraic integers then consists of all number of the form $a + b\sqrt{d}$, where a and b are integers if d is not of the form $4k + 1$, and they are either integers, or are both odd integers divided by 2, if d is of that form.

When d is negative, it is known that prime factorisation is unique for exactly nine values: -1, -2, -3, -7, -11, -19, -43, -67, and -163. Proving uniqueness in these cases is relatively straightforward, but finding whether there are any others is much harder. In 1934 Hans Heilbronn and Edward Linfoot showed that at most one more negative integer can be added to the list. Kurt Heegner gave a proof that the list is complete in 1952, but it was thought to have a gap. In 1967 Harold Stark gave a complete proof, observing that it did not differ significantly from Heegner's – that is, the gap was unimportant. At much the same time, Alan Baker found a different proof.

The case when d is positive is quite different. Factorisation is unique for many more values of d. Up to 50, those values are 2, 3, 5, 6, 7, 11, 13, 14, 17, 19, 21, 22, 23, 29, 31, 33, 37, 38, 41, 43, 46, 47, and computer calculations reveal many more. For all we know, there may be infinitely many positive d for which the corresponding quadratic number field has unique factorisation. A heuristic analysis by Cohen and Lenstra suggests that roughly three quarters of all positive d should define number fields with unique factorisation. Computer results agree with that estimate. The problem is to prove these observations are correct.

Langton's Ant

As the twenty-first century unfolds it has become increasingly apparent that some of the traditional techniques of mathematical modelling are unable to cope with the complexities of the problems facing humanity, such as the global financial system, the dynamics of ecosystems, and the role of genes in the growth of living organisms. Many of these systems involve large numbers of agents – people, companies, organisms, genes – that interact with each other. These interactions can often be modelled quite accurately using simple rules. Over the past 30 years, a new kind of model has appeared, which tries to tackle the behaviour of systems with many agents head-on. To understand how 100,000 people will move round a sports stadium, for example, you don't average them to create a sort of human fluid and ask how it flows. Instead, you build a computer model with 100,000 individual agents, impose suitable rules, and run a simulation to see what this computer crowd does. This kind of model is called a complex system.

To give you a glimpse of this fascinating new area of mathematics, I'm going to describe one of the simplest complex systems, and explain why we don't fully understand it. It is called Langton's ant. Christopher Langton was an early member of the Santa Fe Institute, founded in 1984 by scientists George Cowan, Murray Gell-Mann, and others to promote the theory and applications of complex systems. Langton invented his ant in 1986. Technically it is a cellular automaton, a system of cells in a square grid, whose states are shown by colours. At each time step, the colour of a cell changes according to those of its neighbours.

The rules are absurdly simple. The ant lives on an infinite square grid of cells, initially all of them being white. It carries an inexhaustible pot of quick-drying black paint and another inexhaustible pot of quick-drying white paint. It can face north, east, south, or west; by symmetry we can assume it starts out facing north. At each instant it looks at the colour of the square that it occupies, and changes it from white to black or from black to white using its pots of paint. If its square was white, it then turns 90 degrees to the right and takes one step forward. If its square was black, it then turns 90 degrees to the left and takes one step forward. Now it repeats this behaviour indefinitely.

If you simulate the ant,[93] it starts out by painting simple, fairly symmetric designs of black and white squares. It returns from time to time to a square that it has already visited, but its tour does not close up into a loop because the colour on that square has changed, so it turns the other way on a repeat visit. As the simulation continues, the ant's design becomes chaotic and random. It has no discernible pattern; basically it's just a mess. At that stage you could reasonably imagine that this chaotic behaviour continues indefinitely. After all, when the ant revisits a chaotic region it will make a chaotic series of

Fig 49 Langton's ant's highway.

turns and repaintings. If you carried on with the simulation, the next 10,000 or so steps would appear to justify that conclusion. However, if you keep going, a pattern appears. The ant enters a repeating cycle of 104 steps, at the end of which it has moved two squares diagonally. It then paints a broad diagonal stripe of black and white cells, called a highway, which goes on for ever, Figure 49.

Everything so far described can be proved in full rigour, just by listing the steps that the ant takes. The proof would be quite long – a list of 10,000 steps – but it would still be a proof. But the mathematics gets more interesting if we ask a slightly more general question. Suppose that before the ant starts out, we paint a finite number of squares black. We can choose these squares any way we like: random dots, a solid rectangle, the Mona Lisa. We can use a million of them, or a billion, but not infinitely many. What happens?

The ant's initial excursions change dramatically whenever it meets one of our new black squares. It can potter around all over the place, drawing intricate shapes and redrawing them... But in every simulation yet performed, no matter what the initial configuration might be, the ant eventually settles down to building its highway, using the same 104-step cycle. Does this always happen? Is the highway the unique 'attractor' for ant dynamics? Nobody knows. It is one of the basic unsolved problems of complexity theory. The best we know is that whatever the initial configuration of black cells may be, the ant cannot remain forever inside a bounded region of the grid.

Fig 50 Hadamard matrices of size 2, 4, 8, 12, 16, 20, 24, and 28.
http://mathworld.wolfram.com/HadamardMatrix.html

Hadamard Matrix Conjecture

A Hadamard matrix, named for Jacques Hadamard, is a square array of 0s and 1s such that any two distinct rows or columns agree on half of their entries and disagree on the other half. Using black and white to indicate 1 and 0, Figure 50 shows Hadamard matrices of size 2, 4, 8, 12, 16, 20, 24, and 28. These matrices turn up in many mathematical problems, and in computer science, notably coding theory. (In some applications, among them Hadamard's original motivation, the white squares correspond to -1, not 0.)

Hadamard proved that such matrices exist only when $n = 2$ or n is a multiple of 4. Paley's theorem of 1933 proves that a Hadamard matrix always exists if the size is a multiple of 4 and equal to $2^a(p^b + 1)$ where p is an odd prime. Multiples of 4 not covered by this theorem are 92, 116, 156, 172, 184, 188, 232, 236, 260, 268, and other larger values. The conjecture states that a Hadamard matrix exists whenever the size is a multiple of 4. In 1985 K. Sawade found one of size 268; the other numbers not covered by Paley's theorem had already been dealt with. In 2004 Hadi Kharaghani and Behruz Tayfeh-Rezaie found a Hadamard matrix of size 428, and the smallest size for which the answer is not known is now 668.

Fermat-Catalan Equation

This is the Diophantine equation $x^a + y^b = z^c$ where a, b, and c are positive integers, the exponents. I will call it the Fermat-Catalan equation because its solutions relate both to Fermat's last theorem, chapter 7, and to the Catalan conjecture, chapter 6. If a, b, and c are small, nonzero integer solutions are not especially surprising. For example, if they are all 2, then we have the Pythagorean equation, known since the time of Euclid to have infinitely many solutions. So the main interest is in the cases when these exponents are large. The technical definition of 'large' is that $s = 1/a + 1/b + 1/c$ is less than 1. Only ten large solutions of the Fermat-Catalan equation are known:

$$1 + 2^3 = 3^2 \qquad 17^7 + 76271^3 = 21063928^2$$
$$2^5 + 7^2 = 3^4 \qquad 1414^3 + 2213459^2 = 65^7$$
$$7^3 + 13^2 = 2^9 \qquad 9262^3 + 15312283^2 = 113^7$$
$$2^7 + 17^3 = 71^2 \qquad 43^8 + 9622^3 = 30042907^2$$
$$3^5 + 11^4 = 122^2 \qquad 33^8 + 159034^2 = 15613^3.$$

The first of these is considered large because $1 = 1^a$ for any a, and $a = 7$ satisfies the definition. The Fermat-Catalan conjecture states that the Fermat-Catalan equation has only finitely many integer solutions, without a common factor, when s is large. The main result was proved in 1997 by Henri Darmon and Loïc Merel: there are no solutions in which $c = 3$ and a and b are equal and greater than 3. Little else is known. Further progress seems to depend on a fascinating new conjecture, which comes next.

ABC Conjecture

In 1983 Richard Mason noticed that one case of Fermat's last theorem had been ignored: first powers. That is, consider the equation $a + b = c$.

At first sight this idea is completely pointless. It takes very little grasp of algebra to solve this equation for any of the three variables in terms of the other two. For example $a = c - b$. What changes the whole game, though, is context. Mason realised that everything becomes much deeper if we ask the right questions about a, b, and c. The outcome of his extraordinary idea was a new conjecture in number theory with far-reaching consequences. It could dispose of many currently unsolved problems and lead to better and simpler proofs of some of the biggest theorems in number theory. This is the ABC conjecture, and it is supported by a huge quantity of numerical evidence. It rests on a loose analogy between integers and polynomials.

Euclid and Diophantus knew a recipe for Pythagorean triples, which we now write as a formula, chapter 6. Can this trick be repeated with other equations? In 1851 Joseph Liouville proved that no such formula exists for the Fermat equation when the power is 3 or greater. Mason applied similar reasoning to the simpler equation

$$a(x) + b(x) = c(x)$$

for three polynomials. It's an outrageous idea, because all solutions can be found using elementary algebra. The main result, though, is elegant and far from obvious: if each polynomial has a factor that is a square, a cube, or a higher power, the equation has no solutions.

Theorems about polynomials often have analogues about integers. In particular, irreducible polynomials correspond to prime numbers. The natural analogue for integers of Mason's theorem about polynomials goes as follows. Suppose $a + b = c$ where a, b, and c are integers with no common factor; then the number of prime factors of each of a, b, and c is less than the number of *distinct* prime factors of abc. Unfortunately, simple examples show that this is false. In 1985 David Masser and Joseph Oesterlé modified the statement and proposed a version of this conjecture that did not conflict with any known examples. Their ABC conjecture may well be the biggest open question in number theory at the present time.[94] If someone proved the ABC conjecture tomorrow, many deep and difficult theorems, proved over past decades with enormous insight and effort, would have new, simple proofs. Another consequence would be Marshall Hall's conjecture: the difference between any perfect cube and any perfect square has to be fairly large. Yet another potential application of the ABC conjecture is to Brocard's problem, the first in this chapter. In 1993 Marius Overholt proved that if the ABC conjecture is true, there are only finitely many solutions to Brocard's equation.

One of the most interesting consequences of the ABC conjecture relates to the Mordell conjecture. Faltings has proved this using sophisticated methods, but his result would be even more powerful if we knew one extra piece of information: a bound on the size of the solutions. Then there would exist an algorithm to find them all. In 1991 Noam Elkies showed that a specific version of the ABC conjecture, in which various constants that appear are bounded, implies this improvement on Faltings's theorem. Laurent Moret-Bailly showed that the converse is true, in a very strong way. Sufficiently strong bounds on the size of solutions of *just one* Diophantine equation, $y^2 = x^5 - x$, imply the full ABC conjecture. Although it is not as well known as many other unsolved conjectures, the ABC conjecture is undoubtedly one of the great problems of mathematics. According to Granville and Thomas Tucker, disposing of it would have 'an extraordinary impact on our understanding of number theory. Proving or disproving it would be amazing.'[95]

Glossary

Algebraic integer. A complex number that satisfies a polynomial equation with integer coefficients and highest coefficient 1. For example $i\sqrt{2}$, which satisfies the equation $x^2 + 2 = 0$.

Algebraic number. A complex number that satisfies a polynomial equation with integer coefficients, or equivalently rational coefficients. For example $i\sqrt{2}/3$, which satisfies the equation $x^2 + \frac{2}{9} = 0$, or equivalently $9x^2 + 2 = 0$.

Algebraic variety. A multidimensional space defined by a set of algebraic equations.

Algorithm. A specified procedure to solve a problem, guaranteed to stop with an answer.

Angular momentum. A measure of how much spin a body has.

Arithmetic sequence. A sequence of numbers in which each successive number is the previous one plus a fixed amount, the common difference. For example, 2, 5, 8, 11, 14, ... with common difference 3. The older term is 'arithmetic progression'.

Asymptotic. Two quantities defined in terms of a variable are asymptotic if their ratio gets closer and closer to 1 as the variable becomes arbitrarily large.

Axis of rotation. A fixed line about which some object rotates.

Ball. A solid sphere – that is, a sphere and its interior.

Blowup time. The time beyond which a solution of a differential equation fails to exist.

Boundary. The edge of a specified region.

Chaos. Apparently random behaviour in a deterministic system.

Class E. An algorithm whose running time, for an input of size n, resembles the nth power of some constant.

Class P. An algorithm whose running time resembles some fixed power of the input size.

Class not-P. Not class P.

Class NP. A problem for which a proposed solution can be checked (but not necessarily found) by a class P algorithm.

Coefficient. In a polynomial such as $6x^3 - 5x^2 + 4x - 7$, the coefficients are the numbers $6, -5, 4, -7$ that multiply the various powers of x.

Cohomology group. An abstract algebraic structure associated with a topological space, analogous to but 'dual' to the homology group.

Complex analysis. Analysis – logically rigorous calculus – carried out with complex-valued functions of a complex variable.

Complex number. A number of the form $a + bi$ where i is the square root of minus one and a, b are real numbers.

Composite number. A whole number that can be obtained by multiplying together two smaller whole numbers.

Congruent number. A number that can be the common difference of a sequence of three squares of rational numbers.

Continuous transformation. A transformation of a space with the property that points that are very close together do not get pulled a long way apart.

Coordinate. One number in a list that determines the position of a point on a plane or in space.

Cosine. A trigonometric function of an angle, defined by $\cos A = a/c$ in Figure 51.

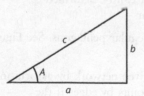

Fig 51 The cosine (a/c), sine (b/c), and tangent (b/a) of an angle A.

Counterexample. An example that disproves some statement. For instance, 9 is a counterexample to the statement 'all odd numbers are prime'.

Cube. A number multiplied by itself and then again by itself. For example, the cube of 7 is $7 \times 7 \times 7 = 343$. Usually written as 7^3.

Cubic equation. Any equation $ax^3 + bx^2 + cx + d = 0$ where x is unknown and a, b, c, d are constants.

Curvature. A measure of how space curves near a given point. A sphere has positive curvature, a plane has zero curvature, and a saddle-shaped space has negative curvature.

Cycle. In topology: a formal combination of loops in a triangulation with numerical labels attached. In algebraic geometry: a formal combination of subvarieties with numerical labels attached.

Cyclotomic integer/number. A sum of powers of a complex root of unity with integer/rational coefficients.

Degree. The highest power of the variable that occurs in a polynomial. For example the degree of $6x^3 - 5x^2 + 4x - 7$ is 3.

Differential equation. An equation relating a function to its rate of change.

Dimension. The number of coordinates required to specify the location of a point in a given space. For example the plane has dimension 2 and the space we live in (as modelled by Euclid's geometry) has dimension 3.

Diophantine equation. An equation for which solutions are required to be rational numbers.

Dirichlet L-function. A generalisation of the Riemann zeta function.

Disc (topological). A region in a surface that can be deformed continuously into a circle plus its interior.

Dodecahedron. A solid whose faces are 12 regular pentagons. See Figure 38.

Dual network. A network obtained from a given network by associating a point with every region, and joining points by edges if the corresponding regions are adjacent. See Figure 10.

Dynamical system. Any system that changes over time according to specified rules. For example, the motion of the planets in the solar system.

Eigenvalue. One of a set of special numbers associated with an operator. If the operator applied to some vector yields a constant multiple of that vector, the multiple concerned is an eigenvalue.

Electromagnetic field. A function that specifies the strengths and directions of the electric and magnetic fields at any point in space.

Elliptic curve. A curve in the plane whose equation has the form $y^2 = ax^3 + bx^2 + cx + d$ for constants a, b, c, d, usually assumed to be rational. See Figure 27.

Elliptic function. A complex function that remains unchanged when two independent complex numbers are added to its variable. That is, $f(z) = f(z + u) = f(z + v)$ where v is not a real multiple of u. See Figure 30.

Euler characteristic. $F - E + V$ where F is the number of faces in a triangulation of some space, E is the number of edges, and V is the number of vertexes. For a torus with g holes it equals $2 - 2g$, whatever the triangulation may be.

Euler's constant. A special number denoted by γ, approximately equal to 0.57721. See Note 67.

Exponent. In a power of a variable x, the exponent is the power concerned. For example, in x^7 the exponent is 7.

Face-centred cubic lattice. A repeating set of points in space, obtained by stacking cubes together like a three-dimensional chessboard, and then taking the corners of the cubes and the centres of their six square faces. See Figures 17, 19.

Factorisation. The process that writes a number in terms of its prime divisors. For example, the factorisation of 60 into primes is $2^2 \times 3 \times 5$.

Fermat number. A number of the form $2^{2^k} + 1$ where k is a whole number. If this number is prime then it is called a *Fermat prime*.

Flat torus. A torus obtained by identifying opposite edges of a square, whose natural geometry has zero curvature. See Figure 12.

Function. A rule f which, when applied to a number x, produces another number $f(x)$. For example, if $f(x) = \log x$ then f is the logarithm function. The variable x can be real or complex (in which case it is often written as z). More generally, x and $f(x)$ can be members of specified sets; in particular, the plane or space.

Fundamental group. The group formed by homotopy classes of loops in some topological space, under the operation 'travel along the first loop and then along the second'.

Gauge symmetry. A group of local symmetries of a system of equations: transformations of the variables that can vary from point to point in space, with the property that any solution of the equations remains a solution provided a compensating change with a sensible physical interpretation is made to the equations.

Gauge theory. A quantum field theory with a group of gauge symmetries.

General relativity. Einstein's theory of gravitation, in which the force of gravity is interpreted as the curvature of space-time.

Genus. The number of holes in a surface.

Group. An abstract algebraic structure comprising a set and a rule for combining any two elements of the set, subject to three conditions: the associative law, the existence of an identity element, and the existence of inverses.

Higgs boson. A fundamental particle whose existence explains why all particles have masses. Its discovery by the Large Hadron Collider was announced in July 2012.

Hodge class. A cohomology class of cycles on an algebraic variety with special analytic properties.

Homology (group). A topological invariant of a space, defined by closed loops. Two such loops are homologous if their difference is the boundary of a topological disc.

Homotopy (group). A topological invariant of a space, defined by closed loops. Two such loops are homotopic if each can be continuously deformed into the other.

Ideal (number). A number that is not contained in a given system of algebraic numbers, but is related to that system in a way that restores unique prime factorisation in cases when that property fails. Replaced in modern algebra by an ideal, which is a special kind of subset of the system concerned.

Induction. A general method for proving theorems about whole numbers. If some property is valid for 0, and its validity for any whole number n implies its validity for $n + 1$, then the property is valid for all whole numbers.

Integer. Any of the numbers $\ldots, -3, -2, -1, 0, 1, 2, 3, \ldots$.

Integral. An operation of the calculus, which in effect adds together large numbers of small contributions. The integral of a function is the area under its graph.

Irrational number. A real number that is not rational; that is, not of the form p/q where p and q are integers and $q \neq 0$. Examples are $\sqrt{2}$ and π.

Irreducible polynomial. A polynomial that cannot be obtained by multiplying two polynomials of smaller degree.

Lattice. In the plane: a set of points that repeats its form along two independent directions, like wallpaper patterns, see Figure 26. In space: a set of points that repeats its form along three independent directions, like the atoms in a crystal.

Lattice packing. A collection of identical circles or spheres whose centres form a lattice.

Logarithm. The (natural) logarithm of x, written $\log x$, is the power to which e ($= 2.71828\ldots$) must be raised to obtain x. That is, $e^{\log x} = x$.

Logarithmic integral. The function $\mathrm{Li}(x) = \int_0^x \frac{dt}{\log t}$.

Loop. A closed curve in a topological space.

Manifold. A multidimensional analogue of a smooth surface.

Maximum. The largest value of something.

Minimal criminal. A mathematical object that does not possess some desired property, and in some sense is the smallest possible such object. For example, a map that cannot be coloured with four colours, and also has the smallest number of regions for which this can occur. Minimal criminals are often hypothetical, and the aim is to prove that they don't exist.

Minimum. The smallest value of something.

Modular arithmetic. A system of arithmetic in which multiples of some specific number, called the *modulus*, are treated as if they are all zero.

Momentum. Mass multiplied by velocity.

Network. A set of points (nodes, dots) joined by lines (edges).

Non-Euclidean geometry. An alternative to Euclid's geometry in which all of the usual properties of points and lines remain valid, except for the existence of a unique line parallel to a given line and passing through a given point. There are two kinds: elliptic and hyperbolic.

NP-complete. A specific class NP problem, with the property that if there exists a class P algorithm to solve it, then *any* NP problem can be solved using a class P algorithm.

Operator. A special kind of function A, which when applied to a vector v yields another vector Av. It must satisfy the linearity conditions $A(v + w) = Av + Aw$ and $A(av) = aA(v)$ for any constant a.

Optimisation. Finding the maximum or minimum of some function.

Packing. A collection of shapes arranged in space so that they do not overlap.

Partial differential equation. A differential equation involving the rates of change of some function with respect to two or more different variables (often space and time).

Particle. A mass concentrated at one point.

Pentagon. A polygon with five sides.

Periodic. Anything that repeats the same behaviour indefinitely.

Phase. A complex number on the unit circle used to multiply a quantum wavefunction.

Polygon. A flat shape whose boundary consists of a finite number of straight lines.

Polyhedron. A solid whose boundary consists of a finite number of polygons.

Polynomial. An algebraic expression like $6x^3 - 5x^2 + 4x - 7$, in which powers of a variable x are multiplied by constants and added together.

Power. A number multiplied by itself a specified number of times. For example, the fourth power of 3 is $3 \times 3 \times 3 \times 3 = 81$, symbolised as 3^4.

Power series. Like a polynomial except that infinitely many powers of the variable can occur – for example $1 + 2x + 3x^2 + 4x^3 + \cdots$. In suitable circumstances this infinite sum can be assigned a well-defined value, and the series is said to converge.

Prime ideal. An analogue of a prime number for algebraic number systems.

Prime number. A whole number greater than 1 that cannot be obtained by multiplying two smaller whole numbers. The first few prime numbers are 2, 3, 5, 7, 11, 13.

Projective geometry. A type of geometry in which parallel lines do not exist: any two lines meet at a single point. Obtained from Euclidean geometry by adding a new 'line at infinity'.

Pythagorean triple. Three whole numbers a, b, c such that $a^2 + b^2 = c^2$. For example, $a = 3, b = 4, c = 5$. By Pythagoras's theorem, numbers of this type form the sides of a right-angled triangle.

Quadratic equation. Any equation $ax^2 + bx + c = 0$ where x is unknown and a, b, c are constants.

Quantum field theory. A quantum-mechanical theory of a quantity that pervades space and can (and usually does) have different values at different locations.

Quantum wavefunction. A mathematical function determining the properties of a quantum system.

Rank. The largest number of independent rational solutions of the equation defining an elliptic curve. 'Independent' means that they cannot be deduced from the other solutions using a standard geometric construction that combines any two solutions to yield a third, see Figure 25.

Ratio. The ratio of two numbers a and b is a/b.

Rational number. A real number of the form p/q where p and q are integers and $q \neq 0$. An example is 22/7.

Real number. Any number that can be expressed in decimals, possibly going on for ever – for example, $\pi = 3.1415926535897932385 \ldots$.

Reducible configuration. A part of a network with the following property: if the network obtained by removing it can be coloured with four colours, so can the original network.

Regular polygon. A polygon whose sides all have the same length, and whose angles are all equal. See Figure 4.

Regular solid. A solid whose boundary is composed of identical regular polygons, arranged in the same manner at every corner. Euclid proved that exactly five regular solids exist.

Rhombic dodecahedron. A solid whose boundary is composed of 12 identical rhombuses – parallelograms with all sides equal. See Figure 15.

Ricci flow. An equation prescribing how the curvature of a space changes over time.

Root of unity. A complex number ζ for which some power ζ^k is 1. See Figure 7 and Note 53.

Rotation. In the plane: a transformation in which all points move through the same angle about a fixed centre. In space: a transformation in which all points move through the same angle about a fixed line, the axis.

Ruler-and-compass construction. Any geometric construction that can be performed using an unmarked ruler and a compass (strictly: a pair of compasses).

Sequence. A list of numbers arranged in order. For example, the sequence 1, 2, 4, 8, 16, ... of powers of 2.

Series. An expression in which many quantities – often infinitely many – are added together.

Set. A collection of (mathematical) objects. For example, the set of all whole numbers.

Sine. A trigonometric function of an angle, defined by $\sin A = b/c$ in Figure 51.

Singularity. A point at which something nasty happens, such as a function becoming infinite or a solution of some equation failing to exist.

Sphere. The set of all points in space at a given distance from some fixed point, the centre. It is round, like a ball, but the term 'sphere' refers only to the points on the surface of the ball, not inside it.

3-Sphere. Three-dimensional analogue of a sphere: the set of all points in four-dimensional space at a given distance from some fixed point, the centre.

Square. A number multiplied by itself. For example, the square of 7 is $7 \times 7 = 49$, symbolised as 7^2.

Stable. A state of a dynamical system to which it returns if it is subjected to a small disturbance.

Standard Model. A quantum-mechanical model that accounts for all known fundamental particles.

Surface. A shape in space obtained by patching together regions that are topologically equivalent to the inside of a circle. Examples are the sphere and torus.

Symmetry. A transformation of some object that leaves its overall form unchanged. For example, rotating a square through a right angle.

Tangent. A trigonometric function of an angle, defined by $\tan A = b/a$ in Figure 51.

Topological space. A shape that is considered to be 'the same' if it is subjected to any continuous transformation.

Topology. The study of topological spaces.

Torus. A surface like that of a doughnut with one hole. See Figure 12.

Transcendental number. A number that does not satisfy any algebraic equation with rational coefficients. Examples are π and e.

Transformation. Another word for 'function', commonly used when the variables involved are points in some space. For example, 'rotate about the centre through a right angle' is a transformation of a square.

Translation. A transformation of space in which all points slide through the same distance and in the same direction.

Triangulation. Splitting a surface into a network of triangles, or its multidimensional analogue.

Trisection. Dividing into three equal parts, especially in connection with angles.

Trivial group. A group consisting only of a single element, the identity.

Unavoidable configuration. A member of a list of networks, at least one of which must occur in any network in the plane.

Unique prime factorisation. The property that any number can be written as a product of prime numbers in only one way, except for changing the order in which the factors are written. This property is valid for integers, but can fail in more general algebraic systems.

Unstable. A state of a dynamical system to which it may not return if it is subjected to a small disturbance.

Upper bound. A specific number that is guaranteed to be bigger than some quantity whose size is being sought.

Variable. A quantity that can take on any value in some range.

Variety. A shape in space defined by a system of polynomial equations.

Vector. In mechanics, a quantity with both size and direction. In algebra and analysis, a generalisation of this idea.

Velocity. The rate at which position changes with respect to time. Velocity has both a size (called speed) and a direction.

Velocity field. A function that specifies a velocity at each point in space. For example, when a fluid flows, its velocity can be specified at each point, and typically differs at different points.

Vortex. Fluid flowing round and round like a whirlpool. May be any size, including very small.

Wave. A disturbance that moves through a medium such as a solid, liquid, or gas without making any permanent change to the medium.

Whole number. Any of the numbers 0, 1, 2, 3, ...

Winding number. The number of times that a curve winds anticlockwise round some chosen point.

Zero (of a function). If f is a function, then x is a zero of f if $f(x) = 0$.

Zeta function. A complex function introduced by Riemann that represents the prime numbers analytically. It is defined by the series

$$\zeta(s) = \frac{1}{1^s} + \frac{1}{2^s} + \frac{1}{3^s} + \frac{1}{4^s} + \frac{1}{5^s} + \frac{1}{6^s} + \frac{1}{7^s} + \cdots$$

which converges when the real part of s is greater than 1. This definition can be extended to all complex s, except 1, by a process called analytic continuation.

Further reading

*Books marked * are technical.*

* Colin C. Adams, *The Knot Book*, W.H. Freeman, 1994.

* Felix Browder (ed.), *Mathematical Developments Arising from Hilbert Problems* (2 vols), Proceedings of Symposia in Pure Mathematics 28, American Mathematical Society, 1976.

* Tian Yu Cao, *Conceptual Developments of 20th Century Field Theories*, Cambridge University Press, 1997.

William J. Cook, *In Pursuit of the Travelling Salesman*, Princeton University Press, 2012.

Keith Devlin, *The Millennium Problems*, Granta, 2004.

Florin Diacu and Philip Holmes, *Celestial Encounters*, Princeton University Press, 1999.

Underwood Dudley, *A Budget of Trisections*, Springer, 1987.

Underwood Dudley, *Mathematical Cranks*, Mathematical Association of America, 1992.

Marcus Du Sautoy, *The Music of the Primes*, Harper Perennial, 2004.

Masha Gessen, *Perfect Rigour*, Houghton Mifflin, 2009.

* Jay R. Goldman, *The Queen of Mathematics*, A.K. Peters, 1998.

Jacques Hadamard, *The Psychology of Invention in the Mathematical Field*, Dover, 1954.

* Harris Hancock, *Lectures on the Theory of Elliptic Functions*, Dover, 1958.

Michio Kaku, *Hyperspace*, Oxford University Press, 1994.

* Jeffrey C. Lagarias, *The Ultimate Challenge: The 3x+1 Problem*, American Mathematical Society, 2011.

* Charles Livingston, *Knot Theory*, Carus Mathematical Monographs 24, Mathematical Association of America, 1993.

Mario Livio, *The Equation That Couldn't Be Solved*, Simon and Schuster, 2005.

* Henry McKean and Victor Moll, *Elliptic Curves*, Cambridge University Press, 1997.

Donal O'Shea, *The Poincaré Conjecture*, Walker, 2007.

Lisa Randall, *Warped Passages*, Allen Lane, 2005.

* Gerhard Ringel, *Map Color Theorem*, Springer, 1974.

* C. Ambrose Rogers, *Packing and Covering*, Cambridge Tracts in Mathematics and Mathematical Physics 54, Cambridge University Press, 1964.

Karl Sabbagh, *Dr Riemann's Zeros*, Atlantic Books, 2002.

Ian Sample, *Massive*, Basic Books, 2010.

* René Schoof, *Catalan's Conjecture*, Springer, 2008.

Simon Singh, *Fermat's Last Theorem*, Fourth Estate, 1997.

Ian Stewart, *From Here to Infinity*, Oxford University Press, 1996.

Ian Stewart, *Why Beauty is Truth*, Basic Books, 2007.

Ian Stewart, *Seventeen Equations that Changed the World*, Profile, 2012.

George Szpiro, *Kepler's Conjecture*, Wiley, 2003.

* Jean-Pierre Tignol, *Galois' Theory of Algebraic Equations*, Longman Scientific and Technical, 1980.

Matthew Watkins, The Mystery of the Prime Numbers, *Inamorata Press*, 2010.

Robin Wilson, *Four Colours Suffice*, Allen Lane, 2002.

Benjamin Yandell, *The Honors Class*, A.K. Peters, 2002.

Notes

1　The German original is: 'Wir müssen wissen. Wir werden wissen.' It occurs in a speech that Hilbert recorded for radio. See Constance Reid, *Hilbert*, Springer, Berlin, 1970, page 196.

2　Simon Singh, *Fermat's Last Theorem*, Fourth Estate, 1997.

3　Gauss, letter to Heinrich Olbers, 21 March 1816.

4　Wiles's title was 'Modular curves, elliptic forms, and Galois representations'.

5　Andrew Wiles, Modular elliptic curves and Fermat's last theorem, *Annals of Mathematics* **141** (1995) 443–551.

6　Ian Stewart, *Seventeen Equations that Changed the World*, Profile 2012, chapter 11.

7　Ibid., chapter 9.

8　Hilbert's problems, and their current status, edited slightly from *Professor Stewart's Hoard of Mathematical Treasures*, Profile 2009, are as follows:

1　**Continuum Hypothesis:** Is there an infinite cardinal number strictly between the cardinalities of the integers and the real numbers? Solved by Paul Cohen in 1963 – the answer depends on which axioms you use for set theory.

2　**Logical Consistency of Arithmetic:** Prove that the standard axioms of arithmetic can never lead to a contradiction. Solved by Kurt Gödel in 1931: impossible with the usual axioms for set theory.

3　**Equality of Volumes of Tetrahedrons:** If two tetrahedrons have the same volume, can you always cut one into finitely many polygonal pieces, and reassemble them to form the other? Solved in 1901 by Max Dehn, in the negative.

4　**Straight Line as Shortest Distance between Two Points:** Formulate axioms for geometry in terms of the above definition of 'straight line', and investigate the implications. Too broad to have a definitive solution, but much work has been done.

5 **Lie Groups without Assuming Differentiability:** Technical issue in the theory of groups of transformations. In one interpretation, solved by Andrew Gleason in the 1950s. In another, by Hidehiko Yamabe.

6 **Axioms for Physics:** Develop a rigorous system of axioms for mathematical areas of physics, such as probability and mechanics. Andrei Kolmogorov axiomatised probability in 1933.

7 **Irrational and Transcendental Numbers:** Prove that certain numbers are irrational or transcendental. Solved by Aleksandr Gelfond and Theodor Schneider in 1934.

8 **Riemann Hypothesis:** Prove that all nontrivial zeros of Riemann's zeta-function lie on the critical line. See chapter 9.

9 **Laws of Reciprocity in Number Fields:** Generalise the classical law of quadratic reciprocity, about squares to some modulus, to higher powers. Partially solved.

10 **Determine When a Diophantine Equation Has Solutions:** Find an algorithm which, when presented with a polynomial equation in many variables, determines whether any solutions in whole numbers exist. Proved impossible by Yuri Matiyasevich in 1970.

11 **Quadratic Forms with Algebraic Numbers as Coefficients:** Technical issues about the solution of many-variable Diophantine equations. Partially solved.

12 **Kronecker's Theorem on Abelian Fields:** Technical issues generalising a theorem of Kronecker. Still unsolved.

13 **Solving Seventh-Degree Equations using Special Functions:** Prove that the general seventh-degree equation cannot be solved using functions of two variables. One interpretation disproved by Andrei Kolmogorov and Vladimir Arnold.

14 **Finiteness of Complete Systems of Functions:** Extend a theorem of Hilbert about algebraic invariants to all transformation groups. Proved false by Masayoshi Nagata in 1959.

15 **Schubert's Enumerative Calculus:** Hermann Schubert found a non-rigorous method for counting various geometric configurations. Make the method rigorous. No complete solution yet.

16 **Topology of Curves and Surfaces:** How many connected components can an algebraic curve of given degree have? How many distinct periodic cycles can an algebraic differential equation of given degree have? Limited progress.

17 **Expressing Definite Forms by Squares:** If a rational function always takes non-negative values, must it be a sum of squares? Solved by Emil Artin, D.W. Dubois, and Albrecht Pfister. True over the real numbers, false in some other number systems.

18 **Tiling Space with Polyhedrons:** General issues about filling space with congruent polyhedrons. Also mentions the Kepler conjecture, now proved, see chapter 5.

19 **Analyticity of Solutions in Calculus of Variations:** The calculus of variations answers questions like: 'Find the shortest curve with the following properties.' If such a problem is defined by nice functions, must the solution also be nice? Proved by Ennio de Giorgi in 1957, and by John Nash.

20 **Boundary Value Problems:** Understand the solutions of the differential equations of physics, inside some region of space, when properties of the solution on the boundary of that region are prescribed. Essentially solved, by numerous mathematicians.

21 **Existence of Differential Equations with Given Monodromy:** A special type of complex differential equation can be understood in terms of its singular points and its monodromy group. Prove that any combination of these data can occur. Answered yes or no, depending on interpretation.

22 **Uniformisation using Automorphic Functions:** Technical question about simplifying equations. Solved by Paul Koebe soon after 1900.

23 **Development of Calculus of Variations:** Hilbert appealed for fresh ideas in the calculus of variations. Much work done; question too vague to be considered solved.

9 Reprinted as: Jacques Hadamard, *The Psychology of Invention in the Mathematical Field*, Dover, 1954.

10 The Agrawal-Kayal-Saxena algorithm is as follows:

Input: integer n.

1 If n is an exact power of any smaller number, output COMPOSITE and stop.

2 Find the smallest r such that the smallest power of r that equals 1 to the modulus n is at least $(\log n)^2$.

3 If any number less than or equal to r has a factor in common with n, output COMPOSITE and stop.

4 If n is less than or equal to r, output PRIME and stop.

5 For all whole numbers a ranging from 1 to a specified limit, check whether the polynomial $(x + a)^n$ is the same as $x^n + a$, to the modulus n and to the modulus $x^r - 1$. If equality holds in any case, output COMPOSITE and stop.

6 Output PRIME.

11 An example of what I have in mind is the formula $[A^{3^n}]$, where the

brackets denote the largest integer less than or equal to their contents. In 1947 W.H. Mills proved that there exists a real constant A such that this formula is prime for any n. Assuming the Riemann hypothesis, the smallest value of A that works is about 1.306. However, the constant is defined using a suitable sequence of primes, and the formula is just a symbolic way to reproduce this sequence. For more such formulas, including some that represent all primes, see

http://mathworld.wolfram.com/PrimeFormulas.html

http://en.wikipedia.org/wiki/Formula_for_primes

12   If n is odd then $n - 3$ is even, and if n is greater than 5 then $n - 3$ is greater than 2. By the first conjecture, $n - 3 = p + q$, so $n = p + q + 3$.

13   I prefer this term to the old-fashioned, but perhaps more familiar, 'arithmetic progression'. No one talks of progressions any more, except for arithmetic and geometric ones. Time to move on.

14   http://www.numberworld.org/misc_runs/pi-5t/details.html

15   My pet hate in this context is 'quantum leap'. In colloquial parlance it indicates some gigantic step forward, or some huge change, like the European discovery of America. In quantum theory, however, a quantum leap is so tiny that no known instrument can observe it directly, a change whose size is about 0.000 ... 01 with 40 or so zeros.

16   Finding a finite dissection of a square into a circle is called Tarski's circle-squaring problem. Miklós Laczkovich solved it in 1990. His method is non-constructive and makes use of the axiom of choice. The number of pieces required is huge, about 10^{50}.

17   The bizarre claims of circle-squarers and angle-trisectors are explored in depth in Underwood Dudley, *A Budget of Trisections*, Springer, 1987, and *Mathematical Cranks*, Mathematical Association of America, 1992. The phenomenon is not new: see Augustus De Morgan, *A Budget of Paradoxes*, Longmans, 1872; reprinted by Books For Libraries Press, 1915.

18   The quadratrix of Hippias is the curve traced by a vertical line that moves steadily across a rectangle and a line that rotates steadily about the midpoint of the bottom of the rectangle, Figure 52. This relationship turns any question about angle-division into the

corresponding one about line-division. For example, to trisect an angle you just trisect the corresponding line. See
http://www.geom.uiuc.edu/~huberty/math5337/groupe/quadratrix.html

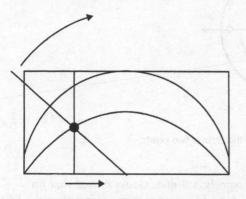

Fig 52 The quadratrix of Hippias (lower curve).

19 Here's an explicit example. Geometrically, if a line meets a circle and is not tangent to it, then it cuts the circle at exactly two points. Consider a line that is parallel to the horizontal axis, distance $\frac{1}{2}$ above it, Figure 53. The equation of this line is very simple: $y = \frac{1}{2}$. (Whatever value x may be, we always get the same value for y.) When $y = \frac{1}{2}$, the equation $x^2 + y^2 = 1$ becomes $x^2 + \frac{1}{4} = 1$. Therefore $x^2 = \frac{3}{4}$, so $x = \frac{\sqrt{3}}{2}$ or $-\frac{\sqrt{3}}{2}$. So algebra tells us that the unit circle meets our chosen line at exactly two points, whose coordinates are $\left(\frac{\sqrt{3}}{2}, \frac{1}{2}\right)$ and $-\left(\frac{\sqrt{3}}{2}, \frac{1}{2}\right)$. This is consistent with Figure 53 and with purely geometrical reasoning.

20 Strictly speaking, the polynomial concerned must have integer coefficients and be irreducible: not the product of two polynomials of lower degree with integer coefficients. Having degree that is a power of 2 is not always sufficient for a ruler-and-compass construction to exist, but it is always necessary. If the degree is not a power of 2, no construction can exist. If it is a power of 2, further analysis is needed to decide whether there is a construction.

21 The converse is also true: given constructions for regular 3- and 5-gons, you can derive one for a 15-gon. The underlying idea is that $2/5 - 1/3 = 1/15$. One subtle point concerns prime powers. The argument doesn't provide a construction for, say, a 9-gon, given one

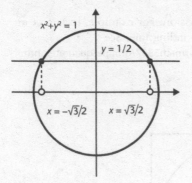

Fig 53 A horizontal line cutting the circle at two points.

for its prime factors – namely, a 3-gon. Gauss proved that no construction is possible for odd prime powers greater than the first.

22 See Ian Stewart, *Seventeen Equations that Changed the World*, Profile, 2012, chapter 5.

23 To make sense of this statement, resolve the quadratic into linear factors. Then $x^2 - 1 = (x + 1)(x - 1)$ which is zero if either factor is zero, so $x = 1$ or -1. The same reasoning can be applied to $x^2 = xx$: this is zero if either the first factor $x = 0$ or the second $x = 0$. It so happens that these two solutions yield the same x, but the occurrence of two factors x distinguishes this situation from something like $x(x - 1)$ where there is only one factor x. When counting how many solutions an algebraic equation has, the answer is generally much tidier if these 'multiplicities' are accounted for.

24 When $n = 9$, the second factor is

$$x^8 + x^7 + x^6 + x^5 + x^4 + x^3 + x^2 + x + 1$$

But this itself has factors: it is equal to

$$\left(x^2 + x + 1\right)\left(x^6 + x^3 + 1\right)$$

Gauss's characterisation of constructible numbers requires each irreducible factor to have degree that is a power of 2. But the second factor has degree 6, which is not a power of 2.

25 Gauss proved that the 17-gon can be constructed provided you can

construct a line whose length is

$$\frac{1}{16}\Big[-1 + \sqrt{17} + \sqrt{34 - 2\sqrt{17}} +$$

$$\sqrt{68 + 12\sqrt{17} - 16\sqrt{34 + 2\sqrt{17}} - 2(1 - \sqrt{17})(\sqrt{34 - 2\sqrt{17}})}\Big]$$

Since you can always construct square roots, this effectively solves the problem. Other mathematicians found explicit constructions. Ulrich von Huguenin published the first in 1803, and H.W. Richmond found a simpler one in 1893. In Figure 54, take two perpendicular radii AOP_0 and BOC of a circle. Make OJ = 1/4OB and angle OJE = 1/4OJP$_0$. Find F so that angle EJF is 45 degrees. Draw a circle with FP$_0$ as diameter, meeting OB at K. Draw the circle centre E through K, cutting AP$_0$ in G and H. Draw HP$_3$ and GP$_5$ perpendicular to AP$_0$. Then P$_0$, P$_3$, P$_5$ are respectively the 0th, 3rd, and 5th vertexes of a regular 17-gon, and the other vertexes are now easily constructed.

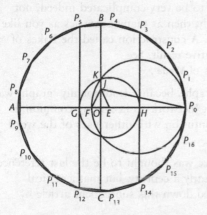

Fig 54 How to construct a regular 17-gon.

26 For the latest discoveries, see Wilfrid Keller, Prime factors of Fermat numbers and complete factoring status:
http://www.prothsearch.net/fermat.html

27 F.J. Richelot published a construction for the regular 257-gon in 1832. J. Hermes of Lingen University devoted ten years to the

65537-gon. His unpublished work can be found at the University of Göttingen, but it is thought to contain errors.

28 A typical continued fraction looks like this:

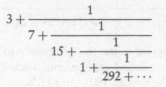

This particular continued fraction is the start of the one that represents π.

29 http://bellard.org/pi-challenge/announce220997.html

30 Louis H. Kauffman, Map coloring and the vector cross product, *Journal of Combinatorial Theory* B **48** (1990) 145–154.

Louis H. Kauffman, Reformulating the map color theorem, *Discrete Mathematics* **302** (2005) 145–172.

31 If the boundaries are allowed to be very complicated indeed, not like a map but far more wiggly, then as many countries as you like can share a common 'border'. A construction called the Lakes of Wada proves this counterintuitive result. See http://en.wikipedia.org/wiki/Lakes_of_Wada

32 The technical term is 'dual graph', because traditionally 'graph' was used instead of 'network'. But 'network' is becoming common, is more evocative, and avoids confusion with other uses of the word 'graph'.

33 Until recently, the *Nature* piece was thought to be the last reference in print to the problem for nearly a century, but mathematical historian Robin Wilson tracked down this subsequent article by Cayley.

34 Working in the dual network, let F be the number of faces (including one big face surrounding the entire network), E the number of edges, and V the number of vertexes. We may assume that every face in the dual network has at least three edges – if it has a face with only two edges then it corresponds to a 'superfluous' vertex of the original network that meets only two edges. This vertex can be deleted and the two edges joined.

Each edge borders two faces, and each face has at least three edges, so $E \geqslant 3F/2$, or equivalently $2E/3 \geqslant F$. By Euler's theorem $F + V - E = 2$, so $2E/3 + V - E \geqslant 2$, which implies that

$$12 + 2E \leqslant 6V$$

Suppose that V_m is the number of vertexes with m neighbours. Then V_2, V_3, V_4, and V_5 are zero. Therefore

$$V = V_6 + V_7 + V_8 + \cdots$$

Since every edge joins two vertexes,

$$2E = 6V_6 + 7V_7 + 8V_8 + \cdots$$

Substituting these into the inequality we get

$$12 + 6V_6 + 7V_7 + 8V_8 + \cdots \leqslant 6V_6 + 6V_7 + 6V_8 + \cdots$$

so that

$$12 + V_7 + 2V_8 + \cdots \leqslant 0$$

which is impossible.

35 'Chain' is misleading, since it suggests a linear sequence. A Kempe chain can contain loops, and it can branch.

36 The proof is given in full in Gerhard Ringel, *Map Color Theorem*, Springer, 1974. It divides into 12 cases, depending on whether the genus is of the form $12k, 12k + 1, ..., 12k + 11$. Call these cases 0–11. With finitely many exceptions, the cases were solved as follows:

Case 5: Ringel, 1954.

Cases 3, 7, and 10: Ringel in 1961.

Cases 0 and 4: C.M. Terry, Lloyd Welch, and Youngs in 1963.

Case 1: W, Gustin and Youngs in 1964.

Case 9: Gustin in 1965.

Case 6: Youngs in 1966.

Cases 2, 8, and 11: Ringel and Youngs in 1967.

The exceptions were genus 18, 20, 23 (solved by Yves Mayer in 1967) and 30, 35, 47, 659 (solved by Ringel and Youngs in 1968). They also dealt with the analogous problem for one-sided surfaces (like the Möbius band but lacking edges), which Heawood had also addressed.

37 The remarkable story of how the bug was discovered, and what happened when it was, can be found at http://en.wikipedia.org/wiki/Pentium_FDIV_bug

38 An excellent site for information about the physics of snowflakes is http://www.its.caltech.edu/~atomic/snowcrystals/

39 C.A. Rogers, The packing of equal spheres, *Proceedings of the London Mathematical Society* 8 (1958) 609–620.

40 Since space is infinite, there are infinitely many spheres, so both the space and the spheres have infinite total volume. We can't define the density to be ∞/∞, because that doesn't have a well-defined numerical value. Instead, we consider larger and larger regions of space and take the limiting value of the proportion of these regions that the spheres fill.

41 http://hydra.nat.uni-magdeburg.de/packing/csq/csq49.html

42 C. Song, P. Wang, and H.A. Makse, A phase diagram for jammed matter, *Nature* 453 (29 May 2008) 629–632.

43 Hai-Chau Chang and Lih-Chung Wang, A simple proof of Thue's theorem on circle packing, arXiv:1009.4322v1 (2010).

44 J.H. Lindsey, Sphere packing in $\mathfrak{R}^3$, *Mathematika* 33 (1986) 137–147.

D.J. Muder, Putting the best face on a Voronoi polyhedron, *Proceedings of the London Mathematical Society* 56 (1988) 329–348.

45 Hales used several different notions for what I am calling a cage. The final one is 'decomposition star'. My description omits some crucial distinctions in order to make the basic idea comprehensible.

46 Suppose the region is a polygon, as in Figure 55. Given any point that is not on the polygon, there exists some straight line from that point that gets outside a big circle containing the polygon, and does not pass through any vertex of the polygon. (There are finitely many vertexes but infinitely many lines to choose from.) This line cuts the polygon a finite number of times, and this number is either odd or even. Define the inside to consist of all points for which the number is odd, and the outside to consist of all points for which it

is even. It is then straightforward to prove that each of these regions is connected, and the polygon separates them.

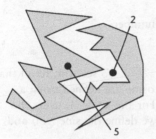

Fig 55 Proving the Jordan curve theorem for a polygon. An odd number of intersections occurs for points in the shaded region (inside), and an even number of intersections occurs for points in the white region (outside).

47 http://code.google.com/p/flyspeck/

48 Andrew Granville and Thomas Tucker, It's as easy as *abc*, *Notices of the American Mathematical Society* 49 (2002) 1224–1231.

49 To expand on this cryptic remark: the formula is

$$\int \frac{dx}{\sqrt{1 - x^2}} = \arcsin x$$

where arcsin (often written $\sin^{-1}$) is the inverse function to the sine. That is, if $y = \sin x$ then $x = \arcsin y$.

50 For example, let k be any complex number, and consider the integral

$$\int \frac{dx}{\sqrt{(1 - x^2)(1 - k^2 x^2)}}$$

This is the inverse function of an elliptic function denoted by sn. There is one such function for each value of k. The set-up is like Note 49, but more elaborate.

51 See Ian Stewart, *Seventeen Equations that Changed the World*, Profile, 2012, chapter 8.

52 The proof can be found in many number theory texts, for example

Gareth A. Jones and J. Mary Jones, *Elementary Number Theory*, Springer, 1998, page 227. On the web, see http://en.wikipedia.org/wiki/Infinite_descent#Non-solvability_of_ r2_.2B_s4_.3D_t4

53 One pth root of unity is the complex number

$$\zeta = \cos 2\pi/p + \text{i} \sin 2\pi/p$$

and the others are its powers $\zeta^2, \zeta^3, \ldots \zeta^{p-1}$. To see why, recall that the trigonometric functions sine and cosine are defined using a right-angled triangle, Figure 56 (left). For the angle A, using the traditional a, b, c for the three sides, we define the sine (sin) and cosine (cos) of A by

$$\sin A = a/c \cos A = b/c$$

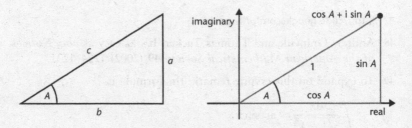

Fig 56 *Left:* Defining the sine and cosine. *Right:* Interpretation in the complex plane.

If we let $c = 1$ and place the triangle in the complex plane, as in Figure 56 (right), the vertex at which c and a meet is the point

$$\cos A + \text{i} \sin A$$

It is now straightforward to prove that for any angles A and B,

$$(\cos A + \text{i} \sin A)(\cos B + \text{i} \sin A) = \cos(A + B) + \text{i} \sin(A + B)$$

and this leads directly to De Moivre's formula

$$(\cos A + \text{i} \sin A)^n = (\cos nA + \text{i} \sin nA)$$

for any positive integer n. Therefore

$$\zeta^p = (\cos 2\pi/p + \text{i} \sin 2\pi/p)^p = \cos 2\pi + \text{i} \sin 2\pi = 1$$

so each power $1, \zeta, \zeta^2, \zeta^3, ..., \zeta^{p-1}$ is a pth a root of unity. We stop there because $\zeta^p = 1$, so no new numbers arise if we take higher powers.

54 Introduce the *norm*

$$N(a + b\sqrt{15}) = a^2 - 15b^2$$

which has the lovely property

$$N(xy) = N(x)N(y)$$

Then

$$N(2) = 4 \, N(5) = 25 \, N(5 + \sqrt{15}) = 10 \, N(5 - \sqrt{15}) = 10$$

Any proper divisor of one of these four numbers must have norm either 2 or 5 (the proper divisors of their norms). But the equations $a^2 - 15b^2 = 2$ and $a^2 - 15b^2 = 5$ have no integer solutions. Therefore no proper divisors exist.

55 Simon Singh, *Fermat's Last Theorem*, Fourth Estate, 1997.

56 Or maybe not. Vladimir Krivchenkov has pointed out that energy of the ground state and the first excited states for the quantum 3-body problem can be calculated by hand. But in classical mechanics, the analogous problem is less tractable because of chaos.

57 Quoted in Arthur Koestler, *The Sleepwalkers*, Penguin Books, 1990, page 338.

58 An animation and further information can be found at: http://www.scholarpedia.org/article/N-body_choreographies

59 Named after the Earl of Orrery, to whom one was presented in 1704.

60 More formally, this is called the Liapunov time.

61 There is a variant that integrates $1/\log t$ from 2 to x, rather than from 0 to x. This avoids a technical difficulty at $t = 0$, where $\log t$ is not defined. Sometimes the notation $\mathrm{Li}(x)$ is used for this variant, and the function defined in the text is called $\mathrm{li}(x)$.

62 The name 'Pafnuty' is unusual. It led Philip Davis to write a quirky but gripping book: *The Thread: a Mathematical Yarn*, Harvester Press, 1983.

63 This follows from Riemann's curious formula

$$\zeta(1-s) = 2^{1-s}\pi^{-s}\sin\left(\frac{\pi(1-s)}{2}\right)\Gamma(s)\zeta(s)$$

where $\Gamma(s)$ is a classical function called the gamma function, defined for all complex s. The right-hand side is defined when the real part of s is greater than 1.

64 Bernhard Riemann, Über die Anzahl der Primzahlen unter einer gegebenen Grösse, *Monatsberichte der Königlich Preußischen Akademie der Wissenschaften zu Berlin*, November 1859.

65 Riemann defined a closely related function

$$\Pi(x) = \pi(x) + \frac{1}{2}\pi(x^{1/2}) + \frac{1}{3}\pi(x^{1/3}) + \frac{1}{4}\pi(x^{1/4}) + \cdots$$

which counts prime powers rather than primes. From this we can recover $\pi(x)$. Then he proved an exact formula for this modified function in terms of logarithmic integrals and a related integral:

$$\Pi(x) = \mathrm{Li}(x) - \sum_{\rho}\mathrm{Li}(x^{\rho}) + \int_x^{\infty}\frac{dt}{t(t^2-1)\log t}$$

Here Σ indicates a sum over all numbers ρ for which $\zeta(\rho)$ is zero, excluding negative even integers.

66 For example, $x + \sqrt{x}$ is asymptotic to x: the ratio is

$$(x + \sqrt{x})/x = 1 + 1/\sqrt{x}$$

As x becomes large, so does $\sqrt{x}$, so $1/\sqrt{x}$ tends to 0 and the ratio tends to 1. But the difference is $\sqrt{x}$, and that becomes larger and larger as x increases. For instance, when x is 1 trillion, $\sqrt{x}$ is 1 million.

67 Euler's constant is the limit, as n tends to infinity, of

$$1 + \frac{1}{2} + \frac{1}{3} + \cdots + \frac{1}{n} - \log n$$

68 Douglas A. Stoll and Patrick Demichel, The impact of $\zeta(s)$ complex zeros on $\pi(x)$ for $x < 10^{10^{13}}$, *Mathematics of Computation* **276** (2011) 2381–2394.

69 http://empslocal.ex.ac.uk/people/staff/mrwatkin/zeta/RHproofs.htm

70 J. Brian Conrey and Xian-Jin Li, A note on some positivity
    conditions related to zeta- and L-functions:
    http://arxiv.org/abs/math.NT/9812166

71 The unit 3-sphere comprises all points with coordinates (x, y, z, w)
    such that $x^2 + y^2 + z^2 + w^2 = 1$. There are several ways to make the
    3-sphere more intuitive. They can all be understood by analogy with
    a 2-sphere, and checked using coordinate geometry. One such
    description ('solid ball with all surface point identified') is given in
    the text, and Figure 57 shows another. To set up the analogy,
    observe that if we cut a 2-sphere along its equator we get two
    hemispheres. Each flattens out into a disc, and this is a continuous
    deformation. To reconstruct the 2-sphere, we just identify
    corresponding points on the boundaries of these two discs. In a
    sense, we have made a map of the 2-sphere using two flat discs,
    much as mapmakers create flat projections of our round planet. We
    can construct a 3-sphere using an analogous procedure. Take two
    solid balls and identify corresponding points on their surfaces. Now
    both have the same surface (because we identified the two surfaces),
    and it is a 2-sphere. It forms the 'equator' of the 3-sphere.

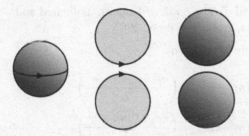

Fig 57 How to make a 3-sphere. *Left:* Cutting a 2-sphere into hemispheres. *Middle:*
Reconstructing the 2-sphere from the two halves by gluing the edges. *Right:* By analogy,
conceptually glue the surfaces of two balls together so that corresponding points are
considered to be identical. This gives a 3-sphere.

72 The usual convention is that we talk of addition and use the
    notation $a + b$ when the commutative law is valid, but talk of
    multiplication and use the notation ab when it might not be. I have
    ignored this convention here because this isn't a textbook on group
    theory and 'addition' seems more natural.

73 Start the count at zero. Every time you pass the bus stop going anticlockwise, increase the count by 1; every time you pass it going anticlockwise, decrease the count by 1. At the end of the trip, add 1 if you arrived going anticlockwise, subtract 1 if you arrived going clockwise. The final count is the total number of times you went round the circle, measured in the anticlockwise direction.

74 Stirling's formula states that $n!$ is approximately $\sqrt{2\pi n}(n/e)^n$.

75 William J. Cook, *In Pursuit of the Travelling Salesman*, Princeton University Press, Princeton, 2012. For current information, see http://www.tsp.gatech.edu/index.html

76 Richard M. Karp, Reducibility among combinatorial problems, in R.E. Miller and J.W. Thatcher (eds.) *Complexity of Computer Computations*, Plenum, 1972, pages 85–103.

77 Z. Xia, The existence of noncollision singularities in Newtonian systems, *Annals of Mathematics* **135** (1992) 411–468.

78 http://www.claymath.org/millennium/Navier-Stokes_Equations/

79 See Ian Stewart, *Seventeen Equations that Changed the World*, Profile, 2012, chapter 14.

80 Leonardo Pisano Fibonacci, *The Book of Squares*, annotated and translated by L.E. Sigler, Academic Press, 1987.

81 Leonardo found a family of solutions

$$\left(\frac{m^2 + n^2}{2}\right)^2 - mn(m^2 - n^2) = \left(\frac{m^2 - 2mn - n^2}{2}\right)^2$$
$$\left(\frac{m^2 + n^2}{2}\right)^2 + mn(m^2 - n^2) = \left(\frac{m^2 + 2mn - n^2}{2}\right)^2$$

where m, n are both odd. The role of d here is played by the number $mn(m^2 - n^2)$, and x is $m^2 + n^2/2$. Choosing $m = 5$, $n = 4$ leads to $mn(m^2 - n^2) = 720$. Moreover, $720 = 5 \times 12^2$. Dividing x by 12 yields the answer.

82 If $x - n$, x, and $x + n$ are squares, then so is their product, which is $x^3 - n^2x$. Therefore the equation $y^2 = x^3 - n^2x$ has a rational solution. Moreover, y is not zero, otherwise $x = n$ so both x and $2x$ are squares, which is impossible since $\sqrt{2}$ is irrational.

Conversely, if x and y satisfy the cubic equation and y is not 0, then

$a = (x^2 - n^2)/y$, $b = 2nx/y$, and $c = (x^2 + n^2)/y$ satisfy the equations $a^2 + b^2 = c^2$ and $ab/2 = n$.

83 That is,

$$\prod_{p \leqslant x} \frac{N_p}{p} \approx C(\log r)^r$$

where r is the rank, C is a constant, and $\approx$ means that the ratio of the two sides tends to 1 as x tends to infinity.

84 The most likely reason is that these are the natural translations from the languages used by the most prominent mathematicians in the two areas.

85 Why b wasn't the number of bananas, I'm not sure. Perhaps because in post-war Britain, bananas were exotic items seldom seen in the shops?

86 Hence a standard mathematicians' joke. A biologist, a statistician, and a mathematician are sitting outside a café watching the world go by. A man and a woman enter a building across the road. Ten minutes later, they come out accompanied by a child. 'They've reproduced,' says the biologist. 'No,' says the statistician. 'It's an observational error. On average, two and a half people went each way.' 'No, no, no,' says the mathematician. 'It's perfectly obvious. If someone goes in now, the building will be empty.'

87 Bohr may have had a serious point. Scientific theories are tested through their predictions, but few of these foretell the future. Most are if/then statements: if you pass light through a prism it will split into colours. The 'prediction' doesn't say when this will happen. So, paradoxically, we can make predictions about the weather without predicting the weather. 'If the warm air from a cyclone encounters cold air then it will snow' is a scientific prediction, but not a forecast.

88 The quotation or a close variant has been attributed to about thirty different sources, including Sam Goldwyn, Woody Allen, Winston Churchill, and Confucius. See http://www.larry.denenberg.com/predictions.html

89 For the latest information, see the Prime Pages: http://primes.utm.edu.

90 Ilia Krasikov and Jeffrey C. Lagarias, Bounds for the $3x + 1$ problem using difference inequalities, *Acta Arithmetica* **109** (2003) 237–258.

91 Jorge F. Sawyer and Clifford A. Reiter, Perfect parallelepipeds exist. arXiv:0907.0220 (2009).

92 R. Fulek and J. Pach, A computational approach to Conway's thrackle conjecture, *Computational Geometry* **44** (2011) 345–355.

93 http://en.wikipedia.org/wiki/Langton%27s_ant

94 The ABC conjecture states: For any $\varepsilon > 0$ there exists a constant $k_\varepsilon > 0$ such that if a, b, and c are positive integers having no common factor greater than 1, and $a + b = c$, then $c \leqslant k_\varepsilon P^{1+\varepsilon}$, where P is the product of all distinct primes dividing abc.

95 Andrew Granville and Thomas J. Tucker. It's as easy as *abc*, *Notices of the American Mathematical Society* **49** (2002) 1224–1231.
    In September 2012 Shinichi Mochizuki announced that he had proved the ABC conjecture using a radical new approach to the foundations of algebraic geometry. Experts are now checking his 500-page proof, but this may take a long time.

Index

9 780465 064892

CPSIA information can be obtained
at www.ICGtesting.com
Printed in the USA
LVHW041614180821
695597LV00014B/272

NEY BEYOND SELENE

Expeditions to the Solar System's 63 Moons

Also by Jeffrey Kluger

Lost Moon: The Perilous Voyage of Apollo 13
(with Jim Lovell; published in paperback as *Apollo 13*)

The Apollo Adventure

JOURNEY BEYOND SELENE

Remarkable Expeditions to the Solar System's 63 Moons

JEFFREY KLUGER

LITTLE, BROWN AND COMPANY

A *Little, Brown* Book

First published in the United States in 1999
by Simon & Schuster Inc., New York
First published in Great Britain in 1999
by Little, Brown and Company

Copyright © 1999 by Jeffrey Kluger

The moral right of the author has been asserted.

All rights reserved.
No part of this publication may be reproduced,
stored in a retrieval system, or transmitted, in any
form or by any means, without the prior
permission in writing of the publisher, nor be
otherwise circulated in any form of binding or
cover other than that in which it is published and
without a similar condition including this
condition being imposed on the subsequent purchaser.

A CIP catalogue record for this book
is available from the British Library.

ISBN 0 316 64842 6

Printed and bound in Great Britain by
Clays Ltd, St Ives plc

Little, Brown and Company (UK)
Brettenham House
Lancaster Place
London WC2E 7EN

*With love to Alejandra, for keeping me mindful of the magic
both at home and on the moons*

Contents

Prologue

Linda Morabito was alone in her lab when she discovered that the moon was exploding. Actually, there was some
question as to whether it was a moon that was exploding at all. A lot
of people—Morabito included—had begun to regard the body more
as a fully certified planet, and a lot of *other* people were coming
around to that way of thinking, too. But the official position at
NASA's Jet Propulsion Laboratory in Pasadena where Morabito
worked was that the maybe-planet was indeed a moon, so that's
what Morabito generally called it. In any event, it was exploding.

Morabito did not know exactly what she was expected to do if
she found out the moon was exploding today. Earlier in the day
would have been different, but earlier in the day she wasn't the only
one here. Indeed, earlier in the day it seemed that everybody was
here: the head of the lab, the head of the department, the head of the
whole space agency flown in special from Washington. That was the
way it always was on picture days, and there was no reason today
would have been any different.

Picture days, it seemed, almost always happened on a Thursday
or a Friday, and for a place like JPL, that made sense. When you're
showing off the first images of some new world taken by a billion-
dollar spacecraft a half billion miles from home, you want to schedule things carefully. Call your press conference too early in the week
and the newsmagazines that don't come out until next Monday start

to see your stuff as stale. Call it too late and they can't get their stories written up by their Friday-night deadlines. No, if you want your pictures to get anywhere beyond the four walls of the JPL imaging room where the hot pixels sent back from the remote robot probe were first assembled into images, you had to call the reporters in, give them what they needed, and have them on their way no later than lunchtime Friday.

Morabito, of course, was not invited to participate in the press conference earlier today—and she didn't expect to be. Nobody from spacecraft navigation ever was. The way the media saw things, the navigation section's job was merely to get a spaceship (in this case, *Voyager 1*) from some terrestrial Point A (in this case, Cape Canaveral) to some cosmic Point B (in this case, Jupiter) in as little time as possible. Never mind that there were 400 million miles between Jupiter and Florida. Never mind that, at its absolute fastest, the little tin ship that was making the trip would never be able to gun its speed much beyond 35,000 miles per hour, a glacial creep that meant it would need years to get where it was going. Never mind that there were a thousand wrong turns the ship could make in the course of its journey, and if it made just one of them it would spin off into the void, never to contact Earth again. No, the reporters didn't care about any of that. What they cared about were the pictures the ship would be beaming home, and it was the planetologists and geologists from the glamorous imaging team—not the drones from the navigation team—who would present them to the reporters.

This picture day the crowd of reporters gathering to receive those images promised to be a big one, but it was not images of Jupiter itself they were coming to see. The planet, after all, was something of a known quantity. Little more than a mammoth, spherical storm of hydrogen and helium, Jupiter had long since been regarded not so much as a planet at all, but as a sort of failed star. It had the size to be a star and the age to be a star, but it never achieved the critical ignition mass to light its internal fires and actually *become* a star. If the solar system had a blown fuse, Jupiter was it.

What circled Jupiter, however, was another matter entirely.

Buzzing electron-like around the giant world were no fewer than thirteen moons. It was 369 years earlier that the four biggest of those satellites—Io, Europa, Ganymede, and Callisto—were discovered. In later decades, better telescopes added Amalthea and Thebe, Metis, Adrastea, Leda, Lysithea, Elara, Ananke, and Carme. However many other moons there might be was impossible to say, but two years earlier, *Voyager 1* and its sister ship, *Voyager 2,* had been launched toward the outer planets to help find out. What intrigued the scientists and the media following the progress of the ships was not merely the exact number of the moons, but the possibility that something might be going on on them.

The moon astronomers knew best—Earth's moon—was, by even the most generous assessments, a carcass of a world: uniformly gray, uniformly dry, uniformly dead. From as far away as Earth, however, astronomers could see that the Jovian moons were a different matter entirely. There were big moons and small moons, patterned moons and plain moons, brightly colored moons and pasty-pale moons. More important, if remote studies with telescopes and spectroscopes were any indications, there were moons that could have atmospheres, water, and even, perhaps, a spark of internal heat. Put them together, and you had moons that could, in theory, harbor life.

It was the life part that interested the reporters most. And it was the internal heat part that was likely to make the difference. No amount of air, water, and organic molecules was going to be able to do all the clever recombining it had to do to create living organisms if you didn't have something to keep them all warm. Out in the interplanetary provinces where Jupiter lies, however—where the sun looks little bigger than a lit match held across the room and offers little more heat—warmth was not such an easy thing to come by. Any energy sufficient to warm a planet-like body this remote would thus have to come from the interior of the planet-like body itself. A place like Earth is a furnace of a world, with a molten core and viscous mantle generating enough heat to keep volcanoes percolating, geysers spouting, and the very continents themselves floating around on all the geological goo like oyster crackers on soup. If a Jovian moon had even a little of this magma-heated metabolism, it just might be able to cook up something living.

That was the theory anyway, and it was one that the press was eager to learn more about as the first of the *Voyager* ships completed its Jupiter flyby and the late-week press conference got under way. For this initial meeting of the media, the focus would be on what *Voyager 1* had been able to learn about the Jovian moon Io—and even the uninitiated appreciated that this was a good choice. Astronomers already knew that Io appeared to have at least a few wisps of atmosphere and a few riverbed-like gulleys, indicating flowing liquid had once been present on its surface. What's more, the surface of the satellite showed a surprising variety of color—with orange and black highlands broken up by ruddy, rusty plains. A moon with this much apparent chemistry going on was a moon that was capable of anything.

It was with much anticipation, then, that the JPL scientists awaited the first batch of Io photos beamed back by the little *Voyager* ship. When the images finally arrived, however, they were, by most measures, a disappointment. Io might have been a dramatic place—tricked out in all the colors Earthbound astronomers had promised it would be—but it also appeared to be a dead place. Nowhere on the surface of the moon was there visible volcanic activity or any other sign of the underground heating the scientists had hoped to find. Io's trace atmosphere and complex soil might yet harbor enough raw materials to give rise to living organisms, but with the moon's internal fires obviously having flickered out long ago, any native life would have flickered out, too.

Putting the best scientific face on this forbidding world would not be an easy thing to do on picture day, and when the media gathered today it was clear they were disappointed. The scientists stressed—and genuinely believed—that with or without heat, Io was still a chemically fascinating place, that its wealth of organic elements could still teach them a lot about how life evolved on Earth and how it might yet evolve elsewhere. The reporters, however, wanted not chemicals, but critters, and at the same time the scientists were congratulating themselves on making their first successful pass over a world that could keep planetary researchers busy for generations, they found themselves tacitly apologizing to the assembled media for not delivering the organic goods. In four months *Voyager 2* would be making its own Jovian flyby, paying special attention to Io's sister moon Europa. If the last moon had been a disappointment, perhaps this next one—with its brightly reflective surface and its odd, icy rind—would have more to offer.

When today's picture day ended, the chemists and geologists went home for the weekend, and, as they always did at points like this in a mission, temporarily turned control of the spacecraft over to Linda Morabito and the rest of the navigation team. Now that the ship had completed its first major rendezvous, it would be the navigators' responsibility to take its bearings, check its headings, and make sure it was pointed true toward whatever destination the planetary scientists had chosen for it next. The tools the navigators would use for that job would be the same Io images the planetary scientists had just shown off to the press—or *almost* the same Io images.

When a spacecraft tearing along at 7.5 miles per second is taking pictures of a moon that's illuminated by nothing more than a smudge of solar light more than 100 million miles away, exposure time is everything. Leave the shutter of your camera open for too short a period, and the picture you'll get will be nothing but an inky smear. Leave it open for too long, and the exquisitely sensitive light-gathering hardware will gather too much, temporarily blinding itself with the flood of incoming illumination and producing merely a white, washed-out sphere where a picture of a richly textured moon should be. At least half of the Io images *Voyager 1* had beamed down were either overexposed or underexposed this way, and while the blackened, underexposed ones did no one any good, the overexposed ones turned out to be surprisingly valuable.

In addition to gathering in the reflected light of the nearby moon, a spacecraft that left its camera's lens cap off too long would pick up lots of tiny pinpoints of starlight. For a navigation engineer like

Morabito, this was a very good thing, since there was no better way to confirm that an unmanned ship was adhering to its planned trajectory than to check its position against the stars. Each time an overexposed picture came down from the ship, it was therefore passed on to the navigation section, where celestial map readers like Morabito would determine exactly where in the heavens the spacecraft was, compare this with where it should be, and decide if a course-correcting engine burn was necessary.

On the evening of the Friday press gathering, Morabito sat in her JPL lab, studying her computer screen as it flashed its navigation images with their hopelessly fuzzy Ios and their wonderfully sharp stars. Morabito was the last navigator from the day shift still at her desk, but others, working a voluntary night shift, might be punching in soon. After just a few minutes, however, it became evident that they probably needn't bother. From even a cursory glance at the stars in the Io images, it was clear that *Voyager 1* was flying true, with each stellar pinpoint showing up exactly where the navigation charts said it ought to be. Picture after numbingly similar picture told the same encouraging story, when all at once Morabito noticed an image that wasn't so similar. Up near the high horizon of one Io picture, off at the two o'clock position, she spotted a curious bulge in the otherwise smooth disk of the moon. It was like nothing Morabito had ever seen before. The bump was too small to be another Jovian moon peeking out from behind Io, too big to be dust or a pixel glitch in the imaging equipment. It was, undeniably, a part of Io itself.

Morabito dug through a pile of photographic prints for another image taken from the same perspective. The bulge was still there. She found another taken from a few degrees away; again the bulge. Indeed, no matter where *Voyager* was as it photographed Io's facing hemisphere, the curious mound in the moon's surface remained. It was almost as if there was some odd atmospheric aneurysm swelling over the landscape in a very specific spot. But Io had no atmosphere to speak of—certainly not one soupy enough to produce such a horizon-transforming cloud.

Morabito worked with the images throughout the evening, calling picture after picture onto her screen and pulling photo after photo out of the stacks that surrounded her. It was only when the night had passed and the sun had risen that Morabito, bleary and still alone in her lab, realized the full magnitude of what she was seeing. This cloud, she now knew, could be only one thing: a plume from an active volcano—and a huge one. It was rounded like a volcanic plume; it was semi-transparent like a volcanic plume; and it rose over a fixed spot on the surface—exactly like a volcanic plume. What's more, it rose astoundingly high. Judging by Io's diameter, the cloud of underground hellfire had to extend more than 160 miles into space. If the same exhaust blast occurred on Earth, it would roar into the sky thirty times higher than the peak of Mount Everest. While the plume was big, however, it was also wispy, made more of gas than ash. It was so wispy, in fact, that even *Voyager*'s finely tuned cameras were not sensitive enough to spot it—at least when they were operating as they should. When they occasionally misfired, however—such as when they kept their shutters open longer than they ought to—enough light flowed into the imaging system to make the smoke and gas visible.

Morabito nodded to herself incredulously. Io was alive, explosively alive, home to what appeared to be the most titanically huge volcano in the solar system. And only Linda Morabito—with her washed-out photos scattered all around her—knew it.

• • •

It was not until Monday that the rest of the Jet Propulsion Laboratory learned what Linda Morabito had discovered over the weekend. And it was not for another week or so that most of them believed it. But slowly, the evidence mounted. Imaging experts digitally enhanced the Io pictures, and the horizon bulge only became clearer. Planetary scientists analyzed its chemical spectrum and found it consisted mostly of sulfur and fine particles—just the stuff a volcano would be expected to give off. What's more, other overexposed images from other parts of the moon suggested smoldering volcanoes there as well. At up to nine different spots on the surface, underground heat appeared to be boiling up and blasting into space. Io, it seemed, was a geological pressure cooker, blowing volcanic holes across its own surface like an overinflated beach ball springing spot leaks.

What this meant for life on the moon was impossible to tell. In the vicinity of the volcanoes, surface temperatures appeared to be approaching a shirtsleeves 60 degrees—more than balmy enough for terrestrial organisms to survive. If the right organic chemicals existed close enough to the volcanoes, it was entirely possible that at least some crude forms of life could have emerged.

The JPL astronomers knew it would be at least a year before they would be able to analyze all of the data *Voyager 1* had beamed back and begin to find out for sure. All they could say until then was that the Jovian system was now known to be a hot system—and hot systems were capable of just about anything. In the meantime, *Voyager 2* was still speeding toward its encounter with Europa, and Europa was likely to make things more complicated still.

Earth-based surveys had long since revealed that the Europan landscape was entirely covered by a bright white crust of ice. Moreover, spectral studies had shown that that ice was composed not of methane or caustic sulfur, but of ordinary water. Melt a little of the ice down over even one hot spot on the moon's surface, and you'd have the first ocean known to exist away from Earth. And it was in the oceans that the only confirmed life in the solar system was known to have begun. If Europa had even a fraction of the heat-giving volcanic activity of its sister moon Io, it could be a practical hothouse for extraterrestrial organisms.

As the *Voyager 2* flyby of Europa approached, JPL scientists planned frantically. For this encounter, they concluded they would not be so choosy about the pictures they'd examine. Overexposed images, underexposed images, images that were little more than a Europan shimmer would all be studied. Planetary scientists would begin scrutinizing pictures of the moon when the spacecraft was still months away from Europa, looking for even the slightest suggestion of a volcanic plume.

Realistically, however, they knew that if there *were* volcanoes on Europa, they probably wouldn't reveal themselves so easily. As thick as the moon's ice layer was, even a relatively big eruption would not make it beyond the surface and into the tenuous atmosphere above. Rather, it would stay beneath the frozen crust, heating the lower layers of ice until they turned first into slush and then, perhaps, into flowing water. The only sign the researchers would get that such thermal turmoil was taking place at all would be in the hard ice that still covered the surface. If the Europan crust was pocked with craters, astronomers would know it was an *old* crust—one that had been repeatedly pounded by meteor storms without ever being resurfaced afresh. If the astronomers found a patch of smooth surface ice, however, they would know that in that area, heat was rising from deep underground, turning the ice viscous and allowing its craters to be filled in and troweled over. It was beneath those fresh plains that oceans—and, in theory, life—might exist.

The day of the Europan encounter, the senior JPL astronomers gathered in the main picture room of the Pasadena complex. Torrence Johnson, the head of the imaging team was there, along with chief planetary scientist Larry Soderblom and chief geologist Brad Smith. Also present was Cornell University astronomer Carl Sagan, who had come to witness the encounter with his usual coterie of graduate students in tow. The images this small group would see today would be projected on half a dozen monitors mounted in heavy steel brackets and bolted to the low-hanging ceiling. The scientists crowded into the room early in the morning, waiting for the moment those monitors would flicker to life. Realistically, they knew it might be a long wait.

From *Voyager 2*'s position deep in Jovian space, the Europa images it beamed to Earth would need a full forty minutes to reach the JPL antennas. From there, the signals would be relayed to a mainframe computer deep in a JPL basement, where they would be rebuilt into a picture. This assembly job would take at least another hour—maybe two or three—and only when it was done would the picture be forwarded to the hanging screens in the little room.

At about 8 A.M. Pasadena time, the scientists received word that

Voyager had encountered Europa and had switched on its cameras. Exactly forty minutes later they got word that the data from the first image had arrived in Pasadena. An hour elapsed, then two, and finally, deep into the third hour, the ceiling monitors began to sizzle with static. The men looked up and watched as a circular image began to resolve itself on the screens, slipped a little, then resolved again. A dumbfounded silence fell over the room.

"What the hell is that?" someone finally asked.

"What's the matter with the picture?" someone responded. Instinctively, though, the men in the room knew that nothing was the matter with the picture at all.

On the screens in front of them was the unmistakable image of a bright white world covered by a sort of sugar shell of ice with barely a single crater anywhere on it. The shell was shot through with a tracery of fine fractures and spider cracks, but apart from those hairline breaks, it was practically pristine. There wasn't a planet or moon in the solar system that didn't show at least a little meteor scarring somewhere on its surface. And yet Europa—which could not have been spared the bombardment all of the other worlds had sustained—had erased virtually all traces of it.

There was only one scientifically sensible explanation. Europa, which the astronomers had hoped might be warm in spots, was warm everywhere. All over the moon there must be enough internal heat radiating up to cause the entire ice crust to soften periodically and then re-form itself. And down *below* the crust, things would only be more dramatic. There, the warmth just might be so great the water would never freeze at all. Beneath Europa's frozen rind might be an ocean that girdled the entire globe. There was only one known place in the solar system where such conditions prevailed, and that was in the icy waters beneath Earth's North and South Poles. And those waters, the researchers in the JPL imaging room knew, fairly teem with life.

"The Antarctic," someone in the room murmured to himself. "I'm looking at the Antarctic."

March 1997

Nobody at the Jet Propulsion Laboratory thought it would take eighteen years to return to Europa. Not that there weren't reasons for such a delay, of course. First of all, there were a lot of other missions JPL was considering flying—missions to Venus, to Mars, to Saturn—and all of those deserved a fair share of the lab's time and money. Then, of course, there was that nasty bit of business with the space shuttle *Challenger* back in 1986. When the shuttle exploded, taking seven unlucky astronauts with it, the space community as a whole seized up and shut down. Surviving shuttles sat idling in their hangars; interplanetary probes sat idling in their dust-free clean rooms. It was only in 1988, when the next of the shuttle siblings at last flew successfully into orbit, that the space program as a whole began to bestir itself, too. Given all that, it was only natural that it took a long time to get back to Europa; a full eighteen years, however, still came as something of a surprise. If there was any consolation for the Pasadena scientists, it was that when JPL did return, it was with a dilly of a ship.

The spacecraft that at last took the lab back to the Jovian system was named, fittingly enough, *Galileo,* after the seventeenth-century astronomer who discovered the planet's four large moons. Like its human namesake, the machine was nothing short of extraordinary, a two-part ship consisting of a suicide probe that would plunge into the atmosphere of Jupiter and a far larger ship that would orbit the world, spending at least two years photographing its atmosphere and its swarm of moons. It was not an easy matter traveling the 400 million–odd miles out to the giant world, and *Galileo,* which was launched in the autumn of 1989, suffered its share of breakdowns on the way. In December 1995, however, it arrived at Jupiter, firing off its probe and then settling into planetary orbit. The ship spent its first year in the Jovian system productively, getting its bearings, calibrating its instruments, and sending back a wealth of photos and data as it flew twice by Ganymede, once by Callisto, and once—fleetingly—by Europa. In early 1997 it was ready for its first close pass over Europa, one that would at last allow it to take a good, detailed look at the moon. If any survey of the marbleized world was going to prove the existence of the ocean the *Voyager* astronomers had been rhapsodizing about since 1979, it would be this one.

JPL scientists knew that little would have changed on the four-billion-year-old Europa in the flicker of cosmic time since the lab had last dispatched a ship there, but at JPL itself much was different. Over the course of the last decade or so, the increasing miniaturization of both the space agency's budget and of the equipment it bought had wrought dramatic changes. In the lab's main mission control building, the amphitheater-like room from which the flights of *Voyager* and so many other spacecraft had been run was now dark and shuttered. In its place was a far less grand, far more efficient warren of flight control cubicles in a modest office building on the JPL campus. The mammoth, immobile consoles that had filled the old auditorium had been done away with, too, replaced by prim little desktop computers plugged into the new workstations.

Most important—at least to the researchers on the imaging team—the communal TV rooms where groups of scientists used to gather to whoop and gape as the first grainy pictures from deep space flickered onto black-and-white monitors had been closed up as well. The next time a spacecraft visited a planet or moon, scientists who wanted to view the snapshots it sent home would simply retreat to their individual offices, and there, with the help of their own computers, click a key and call the images up. The pictures, to be sure, would be crisper; their colors, certainly, would be sharper. The scientists, however, would be all alone when they saw them, glimpsing a brand-new world from the same solitary seat at which they answered their mail and made their calls and ate their brown bag lunches. Whether the trade-off was a good one, no one could say.

It was on March 20 when word went around that the latest images from the Europa flyby were on their way. According to the telemetry readings, *Galileo* had barnstormed the moon at a distance of barely 364 miles—closer than any spacecraft had ever come before. The on-board cameras, according to the same data, had successfully snapped hundreds of pictures before the ship flew on, though just how good those pictures would be would not be known until computers on the ground had had a chance to receive and translate the spacecraft's transmissions. Early in the afternoon, it was said, most of this processing work would be done, and by 1 P.M., the hallways and common spaces in the JPL office buildings began to empty. Torrence Johnson, who, as a young engineer, had sat with Sagan and the rest when the Voyager pictures were first fired home, was once again head of the image analysis team. With thoughts of that fine, communal day running through his mind, he watched as the scientists all around him vanished into their offices and then, after a moment, returned to his own.

Sitting down at his computer, Johnson, like dozens of *Galileo* team members at dozens of computers around JPL, did nothing at first but wait, fixing his eyes on the monitor for a sign that the pictures were ready. Almost immediately, a small tone sounded and a tiny icon appeared on the screen. Johnson moved his cursor to the icon and clicked it quickly. In the offices and cubicles all around him, solitary scientists did the same.

On all of the computer screens, a fuzzily luminous landscape began to appear. From the poor definition and the out-of-focus edges, the picture could have been taken almost anywhere—Europa, Callisto, arctic Canada. After an instant the image scanned again and the contours sharpened. No Canada anymore, this was clearly another world. One more scan and that world was clearly Europa.

Johnson ran his eyes over the familiar eggshell landscape of the smooth, hard-frozen globe—a landscape he had first seen nearly a generation ago—and something quickly caught his attention. Near the center of the image, unnoticeable at first, was what appeared to be a small, jagged peak poking up through the surface. It was little more than a shard really, easy to overlook, but the fact that it was there at all was curious. Europa, as nearly as anyone could tell, had no hard topography—at least not any that was visible. Whatever hypothetical features defined its surface lay deep beneath its hypothetical ocean.

Johnson studied the image curiously and understanding quickly flooded in. If he wasn't looking at a mountain and he wasn't looking at a hill, there was only one thing he *could* be looking at: an iceberg. The size was right, the shape was right, the crazy angle at which it appeared to be bobbing was right. Ice, of course, could exist anywhere; ice*bergs,* on the other hand, could exist only in certain places—places that had oceans. Somewhere beneath the hard face of Europa, the frozen mantle must have melted into liquid, and that liquid was apparently churning and surging, cracking the ancient crust above it.

More pictures flashed on the screen now and more icebergs appeared, all glinting the color of fresh snow. There were big bergs and small bergs; upright bergs and tipped bergs. There were icebergs that had shattered into little more than shards and icebergs that had calved away in chunks big enough to fill San Francisco Bay. All of the countless fragments appeared to be locked in place, as if the surface had gone soft just long enough to allow the ice to chip and tip, after which it had frozen up again. But an ice sheet this brittle would certainly crack again, and when it did, the sea would surely spill through it.

Johnson let the full album of images play across his screen, then slowly got up and walked into the hallway, blinking a bit dazedly at what he had just seen. From identical offices up and down the hall, other scientists emerged, wearing identical expressions. Johnson headed straight for the office of Bill O'Neil, the *Galileo* project manager, and caught him before he could even rise from his desk.

"We've got it, Bill," he said simply. "This is the real thing."

O'Neil turned to Johnson and simply smiled. On his screen, Johnson could see, the icebergs were sparkling, too.

Nobody ever called them pilots, and that always galled them a bit. Nobody ever called them explorers, and that rankled, too. When you're building ships in as unlovely a place as Pasadena, trucking them across the country to as unremarkable a place as Florida, then picking them up and flinging them into space—steering them to planets and moons billions of miles away—you'd like to be known as the celestial adventurers you are. Historians, however, never saw things that way.

The problem was that when it came to celestial adventurers, historians had a bellyful. They called them astronauts—and astronauts, as anyone knew, did remarkable things. Unlike the self-styled adventurers at the Jet Propulsion Laboratory in Pasadena, astronauts didn't just build ships, but climbed inside them. Unlike the adventurers at JPL, astronauts didn't fly their spacecraft by remote control, but from the inside of a cockpit, sailing through space by stick, rudder, and the seat of their silvery pants. Unlike the JPL adventurers, astronauts knew that if the ship they were flying turned rabbit on them, spinning out of control and tumbling off into space, they'd never be coming home again. With all that, it was no wonder historians devoted most of their verse to the astronauts alone, remembering the engineers at the little laboratory in Pasadena as mere exploratory afterthoughts.

But the engineers at JPL were more than afterthoughts. While NASA's astronauts never got further than the moon, the robot probes of JPL *started* with the moon. Before human pilots had dipped barely a toe into space, Pasadena engineers were firing off machines designed to crash-land on the lunar surface, soft-land on the lunar surface, serenely orbit the great girth of the ancient moon. When astronauts finally did reach the moon themselves—settling down into the soil with their air tanks and water bags and all the other pulsing, bubbling equipment they needed to keep themselves alive—the lean, fleet ships of the Jet Propulsion Laboratory had already gone elsewhere, heading off into the true trackless depths of the unexplored solar system. What they found there surprised even their inventors.

Earth's own moon—known as Selēnē by the Greeks, Luna in the Romance languages, and simply "the Moon" in uncluttered English—was never really worthy of such a flurry of names. Smaller than Mars, deader than Mercury, it is one more rock in a solar system of larger rocks. The other sixty-two moons circling six of the other eight planets are another matter.

They are strange, almost whimsical places, those other moons, places where the ordinary rules governing heavenly bodies seem not to apply. They're places where volcanoes spew sparkly snow, where rivers run with scalding ammonia, where geysers spout carbonated water, where lakes brim with organic ethane, where fires that burn on one world dust the cliffs of another with ash, where whole globes may shatter into shards and then reassemble themselves. And they are places where life, even now, might be taking hold.

No two of the five dozen or so moons are exactly the same. There are some that are big—bigger than Mercury. There are others that are small—smaller than Manhattan. Some of the moons are red, some are orange, some are black, some are white. One moon is as reflective as a silver platter; another is tarry and dark; a third is both shades at once, with a leading hemisphere black as asphalt and a trailing one white as snow. Like most worlds, the moons are generally round and generally marked by mountains and valleys—but not always. Some are buffed smooth as eggshells; one is as oblong as a great potato. There may not even be exactly sixty-three moons in the solar system at all. There may, in fact, be many more.

Whatever their exact number, the sixty-three or so moons do exist—a second, unknown solar system within the existing known one. For the last thirty-five years, a whole uncelebrated class of explorers has been visiting all of these worlds, sending out probes to run cosmic reconnaissance for the human species back home. The discoveries made by the unmanned ships—as well as by the few manned ones that have followed them as far as Earth's own moon—have been nothing short of remarkable. The stories of how JPL engineers learned to send their robot spacecraft on their extraordinary journeys have been even more so.

Part I

Near

1

A Splendid Suicide

Pasadena, Calif., February 2, 1964

William Pickering had reason to believe that Lyndon Johnson was mad at him. Certainly, the president had never told the chief of the Jet Propulsion Laboratory as much, and it was unlikely he ever would. But what Johnson said and what Johnson thought were often two different things, and there was little doubt that right about now, the president was fed up.

It was only in the last few minutes that Johnson's mood would have turned sour. Up until then both he and Pickering—along with the hordes of reporters following Pickering's work on both of the country's coasts—were in a fine frame of mind. Pickering had not necessarily expected this kind of attention at this kind of hour. Out at his laboratory in Pasadena, after all, it was already after one in the morning. That meant that back in Washington, where the president and much of the eastern press establishment lived, it was after four. Nonetheless, more than a hundred reporters had crowded into NASA's Washington media room to follow a closed-circuit radio account of the goings-on at Pickering's lab, and two hundred more were jostling to get into JPL's own Von Karman Auditorium to listen to the same transmission.

Up until a few minutes ago Pickering was delighted by such unexpected coverage. A few minutes ago, however, things hadn't yet gone all to hell. Now they had, and Pickering would have been just as happy if the reporters would forget all about his little project for

the moment and come back some other time when he had better results to report.

The problem was that most of these people had been coming
back to Pickering for a while now. For those who were still counting,
this was the sixth test JPL had run like this, and it looked as if it was
going to be the sixth one to come to grief. It wasn't just the cost of
Pickering's little adventures that got people's backs up—though the
little adventures didn't come cheap. NASA was paying about $9 mil-
lion for each of the silvery, sixty-six-foot Atlas rockets Pickering periodically needed; another $5 million or so for the twelve-foot,
second-stage Agena rocket that went on top of it; and a good $8 mil-
lion more for the most important piece of hardware of all: the ingenious, ten-foot tall, vaguely cone-shaped *Ranger* moonship that went
on top of both of them. Add the cost of manpower, and Pickering
was burning up more than $25 million every time one of his improb-
able contraptions left the ground.

What really bothered the press, the public, and the president
wasn't just the cost of these Rolls-Royce spacecraft, and it wasn't
just the fact that they were all failing. What really raised hackles was
that they were failing so publicly. All five of the previous *Ranger*
launches had been globally covered events; and all five had been followed by some globally covered disaster, which itself was followed
by a globally covered red-faced press conference in which Pickering
and the rest of the JPL brass tried to stammer through an explanation of what went wrong and pledge—unconvincingly—that they
knew what had to be done to put things right.

For anyone who knew anything about space travel, putting
things right should not have been all that difficult. Pickering's ships
weren't intended to carry people to the moon, after all; that was for
the *Apollo* teams to pull off sometime before the end of the decade.
These ships weren't even intended to make an unmanned but controlled landing on the moon; that was for the engineers building the
still-to-be-unveiled *Surveyor* spacecraft to figure out in the next two
or three years.

No, the *Ranger*s were supposed to beat both of these craft to the
moon to take some preliminary readings, snap some preliminary

photos, and do both the easiest way they could. That meant a crash landing. Launched from Earth on a dumb, cannonball trajectory, a *Ranger* ship would fly all but mutely out to the moon, doing virtually nothing at all for the sixty-six or so hours it needed to get there. Then, in the middle of the third day of its life, when the lunar sphere had grown from a distant aspirin tablet to a genuine arm's-reach world and the speeding spacecraft was roughly fifteen minutes away from annihilating itself on its surface, it would switch on a bank of six cameras and begin taking pictures of the fast-closing landscape, beaming them back to NASA's Goldstone Tracking Station in the Mojave Desert at a rate of better than five images every second. Less than a single second before impact, the *Ranger* would click its shutters for the last time, and then, like a sea captain flinging a bottled message into the ocean the instant before his ship is shot out from under him, toss this final image faithfully home, too. An instant later, it would commit a spectacular suicide on the very patch of lunar soil it had just photographed so faithfully.

Pickering and NASA weren't the first ones to dream up such a kamikaze spacecraft. In 1959—more than three years before JPL's hapless *Rangers* started taking to the skies—Soviet scientists had launched exactly this kind of ship, skeet-shooting the moon with their ingenious *Lunik 2,* a probe that successfully blasted itself to bits in the northern lunar hemisphere and left behind a metal pennant embossed with a coat of arms to commemorate the event. The Soviets had been crowing about their lunar bull's-eye ever since it happened, and in the past two years—after the *Rangers* began flying and failing—Soviet premier Nikita Khrushchev had had himself a ripping good time reminding the world of Russia's successes and clucking in mock sympathy at the struggles of the bumbling Americans.

"The Soviet pennant on the moon has been awaiting an American pennant for a long time," the bumptious Russian liked to say. "It is starting to become lonesome."

Pickering, like the rest of the people at JPL, had gotten sick of hearing about the Soviet pennant, mostly because there probably *was* no pennant—not anymore, anyway. Einstein wasn't kidding when he said that the line between energy and matter is a murky one, and a projectile tearing through space at 6,000 miles per hour would become an energetic object indeed. A mere twenty-pound shot put striking the surface of the moon at such sizzling speed would release a blast of energy equivalent to a suitcase full of TNT. An exponentially bigger, 800-pound spacecraft would be exponentially more powerful, not just exploding, but liquefying—perhaps even vaporizing—the moment it hit the surface, destroying itself and any cargo it may have been carrying. Crash-landing spaceships didn't end their lives in a heroic hail of debris as Khrushchev suggested, but in a sort of ignominious splatter. American scientists tried to explain that to the American public at least a thousand times, but the American public was having none of it. When it came to little *Lunik,* all anybody saw was that glinting, coin-like coat of arms winking down at Earth from Moscow's shiny metal pennant. Heads or tails, the Russians won.

Now, nearly five years after the launch of *Lunik 2,* it looked as if they were going to win again. In the early morning hours of February 2, 1964, *Ranger 6* was more than 240,000 miles from the surface of Earth, just over 1,500 miles from the surface of the moon, and moving at close to its 6,000-mile-per-hour maximum velocity. A speed like that and a distance like that meant that the spacecraft was just seconds away from entering its final, mortal plunge and fifteen or so minutes away from lunar impact. Pickering, like the rest of the engineers at JPL, had spent the last sixty-six hours looking forward to this single quarter hour. *Un*like the other engineers, he would be watching it unfold almost totally alone.

NASA had recently built JPL a spanking new mission control center—or Space Flight Operations Facility, as the space agency liked to call it—a three-story, windowless bunker designed specifically for flights like this. The final preparations of the new SFOF had taken longer than planned, however. While most of the furnishings and electronics had already been installed in the place, the desks and consoles that would fill the mission control room proper had yet to arrive. *Ranger 6,* like all of the earlier *Ranger* flights, would thus be run from JPL's existing, comparatively spartan control center. Pickering himself, however, had decided that in the final minutes of this particular mission it might be wise for him to keep his distance from this older facility, lest his flight controllers, already stooping under the weight of five failed lunar trips, feel even greater pressure as their silent, watching boss looked over their shoulders on the sixth. Instead, he and one trusted lieutenant—Homer Newell, chief of NASA's space sciences division—would retreat alone to the new SFOF's glassed-in viewing gallery, overlooking the eerily empty mission control room, while a public address announcer who was in constant touch with the Goldstone Tracking Station in the Mojave Desert kept them abreast of the progress of the flight.

At just before 1 A.M. Pasadena time, Pickering and Newell took their seats in the new building. At that point, the mission clock at the front of the empty amphitheater read twenty minutes to impact. At the stroke of 15:00, the two men knew, the photography phase of the mission should begin, as *Ranger 6* switched on the power to its video system and began warming up its camera array. Exactly two minutes after that, the six video eyes should actually open and begin transmitting images of the lunar landscape that just 780 seconds later would claim the ship. Alone in the brilliantly lit VIP gallery, Pickering and Newell spent five minutes uneasily glancing around the room, shifting their eyes from the clock to one of the speakers built into the freshly painted and papered walls, and then back again. At exactly the appointed moment, the speaker came obligingly to life.

"Fifteen minutes to impact," the public address announcer said, echoing simultaneously through the old mission control, the new SFOF, and the press rooms in Pasadena and Washington. "Telemetry confirms that the cameras are warming."

Pickering exhaled a deep breath, turned to Newell and gave him a tight nod. Beyond the glass of the VIP booth, on the ghost-ship floor of the empty control room, he could almost see the spectral flight controllers hunched over their invisible consoles. A half minute passed.

"Fourteen-thirty to impact," the public address voice said. "Cameras still warming."

Pickering managed a small smile. The voice coming through the speaker belonged to Walt Downhower, a *Ranger* scientist, and a popular one. Pickering himself had tapped Downhower for this historic, if largely symbolic assignment and he was pleased for both of them that things were thus far going well. Already, *Ranger 6* had lived much longer than any other *Ranger* before it; if it survived another fourteen and a half minutes, it would also have lived much better.

"Cameras still warming," Downhower said again after thirty more seconds. He repeated the call again after another thirty, and once more after yet another. Pickering flicked his eyes at Newell and then at the advancing clock on the forward wall, watching it tick down to the 13:00 mark. Just … about … now, the half dozen apertures in the half dozen lenses of *Ranger 6*'s imaging system should be flying open. Pickering listened for the call confirming that this had indeed happened, but when he next heard the voice over the public address system, he froze.

"No indication of video," Downhower announced flatly.

Newell turned to Pickering and started to speak, but Pickering raised a silencing finger and held it there. Fifteen seconds later Downhower spoke again.

"Still no indication of video signal," he said. "Cameras are receiving power but Goldstone is receiving no images."

Another fifteen seconds passed. "Still no indication," he said once more.

Pickering lowered his finger and slowly closed his eyes in disgust. If he was getting this news, so too was the press and—somewhere in the White House—so too was the president. At this instant, Pickering knew, the men in the old mission control room just a building or two away would be working furiously, trying to find out just what had blinded the plunging ship and just what could be done to restore its sight in the dozen or so minutes it had left to live. A dozen minutes could be a long time in the life of a spacecraft—even a spacecraft in a 6,000-mile-per-hour death dive—and given the expertise of the men in the room, it was entirely possible they would succeed in getting the problem solved. With a trail of five dead *Rangers* behind them, however, it was also entirely possible that they wouldn't. Either way, in less than half an hour, Pickering was going to have another press conference to address.

Late 1946

William Pickering always enjoyed the story about the time Wernher von Braun shot a missile at himself. You could say a lot of things about Wernher von Braun—and most people did—but a man who would shoot a missile at himself was a man who deserved at least a little of your respect.

It was just after the end of World War II when Pickering got to know von Braun, at the White Sands missile range in New Mexico. At the time Pickering could not have had less in common with the beefy German. Pickering was born in New Zealand in 1910 and lived there only until 1929, when he moved to the United States to study electrical engineering at Pasadena's California Institute of Technology. New Zealand, unlike most of the other Western and Pacific democracies, had taken its time getting itself electrified, mostly because the dynamos necessary to do the job required oil, and oil required money and New Zealand—which had long gotten by perfectly fine selling the world its butter, wool, lamb, and fruit, thank you very much—wasn't about to bankrupt itself now just to string the home islands with electrical cables. But if New Zealand lacked oil, it did not want for fjords and falls, and once the government was persuaded that it was possible to light the country using nothing more than the power of flowing water, the future of the hydroelectric industry started to look bright.

Pickering, who always had kind of a knack with machines, figured there was a lot of money to be made helping to plug the country in this way, provided he got the right education, and that meant Caltech. The plan was a good one, and it might well have worked out, but barely a month after the visiting student arrived in the United States, the American stock market took a nasty and much-publicized tumble. If the American economy was falling, the economies of much of the rest of the world would be falling further, and in New Zealand, that meant you could say goodbye to electrification. Pickering spent the next six years at Caltech, earning a Ph.D. in electrical engineering and all the while keeping an eye on the job market back home. After graduation, with employment prospects just as bleak in New Zealand as he feared they would be, he decided to settle permanently in southern California.

Even in southern California, it turned out, there weren't that many jobs to be had, but Caltech was willing to help out its own. No sooner had Pickering earned his doctorate than the school offered him an instructor's position, teaching new students essentially the same science he had just been taught. The job didn't include much of the hands-on engineering work Pickering loved, but Caltech sweetened the deal: Whenever the new instructor had time, he was welcome to drive out to the nearby Arroyo Seco riverbed and lend a hand designing some prototype rockets a team of Caltech engineers and graduate students had been fooling around with lately. The place they'd been conducting their research wasn't much—little more than a stretch of dead canyon wash that looked to be more moonscape than landscape. Nonetheless, the instructors and students working there had taken sort of a liking to the little patch of wilderness, and had lately christened it—with decidedly more grandeur than it deserved—the Jet Propulsion Laboratory.

Pickering, who had never much thought about rockets, made a few visits out to the so-called JPL and found that he had something of a flair for the machines being built there. Over the course of the next ten years, he and the other desert designers found they were spending more and more time in their remote riverbed, dedicating themselves to the straightforward task of trying to build new generations of rocket motors that didn't do what most existing rocket motors tended to do, which was blow up. The team had middling success with these efforts, designing both solid- and liquid-fueled missiles that sometimes flew and sometimes didn't, but in neither event went very far. What made this kind of incremental progress hard to take was that even as the Caltech researchers were assembling their firecracker ships, they knew there was one place in the world where the rockets were flying far, indeed: Germany.

During the final years of the Second World War, scientists in both the East and West had watched with a mixture of horror and respect as the German Reich introduced the other nations of the world to its newly invented V-2 rocket. From somewhere in greater Germany—

intelligence had pretty much pinned it down to the northern village of Peenemünde—the military was assaulting hated Britain with a forty-seven-foot-tall, 29,000-pound missile capable of flying over the Baltic Sea to the heart of London and doing a regiment's worth of damage without risking a single German life. The V-2, to be sure, was a brutish machine, a rude and random terror device that may have been good at blowing things up, but did so with little more tactical precision than a hurled stone. To rocket scientists like Pickering, however, such an inelegant device was also a magnificent device. Here was a missile that flew where it was supposed to and *when* it was supposed to, and did both without consuming itself in an unplanned fireball before it got there. If you could somehow put the homicidal purpose of the V-2 hardware out of your head—no small feat, to be sure—the hardware itself inspired nothing but respect. Von Braun, the man who was said to be the rocket's chief designer, inspired almost as much—or at least he started to after the story got around about the business of firing a missile at himself.

During the earliest tests of the V-2, the rumors went, the Peenemünde scientists generally aimed their unproven rockets out over the Baltic. From the safety of their bunkers, they could see the missiles vanish over the horizon and assumed—reasonably—that somewhere in the vast waters off the German coast they were splashing down and sinking. But when the engineers moved their tests inland and aimed instead at firing ranges deep in Poland, they got a nasty surprise: Their rockets went up all right, but inexplicably, they never came down. Somewhere in their trajectory, the fifteen-ton machines were simply ceasing to exist. Von Braun polled his engineers and got all manner of theories to explain the phenomenon, from guidance problems to structural problems to premature detonation of the warhead. What he didn't get, however, was a solid answer. Exasperated, he came up with a way to get one: Scribbling out the coordinates for a particular spot far off in the test range, he handed the sheet of paper to one of his launch technicians and told him to fire the missile at precisely that point in precisely two hours.

Will you be here for the launch? the technician was said to have asked.

No, von Braun reportedly told him.

Will you be back in Berlin?

No, again, came the answer from the chief designer.

Where then? the technician pressed.

Von Braun, the story went, then simply smiled and pointed at the target coordinates he had just scribbled down.

Two hours later the man who invented the V-2 stood at the far end of the ballistic rainbow one of his machines was about to paint in the sky, picked up a field phone, and ordered his team to fire. A few minutes later he saw the black-and-white body of the five-story rocket appear over the horizon, climb high over his head, and start to fall toward him. Von Braun watched and watched as the missile fell and fell, until, shortly before it should have reached the ground, it simply incinerated, breaking into pieces and melting to hot, plummeting slag. Von Braun nodded to himself and walked from the field. The loss of his missile, he now knew, was not due to a problem with its payload or its gyros, but rather to a mere meltdown of its metal skin, which was evidently too fragile to handle the white-hot heat the vehicle generated as it tore through the atmosphere. Von Braun consulted with his engineers, toughened his rocket's hide, and soon had a missile that was fit to fly.

Pickering was intrigued by such a line-of-fire scientist, and in 1946 he learned that he was going to get a chance to observe him up close. Shortly after Germany fell to the Allies, word got around that an arsenal of V-2s had been bundled up like hellish cordwood and shipped to White Sands for research. More remarkably, a detachment of German scientists headed by von Braun had been bundled up in their own way and shipped along with them. Most remarkably of all, Washington was going to ask a few American engineers to travel to the New Mexico desert to study with the Germans, and Pickering was going to be among them.

Over the course of the next decade, with von Braun working exclusively at White Sands, and Pickering—who by now had been named chief of the modest Jet Propulsion Laboratory—traveling back and forth between Pasadena and New Mexico, the Germans and Americans cranked out an impressive line of rockets, some capable of traveling more than 250 miles into the sky. More than 250 miles into the sky, of course, was well into what astronomers thought of as space, and the White Sands engineers liked to think of it that way as well, flattering themselves that they were designing not just rocket ships but starships. The problem with this idea was that while the New Mexico missiles could climb high, they couldn't climb very fast, and without enough propulsive muscle to attain the 17,500 miles per hour necessary to achieve Earth orbit, every putative starship that went up simply fell back down. In 1956, von Braun and Pickering thought they had at last come up with a way to get this problem licked.

Working at his JPL riverbed, Pickering and other Caltech researchers had developed a three-foot tall, solid-fueled rocket that they used principally to study aerodynamics. Out at White Sands, von Braun had developed a slimmed-down and souped-up V-2 that he called a Redstone. In order to study the kind of reentry heating problem that had claimed von Braun's early V-2s, the two scientists had recently combined their inventions, stacking a cluster of eleven of Pickering's little rockets on top of von Braun's big one, and then mounting another three-engine cluster on top of that. The purpose of the improbable assembly was to carry the top stage of the missile as high as it could go, so that when it tipped over and fell back to Earth, it would generate air friction similar to what a vehicle would experience when it returned from space. What the scientists hadn't counted on was just how high and how fast that three-engine top stage would climb. Put one more single-engine stage on top of it, they realized, and it just might be possible to kick a small satellite into orbit.

In as insular a community as the missile community, whispered word of a possible, practical space rocket doesn't stay whispered for long, and in early 1957, General John Medaris, the military commander of the White Sands base, approached Pickering.

"You sure about the design of this rocket of yours?" the general asked.

"Absolutely," Pickering answered.

"It could really get us into orbit?"

"It really could."

"Soon?"

"Very soon."

Medaris nodded slowly. "Impressive," he said. "I wish I could let you fly it."

Pickering looked dumbfounded. "You *wish?*"

"This comes straight from Eisenhower. The Redstone's a military missile, and if we're going to go into space without spooking the rest of the world—particularly the Russians—we're going to have to do it aboard a civilian rocket that can do the same job."

"We don't *have* a civilian rocket that can do the same job," Pickering said.

"Then," Medaris answered, "we'd better invent one."

Whether the scientists and military men of the Soviet Union were actually spooked by the rocketry prowess of the United States was impossible to know. If they were, they didn't let it distract them from their work. On October 4, 1957—just nine months after Pickering's meeting with Medaris—Moscow stunned the West with the announcement that it had just placed a 184-pound robot ball in a 560-mile circular orbit around Earth. The little artificial moon wasn't much—a twenty-inch metallic melon containing instruments no more sophisticated than a thermometer, a radio, and a battery to power both. Nonetheless, the satellite—or the *Sputnik,* as the Russians urged the world to call it—was indeed aloft, speeding around the planet once every ninety minutes and making a conspicuous transcontinental passage over the United States on most of those trips. The rocket used to loft the craft, in the event anyone was concerned, was the huge and fearsome R-7, originally built for the sole purpose of carrying intercontinental ballistic missiles.

In the United States, the scientists answered back fast—after a fashion. Even before Eisenhower dispatched Medaris to talk to Pickering, the president had instructed the Pentagon to order its highly respected Naval Research Laboratory to try to come up with a rocket capable of flying into space. Despite the lab's military pedigree, virtually all of the machines it had ever developed had been used for non-battlefield purposes like weather forecasting and atmospheric study, and Eisenhower was convinced that its space missiles, too, would thus be seen as instruments of peace. In the months Pickering and von Braun's four-stage rocket was sitting idle in its hangar, the Navy scientists had apparently come through, designing a slender reed of a rocket they named Vanguard. The new missile was a disarmingly frail-looking thing next to Pickering and von Braun's burly machine, and next to the Soviets' giant R-7 . . . well, that was a comparison it didn't pay to make.

Appearances notwithstanding, the Vanguards were said to work, and on December 6, 1957, just two months after *Sputnik* was launched, the Navy scientists rolled one of the rockets out to a launch pad at the edge of an old missile range in Cape Canaveral, Florida, uncrated a tiny, shiny satellite of their own, and mounted it lovingly atop the rocket. With news crews covering the event, the engineers backed away, lit the Vanguard's fuse, and watched first in pride and then in horror as the missile smoked a little, steamed a little, rose a few promising feet off the ground and then suddenly ate itself in a bright, white explosion. The spherical satellite, evidently made of sterner stuff than the rocket it relied on, fell to the ground, bounced a few dozen yards away, and came to rest in front of the nearby blockhouse, beeping its idiot signal to the humiliated engineers closed up inside.

The national press howled with laughter at the Canaveral debacle, filling the next morning's newspapers with stories about the Navy and its spectacular new "Kaputnik" satellite. The government only deepened its public relations hole later in the day when a spokesman held a press conference intended to explain away the disaster.

"What do you think was the cause of the explosion?" a reporter asked.

"I'm not sure I'd call it an explosion," the spokesman said.

"What would you call it?" asked the incredulous reporter.

"Rapid burning."

However the government chose to describe its public humbling, it was a humbling just the same, and to Pickering and von Braun, it was an unnecessary one. To the scientists' surprise, General Medaris—the very man who had grounded their far more fit booster in the first place—agreed with them. Several days after the Vanguard disaster, word went around White Sands and JPL that Medaris had been called to the White House to discuss the accident with Eisenhower. When he got there, he wasted little time telling his commander in chief what he thought he needed to know.

If you want a satellite, the general was said to have said, *we can put one up for you.*

What would you use to launch it? Eisenhower reportedly asked.

The Redstone, Medaris answered, betraying not a trace of apology for his military missile. *A Redstone first stage with solid rocket clusters on top of it.*

And you're sure this would work? the president wanted to know.

As sure as we can be.

How long would it take you to launch?

Medaris paused. *Eighty days,* he said. *Give us eighty days and you'll have a satellite.*

At this, the stories went, the president looked at Medaris for a long and stern moment and then nodded. *Eighty days,* he said crisply. Medaris didn't have to be a military man to know that this was an order.

With Eisenhower's okay, Pickering, von Braun, and their rocketeers scrambled to get ready. Of course, just what kind of satellite they'd be firing into orbit if they indeed made the president's deadline was not clear. From the beginning Pickering and von Braun frankly didn't care if they put a bocce ball into space, so long as they put it there successfully. For Washington, however, it was important to maintain at least the appearance that this was more a scientific enterprise than a political one, and the order thus came down that a payload of some kind would be necessary. The engineers complied, cannibalizing a cosmic ray detector that had been built by University of Iowa astronomer James Van Allen for the Vanguard satellite and reconfiguring it to fit on top of their booster. Van Allen, they knew, was a reliable scientist whose instruments would be likely to operate as advertised, and a cosmic ray detector was exactly the kind of otherworldly instrument the politicians would be looking for on this mission. The satellite would be appropriately christened *Explorer 1.*

The booster, shedding its military past if only in name, would be known by the heroic handle Jupiter-C.

The last week in January—weeks before the expiration of Eisenhower's eighty days—Pickering and von Braun, with Van Allen at their side, mated the little *Explorer* to the giant Jupiter and rolled the rocket out to its Cape Canaveral pad. Several days later they announced, probably on the evening of January 31, they would at last attempt to launch it.

When that evening arrived, the beaches along the Canaveral coast were dotted with newsmen, most of whom had been on hand for the headline-making Vanguard pratfall, and none of whom would dare miss this one as a result. Mingling with the reporters were thousands of locals and tourists who had arrived in cars and on foot, and were determined to stay put until the new rocket either flew, or—far more likely—spectacularly failed to fly. Most of the members of Congress, it was said, were following the launch preparations as well, and though few of them had managed to make it here, many had sent press aides who were filing regular dispatches by phone. Even Eisenhower himself, who was on vacation in Augusta, Georgia, and had set the evening aside for dinner and a few hands of bridge with friends, had given explicit instructions that he was to be informed immediately if the Jupiter-C actually left the pad.

To the surprise of nearly everyone on hand, at 10:48 P.M. the Jupiter-C left indeed, bursting to life and rising from the ground in an explosion of hellfire that made its V-2 granddaddy look like a sparkler. Minutes later it vanished into the Florida sky and began speeding toward space, carrying the modest *Explorer 1* with it. Though the crowd that witnessed the rocket's launch was a big one, Pickering, von Braun, and Van Allen were not part of it. After the Jupiter-C took off, it would be impossible to know if *Explorer 1* actually reached orbit for at least ninety minutes—about the time it would take the satellite to complete its first circuit of the globe and at last be in range of the desert tracking station in California's Earthquake Valley. Throughout that time, the project's military sponsors had decided, it would be wise if the three engineers who designed the hardware were not off somewhere on the coast of Florida, but were

within arm's reach at the Pentagon—the better to address a hastily assembled press conference if the mission succeeded, or to tell the De-
fense Department brass what went wrong if it didn't. The men
responsible for creating the rocket that was attracting so much attention tonight were thus nowhere near their machine, but were instead
closed up in a windowless government conference room with General
Medaris and several other officers, following the events by radio.

The three scientists endured the launch in silence—nodding reas-
suringly to the Pentagon men when word arrived that the Jupiter-C
had left the ground successfully—and spent the next eighty-nine or
so minutes smoking cigarettes, glancing at watches, and milling
pointlessly about the room. Finally, at exactly the ninety-minute
mark in the mission—with the stares of the other men in the room
boring into him—Pickering picked up the phone and called his lab.
Frank Goddard, an assistant JPL administrator, took the call.

"Frank?" Pickering said, avoiding the glances of Medaris and the
others. "Anything from the desert?"

"No word yet," Goddard answered.

No word yet, Pickering mouthed to the other men, and tried a
nonchalant smile.

"Are we accurate on time of acquisition?" Pickering asked.

"Should be happening right about now," Goddard answered.

Any minute now, Pickering mouthed.

A silence fell over both the phone line and the room.

"Some launch, huh?" Pickering said into the phone.

"Mm-hmm," Goddard answered.

Mm-hmm, Pickering nodded. He tried the smile again; there
were few in return. Sixty seconds or so passed.

"How late are we, Frank?" Pickering asked.

"Less than two minutes past planned acquisition."

Pickering covered the phone. "Less than two minutes," he said to
the room, and waved a little dismissively. "That's nothing at all.
These things are always rough."

Pickering was telling the truth as far as it went, but he also knew
that while predicting a spacecraft's flight profile was an approximate
science, it could only be *so* approximate. *Explorer 1* might be three

or four minutes late in showing its face, but once it got up to five or six or seven, it became increasingly likely that the satellite was lost. No one said anything for another fifty or so seconds.

"Nothing, Bill," Goddard said at the toll of the third minute.

"Nothing, Bill," he repeated after the fourth.

A fifth minute passed, then a sixth, and then a seventh, each marked by its own grim announcement. Finally, as the eighth minute approached, Pickering closed his eyes. The satellite, he was now certain, had failed. Goddard might not yet be willing to admit it, but Pickering would nevertheless have to announce it, conceding to the generals in the room that, despite his best efforts, the Jupiter-C was just another Vanguard. Suddenly, however, on the other end of the phone line, someone passed a paper to Goddard. Someone else passed him another one, and Pickering suddenly heard a whoop in his ear.

"Acquisition!" Goddard announced. "Acquisition! *Explorer*'s talking to Earthquake Valley!"

Pickering pumped a fist in the air. "They've got it," he shouted to the room at large. "We're in orbit!"

In the windowless Pentagon room, pandemonium—or the most that could ever pass for pandemonium in a windowless Pentagon room—broke out. Generals shook hands, slapped backs, and lit cigars while Pickering held one ear, pressed the phone closer to the other, and tried to absorb as much information as he could about the data that was streaming down to Earth from his brand-new moon. Finally, one of the generals caught Pickering's eye, motioned to his watch, and indicated it was time for him to hang up the phone. Pickering did, and almost immediately, he, von Braun, and Van Allen were hustled out of the room, out of the building, and into the back seat of a black government car. With Medaris and two other Pentagon officials pressed in with them, they were driven through the rainy winter night to the nearby National Academy of Sciences.

When they arrived at the academy, Pickering was surprised to see the windows dark and the building all but deserted. Nonetheless, the generals hurried the scientists out of the car, into the building, and down a long, empty hallway until they reached a set of doors marked

"Auditorium." Behind the doors Pickering thought he could hear a hum and murmur. Medaris nodded to the scientists encouragingly and gave the doors a push. Instantly, the hum and murmur turned into a roar. Inside, Pickering could see, were more newsmen than he'd ever imagined could gather in one place at one time. When the reporters caught sight of the party that had just arrived, television lights flicked on like klieg lamps, flashbulbs started to pop like flash paper, and a low, guttural ovation began and built.

Pickering, von Braun, and Van Allen froze in the doorway until Medaris nudged them gently from behind. The newsmen instantly descended on them. Moving in the middle of the crowd toward the front of the room, the scientists realized that they were not quite being borne aloft, but nor were they moving entirely under their own power. Once they reached and mounted the auditorium stage, someone positioned them in a tight group and another fusillade of cameras went off. Someone then changed their pose and the cameras fired once more. Finally, someone handed them a model of the three-foot-long *Explorer* satellite, the satellite that, even now, was circling Earth every hour and a half at speeds approaching five miles per second. The scientists glanced at the model for an instant, uncertain what they should do with it, and then, as if with one mind, they held it aloft. It was this picture that drew the most flashes of all, and it was this picture that would appear in sunrise editions of newspapers all over the world the next day. Pickering surveyed the crowd as the cameras were firing.

Never the same, he thought. *My life's never going to be the same.*

•   •   •

Later that year the government took steps that ensured Pickering would be right. With the country having at last established a toehold in space, Washington moved fast, announcing that it was dissolving its forty-year-old National Advisory Committee for Aeronautics, or NACA, a sleepy federal agency that had been charged with the responsibility of keeping the country abreast of developments in the field of aeronautics. In its place would be the newer, sleeker National Aeronautics and Space Administration, or NASA. Hoping to get itself up and running in a hurry, NASA decided to forgo the tedious business of establishing its own research centers and staffing them with its own scientists. Rather, the agency would simply go cherry-picking among the country's existing missile ranges and rocketry labs, inviting a select few to join the new agency. Those that accepted would continue to be part of the private university or military branch that currently controlled them, but from now on, part of their funds and much of their work would come from NASA.

With the government writing the checks for such an institutional shopping spree, the space agency wasted little time, snatching up facilities from all over the country. Included in NASA's haul were the Air Force's Cape Canaveral missile range in Florida, the Army's ballistic missile test agency in Huntsville, Alabama, and two old NACA centers in Hampton, Virginia, and Greenbelt, Maryland. In June 1959 the space agency also made an offer to Caltech's Jet Propulsion Laboratory in Pasadena, California. JPL was an unprepossessing place—a modest cluster of buildings that had been slapped up in the desert almost haphazardly over the years. But with the right infusion of NASA cash, it might be possible to turn the scruffy little lab into a first-rate spaceport.

Caltech, in consultation with Pickering, accepted the invitation, with but a single condition: While most of the new labs NASA had recruited would be concentrating their efforts on putting astronauts and satellites into near-Earth orbit, JPL wanted to go deeper. The desert lab hoped to build unmanned machines that would not simply circle the home planet, but fly far beyond it, surfing the worlds of the far-flung solar system. If NASA could promise that, the lab would come aboard.

NASA indeed promised, JPL received its membership papers, and before the end of the year Pickering assumed the reins of the newly federalized space center. His engineers' first target, he decided, would be the moon. The ship that would get them there would be called *Ranger.*

August 23, 1961

Even deep inside Cape Canaveral's Hangar AE it was possible to hear the frogs croaking. When you were outside the giant, stamped-tin building that stood in the middle of the old Air Force base, you expected to hear frogs. Indeed, from almost anywhere on the 15,000 acres of mud and sand that was all the cape really was, it was impossible to get away from them. But once you went inside—once you slammed the door of the building where you spent so many of your workdays and tried to concentrate on what you were doing—you expected a little peace. In Hangar AE you rarely got it.

It didn't come as a surprise to most people that the walls of the giant Canaveral Quonset hut weren't quite up to the job of shutting out the frogs. As mission control rooms went, after all, the flimsy building wasn't much. Yes, it had consoles and monitors and headsets like any other control center; some time ago someone had even thought to throw a huge swatch of industrial carpeting down on the raw, concrete floor. But an airplane hangar was still an airplane hangar, and when it came to accommodations NASA didn't appear to be prepared to offer more.

Certainly, the agency was *capable* of offering more. Earlier this year Alan Shepard and Gus Grissom had flown their historic *Mercury* capsules atop their historic Redstone rockets, becoming the first and second Americans ever to travel into space, and during those flights the whole world had gotten a peek inside the glittering mission control room where teams of earnest young engineers nursed the spacecraft off the pad, into space, and safely down again. The men in Hangar AE, however, weren't launching men into space, but robot probes, and for a job like this, a stamped-tin Quonset hut would have to do. Tonight, that job consisted of launching the *Ranger 1* space probe toward the moon.

It had been more than two years since William Pickering and his Jet Propulsion Laboratory had joined the fledgling NASA, and much of that time had been devoted to preparing for this one night. Pickering had spent most of the past twenty-four months shuttling between

JPL and NASA's new Washington headquarters, seeing to both the design of the new ship and the funding that would get it built. His wife, Muriel, had spent most of *her* time learning to tolerate his frequent absences. Muriel was the sister of one of Pickering's old Caltech roommates, and the two had met early on in Pickering's undergraduate career, when she came to campus to visit her brother. Having spent so much time around apprentice engineers, Muriel had grown accustomed to their near-mad absorption in their work, but even this had not prepared her for the pace her husband had been keeping since 1959.

"I'm home a lot more than you think," Pickering used to insist unconvincingly. "On average, I only go to Washington once a month."

"Then most months," Muriel would answer archly, "you beat the average."

Muriel's objections notwithstanding, on this evening, all the coast-to-coast commuting appeared to have paid off. Strictly speaking, the *Ranger 1* moonship Pickering was preparing to launch tonight was not headed straight for the moon, and barring some gigantic miscalculation in trajectory, it would never even come close to it. Where it was headed was much farther into space. From the beginning, the engineers designing the *Ranger* spacecraft realized that the machine they were building had to be a hardy piece of hardware. Leaving the ground aboard the bone-jolting Atlas-Agena rocket, it would be required to survive a rapid acceleration to 25,000 miles per hour and a three-day journey in the killing, −200 degree cold of translunar space, arriving in the lunar vicinity with none of its countless delicate systems either frozen or broken or shaken to pieces. It was an improbable bit of shipbuilding, and in order to prove it could be done, the engineers needed a shakedown cruise, a test flight with a prototype *Ranger* that didn't have to trouble itself with making it to the moon, but merely with getting into deep space, flying about a bit, and making sure its systems worked as they should.

Tonight's flight, then, would be less a finely aimed arrow-shot at the shiny center of the lunar apple a quarter million miles or so away than a crude hammer throw more than 500,000 miles into the void, during which the controllers would send the ship a series of commands and check to see how it responded. After a day or two of these deep-space calisthenics, they would simply abandon the craft to the cosmos.

Pickering, of course, was here for tonight's launch, flown in special from his JPL headquarters out in Pasadena, and from the moment he arrived he could see that his hangar was a busy place. Though the launch of *Ranger 1* was still hours away, the giant Atlas missile was already on its pad, sweating clouds of liquid oxygen vapor. Tucked inside its nose cone like a kangaroo joey was *Ranger 1* itself, transmitting its vital signs back to the control consoles in the hangar, where the nervous scratchings of oscilloscope beams served as a sort of vehicular EKG, reassuring the engineers seated at their monitors that the cargo was healthy.

The signals streaming back from the launch pad were not being monitored in just one control center, but in three. There was the launch blockhouse—located about half a mile from Hangar AE— where Wernher von Braun and a few select members of his Peenemünde team were waiting for the signal to light the engines that would loft *Ranger 1* into the sky. There was Hangar AE itself, where controllers would see to the welfare of the ship from the time the rocket's engine bells cleared the launch tower to the time the spacecraft arrived in space. Finally, there was a third, equally modest control room out in Pasadena, where a remaining group of engineers would oversee the rest of the mission.

Though Pickering supervised the work in all of these facilities, in Hangar AE he was not the most important man present. That person was Jim Burke, the scientist Pickering had tapped to head the *Ranger* program as a whole. Like Pickering, Burke had learned his craft at Caltech, and had first practiced it for the military, building and test-flying battlefield rockets like the Loki, a nasty little needle of a thing that was designed for the job of targeting enemy warplanes and blowing them out of the sky. Burke was an easy, congenial man, and whether or not he enjoyed such a grim line of work was impossible to know. When he showed up at JPL apparently looking for a happier line, however, the lab's bosses quickly hired him on, evidently trusting him enough to put the entire *Ranger* program in his care.

Burke took to his new job readily—all the more so because he'd be performing it among friends. Joining him in the new enterprise would be Bud Schurmeier and Cliff Cummings, two of his Caltech classmates, with whom he'd spent years studying, skiing, and recreationally flying. Cummings would be one of Burke's superiors, overseeing the development not just of the *Ranger* probes, but of any other unmanned moonships JPL might build down the line as well. Schurmeier would be one of Burke's lieutenants, overseeing quality control and development of the *Ranger* spacecraft's on-board systems. Pickering was pleased that three people who would be working so closely with one another were already good friends, but what mattered more to him was that they were also good engineers. Tonight, with *Ranger 1* smoldering on its Cape Canaveral pad, he'd need all their skills.

It was shortly before sunrise when the *Ranger* countdown clock at last reached 0:00 and the Atlas missile carrying the fragile moonship rumbled to life. The controllers inside the hangar could hear a low, almost subauditory roar growing outside and, as if on cue, turned to look over their shoulders. In the control center where the *Mercury* flights were run, huge TV screens at the front of the room afforded a from-the-blockhouse view of the rocket carrying the astronaut from the moment it was fueled to the moment it vanished into the sky. In Hangar AE, where engineers launched robots, not men, the front of the room was nothing but a blank aluminum wall. On the walls at the back of the room, however, high up near the ceiling, was a row of transom-like windows. During daytime launches the fire from even the brightest rocket was not enough to add much light to the bleached Florida sky. But during a night launch the erupting missile more than two miles away caused the entire bowl of the heavens to light up an iridescent orange. A glance at the windows from inside the hangar could thus help reassure the controllers that their rocket was igniting as it should. Now, looking backward, the *Ranger* team saw with satisfaction that the glass was indeed starting to flare the color of fire.

"Liftoff confirmed," Burke said into his headset, turning his eyes

back toward his screen, where the cathode scribble told him by analog inference what the transom had told him directly. "Launch tower
has been cleared."

If the missile machinery was performing as it should, it should
not be long before Cape Canaveral would once again be dark and
quiet. The Atlas rocket would do most of the heavy lifting on this
flight, carrying the moonship all the way up to a near-orbital altitude
of eighty or ninety miles. The spent missile would then fall away, and
the little Agena booster—a relative popgun of a thing housing the
Ranger itself—would fire briefly, climbing to an altitude of 115 miles
and entering Earth orbit. After less than a single circuit of the planet,
the Agena would light again, pushing the *Ranger* up and away from
its parent world and out into deep space. The *Ranger* would then
separate from the used-up Agena and fly on by itself. From the moment the Atlas left the pad, it was clear that the rockets were following this flight profile precisely, and less than fifteen minutes after the
engines were lit, word came back to the hangar that the Agena and
Ranger were in Earth orbit.

Pickering, Burke, and the other controllers allowed themselves a
moment to stand and stretch. From here they could do little but wait.
Their spacecraft had been fired east and would thus have to make
nearly a full lap around the back side of the world before it could
emerge from the west and head out into deep space. The problem
was, on the back side of the world there was no way to stay in touch
with the ship, and it was while *Ranger 1* was in this communications
blackout that its Agena engine would have to ignite the second time
to kick it out on its final trajectory. If the Goldstone Tracking Station
on the front side of the globe reacquired *Ranger 1*'s signal relatively
quickly—say in forty-five minutes or so—the controllers would
know that the Agena had indeed hit the gas and accelerated the
moon probe out into deep space. If it took longer for the ground to
hear the spacecraft's call—perhaps an hour or so—it would mean
that the engine had failed and the craft was marooned in a pokey orbit around Earth.

As with *Explorer 1,* Pickering knew that everything would hinge
on the word that came back from the Mojave Desert. *And* as with

Explorer 1, when that word finally came, the news at first was not good.

"Sixty seconds to planned acquisition," the tracking officer announced to the room at large when the forty-four-minute mark in the mission arrived. "Goldstone reports no contact."

Sixty seconds later he repeated that call, and at every sixty-second interval after that for the next several minutes. With each announcement the mood in the hangar grew darker and darker. Burke slumped in his chair; other controllers alternately paced and sat. Finally, just over an hour after launch—when the tracking officer had made his announcement more than fifteen times—the call that the controllers feared would come, came.

"*Ranger 1* in low-Earth orbit," the Goldstone man said over the phone to the tracking officer. "Agena apparently failed to burn."

"Roger," the officer repeated for the benefit of the rest of the room. "Agena failure. *Ranger* in low-Earth orbit."

Burke cursed silently to himself; the men on either side of him, unconstrained by leadership positions, muttered their imprecations out loud. Pickering, by contrast, felt surprisingly untroubled. The Agena was a moody rocket and everyone knew it. If it failed to perform as it was supposed to for *Ranger 1,* the telemetry that had been streaming down from the ship would tell the engineers what they'd have to fix in time for *Ranger 2.* For now, the *Ranger* scientists undeniably had a spacecraft in space—never mind that it was near-Earth space—and at the moment, it was heading their way. If Goldstone had acquired *Ranger 1* in California, that meant it was currently less than 3,000 miles away and heading east. Crossing the country at 17,500 miles per hour, it should be over Florida in little more than ten minutes. No one yet knew how well the internal systems of the *Ranger* had survived the punishing launch, but it was important to find out. Pickering thought he knew of a way.

On the Canaveral grounds in front of Hangar AE was a rickety, fifteen-foot light tower equipped with a small, hand-operated antenna. Before launch the antenna was pointed toward the pad, allowing it to pick up the signals from *Ranger* that, in turn, produced the vital-sign scribbles on the controllers' screens. After launch, when the spacecraft should be heading moonward, the antenna would be useless since *Ranger* would be aiming its transmissions not toward some hand-cranked receiver standing in the marshes of Florida, but toward the giant dishes of the Goldstone antennas that had been built to receive just such deep-space transmissions. But *Ranger 1* and its fizzled Agena were not in deep space now, and the little antenna that was never supposed to see them again would once again be within range of their signals.

Even as Pickering was contemplating this, Burke seemed to realize it as well. Glancing at his watch, he jumped to his feet and motioned to a nearby controller, and the two of them sprinted outside into the breaking Florida morning. Wading through the swamp grass, they scampered up the metal ladder of the tower and mounted the platform. Burke seized the handle of the antenna—which was still pointed at the now-empty launch pad—and swung the entire assembly back ninety degrees, pointing it roughly skyward. The controller with him then donned a headset, plugged its cord into an outlet on the platform, and squinted into the sky. If Burke's antenna was picking up the spacecraft's signals, the headset would detect them.

"Got it?" Burke asked.

The controller strained to listen, but heard only a soft hiss.

"Nothing," he said.

Burke fiddled with the handle a bit, moving it this way and that.

"Now?" he asked.

"Nope," the controller said, and then instantly reversed himself. "Yes! Got it. It's coming in."

In the headsets, the engineer could hear something that sounded for all the world like a high-pitched Morse code. What it actually was, however, was the sound of *Ranger 1*'s on-board computer beaming down information in the only language the ground-based computers could understand: the mind-numbing strings of ones and zeros that make up a digital code. A larger transmitter aboard the ship would have been able to send the data faster, but *Ranger*'s tiny system could only manage an electromagnetic trickle—so slow a trickle, in fact, that a technician who was familiar with the systems could actually listen for familiar digital patterns and determine if the ship was healthy or not.

"How's it sound?" Burke asked.

The controller closed his eyes and listened. "It sounds good," he said. "It sounds *really* good."

Burke smiled broadly, reached to his right, and opened a night watchman–style box built into the platform. Lifting the receiver, he was instantly in touch with JPL. For the next several minutes the controller with the headset continued to listen to the stream of *Ranger 1* data, reading all he could back to Burke, who, in turn, read it back to Pasadena.

When the spacecraft at last passed out of range and Burke hung up the phone, he hung it up happily. The ship he and the other JPL engineers had built was operating as it was built to operate, and even though this particular moon probe would be staying close to home, the signals indicated that it clearly would have been up to a lunar trip. Instinctively, Burke looked upward as if he could see the invisible probe passing over, though even if *Ranger 1* had still been directly above him, the brightening sky would have made that impossible.

With the break of day, Burke noticed, the frogs had at last stopped croaking.

•   •   •

After 111 egg-shaped orbits of the Earth—covering three million miles and nearly seven days—*Ranger 1* came ignominiously home, tumbling into the atmosphere, bursting into meteoric flame, and sizzling into the ocean. As Burke and Pickering had anticipated, the JPL engineers were able to read through the hieroglyphs of *Ranger* telemetry and dope out the cause of the Agena breakdown pretty easily.

This Agena—and, indeed, all of the Agenas—were birthed by their inventors with a tiny flaw in a tiny valve that controlled the flow of oxygen to the engine's combustion chamber. The valve, the *Ranger* team found, was placed too close to the chamber, causing it to be exposed to a searing blast of heat every time the engine fired. In the case of most Agenas, which were only fired a single time to put a payload into Earth orbit, this didn't present a problem, since all the valve had to do was open once and close once, and the rocket could then be discarded. In the case of the *Rangers'* Agena, which had to fire twice—once to kick the ship into Earth orbit and once to push it out—flash-frying such a delicate component the first time the engine was lit made it likely that there would never *be* a second time. And indeed, when *Ranger 1*'s Agena tried to start itself back up as it orbited the far side of the Earth, the cooked valve was able to flutter open only a bit—allowing just a little oxygen to drizzle into the combustion chamber—before shutting right back down again. A tiny piece of insulating foil would have been all it took to correct the flaw, but *Ranger 1*'s Agena didn't have that foil, so the intended deep-space ship went almost nowhere at all.

Three months later, on November 18, 1961, the same group of controllers gathered in the same hangar to launch essentially the same spacecraft, though this one was called *Ranger 2*. To their dismay, they got essentially the same results. Once again the Atlas performed well; once again the Agena placed the *Ranger* in Earth orbit; and once again the second burn of this second stage went awry. This time the engine lit as it should have—thanks to the tiny piece of protective foil—but the Agena's gyroscopes failed, causing the ship to wobble and tumble like a badly thrown football. By the time the engine shut down, the spacecraft had completely lost its trajectory footing and found itself stuck in a shaky Earth orbit barely ninety-four miles high. Ninety-four miles is practically treetop height by spacecraft standards, and an orbit so low could not be sustained for long. Little more than a day after *Ranger 2* was launched, it came crashing back to Earth.

With two spacecraft kills on its rap sheet, JPL might have benefited from taking a little time to collect itself. It was hardly wise to send a *Ranger 3* off to the dry seas of the moon when *Rangers 1* and 2 had wound up in the oceans of the Earth. Pickering, however, saw things differently. If there was one lesson missile men learned at places like White Sands, it was that the only way to design a rocket that works is to launch lots of them that don't. When one finally goes where it's supposed to go, just remember how you built it and do it that way again. Shortly after the loss of the second *Ranger,* Pickering called Burke, Cummings, Schurmeier, and other ranking members of the lunar team into his office and reminded them of that fact.

"It looks like we're going to have to sneak up on this problem," he told them. "We're not here to design a spacecraft that we know will fly. We're here to design one that we *think* will fly."

"Seems like we ought to be more certain than that," somebody offered.

"When it works," Pickering said, "we'll be certain." The JPL chief smiled at his team and adjourned the meeting shortly after it began. The lab, it was clear, was going to aim for the moon, and the very next *Ranger* was going to be the one to make the trip.

•   •   •

The engineers at JPL expected to accomplish a lot of things when they designed *Ranger 3,* but building a wooden space probe was not among them. As the liftoff of JPL's next moonship approached, however, that's just what they were preparing to do.

As happy as lunar mapmakers had always been with the *Rangers'* mission profile, lunar geologists had always been less so— and with good reason. A spacecraft beaming back pictures of the moon might do a lot to help you map the lunar surface, but it's absolutely useless when it comes to mapping the *sub*surface. What the geologists had long dreamed of was equipping the ship with at least a crude seismograph that would be able to take at least a few underground soundings of the moon after the ship hit the surface. That seemed impossible, of course, because when a vehicle hits the ground hard enough to vaporize itself, it's hardly likely that any instrument package—let alone one that contains a delicate watchmaker's contraption like a quake detector—is going to live through the impact. Nevertheless, with the announcement that *Ranger 3* was indeed going to head out to the moon, JPL engineers came up with a way to help at least a few of its instruments survive its calamitous landing.

Attached to the top of the new *Ranger 3*—like a sort of giant pom-pom on the stocking cap-shaped body of the spacecraft—was a thirty-six-inch, shock-absorbing sphere. Packed inside the sphere was a sealed instrument package floating inside a sealed envelope of oil. The way the scientists who designed this unlikely device envisioned things, when the spacecraft was just fifteen miles above the moon and falling at 6,000 miles per hour, the sphere would pop free and a small rocket motor attached to its bottom would ignite, slowing its speed to just 80 miles per hour. A good ten minutes after the spacecraft itself had finished snapping its pictures and annihilated itself on the surface, the ball would at last plop to the ground, bouncing several times in the reduced one-sixth gravity, rolling a few hundred feet, and coming to a stop. The floating instrument package inside, which was deliberately designed to be slightly bottom heavy, would then orient itself, switch on its systems, and spend the next thirty to sixty days looking for moonquakes and transmitting what it learned back to Earth.

The contraption was simultaneously ludicrous and brilliant, but it had one small problem: No one had any idea what the ball should be made of. Metals—even the lightest metals—were too heavy and fragile. Plastics and rubbers—even the strongest ones—were too brittle. Someone suggested surrounding the sphere with shock-absorbing air bags that would inflate shortly before impact, but while that might work on a world like, say, Mars, where residual atmosphere could slow the giant balloon down before it hit the ground, on the airless moon even the strongest inflatable material would simply burst on impact.

Finally, Jim Lonborg, one of the lab's more imaginative designers, hit on a possible solution: balsa wood. It was light, it was strong, and most important, it absorbed shock easily—everything the designers were looking for in their seismic ball. Lonborg approached Burke a little warily and began to pitch his idea, but even before he finished, Burke cut him off, laughing.

"Try it, try it," Burke said to his designer. "I've spent half my life building model airplanes. You don't have to sell me on balsa wood."

That day the wooden seismograph sphere went into development at JPL. Not long after, the first working model was mounted atop the *Ranger 3* spacecraft and tagged for flight to the moon.

Just before the spacecraft was shipped from JPL to Cape Canaveral, it underwent one other, far less conspicuous change. At the bottom of every *Ranger* ship were six metal feet designed to help connect the spacecraft to the Agena rocket; attached to each foot were titanium washers that kept the connection snug. Titanium was a good choice for a component like this since the near-indestructibility of the metal would help the tiny rings survive the vibration of launch without coming loose. In fact, so strong were the washers that engineers began to wonder whether they might even survive the crash on the moon itself. In the event they did, it was decided, something else would survive with them.

Inscribed on the washers of this first moonbound craft were the names of Jim Burke and all of the other key engineers who had spent the last three years midwifing its design. The scientists had done the etching themselves—taking turns passing around a vibrating electric pencil in the privacy of their lab as self-consciously as if they had been passing around a bottle of moonshine—and a number of them had taken the opportunity to personalize things even further. Some had inscribed their wives' names; some had included their children's; one had even immortalized his dog. At first, Burke had reservations about allowing the engineers this manifestly unscientific indulgence, but after a while he relented. Planting their nation's colors on the moon might motivate scientists, he realized. But planting their family colors would galvanize them.

•   •   •

Less than five days after it left Pasadena, the air-conditioned van carrying JPL's new, and personalized, *Ranger 3* arrived at Cape Canaveral and backed into a hangar. The Florida launch team would have about six weeks to prepare for what Pickering hoped would be a January 26 launch, and throughout the last weeks of 1961 and the first weeks of 1962, it seemed that they might make that deadline. But barely a week before the time set for liftoff, it began to look as if the new spacecraft would not be going anywhere at all. This time the problem wasn't the *Ranger* itself, and for once it wasn't the cantankerous Agena. This time it was the giant Atlas booster.

The largest organ in the eighty-foot rocket's anatomy was a six-story fuel tank, designed to be filled with explosive kerosene. Resting on top of the tank was a far smaller one holding liquid oxygen that would mix with the fuel and provide it with the flame-feeding atmosphere that would allow it to burn. The stacking of the tanks was a sensible arrangement in the body of so long and narrow a booster, with one small problem. Liquid oxygen is cold—nearly 340 degrees below zero worth of cold—and placing a tank of the frigid stuff on top of a second tank of kerosene was a sure way to chill that liquid, too, turning the already viscous stuff into an unmanageable goo.

To address the problem, the Atlas designers had added a layer of polyurethane insulation to the inside wall of the kerosene tank and covered that with a microthin layer of stainless steel. The design seemed like a good one until barely seven days before liftoff, when pad engineers discovered that in *Ranger 3*'s Atlas, the stainless steel layer had inexplicably sprung a leak and the insulation inside was sopping up kerosene like a giant sponge. The problem was clearly the fault of North American Aviation, the booster's San Diego–based designer and builder; but now North American's problem had wound up in the collective laps of the JPL engineers.

The moment Burke heard about the leak, he phoned Cape Canaveral, alerted the launch team that he would be flying in from Pasadena, and asked to meet with three people: Kurt Debus, the Peenemünde veteran who ran the Florida launch complex; Major Jack Albert, an Air Force officer whose job it was to certify all Atlases either fit or unfit to fly; and John Tribe, a gifted young booster engineer who had nowhere near the seniority of the other three men, but made up for it in raw missile smarts. The group gathered in Debus's office, and once they were there Burke got right to business.

"So what do we do with this leaky missile?" he asked straightaway.

"Not much we can do here," Albert said. "This is North American's job."

"Which means we pull it back from the pad—" Burke began.

"—and ship it back to San Diego," Debus finished.

"And how long for the fix?" Burke pressed.

Debus turned his palms up. "Two months. Maybe three."

Albert and Burke whistled softly. With two *Ranger*s already lost and NASA starting to ask when it could expect some results from the millions of dollars it was spending for each of these temperamental ships, two months might as well have been two years. Burke started to speak, but Tribe interrupted him.

"Don't ship that rocket anywhere," he said.

"North American isn't going to come here and fix it for us," Albert objected.

"They don't have to," Tribe said with an emphatic nod. *"I'll* fix it here."

The three senior men stared at the junior engineer for a moment. "John," Burke said, "you can't do that without taking the tank out of the rocket, and it takes the designer for that."

"We don't need the designer," Tribe said, "if I go inside the tank."

The plan Tribe proposed was a radical one. At the bottom of the Atlas's kerosene tank was an eighteen-inch hole where a bolt ring held one of the engines in place. If the tank was drained and the bolt removed, an engineer—at least a wiry engineer like Tribe—could wriggle inside. The atmosphere he found there would be poisonous, since no amount of draining could remove the toxic vapors six stories of kerosene would leave behind. But with the help of a gas mask, a miner's flashlight helmet, and a carpenter or two, he could build a wooden, ship-in-a-bottle scaffold, climb the sixty feet to the top of the tank, and perform whatever repair work needed to be done. When it was complete, the scaffold could be disassembled and extracted and the tank refilled.

"How long would this fix of yours take?" Burke asked.

"A week or so," Tribe said, and then added pointedly, "seven weeks less than the contractors."

Even if Pickering and Burke could argue with Tribe's plan, they couldn't argue with his math. Within days *Ranger 3*'s Atlas was rolled into a hangar, its engines were detached, and John Tribe and his pair of helpful carpenters were crawling around in its innards. One week later, as promised, the tank was once again fit to be fueled, and the surgically repaired Atlas was rolled out to the pad. On January 26 it took off for the moon. Like the *Ranger*s that preceded it, it would never get where it was going.

•   •   •

When flight planners plot a path to the moon they are always conscious of the fact that they're aiming at a moving target—and a fast-moving one. The moon needs less than twenty-eight days to inscribe its 1.5-million-mile circle around Earth, which means that it covers more than 53,000 miles every day, or more than half a mile every second. For that reason, terrestrial rocketeers can't simply point a spacecraft to the spot in the sky where the moon is today, but rather, where they calculate it will be in about three days. If the spacecraft takes it into its head to try to speed things up, all it winds up doing is arriving at the designated rendezvous point before the moon itself shows up, and then whizzing past into trackless space. If it moves too slowly, the moon will get there first and be long gone by the time the slowpoke ship arrives. For this reason, launch directors trying to fire a spacecraft on a lunar trajectory had to make it a point to keep an especially tight rein on their speeding rockets.

On January 26, 1962, the day *Ranger 3* was launched, it was clear the Atlas was fighting the bit. Just minutes after liftoff, the incoming telemetry indicated that the engines were burning far too hot, adding unwanted speed to its climb. Burke told his controllers to radio up commands instructing the booster to throttle back a bit, but the booster ignored them and raced on. When the out-of-breath spacecraft arrived in orbit, it was oblivious to the excess velocity it had attained, and before the controllers could countermand its programming, it lit up its Agena and headed out to the moon at the same too-high speed.

Watching from his console in the center of the room, Burke was stunned at what he was seeing. The first and most obvious thought was that the problem had something to do with the improvised fix John Tribe had dreamed up for the Atlas, but Burke knew this couldn't be. Tribe's fix was a simple thermal fix; this problem was caused by something far more complicated. Just what that thing was, though, was a puzzle to be solved later. More pressing now was determining just how much damage it had done. If *Ranger 3* was only going fast enough to miss the moon by a few thousand miles, it might be possible to fire the probe's own pint-sized engine at some point during the three-day trip, slow things down a bit, and hit the lunar target after all. If the ship was going to miss the moon by 10,000 or so miles, however, no amount of corrective action would be much help. It would take the JPL tracking officers several hours to make these calculations precisely, and since the Hangar AE team was scheduled to hand control of the spacecraft over to the Pasadena team at this point in the mission anyway, Burke promptly hopped a military plane for the West Coast.

When the *Ranger* project manager at last arrived at JPL late that day, he headed not for the control center, but for a nearby support room where Marshall Johnson worked. Johnson was JPL's trajectory shaman, and all day today he was holed up with his cabinet-sized IBM computer, feeding it punch cards and waiting for it to spit back data on the flight path of the racing *Ranger*. As soon as Burke arrived in Johnson's work area and set eyes on the scientist, he could see that the news was not good.

"Where do we stand?" Burke asked.

"Spacecraft is high and fast," Johnson said.

This did not help Burke; he knew *that* already. "And?" he asked.

"And we're going to miss the moon by twenty thousand miles."

Burke quietly cursed—and then did what every missile man does when he finds that his vehicle is not going to be able to perform its primary mission: He began thinking of a secondary mission. As it turned out, Pickering—who had gotten Johnson's report even before Burke—was already coming up with one. While *Ranger 3*'s trajectory ruled out a lunar impact, it did not rule out remote photography. In two and a half days the spacecraft would move closer to the moon than any American spacecraft ever had before, and with a carefully planned firing of its engine, that gap could be closed even further—if never completely. All the *Ranger* team would have to do was nudge the ship as close to its lunar target as they possibly could, time its cameras to fire just as it was flying by, and they could still deliver an album of lunar photos the likes of which American scientists had never before seen.

Burke, Pickering, Johnson, and the trajectory engineers sat down in the *Ranger* control room to plan the engine maneuver and discovered, to their delight, that it would be a fairly uncomplicated one. If they simply ordered *Ranger 3* to roll a bit so that its flat base faced forward, and then fired its main engine for a few seconds, they could change its speed by eighty miles per hour. Eighty miles per hour might not have been much for a ship that was currently traveling about three hundred times that snail's pace speed, but factored out over the two-plus days of flight time, this was enough to bring it at least 6,000 miles closer to the moon and improve its picture-taking resolution dramatically.

Late that day Burke's assistants ran the numbers over to a communications officer, who radioed them out to an engineer at Goldstone, who, in turn, radioed them up to *Ranger 3*. A few minutes later word came back that the wayward spacecraft had done what it was told, the engine had fired, and the ship, even now, was easing itself moonward.

At JPL, the relief was palpable. Not only had the *Ranger* team salvaged at least part of their spacecraft's picture-taking mission, they had also planned and executed one of the most delicate bits of deep-space flying ever attempted. The flight team exchanged congratulations all around, but the self-kudos turned out to be short-lived.

The morning after the engine burn Burke got up early and went straight to the control center to meet with his flight dynamics officers and check on the overnight trajectory of his ship. When the men looked at their data, they were stunned. According to the numbers coming down from the spacecraft, *Ranger 3* wasn't moving closer to the moon, but farther away from it. More disturbing was the eerie *way* it was moving away. The ship wasn't wandering randomly off course the way it would if a gyro had blown or a thruster had inadvertently fired. Rather, it was moving at the precise angle and speed it had been instructed to, just 180 degrees in the opposite direction. It was as if JPL had told the ship to make a very precise right, and it had instead made an equally careful left.

Burke suspected he knew just where the problem lay. Collaring two of the trajectory analysts, he tore over to the spacecraft assembly building and began ripping through the programming books that contained the spacecraft's navigational codes. In short order he found what he feared: A single digit out of all the flyspeck numbers that flecked the pages had been inverted, causing a +3 to be entered into the spacecraft's brain where a −3 should have been. No matter where the ship was told to fly, it would thus fly exactly the other way. Burke dropped his head into his hands.

"Talk to Goldstone," he said wearily to one of the assistants. "Tell them to tell the spacecraft to change this"—he pointed to the +3—"to this"—he pointed to the −3. "And tell them," he added, "to do it right now."

The assistant obeyed, but even *right now,* Burke suspected, was probably much too late. The next day *Ranger 3* sped past the moon at a distance of 23,000 miles and tumbled into orbit around the sun, becoming, in essence, a tiny, 800-pound planet. Shortly afterwards, it vanished from JPL's tracking screens altogether without ever having returned so much as a single picture.

•　　•　　•

The Soviet Communist Party meetings in Vladivostok were usually not very lively affairs. The apparatchiks would troop into the dreary meeting hall for two or three days of speeches, listen as this or that new five-year plan was unveiled, and then troop home again, little more enlightened than when they came. At some point, Premier Khrushchev usually made an appearance, and this generally picked things up a bit, but even Khrushchev had trouble giving what was essentially a prefab pep rally any appreciable pep. On January 30, 1962, however, the premier was in a frisky mood. Another American spaceship—this one they had called *Ranger 3*—had just gotten lost on its way to the moon, and this time it appeared to have gone so awry the American scientists were unlikely ever to hear from it again. When Khrushchev took the stage for his scheduled address, he couldn't resist speculating about the reason.

"Why does the United States try to launch a rocket to the moon but fail to hit the target?" he asked with playful bewilderment. A few members of the audience laughed self-consciously.

"It's not that they haven't got the dollars," he said, "in fact, they have lots of them." A few more audience members laughed.

"It's not that they haven't got any scientists," he went on, "for they have those, too. Then what are they lacking?" he asked. Before anyone in the audience could shout out an answer, he jumped in himself.

"They lack the regime which exists in our country," he sang out triumphantly. "Socialism," the premier declared, "is the only reliable launching pad from which to launch spaceships. Here there is no anarchy of production, no fierce competition, no class antagonism which hinder the necessary concentration of effort necessary to solve scientific problems."

It was familiar stuff to the people in the hall, but it was exuberantly delivered. Khrushchev, in fine form, went on this way for a while, whipping the otherwise somnolent crowd into what passed, at a function like this, for a frenzy. The next day, the complete text of the hugely successful speech was released through TASS, the government-controlled news agency. The day after that, Washington circulated it to NASA; the following day, NASA passed it on to JPL. In Pasadena, most of the *Ranger* officials made a point not to read a word of it.

•   •   •

For the JPL scientists, with three dead *Ranger*s on their hands and a tab for the failed moon flights that, including manpower and R&D, was already approaching $100 million, the answer to bad P.R. was to launch again, and to do it soon. So far, nobody in Washington was talking about pulling the plug on the entire troubled program, but that was certain to change. The longer the *Ranger* project went without any tangible results, the more fixed its reputation as a waste of limited space funds would become. The next time NASA's budget came up for review, it would be those deadweight projects that would be the first to go.

It didn't take long to pinpoint the flaw in the guidance program that had led to the excessive speed in *Ranger 3*'s Atlas, and it took even less time to fix it. As soon as it *was* fixed, *Ranger 4,* freshly trucked in from Pasadena, was hoisted atop the new, modified rocket. Just two months after that, on April 23, 1962, it blasted off its Canaveral pad and headed toward the moon.

At first, the flight of *Ranger 4* looked not just promising but downright flawless. The repaired Atlas flew straight and true; the temperamental Agena fired perfectly to put the moonship in Earth orbit; an hour later it refired on cue to kick it out into deeper space. At their consoles in Hangar AE and JPL, the controllers looked at one another with wary, almost incredulous smiles. This was exactly—*exactly*—the way they had built this cursed ship to operate. Indeed, so well was it working today that a short while later, when Marshall Johnson's preliminary trajectory data came down, the flight dynamics officers realized that their jobs were almost done. *Ranger 4*'s aim was so good that without so much as a tweak from its thrusters or a breath from its main engine, it would strike the moon on precisely the day it was supposed to, at precisely the right time and precisely the spot. Even as this cheering news was being passed around JPL, however, the spacecraft was preparing to breathe its last.

All of the *Ranger* moonships were electrically connected to the rocket that carried them by a simple arrangement of plugs and bolts. Ringing the top of the Agena were half a dozen ordinary-looking electrical sockets, and protruding from the bottom of the *Ranger*s were half a dozen pairs of corresponding prongs. When the moonship was mounted on top of its Atlas-Agena, it was thus literally plugged into the rocket, and then secured in place by a series of explosive bolts. After launch, when the *Ranger* had been fired out of

Earth orbit and the Agena's work was done, the bolts would deto-nate, the ships would separate, the spent Agena would tumble away, and the spacecraft's on-board batteries and solar panels would take over the job of providing it with power.

That was the way the system was supposed to work anyway, and on *Ranger 4* that was the way it *started* to work. But when the space-craft and Agena separated, a few flecks of foil were set adrift be-tween them. By pure, free-floating randomness, one of the flecks happened to settle atop two of the electrical prongs protruding from the bottom of the *Ranger,* and as soon as it did, it completed a killing circuit that instantly tore through the entire body of the ship. Like an electrocuted man, *Ranger 4* shook, cooked, and slowly began to siz-zle and die.

On the ground, the calamity played out quickly. One moment a healthy stream of telemetry was pouring down to the happy controllers in the Earth-based tracking stations; the next moment there was noth-ing at all but a tiny signal from the tiny transponder buttoned up inside the balsa wood ball. Since the battery that was intended to power the seismograph on the surface of the moon was separate from the ship's power source, it was unaffected by the system-wide short circuit. As the suddenly cold, suddenly dead *Ranger 4* sped away from Earth at 25,000 miles per hour, this little, muffled voice was all that was left to mark its transit.

Pickering was in the JPL control center when the spacecraft telemetry suddenly vanished, and for the better part of the day he, Burke, and the rest of the controllers concocted different combina-tions of commands that they hoped might awaken the sleeping ship. Throughout that time JPL issued only the vaguest mission status statements to the reporters crowded into Von Karman Auditorium, but at 6 P.M. or so the media people would expect—and be entitled to—some kind of end-of-the-day wrap-up. It fell to Pickering to tell them that the ship they were following so closely had flatlined. Walk-ing into Von Karman, he pushed his way without comment through the reporters who had been gathered there for most of the day, mounted the stage, and drew a weary breath.

"It looks like we have a fairly significant problem with the spacecraft," Pickering said a little blearily. "We don't really know how major the problem is."

"What does the telemetry *say* the problem is?" a newsman asked.

"We're not receiving any telemetry," Pickering confessed.

"Is the spacecraft dead?" another asked.

"Not completely."

"Has its guidance system been able to align itself with the sun?"

"No."

"Have its computers been able to communicate with Earth?"

"No."

"What does that mean?"

Pickering paused. "That," he said with a sigh, "indicates that the trouble is deep within the brain of the system."

Pickering was being as honest with the reporters as he could be, and over the next day, as his unconscious ship showed no signs of awakening, he became increasingly candid, conceding to the media that this *Ranger,* too, was likely to be a failure. A dead pod heading for a pointless lunar impact was not the stuff of headlines, and as *Ranger* 4 proceeded all but mutely through its two-and-a-half-day journey, Pickering anticipated that the media's interest in it would vanish all but entirely. Unexpectedly, however, the flight of the disabled ship became perversely fascinating to the press. The heartbreakingly hopeful, chick-like peeping of the balsa wood ball's 50-milliwatt transmitter continued to register on Goldstone receivers even as the ship receded farther and farther into the void. As it did, both the reporters and the public, suckers for hard-luck cases, demanded more and more information about its progress. Nobody pretended that the mission would return any good science, but as long as the trajectory held steady, the fact remained that before the end of the week, an American spacecraft—even if it was an American spacecraft sunk deep in an electronic coma—was going to strike the surface of the moon, and that was an event worth noting.

Pickering and the JPL public relations team did what they could to satisfy this surprising national interest, tracking the steady chirps of the seismograph ball and issuing regular updates on the spacecraft's status. Trajectory data gathered by the Goldstone station in the Mojave indicated that the ship would strike the moon shortly between 4:30 and 5:00 A.M. on April 26. A slight list that had appeared in its flight path would now cause it not to hit the lunar sphere head-on, but to soar just past its western edge. This approach would nevertheless be so close that the ship would still be captured by the moon's gravity, arc around its far side, and strike the surface just out of view of Earth. If the peeping of the transponder stopped at 4:47 A.M. when the ship vanished over the western horizon and did not resume some forty minutes later—the amount of time it would take it to reemerge over the eastern side if it did not crash—the JPL men would know they had hit their lunar target. Two and a half days after *Ranger 4*'s launch, Pickering, Burke, Cummings, Schurmeier, and several other JPL officials flew to Goldstone to follow their spacecraft's descent.

When the JPL team arrived in the desert, the sun was still not up and the Goldstone facility was lit by little more than the blinking red lights at the base of its white, wide-mouth antennas. In an auditorium inside, technicians had set up a large television monitor filled with a live image of the moon, captured by a sixty-inch telephoto lens aligned with one of the receiving dishes outside. *Ranger 4*—a distant, ten-foot dust mote—could not possibly show up on the screen, but a superimposed crosshair was able to pinpoint its location as it fell closer and closer to the edge of the moon. A speaker placed next to the indoor monitor broadcast the peeping of the transponder as it called tenaciously back through the growing, deep-space static. A similar system had been set up in a nearby press room for the small crowd of newsmen who had also managed to find their way to this normally forgotten outpost at this normally forgotten hour.

As the JPL scientists settled into their seats in the Goldstone auditorium, *Ranger 4* was just a few thousand miles from the moon. As it drew closer and closer, the gravity lasso that had snared it would pull harder and harder and the ship would begin to move faster and faster. The peeping of the transmitter would continue at the same speed and frequency no matter how fast the *Ranger* was going, but the shape of the electromagnetic signal that was carrying it would change dramatically. Since radio waves tend to stretch and flatten as the transmitter emitting them moves away from the antenna receiving them, all the Goldstone engineers had to do was measure this change and they could measure the spacecraft's velocity, too. As the peep-peep-peep sounded hypnotically through the room and the crosshair on the TV closed in on the lunar limb, an engineer approached with a readout of the transmitter signal and handed the single sheet of paper to Cummings. The transmission, it was clear, was indeed flattening out and the spacecraft, by implication, was indeed in its death plunge.

"Oh, baby," Cummings muttered as he scanned the numbers.

At the front of the room, a large mission clock indicated that loss of spacecraft signal was now just two minutes away. Two minutes melted to one and one melted to just thirty seconds. Nearly 240,000 miles away, *Ranger 4* was climbing to its maximum speed and falling to within a few miles of the surface. Then, just fifty seconds after the clock in the front of the Goldstone auditorium flashed 4:47 A.M.—almost the exact moment the trajectory experts had predicted—the peeping from the spacecraft vanished as suddenly as if it had been cut off by a knife and the room was filled with the soft hiss of translunar space. Pickering glanced briefly at his watch and nodded a silent farewell at the screen. Even without waiting the formal forty minutes, he knew he'd never hear from his *Ranger* again.

• • •

Almost from the moment the dead *Ranger 4* struck the lunar surface, the JPL press apparatus began cranking itself up to start recasting the failed mission as at least a qualified success. To the delight of the lab, the press seemed more than willing to agree with this happy interpretation of things. Paper after paper hailed the accuracy of the rocket and the precision of the flight, and overlooked, at least for the time being, the death of the payload.

But if the American news community was willing to turn a deaf ear to failure, Khrushchev wasn't. Shortly after *Ranger 4* died, the

Soviets were enjoying their National Railway Day celebrations, when the party leader took another opportunity to tweak the West. Delivering a speech to a group of transport workers, he suggested that maybe, just maybe, the Americans' latest spacecraft hadn't hit the moon. After all, he said with a conspiratorial twinkle, three straight *Ranger*s had already failed, and with a fourth one now added to the toll, perhaps the scientists on the other side of the ocean just cooked up a story of a lunar impact to save global face.

This time Pickering was riled. Attack his *Ranger*s if you want; attack his Atlas; you could even attack his engineers. But don't question his truthfulness. Pickering was a missile man first, and when a missile man tells you he hit his target the way he was supposed to, you can be absolutely certain he did. The day after Khrushchev's speech, a curt Pickering delivered an address of his own. Calling a morning press conference at JPL, he waited until the reporters were assembled in the media briefing room and then strode in silently. Mounting the stage, he began his address without a word of preamble.

"On April 26, 1962, at 4:47 and 50 seconds A.M.," he said, "*Ranger 4* was tracked by the Goldstone receiver as it passed the leading edge of the moon. At 4:49 and 53 seconds it crashed on the moon at a lunar longitude of 229.5 degrees east and a lunar latitude of 15.5 degrees south."

Pickering stepped away from the microphone and looked out at the reporters in front of him for a long, resolute second. No one, it was clear, was likely to ask a follow-up question.

•   •   •

In the weeks following the death of *Ranger 4,* Jim Burke and William Pickering went in dramatically different directions: Burke decided to go to Europe; Pickering decided to go to Venus. By any measure, Pickering's decision was the safer one.

Though the public and the press did not always appreciate it, JPL was in the business of doing more than just flying to the moon. Nearly eighteen months earlier, in February 1961, the Russians had launched the *Venera 1* spacecraft on a 224-million-mile journey to

Venus. The spacecraft was not much—a tin can of a thing equipped with some primitive cameras and deep-space sensing equipment. The trajectory it flew was cruder still: Fired out of the clean, orbital circle Earth inscribes around the sun, it spiraled slowly in toward Venus, but was never intended to land on it. Rather, it crossed Venus's orbit and flew fleetingly past the cloud-covered world at a whopping distance of 62,000 miles. The pictures the ship could snap were thus little better than what could be captured from the window of a speeding car—and a remote speeding car at that. Nevertheless, snap pictures it did, and a few days later these fuzzy portraits were proudly flashed around the world.

On its face, the feat ought to have seemed impressive, and to people unacquainted with the business of space travel, it indeed was—certainly more impressive than flying to some close-to-home target like the moon. Scientists who made it their business to undertake such flights, however, saw things differently. If human beings were ever going to follow robot probes into space, flying to particular, predesignated spots on the surface of new worlds, they were going to have to master a brand of pinpoint piloting never before even contemplated. To learn this kind of flying, you're a lot better off rehearsing with the close-up, bull's-eye missions the *Ranger* ships were flying than the long-distance miss *Venera* achieved. While American scientists understood this, however, the American public often didn't seem to. And it was the American public that ultimately paid the bills, applied the pressure, and decided if NASA was going to exist at all. If the Soviet Union was flying to Venus, Americans would insist on going, too—and it would be William Pickering's JPL that would be expected to take them there.

As distractions went, the prospect of attempting a mission to Venus was a surprisingly welcome one. After the spectacular failure of *Ranger 4,* JPL was in no hurry to launch a *Ranger 5.* The lab's scientists sniffed out the cause of the *Ranger 4* flop pretty easily and came up with an attractively straightforward solution. On future missions the electrical connection between the spacecraft and the booster would be redesigned, with the plugs repositioned to protrude from the top of the rocket and the sockets now built into the bottom of the spacecraft. Any stray piece of foil that contacted the plug prongs would thus only short out the empty Agena, which by that point in the mission would have been jettisoned as junk anyway.

Uncomplicated as this reengineering work seemed, conducting it would nevertheless take six months or so, meaning that no *Ranger* could set out for the moon until the late fall of 1962. And that left a brief window for Venus. In order to get this first interplanetary mission off the ground as fast as possible, JPL engineers decided not to develop a whole new type of spacecraft to make the trip. Instead, the Venus ship would be little more than a modified *Ranger* ship with longer-range cameras, deeper-space sensing instruments, and no seismograph at all. The name, too, would be changed to a far-more vagabond *Mariner*.

In July the first *Mariner* was mounted atop an Atlas booster and fired from Cape Canaveral's launch pad 12. It promptly did what all of its *Ranger* littermates before it had done: It failed. As with *Rangers 1* and *2*, the *Mariner* flopped because the booster flopped, losing its way above the ocean and pointing its nose down toward the Atlantic instead of up to space. *And* as with the first *Rangers*, JPL responded quickly, hammering together another *Mariner,* rolling out another Atlas, and resolving to try again. But even with the scientists working so feverishly, it would still be at least another month before *Mariner 2* could get off the ground, and for Jim Burke, this seemed like a good time to get out of town.

Though Pickering, as head of JPL, had to concern himself with all of the lab's projects, Burke, as the *Ranger* project manager, was strictly a moon man, at least until he was assigned to another beat. For the foreseeable future, there would thus be little for Burke to do but knock around his lunar shop, waiting for the chance to fly again. For the month of August, Burke therefore decided, he and his wife would take a long-delayed trip to Europe.

Pickering might have objected to Burke's decision to leave the country, but if he did, he was not inclined to say so. At least some of the other engineers at JPL almost certainly objected—if only because the absence of the project chief when the project itself was in such distress just didn't look right. Burke, however, was a practical man, one who appreciated that vacation time was rare, and he was not inclined to sacrifice what little he could find merely for the sake of appearances. As the *Ranger* team sat largely idle and the Venus team prepared to roll *Mariner 2* out to the concrete griddle of launch pad 12, Jim and Lynn Burke flew off to Europe, spending most of the month of August touring the countryside of Denmark, the cities of Germany, and the coolly canopied trails of the Black Forest. Toward the end of the month, the Burkes arrived in London for the last leg of their trip. On August 27, at 7:53 A.M. Greenwich mean time—just as they were sitting down to breakfast in their downtown hotel—the nighttime skies over far-distant Florida lit up, and *Mariner 2,* bound for Venus, left the ground.

Several hours later, as Burke and his wife were climbing into their hotel elevator, the white-gloved hand of a bellboy grabbed the door and stopped it from closing.

"Mr. Burke?" the boy said tentatively.

"That's me," Burke said.

"Telegram for you."

The boy handed Burke the flimsy yellow paper and released the door. The elevator started to rise, and Burke tore into the envelope and pulled out the single sheet inside. The message was from Jack James, the head of JPL's Venus project.

"*Mariner 2* launched successfully," the three-line letter read. "Midcourse maneuver pending. We're on our way to Venus."

Burke whooped, clapped, hugged his wife, and only then, showed her the telegram. When she read it, she grinned broadly and hugged him back. The youngish man operating the elevator stared correctly ahead and Burke, impulsively, shoved the telegram toward him. The operator flicked his eyes downward, scanned the three lines, and nodded politely.

"That's wonderful, sir," he said, without ever looking at Burke.

Half an hour later Burke was sitting on the bed in his room, trying to make a transatlantic phone connection to California, when there was a knock at the door. His wife answered and stood aside as a waiter entered with a chilled bucket of champagne and a vase of flowers on a linen-covered cart. The accompanying card, embossed with the name of the hotel manager, read simply, "With our compliments—and our congratulations."

●　　　●　　　●

Little more than a week later a well-rested Jim Burke returned to a jubilant JPL. Little more than two months after that, on October 18, 1962—as *Mariner 2* was passing what was roughly the halfway point on its still-smooth, four-month transit to Venus—*Ranger 5* was launched toward the moon's Ocean of Storms. Deep in the spacecraft's power switching and logic unit was a single bolt that was responsible for the single job of crimping a pair of terminals together, providing electricity to most of the instruments aboard the ship. Unknown to anyone on the *Ranger* team, sometime in the weeks leading up to launch, that bolt had worked its way perhaps a quarter of a turn loose. No one would ever discover why—perhaps it was a careless technician who was thinking about quitting time when he should have been thinking about his work; perhaps it was a single bump on a road the spacecraft encountered on its 3,000-mile truck trip from Pasadena to Canaveral. Whatever the cause, the problem was inconspicuous enough that no one was aware it existed at all until an hour after launch, when a horrified Jim Burke looked at the strip charts and telemetry streaming into the control room and saw the spacecraft power beginning to plummet.

Less than nine hours later it winked out completely, and *Ranger 5*, like *Ranger 4* before it, lapsed into electronic unconsciousness. On October 21 the insensible ship—by now somersaulting uncontrollably through space—flew past the moon and entered permanent orbit around the sun. This time, Burke and the other men of the program knew, there'd be hell to pay.

●　　　●　　　●

William Pickering did not know how he was going to fire Jim Burke. Actually, he didn't even know *if* he was going to fire Jim Burke. He wasn't going to call it a firing, that was for sure. Maybe he'd call it a reassignment. Maybe he'd call it a project swap. And maybe—if he was improbably lucky—he wouldn't have to do it at all. A lot depended on the commissions.

If you had asked any of the engineers and other workers at JPL what to expect after the utter failure of *Ranger 5,* none of them would have failed to anticipate a commission. JPL, after all, was equal parts academic, governmental, and military lab, and if the folks who made up those three groups wouldn't be thinking commission at a time when things had gone so dismally wrong, no one would. What no one would have anticipated would have been *three* commissions.

Almost the moment *Ranger 5* was irretrievably lost, Pickering announced that an internal JPL panel would be convened to look into the causes of the failure and the other four failures that had preceded it. Shortly afterwards, NASA headquarters in Washington trumped JPL by announcing a second board of inquiry of its own. After that, Congress raised them both by calling for a third. Of the three, it was NASA's that looked the most menacing.

While congressional hearings might be a high-profile show, they would be little more than that. *Ranger* engineers might be summoned out of their labs, buttoned into coats and ties, and flown off to Washington to explain to lawmakers why their moon flights kept failing, but in the end, the lawmakers knew full well that they didn't understand a thing about how the space scientists did their work, and if they started mucking around with how the project was run, they stood a good chance of making things worse.

Similarly, the JPL investigation promised to have little bite. There were few people more sympathetic to the trials of an engineer than a group of other engineers, and few engineers they'd be more inclined to sympathize with than the affable Jim Burke, Cliff Cummings, Bud Schurmeier, and the other men of *Ranger.* Indeed, one of the key JPL engineers who would be serving on the commission had himself encouraged Burke to take his vacation back in August. A tribunal of fellow scientists like this might be a lot likelier to get to the bottom of what ailed *Ranger* than a tribunal of congressmen, but whether they would be inclined to do anything about it—firing the friends who needed to be fired and disciplining the ones who needed to be disciplined—would be another thing entirely.

The NASA committee was a different matter altogether. The space agency had an entire exploratory program to run—manned and unmanned missions, near-Earth and deep-space missions—and it was not about to tolerate one small project in one branch office jeopardizing the whole operation. *Ranger* had promise and NASA had been backing it for a while now, but if that promise had faded, the agency would not hesitate to cut the program loose. The new panel was headed by Albert Kelley, a onetime Air Force commander, and at JPL, the day the Kelley commission finally convened and began to hold its hearings was a troubling day, indeed.

Like a storm front, the commission gathered first in the east and moved slowly west, stopping to collect affidavits from people at various NASA facilities and various NASA contractors. Kelley looked into the training of the engineers, the quality of the parts they made, the management skills of their supervisors, the incentives provided for good work, the sanctions imposed for bad work, and, finally, the guts of the spacecraft the scientists put together. The commission started its work in late October, conducted its investigation for the remainder of the month, and finally arrived at JPL sometime in November, spending two weeks interviewing every engineer or administrator who had anything to do with *Ranger*, including Pickering and Burke themselves. By the latter part of the month, Kelley and his men closed their files, gathered their notes, and flew back to Washington to write their report.

For several days not a word was heard from the East. Then several more. Finally, on December 7, Pickering received a bound folder and a phone call. Later that day he made a call of his own, to Burke, telling him he needed to see him.

"Coffee?" Pickering asked when Burke arrived in his office, slightly out of breath from hurrying across the JPL campus.

Coffee, Burke figured was a bad sign. He nodded no.

"Cigarette?"

A cigarette was a worse sign. Again Burke nodded no.

Pickering drew a breath. "Jim," he began, "I've got problems."

Burke nodded, smiling sympathetically and even adding a don't-I-know-it eye roll. Pickering went on.

"I've got congressional committees jumping on me, Kelley's people jumping on me. We've all appeared before them and we know damn well what the situation is."

"We do," Burke said.

"And the conclusion is that we've got to do things differently."

"Okay," Burke said hopefully.

"And change our ways."

"Okay."

"And that means changing people." Pickering gestured vaguely toward a document on his desk. "The report's come down. You and Cliff Cummings are going to have to do something else."

Burke, who had been sitting expectantly forward, now slumped back a bit. Pickering went on hurriedly. "That doesn't mean anyone's leaving the lab," he said. "It doesn't even mean anyone's leaving *Ranger*. It just means we'll put you on a part of the project you're more . . . suited to."

Looking away for a moment, Burke collected himself. He and his friend Cummings were to be sacrificed. His friend Schurmeier was to be spared. He looked back at Pickering. "Do you want to hear from me who I think the new project manager ought to be?"

"Sure," Pickering said.

"I think it ought to be Bud Schurmeier."

"Well, you scored," Pickering said. "Because that's who it's going to be."

Jim Burke, the former head of the *Ranger* program, accepted his new position as lieutenant to his former lieutenant without complaint. From now on he would focus exclusively on the narrow problem of designing new connections between the *Ranger* and the Atlas-Agena and within the *Ranger* itself, making sure there would be no more problems like the ones that plagued *Ranger*s 4 and 5. Cummings would leave the program entirely. After claiming five robot spaceships, the *Ranger* program at last claimed its first scientists. Exactly one week later, the overachieving *Mariner 2* sailed grandly by Venus at a relative cloudtop height of just 23,000 miles—coming more than twice as close to the distant planet as the vaunted Russians ever got.

• • •

In the fourteen months after Pickering busted Burke down from general of the *Ranger* division to mere field officer, the program reinvented itself. With Schurmeier now driving the team and Pickering driving Schurmeier, the design and manufacture of *Ranger* 6 proceeded with a near-fanatical attention to technological detail. Manufacturing processes that were once considered routine were monitored by squads of quality-control engineers. Spacecraft assembly hangars that had once been considered more than adequately free of contaminants were now stripped bare and scrubbed raw before so much as a single strut or solar panel was brought inside. Design teams tore into the *Ranger* blueprints, disassembling the schematic ship and not putting it back together until they were sure they had a configuration that would allow it to work right. There would be no loosely torqued bolts on *this* spacecraft, no +3 in the programming where there should have been a −3. There wouldn't even be a balsa wood ball anymore. *Ranger* 6 would be a clean, light, lithe ship, one that would be mounted atop the most exhaustively checked Atlas-Agena ever launched from Cape Canaveral, and sent to the moon to do two things only: take pictures and die.

On the morning of January 30, 1964, that combed and scrubbed rocket took off from Cape Canaveral carrying the rebuilt *Ranger* 6 with it. With Schurmeier in the center seat in Hangar AE, the Atlas and Agena—like few Atlases and Agenas before them—performed precisely as they should, carrying the *Ranger* to just the right altitude, pushing it out of Earth orbit at just the right moment, and sending it toward its lunar intercept on just the right trajectory. So precise was that flight path that if the guidance officers had shut down their consoles that day, their spacecraft would still have arrived in the lunar vicinity less than three days later and flown just six hundred miles wide of the moon, a miss distance that could easily be corrected by the barest pulse from the spacecraft's main engine the morning after launch. Indeed, virtually everything about the early part of the flight looked to be perfect, with one tiny exception.

In the middle of the rocket's climb to Earth orbit—the instant after the Atlas booster shut down and the instant before the Agena lit—a curious bit of telemetry had flashed onto the screens in the control room. The television cameras aboard *Ranger 6*, it appeared, had switched inexplicably on, blinked around briefly in the darkness of their Agena shroud, and, finding no moon there, had switched off again, returning to the long sleep they would maintain until the very end of the mission sixty-six hours later. The entire event had lasted barely eight seconds, and the engineers had not given it a whole lot of thought after that. An anomaly during launch might be troubling, but an anomaly that corrected itself—especially during an otherwise flawless mission—was easy to forget about.

Two and a half days later, however, forgetting about it was not so easy. On February 2, 1964, at 1 A.M., as William Pickering sat in the viewing room of JPL's spanking new—though utterly empty—Space Flight Operations Facility with Homer Newell at his side, hundreds of reporters crowding into the nearby Von Karman Auditorium, and the specter of an angry Lyndon Johnson following things from his White House office, word came back that *Ranger 6*, which was less than fifteen minutes away from crashing onto the surface of the moon, had been struck suddenly blind. The anomalous cameras that had fluttered so suspiciously during launch had apparently blacked out entirely.

It was Walt Downhower who brought Pickering the bad news, flatly intoning, "Still no video, still no video" over the loudspeaker in the SFOF. And as Pickering learned it, the press and the president learned it, too. Now, with the spacecraft approaching 6,000 miles per hour, the moon only a thousand miles below it, and Pickering and Newell alone in their viewing room, Downhower spoke up again.

"At this time," he said, "we are still receiving no high-power video."

"I don't believe what I'm hearing," Pickering muttered. Newell nodded. "I don't *believe* it," Pickering said again. Over the P.A., the high-pitched tones of *Ranger 6*'s telemetry—sounding eerily like the wind the spacecraft would be whipping through if there were any wind where it was flying—was easily audible.

In the old SFOF the *Ranger* engineers were working frantically to get the spacecraft's electronic eyes to open. The newly elevated Schurmeier, who had returned from Cape Canaveral shortly after the *Ranger 6* launch, ordered his imaging chief to send a backup signal to the spacecraft, instructing it to override all existing commands and engage its cameras at once. The signal went out.

"Still no full power video," Downhower's voice said.

Schurmeier told the ship to override the override and try again.

"Full power video still negative," Downhower said.

In Von Karman Auditorium the assembled reporters began to glance at one another. At NASA headquarters they did the same. The mission clock read eight minutes to impact; the cameras had already missed at least five precious minutes of photography time. Around the press room eyes began to roll. With six different cameras aboard this ship, and each one mounted on two different assemblies, probability alone made it all but arithmetically impossible that every part of this superredundant system would fail at once. And yet that's just what seemed to have happened. From a JPL team that had already killed five full spacecraft, however, what more could be expected?

"These guys . . . ," one reporter muttered in disgust.

"Unbelievable," another responded.

"Still awaiting full power video," Downhower answered both of them.

A quarter of a million miles away *Ranger 6* accelerated to 6,000 miles per hour and its altitude shrank from a thousand miles to hundreds. In the empty viewing room Pickering saw the mission clock cross below eight minutes, then seven, then six. If the cameras came to life now, Pickering found himself thinking—this *instant*—he could still salvage maybe two thousand pictures. That number, however, was shrinking as fast as the altitude figure, and on the Sea of Tranquillity, the ten-foot *Ranger* would now be almost close enough to the ground to cast a pinpoint shadow. "Five minutes to impact,"

Downhower said. A minute later it was four, then three, then two.

"One minute to impact," Downhower at last said with finality.

In the press rooms on both coasts, the reporters fell silent; in the viewing room at JPL, Newell and Pickering stared at the mission clock; at his microphone Downhower stopped counting. For sixty seconds nothing was heard in either of NASA's press rooms but the plaintive piping of *Ranger 6*'s telemetry. Then, all at once, it stopped. For an instant there was absolute silence in the NASA halls. Then Downhower spoke up one last time.

"We have first indication of impact," he said uninflectedly. "There is no indication of switchover to full power video throughout the terminal event."

In the new SFOF, Pickering cast one last glance at his mission clock. It read 2 days, 17 hours, 35 minutes, and 29 seconds—the length of *Ranger 6*'s entire life, and not a single picture was returned to mark its passing. Suddenly, however, the P.A. system that had been carrying the spacecraft's telemetry emitted a brief but promising burst of static. Pickering turned expectantly toward it. The reporters and controllers did the same. Somewhere high above them, in the electromagnetic eddies flowing through the upper atmosphere, a signal streaming from who-knew-where collided with the one streaming back to NASA and caught a ride aboard it straight down to the very rooms where things had just gone so silent. When it arrived, that silence was broken by a near-surreal woman's voice.

"Spray on Avon cologne mist," the woman said, "and walk in fragrant beauty."

At the communications station in mission control, the radio man leaped for his console and cut the connection, causing things once again to go quiet. At the control consoles around him the *Ranger* team stared openmouthed. In the press rooms the reporters did the same. In his viewing room Pickering could not help appreciating the dark appropriateness of the moment. His horrific day now had a perfect closing coda. The humiliation of his lab was complete.

•   •   •

JPL engineers did not pretend it would be easy to figure out what had claimed *Ranger 6*'s sight, and they were right. For weeks after the death of the spacecraft, they pored over the strip charts and telemetry records of the flight like heart surgeons reading the final EKG tracings of a man who had died on their table, but they could not find a single scribble that would explain the failure. The camera anomaly during launch was the obvious place for them to start their post-mortem, but to the relief of Pickering and the other controllers who had elected not to look into the problem further, that appeared unlikely to yield anything. Spacecraft were complicated electronic organisms, and, like anything else with so elaborate an anatomy, they were subject to the occasional twitch or hiccup. The twitch *Ranger 6*'s cameras had experienced might have been an untimely one, but it was apparently a harmless one.

Or at least that's the way it seemed. After the initial investigation of the flight had hit its technological dead end, however, Alex Bratenahl, an obscure JPL engineer, approached the *Ranger* team and asked if he might have a look at their launch-day films. Strictly speaking, Bratenahl had no business making such a request. His field of expertise was solar and plasma physics, and he did his work largely in the lab's astronomy buildings—a scientific world away from the *Ranger* team. A theoretical researcher like Bratenahl could work at JPL for years and never have to trouble himself with the crude, workbench business of designing or fixing spaceships. Nonetheless, when he asked for the films, he got them, and within a day after first threading them into his projector he had *Ranger*'s problem licked.

When a gigantic machine like an Atlas-Agena flies through the atmosphere at supersonic speeds, it stirs up all manner of acoustic and aerodynamic hell. One of the most dramatic moments occurs in the middle of that ascent, when the rocket shuts down its main engines, drops its first stage, and prepares for the ignition of the second-stage Agena. At that instant, gases and erupting fluids around the missile cause a sudden, flaming shock wave to engulf it, flowing up toward its nose and then back down again. The shroud of fire lasts less than a second, and is thus unable to do any damage to the skin of the ship itself. But beneath the skin, Bratenahl suspected, it could do quite a bit of harm.

Much of what made up the rocket's veil of flame was plasma, a charged gas that, to the uninitiated, looks like any other gas, but to a plasma physicist is a very different chemical beast. If atmospheric conditions are just right, the almost equal number of positively charged ions and negatively charged electrons in the curious vapor give it the ability to conduct electricity as efficiently as a copper cable. If an Atlas rocket climbing toward space was surrounded by a cloud of the sizzling stuff, and if there was even a microscopic break in the metal shell that enclosed the booster, stray plasma could stream inside, filling the payload chamber with a high-energy cloud. Any piece of electronic equipment the rocket was carrying—including, say, a robot spacecraft on its way to the moon—could then easily switch itself on and roast alive in the supercharged mist. The telemetry data indicating that *Ranger 6*'s cameras had awakened during launch was probably accurate, Bratenahl concluded. The signal that followed eight seconds later indicating that they had shut back down, however, was probably false. The video system hadn't simply turned itself off, it had overheated and died.

The next day Bratenahl returned the films to the *Ranger* team. "Insulate your cameras better," he told them simply. "That should solve your problem." The engineers immediately set about doing precisely what Bratenahl suggested.

But just insulating the *Ranger* spacecraft from injury did nothing to insulate the *Ranger* scientists from criticism, and at JPL they expected a whole different kind of firestorm. To their surprise, this time the criticism was muted. Once again Pickering and NASA convened commissions to look into the latest disaster, and once again the commissions grumbled and fumed and cautioned JPL scientists to watch their steps. Oddly, though, the warnings seemed halfhearted, almost as if NASA no longer expected such agency finger-wagging to have any effect. This sense of surrender was far more ominous than the outrage that had followed the loss of *Ranger 5*. Back at NASA headquarters, Pickering feared, the fate of his *Ranger* program was probably hanging by the most tenuous of threads. Not long after, NASA administrator James Webb called to inform him that that thread had snapped.

"Bill," Webb said when Pickering answered the phone in his JPL office, "we need to talk about a mistake I may have made."

"A mistake *you* made?" Pickering asked.

"You bet," Webb said. "I think I may have gone and given you and your *Ranger* boys more than you can handle."

Pickering closed his eyes. This, of course, was how the manuals said it was done. If you merely plan to reprimand a man, you tell him straight out that he screwed up; if you plan to sack him altogether, you extend him the genteel gesture of pretending that the fault lies with you.

"I can't say I agree with you," Pickering said a little hoarsely.

"I didn't expect you to," Webb answered. "But you're a little too close to the program to see things clearly. From here, it seems like you're in over your heads."

"I think we can climb out of it," Pickering answered.

"You've had time to climb out of it and you haven't yet." Webb took a brief pause and finally said what he came to say. "How about if we just let you fellows out of this *Ranger* contract and turn the whole project over to another contractor?"

Pickering heaved a deep sigh. He had feared he was going to hear something like this; actually, he *knew* he was going to hear something like this. So certain was he that when he did hear it, he felt a perverse sense of freedom. The institutional axe he had been fearing since JPL's *Ranger*s started dropping out of the skies had at last fallen. The lab had failed, Pickering himself had failed, and the profound struggle of making believe that all of the recent failures were mere setbacks, mere engineering challenges, was at last over.

Or it should have been over. The problem was, Pickering believed the make-believe. As he had been taught more than twenty years ago in the dry riverbed that JPL once was, and as he himself had taught his team at the sprawling space center the lab had become, the very heart of the missileman's business is failure—repeated, relentless, humiliating failure—until once, just once, you manage to succeed, and when you do, you know forever how the

machine you're trying to invent ought to be built. With his long-en-
dangered *Ranger* now apparently lost, Pickering had nothing further
to lose by explaining this very thing to the very man who had taken
the spacecraft away from him. For several minutes Pickering held
Webb's ear, telling him that love or hate his *Ranger* program, you
had to admit it had been run exactly the way such a program had to
be run. In the course of explaining all this, he realized that despite
himself, he was also pleading for a chance to try just once more.
Pickering's engineers had fired off six *Ranger*s so far, and with everything that had gone wrong with the program, two of them had flown
straight from Pasadena to the moon. The next one, the JPL boss was
certain—*the very next one*—would land there, too, and before it
crashed, this one would do the job it was built to do.

Webb said nothing throughout Pickering's speech, nothing when
he finished, and nothing for a moment or so after. Only the buzz of
the open connection let the JPL chief know the NASA chief was there
at all. Finally, the instant before Pickering could say anything else,
Webb at last spoke.

"One more flight," he said simply. "You've got one more flight."

•   •   •

Whether the scientists at JPL knew of Pickering and Webb's conversation was impossible for Pickering to know. They wouldn't be likely
to ask their boss if he had been forced to plead for their jobs, and he
certainly wouldn't be inclined to tell them if he had. If the question
was going to come up at all, it would likely come up several nights
later, at the Miss Guided Missile dance. The Miss Guided Missile
dance was an annual late-winter tradition at which the JPL engineers
would gather at a local hall, spend the evening dancing and socializ-
ing, and at the end bring a young secretary or administrative assistant up onstage, place an aluminum foil crown on her head, and
declare her queen of the evening. Most years Pickering would have
been just as happy to give the whole affair a pass, but as head of the
lab he had the responsibility of performing the ceremonial crowning.

The night of this year's dance Pickering arrived late, hoped to leave early, and spent most of the time he was there looking less than enthusiastically forward to the little coronation he would have to perform. Nonetheless, at the appointed hour, a fanfare sounded and someone seized the auditorium microphone and called the lab chief to the stage. Pickering smiled wanly and, rehearsing his little piffle of a speech, began making his way to the front of the room.

As he did, he noticed a reserved, though sustained applause breaking out. The clapping was accompanied by something like a cheer, but a low, contained one—less celebratory than respectful. The cheer built slowly until it was closer to a whoop, then a roar, then an almost ecstatic howl, accompanied by a crack of far louder applause. Pickering looked dazedly around him and pushed his way through the crowd as the sound continued to build. When he at last climbed the stage, the ovation—no denying it was an ovation now— crashed around him. Pickering shadowed his eyes with his hand and tried to squint out over the room, but the bright lights—and the light mist—that were suddenly obscuring his vision made it impossible to see. The applause played out for another minute or two, until finally, the man who had sent *Ranger*s 1 through 6 toward the moon and had just bought his team enough time for a *Ranger 7* leaned toward the microphone.

"We're going to fix it," he said simply. "We're going to make it work."

Nobody was thinking about Miss Guided Missile anymore.

•   •   •

William Pickering's last-chance *Ranger* was assembled at JPL in the spring and early summer of 1964. If the engineers were giving much thought to the possibility that they might never build one again, they didn't show it. There were no final names engraved on any final washers, no teary applause when the little craft at last rolled out. Rather, the ship was simply finished up, checked out, and, like the six *Ranger*s that came before it, loaded unceremoniously into an air-conditioned van for the seventy-two-hour trip to Florida.

The target for the spacecraft this time would be the Sea of

Clouds, a vast plain in the moon's southern hemisphere just to the east of the Ocean of Storms. The announced purpose of all of the *Ranger* flights was to look for landing sites for later manned ships, and the Sea of Clouds, which had been troweled almost completely smooth by lunar lava billions of years ago, was a good candidate. In order to ensure that the pictures this and any other *Ranger* space-craft took survived—assuming one finally took them—the JPL team planned to process the images in several ways.

All of the photographic data that gushed back to Earth in the final minutes of the spacecraft's life would be caught by the twin, eighty-five-foot maws of the Goldstone antennas. The first and best images these signals would produce would be created by channeling the impulses into an ordinary video system. There a cathode-ray tube would convert the signals into electron pulses and fire them in a rapid scan pattern onto a phosphorescent screen. To the unaided eye, all that would be visible on the screen as this took place would be a rapidly moving pinprick of light, looking less like a picture than a single moving pixel. Undetectably, however, that pixel would be flickering and blinking, tattooing out a pattern that directly corre-sponded to the shades of black, white, and gray that make up a fully developed television image. A 35mm camera would be positioned in front of the screen with its shutter open throughout this process, so that when the scan was done, a pointillist image would have been constructed on one of its frames. The camera would then automati-cally advance to the next frame and the next and the next until it had captured no fewer than 4,200 images in just thirteen minutes. The film would then be stored lovingly in a refrigerator and flown to Hol-lywood to be developed in the best movie industry lab the space agency could find.

This processing would take a bit of time, and if the mission suc-ceeded, NASA knew that that would be too *much* time for the press. In order to satisfy the see-it-now needs of reporters, who would ex-pect pictures to be available within hours, the space agency would also create a quicker, somewhat cruder sets of prints by recording *Ranger*'s data transmissions on tape, playing them back through an-other video system, and capturing these images with another 35mm camera. If an even quicker look was needed, an ordinary Polaroid could snap a picture of either these recorded transmissions or the raw live ones. Both the Polaroids and the backup set of 35mm negatives would then be flown directly to JPL.

For any lay viewer, the difference between these secondary prints and the better, primary ones would be all but nonexistent. To lunar cartographers, however, the infinitesimal data deterioration that occurs whenever a cruder, second-generation image is created could make the difference between spotting a pothole or boulder on the Sea of Clouds and not spotting it, and this, in turn, could make a huge difference to a lunar crew a few years down the line.

That, in any event, was how the system was supposed to work. But with no *Ranger* having had a chance to try it out so far, no one knew if it would indeed operate the way it was supposed to. Pickering was determined that this time the ground-based cameras would have work to do.

Liftoff for *Ranger 7* was set for July 28, 1964, shortly before 10 A.M., JPL time. Jim Burke, who had overseen five liftoffs as the project manager, would work this launch from an ordinary yeoman's seat in JPL's spanking new, at-last up-and-running Space Flight Operations Facility. Pickering would watch from his no-longer-deserted perch in the SFOF viewing area. A few hundred yards away, newsmen and other JPL employees would once again follow things in Von Karman Auditorium. At 9:50 A.M.—precisely when scheduled—they got the first bit of what they came for, when *Ranger 7* successfully left the pad. Less than fifteen minutes later it left the atmosphere, and less than ninety minutes after that it headed out toward the moon.

During the sixty-six hours the spacecraft was in its translunar coast, the press coverage of the flight was uncharacteristically—but perhaps understandably—sparse. From all signs, *Ranger 7* was on target for its lunar impact, but *Ranger 4* had been, too. Its telemetry stream indicated that all of its systems were functioning, but for a while at least, so had *Ranger 5*'s and 6's. The reporters, it was clear, would be willing to give the story a little coverage, but this time *just* a little. Only when the ship arrived above the Sea of Clouds at about

6:25 on the morning of July 31 and prepared to take its pictures, would the oft-fooled press pay it closer attention.

As the sun prepared to rise on the West Coast that Friday morning, the reporters at both JPL and NASA's Washington headquarters were indeed in their press rooms. *Ranger 7* itself, however, was not precisely where it was supposed to be. With only a few thousand miles to cover and just over an hour of flight time to go, the spacecraft was clearly headed for the Sea of Clouds, but a subtle drift in its flight path had nudged it a bit off course, causing it to approach its impact site at a slight angle, one that sharpened shadows and distorted perspectives slightly. The mission directors were fully aware that this might happen and had built an opportunity for an eleventh-hour trajectory adjustment into the flight plan. At shortly after 5 A.M., that eleventh hour arrived, and Bud Schurmeier now had to decide whether to fire his ship's thrusters and change its orientation slightly or simply let it fly on.

At his console in the SFOF, Schurmeier huddled with Gerard Kuiper, the head of the experiment team that would be analyzing *Ranger 7*'s images. Kuiper, he knew, had been huddling with his own scientists, and he had a pretty good idea what message he came bearing.

"We've been talking it over in the back room," Kuiper said, "and we really think we need to nudge this ship."

"Which means disturbing its guidance system," Schurmeier said.

"I know," Kuiper answered.

"And firing up its thrusters."

"Mm-hmm."

"And you think it's worth it?"

"I do," Kuiper said.

Schurmeier nodded thoughtfully. "I don't," he said. "I don't want to touch this spacecraft unless it's absolutely necessary."

"If we don't get the right kind of pictures," Kuiper objected, "we miss half the point of the mission."

"If we don't get any," Schurmeier answered, "we miss it all. We'll go with the trajectory we've got."

Kuiper nodded once and Schurmeier keyed open his microphone. "Terminal maneuver will be a no go," he said with finality. In the viewing room Pickering smiled approvingly; at his console on the SFOF floor, Burke agreed. On the wall above them, the mission clock read barely one hour to impact.

For the next forty-plus minutes *Ranger 7* would do nothing note-worthy—not until precisely eighteen minutes before impact when the on-board timer would order the cameras to begin charging them-selves up. Three minutes later, just as they had on *Ranger 6,* they would be instructed to open their eyes and send their pictures home. When the mission clock read thirty minutes to impact, the men in the SFOF began nervously studying the trajectory plots. When it read twenty minutes, the ones who had been milling about returned to their seats. When it read nineteen minutes to impact, they leaned to-ward their consoles. At the eighteen-minute mark, Schurmeier squinted at his screen, drew a breath, and then released it in a great gust as the tiny data point on the monitor indicated that the cameras were indeed receiving juice.

"Cameras in full power mode," he announced. The men at the consoles made celebratory fists; the men in the viewing room flashed them a thumbs-up.

In Von Karman Auditorium, George Nichols, a *Ranger* engineer, would be performing the public address job Walt Downhower had handled on *Ranger 6.* Listening through his headset, he heard Schurmeier's call and repeated it to the room. "Ranger Control re-ports that cameras are in full power mode," he said. The press im-mediately broke into applause and Nichols quickly hushed them. "We point out that having the TV system in full power mode is not proof that the cameras are operating," he called out. "That will not come for another three minutes."

For the people in the SFOF and the two NASA press rooms, that time would play out slowly. Before the necessary 180 seconds elapsed, the ship would have to free-fall at least three hundred more miles, plummeting from 1,800 miles above the Sea of Clouds to just over 1,500. The first minute ticked away, then the next, then the next. Then, at almost precisely the instant the mission clock turned over from 15:00 to 14:59 and the spacecraft plunged through the 1,520-mile altitude mark, the antennas in the Mojave Desert twitched to life.

"Video signal received!" a Goldstone technician shouted into his headset to JPL. "Video signal received! It looks strong and it looks clean."

Schurmeier grinned broadly. "Strong and clean video being received by Goldstone!" he repeated to the SFOF.

In Von Karman, Nichols made the call, too. "At this point," he said evenly, "a video signal is being received and is reported to be strong and clear." A roar went up in the room, and Nichols again tried to override it. "This-does-not-mean-a-picture-is-being-received," he shouted in a hoarse staccato. "It will take some analysis and a check of the recorders to determine that."

Nichols was right, of course. If the signal flowing from the spacecraft was weak or garbled, the desert antennas would be receiving little more than cosmic nonsense. But even as Nichols was making his cautionary announcement, equipment began to stir deep in a Goldstone lab. On a bank of black-and-white television monitors, tiny, phosphorescent fireflies had appeared and begun tracing a crazy path across the glass, and a series of cameras mounted in front of them began to snap and snap and snap. At one monitor a Polaroid camera clicked, whirred, and spit out a single card of photographic paper. An engineer standing by instantly grabbed it and ripped the backing away. In front of him he saw what the spacecraft's cameras had seen, and what the spacecraft's cameras had seen were hills and rills, craters and gullies, canyons and plains and lunar seas. What the cameras had seen was the moon.

"We've got pictures!" the technician called out.

"Pictures are in!" a Goldstone communicator relayed to Schurmeier.

"Goldstone has pictures!" Schurmeier repeated to the SFOF.

In Von Karman, Nichols braced. "Preliminary analysis of the signal," he said, "is that we are seeing pictures." This time, he knew, there would be no silencing the reporters' cheers. All over the twin NASA press rooms, shouts went up, hugs were exchanged, notebooks were tossed in the air. In the SFOF viewing room, Pickering

and the other administrators reacted with similar—if more reserved—glee, rising almost as one, shaking hands, flinging arms
around shoulders, and applauding down at the mission control team
working below them. The team members themselves looked over
their shoulders at their appreciative superiors and broke into huge
and helpless smiles.

For the next fourteen minutes the 807-pound *Ranger* 7 continued to fall and its bank of six cameras continued to fire. Dozens,
then hundreds, then thousands of electronic images deluged the
thirsty Mojave antennas. As film spun through the 35mm cameras
and picture after picture dropped from the Polaroids, the whistle of
the accelerating spacecraft's trajectory seemed to grow louder and
louder in the headsets of the controllers. The ship plunged from an
altitude of 1,000 miles, down to 500, and then on down to just a few
dozen as its speed climbed and climbed and climbed. From the surface of the moon, what had appeared only moments ago to be a single, stationary star would now appear to be a moving, glinting,
growing mass. Finally, at exactly 6:40 A.M. Pasadena time, after returning 4,316 pictures, *Ranger* 7 completed its plunge, slammed into
the ancient surface of the Sea of Clouds, and, as the physicists predicted it would, liquefied, vaporized, and ceased to exist.

More than 240,000 miles away, a team of astronomers in a
British observatory were certain they saw the flash.

•   •   •

Less than three hours after the spacecraft that had been known as
Ranger 7 passed into history, William Pickering retreated to his office and invited Jim Burke, Gerard Kuiper, Bud Schurmeier, and a
few other *Ranger* scientists to join him there. A few minutes after
they arrived, a delivery boy showed up with sandwiches and sodas.
A few minutes after that, there was another knock at the door.

Pickering rose and opened it, and a young technician from the
JPL photo lab entered carrying a portfolio-sized manila envelope.
The technician handed the envelope wordlessly to the JPL chief, who
nodded his thanks and, with a smile, beckoned the boy to leave. Af-

ter he had gone, Pickering opened the little metal clasp at the top of the envelope, lifted the flap, and slipped his hand inside, sliding it beneath a stack of slightly warm photographic paper. Easing the stack out with the care of a waiter maneuvering a heavy tray, he nodded to the men in the room, who swept the cups and food wrappers off the table where they had been eating. Carefully then, Pickering laid the moon before them.

The senior members of the *Ranger* team looked down at the images on the table in something close to utter silence. Someone might have murmured "astounding"; someone else might have murmured "remarkable." Whatever was or wasn't said, the pictures their ship had brought them were extraordinary. There were great plains and small flatlands, jagged peaks and smooth hills. Over here was a chasm, over there was a crater, over there was a boulder as small as a car. The scientists had looked at thousands upon thousands of lunar images before, but all had been taken by telescopes anchored on Earth. These pictures were taken by the first American camera to cross the translunar distance and do its work up close. Slowly, slowly the scientists flipped through the stack of images, and then when they were done, they flipped through them again.

The quiet scene in Pickering's office took an hour or so to play out and was nothing like the bedlam that had broken out just a little earlier in Von Karman Auditorium, when the same men made their first appearance in front of the press after the spacecraft's impact. Even before the scientists arrived, the newsmen had risen and offered an anticipatory round of applause to their tiny TV images when the SFOF camera caught them in its sights and flashed them onto the press room monitor. When the scientists arrived in Von Karman, the applause turned into a cheer. The moment the *Ranger* bosses mounted the stage, the already boisterous crowd hurled questions at them.

"How good did the Goldstone people say the pictures are?" someone asked.

"Several times better than any pictures of the moon we've ever seen before," Pickering said judiciously.

"Have they seen anything unexpected?" someone pressed.

"If you mean were there any little green men," Pickering said, "the answer is no." The room erupted in laughter, though at this point Pickering suspected he could have read aloud from a grocery list and this crowd would still have responded happily.

"What do you think about the future of JPL now that this mission has succeeded?" a voice called out.

This time it was Pickering's turn to laugh. "I think," he said, "it's improved."

Later that day a phone call came in to Pickering from Lyndon Johnson. The president congratulated the *Ranger* team on its work, inquired about the newly returned pictures, and allowed as how he wouldn't mind seeing a few of them himself. Lyndon Johnson allowing that he wouldn't mind seeing something meant that he jolly well expected to see it, and later that night a somewhat dazed Pickering found himself on a red-eye to Washington. By eight the next morning, he was standing in the Cabinet Room in the White House, with the president, a handful of senators, and Jim Webb, the beaming NASA administrator. With the help of an easel, a pointer, and blowups of the same pictures Pickering himself had seen the previous afternoon, the JPL chief showed the president the curious place his space agency's spaceship had visited the day before. Johnson listened attentively, clucked over the pictures, and once again offered Pickering his congratulations. An hour later Pickering was on a plane headed west. Six hours later he was back in his lab, as if his trip to the White House hadn't taken place at all. For all he knew, in the president's mind, it never had.

Seven months later, in February 1965, *Ranger 8* flew flawlessly to the moon, sending back thousands of pictures of its own before annihilating itself against the ancient face of the Sea of Tranquillity. A month after that *Ranger 9* followed, successfully self-destructing in the Alphonsus crater, after returning a third album of images. Following that, Pickering knew, no more *Ranger*s would be flying. Now that NASA had proven it could get where it wanted to go, it had far bigger plans to pursue. There were the *Surveyor* ships that would soft-land on the moon in a single, functioning piece; there were the *Viking* probes that would accomplish the same feat on Mars. There were the *Mariner*s to Venus, the *Pioneer*s to the sun, and the other, as yet unnamed ships that might go as far out as Jupiter. And then, of course, there were the grand *Apollo* skyliners that would take not hardware, but humans to the same moon *Ranger* had so recently claimed.

With such magnificent ships to build, the agency didn't have much time for little spacecraft able to do no more than fly out to the moon and belly flop in its soil. *Ranger* may have been the first probe to make such a deep-space trip but it wouldn't be the last, and before the final, vaporous remains of *Ranger 9* had even had a chance to scatter in the nonexistent lunar winds, Pickering knew it was time to turn his mind to other things. It would be awhile, he realized, before he'd be getting another call from the White House.

Early 1966

Jim Burke was finding it hard to believe what the numbers coming out of the computer were telling him. As an engineer Burke had long since learned to trust the numbers he worked with every day, even if what they suggested and what common sense suggested were two different things. Today, however, the numbers were suggesting something fantastic.

In recent months Burke had grown increasingly familiar with fantastic numbers. It had been about a year since the end of the *Ranger* program, which meant it had been about a year since he had started working for Homer Joe Stewart. Homer Joe Stewart, as anyone at JPL could tell you, was a bit of a visionary, and thus it was fitting that when he got his own JPL section to head, it was an unconventional one. Stewart's group wasn't involved in something as mundane as developing hardware or flight plans for current missions, but cooking up schemes for future ones. If there was a planet no one had even considered visiting yet, it was Stewart's group that concocted a way to get there. If there was a spacecraft no one had even dreamed of building, it was Stewart's group that would dream it up. For an engineer like Burke, who was accustomed to assembling here-and-now machines, this kind of blue-sky thinking was a new thing. Gradually, however, he'd gotten used to the way Stewart's department did things, knowing that if the ideas were good enough and the ships imaginative enough, at least a few of the whimsical flights might actually get flown. The data coming from the computer today was the stuff of whimsy, indeed.

Burke had been called down to the computer room by Jim Long, a more junior member of the department who had been exploring the idea of missions to the solar system's outer planets. Long, it was evident, loved playing about with long-distance flights like these. The billions of miles the ships would have to cover and the dozen or so years it would take them to make their journeys made *Ranger* and *Mariner* missions seem like day trips, and Long used to spend untold hours hunched over his solar system maps, happily plotting the imagined ships' improbable trajectories. When Burke arrived in the computer room a few minutes ago, Long looked even more pleased than usual, and before the senior man could say anything the junior man thrust a sheaf of papers at him.

"Look at this," Long said. Burke started to respond, but Long cut him off. "Just look at it."

Burke glanced at the page and saw that it was dense with numbers and planet names. He looked a little closer and saw that what he was seeing were orbital calculations, but not present-day ones; these were projections for the orbits of the planets more than a decade in the future. He looked closer still, and the full impact of what he was seeing hit him. He whistled to himself softly and turned to Long with a smile.

"When will this happen?" Burke asked.

"Nineteen seventy-nine," Long answered.

"You're sure of this?"

"Completely."

Burke smiled again. What Long was telling him was remarkable. With nine planets in the solar system and each orbiting the sun at different radii and different speeds, the local family of worlds looked, to the lay eye, like nothing more orderly than a nine-ball roulette wheel. But even out of chaos can come order, and the solar system was no exception.

Every 176 years, random orbital motion caused the four large, outer planets—Jupiter, Saturn, Uranus, and Neptune—to organize themselves in a near-perfect line. The tidiness wouldn't last long; look away for too long and when you looked back, things would have already started to deteriorate again. But if you acted fast—if you acted with foresight—you could launch a spacecraft from Earth that would reach the outer system just when the four gas giants had begun to form ranks. With the right navigational plan, you could then tour them all, using Jupiter's gravitational energy to fling you on to Saturn, Saturn's to get you to Uranus, and Uranus's to get you to Neptune. Neptune would then toss you out of the solar system altogether, sending your spacecraft not just into the void between the planets, but into the void between the stars. The idea wasn't a completely new one; JPL mission planners had been talking about the possibility of just such an orbital alignment for a long time. Now, however, it looked as if it was certain to happen—and happen soon.

Burke looked up at Long again.

"When do you think we'd need to be ready to launch?" he asked.

"Nineteen seventy-seven," Long said.

"Then we need to start planning now."

2

Green Glass and White Rocks

Thursday, July 29, 1971, 9 P.M.

Dave Scott was two-thirds of the way to the moon when he at last caught the smell of pork and scalloped potatoes cooking in his command module. The *Apollo 15* crew had smelled a lot of smells in the eighty or so hours they'd been away from Earth— the new rubber of their spacesuit liners, the fresh canvas of their couches, the vaguely acrid scent of oxygen that had traveled through hundreds of feet of plastic and metal tubing before at last flowing into their tiny cockpit—but none of them had been as welcome as the pork and potatoes they were smelling now.

To be sure, it was something of a stretch to say the pork and scalloped potatoes were cooking at all. Cooking required an oven, and an oven required room, and room was not something three men could hope to find in their cramped little pod of an *Apollo* spacecraft. In another year or two NASA would be launching its much talked-about Skylab space station. While the huge orbiting liner wouldn't travel very far from Earth—permanently circling the planet just 235 miles up—it would keep its lucky crew of three in the kind of comfort only a big ship with lots of interior acreage could provide. You could bet Skylab would have an oven on board, and when it came time for dinner on *that* ship you'd get genuine smells of genuine food from one end of the forty-five-foot craft to the other.

Aboard *Apollo* things were different. All of the food for a two-week lunar journey was stored in a foot locker–sized larder tucked away at the foot of the command module. The 126 meals—three a day for three men for fourteen days—were freeze-dried, shrink-wrapped, and then color-coded red, white, or blue, indicating the commander, command module pilot, or lunar module pilot. When mealtime rolled around, the crew members would simply scoop out a few packets, snip the ends off with a pair of shears, and then inject hot water into the opening with a hose and gun assembly stashed in the lower equipment bay. With much of the food—processed and puréed to an all but unidentifiable paste—it wasn't always easy to taste what you were eating, much less smell it. But for some reason, pork and scalloped potatoes were a different story. Hydrate the meal with just enough water, and the fragrance coming out of your packet—if not the look of the food inside it—could almost make you believe you were eating an actual Earth meal in an actual Earth kitchen.

Dave Scott had been looking forward to pork and potatoes night for a while now, and as the menu rotation would have it, it fell on the best of all possible evenings. In just eighteen hours—or what would be mid-afternoon on Friday if the crew was still in Houston—he and Jim Irwin would separate their spindly lunar excursion module from their cone-shaped command module, fire their descent engine, and head down to the surface of the moon. A few hours after that the LEM would touch down on a prairie known as the Hadley Plain, hard by the toothy Apennine Mountains, just near the spot where the waters of the Sea of Serenity would overlap the waters of the Sea of Rains if the Sea of Serenity and the Sea of Rains, in fact, had any water. It was a spot Scott had lobbied hard to visit, and tomorrow at this time he was going to arrive there. For a man hoping to study lunar geology, there was no better place to be.

For all their trailblazing achievements, *Apollo*s *11* and *12*—the first two lunar landing missions—had played it relatively safe. The target for those missions, both of which flew in 1969, had been the Sea of Tranquillity and the Ocean of Storms, and if you were looking for a good, safe spot to set down a lunar lander, you couldn't pick better runways. For one thing, both places were utterly flat. It wasn't for nothing that ancient astronomers had mistaken the vast, unbroken stretches of extraterrestrial terrain that mottled the moon for seas, and for acre after acre, mile after mile, they rolled on as huge and featurelessly as the bodies of water for which they had been misnamed.

For another thing, both places were well mapped. In 1965, *Ranger 8* had ended its brief life in the Sea of Tranquillity, transmitting more than seven thousand pictures of the surrounding landscape before destroying itself in the ancient hardscrabble. In 1966 and 1967, *Surveyors 1* and *3* soft-landed in the Ocean of Storms, touching gently down on a plume of rocket exhaust and returning nearly 17,000 pictures of the land around them. Shortly after, *Surveyor 5* followed, returning to the Sea of Tranquillity, soft-landing not far from where *Ranger 8* had crash-landed, and broadcasting home more than 18,000 photographs to supplement the ones already in hand. In addition to all that reconnaissance, NASA had dispatched a fleet of five *Lunar Orbiter* ships to circle the moon between 1966 and 1967, conducting further photo surveys of all of the largest lunar seas—Tranquillity and the Ocean of Storms among them. After so much high- and low-altitude surveying, there was barely a rock or divot in the surface of either of the two plains that hadn't had its picture taken over the last few years. While NASA flight planners couldn't guarantee that an errant crater or boulder field wouldn't still pop up to surprise an approaching astronaut crew, they were pretty sure that they were now utterly familiar with the lay of the lunar land.

Sending astronauts to such safe and nondescript places, however, came at a price. From the moment modern telescopes and robotic spacecraft confirmed that the moon did not appear to have ever held so much as a cupful of water—or, for that matter, a lungful of air—it seemed clear that one thing its seas did hold was lava. With no wind or rain to sculpt the land, there was only one other force besides water that could trowel the lunar seas down as flat as they were, and that was molten rock pouring over the surface from magma chambers far below.

The problem with studying land like this is that if you're looking to peek back into lunar history, a lava plain can take you only so far. Geologists seeking to determine the age of surface formations lived by a simple, sacred doctrine known as the Law of Superposition. For all of the arcane chemistry and complex mineralogy that is part of the geological art, the Law of Superposition was a wonderfully straightforward idea. The deeper a geographic feature is buried, the law decreed, the older it is; the shallower it lies, the younger it must be. A geologist looking at a photograph of a rock lying on the lip of a crater, which was itself dug out of the foot of a hill, knew instantly and intuitively that the hill was pushed up first, the crater blasted out of it later, and the rock—through some other dynamic event—deposited on it last.

Applying this same reasoning to lunar seas, the scientists realized that these huge flatlands—which covered everything around them and yet were themselves marred by almost nothing at all—had to be among the youngest features on the by-no-means young moon. If you wanted relatively recent geology, the seas were a perfectly adequate destination. But if you wanted to collect the true keystones to the lunar past, you had to head for the lunar hills.

In 1970, *Apollo 13* had tried a landing in the moon's highlands, setting out for the hills of the Fra Mauro formation in the moon's southwest quadrant. After a translunar explosion blew the ship and the mission all to hell—forcing commander Jim Lovell and his rookie crew to turn around and fly their mortally wounded spacecraft back home—*Apollo 14* aimed for, and this time reached, the same site. Now, *Apollo 15* was going to raise the topographic ante, bypassing the plains and the hills, and heading straight for the base of the Apennine Mountains, three-mile-high peaks that were among the starkest features anywhere on the moon.

There was no telling exactly how a mountain range as dramatic as the Apennines formed, but geologists had a pretty good idea. Long ago, they hypothesized, the moon was a fairly undifferentiated place, with perhaps a few wrinkly hills breaking up its otherwise egg-smooth surface, but not much else. Now and then, however, a mammoth boulder cannonballing through space would take aim at the airless—and thus defenseless—moon and collide with it broadside. Vaporizing on impact, a meteorite this big wouldn't leave a simple crater, but instead would gouge out a valley that might measure hundreds of miles across. The injury the projectile inflicted would not be just wide but deep, causing a great hemorrhage of the moon's molten lifeblood to flow out over the surface. The bleeding would continue for tens of millions of years before it was stanched, and when it finally was, it would leave behind a smooth, new lunar sea across most of the valley the collision had created.

Around the edge of the newly formed basin, things would not remain so flat. The land the meteorite excavated when it hit would have to go somewhere, and in general, where it would go would be out and up, crumpling toward the sky in huge, instant mountain ranges that would surround the sea. Since the new peaks would be spared the deluge of lava that flowed into the valley below them, their flanks would still be made almost entirely of ancient crustal material. A crew of astronauts who put their spacecraft down near the mountains would thus merely have to look for a little high-lying rubble and they would likely have themselves an authentic bit of the original moon.

It was precisely this kind of debris the *Apollo 15* crew would be hoping to find when they landed at the foot of the Apennines, and the material they would be keeping a special eye out for would be something called anorthosite, a bright, white crystalline material made up principally of the mineral plagioclase, which was itself made up of aluminum and calcium. Scott and Irwin had seen plenty of anorthosite on geology treks they'd taken into California's San Gabriel Mountains over the course of the last year; now, they'd be looking for the same common debris on the surface of an entirely uncommon world. If they were able to bring a little bit home, and if the age of the sample was older than the age of the lunar sea itself, they would both prove the impact theory of basin formation and confirm the age of the moon as a whole.

As a novice lunar geologist, Scott looked forward to digging through the soil to hunt for the ancient samples. As a veteran astronaut and test pilot, he looked even more forward to the business of getting there in the first place. Steering his little, foil-skinned LEM through the spires of the Apennines and setting it down at the base of the mountains would be a nifty piece of piloting—a *sporty* piece of piloting, as Scott liked to say—and he was only too pleased to try his hand at it.

Tonight, at 9 P.M., he was just eighteen hours away from getting his chance. Floating above his canvas seat on the left-hand side of the spacecraft, he scooped a spoonful of pork and scalloped potatoes from his plastic food pouch and chewed slowly and silently. In the center seat, Al Worden—the command module pilot, who would stay behind the wheel of the lunar orbiter while his crew mates were down on the surface—did the same. Off to the far right, Irwin mirrored them both.

Scott turned idly to his left and happened to look out the window on the port side of the cabin at just the moment the wasted landscape of the Sea of Serenity rolled by sixty miles below. Next to it lay the vast, gray stain that was the Sea of Rains, or Mare Imbrium, as the scientists both on the ground and in the spacecraft less colloquially called it. Somewhere on Imbrium was the spot where Scott and Irwin would be walking in less than a single Earth day. Scott scanned the terrain and broke into a broad grin.

"Hey, Jim," he called to Irwin. "You know what I'm doing tomorrow?"

"Can't guess," Irwin answered, with a smile of his own.

"I'm gonna go shoot a landing"—Scott looked out his window, found his spot, and pointed—"right there. I'm gonna put that baby right down there."

"That's right," Irwin said, a little less certainly.

"Yes, sir," Worden agreed.

On the starboard side of the cockpit, Irwin shifted his body slightly, peered out his own window, and watched the Mare Imbrium landing site Scott had just spotted slide further into view. Unlike his commander, who had flown in both the *Gemini* and *Apollo* programs, Irwin was new to the outrageous business of space travel. And unlike Worden, who was also a novice, he would not be spending the entirety of his maiden flight in the relative safety of the command module. Rather, he would be relying on his rookie skills—and his rookie nerves—to see him all the way down to the surface of the moon.

"You know what I want to do first thing when we get back home?" Irwin said after a moment's reflection.

"What's that?" Scott asked.

"Have a beautiful night in Tahiti."

"You're on, buddy," Worden said. "You're on."

"You bet," Scott agreed.

"No, really," Irwin persisted, all playfulness gone. "I'm not kidding. We really ought to think about it."

"Absolutely," Scott agreed. "We'll book a big airliner."

Irwin looked around himself at the little nutshell of a command module that was keeping him and his crew mates alive; then he cast his eye out his forward window, where he could just see the roof bolts of the fragile lunar module that would soon be transporting his mortal soul to the foot of the Apennine Mountains and back again. A veteran like Scott might never quite get his fill of flying, but after this week a first-timer like Irwin would have had enough to last him quite a while.

"Without the aviation," he said flatly to Scott. "Without the aviation."

July 30, 1971, 5 P.M.

The *Apollo 15* moonwalkers were under no illusions that flying their lunar module down to the surface of the moon would be a simple matter. What made it even harder was the fact that they'd be flying it blind.

Earlier today, at just 3:00 P.M. Houston time, Scott and Irwin had climbed into the LEM, sealed their roof hatch, and cast off into orbit. When they did, they were moving at about 5,000 miles per hour—a velocity that would have kept them in lunar orbit indefinitely. In order to descend to the surface, they would have to reduce that speed by more than half; in order to do that, they would have to position their LEM so that its descent nozzle faced forward, then hit the ignition and let the chemical fire that poured from the engine bell apply a propulsive brake. The problem with this simple maneuver was that since the LEM was designed to land in a standing position, its descent nozzle was built into its underside. Pointing the engine forward thus required rotating the ship 90 degrees backward, and that left the astronauts lying flat on their backs—flying feet first, head last, eyes looking up at the stars.

For almost any pilot, this was an awkward—not to mention undignified—way to fly, yet for most of the sixty-mile descent to the moon, that was just what the mission plan demanded. It was only when the crew was 7,000 feet above the surface and three minutes from touching down that the on-board computer would at last order the clusters of thrusters arrayed around the ship to bloom to life, rolling the LEM to an upright position. The commander, with his feet back beneath him and the lunar landscape flooding his window, would then grab his maneuvering stick, scan the terrain for familiar surface features, and at last begin the high-wire flying he came here to perform.

Lunar module crews—who had spent months studying terrain maps before setting out for the moon—typically had a lot of landmarks to help them steer their ships in, and Scott and Irwin were no exceptions. In addition to the mammoth Apennine Mountains, which would serve as their first and most prominent harbor marker, they would be looking for a depression known as Spur crater, a deep abyss that marked the edge of their prime landing site. Also helping to frame the target area would be Hadley Tower, a sheer 11,000-foot mountain at the edge of the Apennines; the North Complex, a small cluster of prominent craters; and Hadley Rille, a sinuous lava channel snaking through the surrounding Hadley Valley.

Along with these friendly landmarks, there was also one that wasn't so friendly. On the other side of Hadley Rille was a single, low hill easily visible from high above. If Scott steered his ship too close to that unmistakable landscape bump, he would know that he had overshot his prime target, was drifting well beyond the area NASA considered safe, and was in danger of putting the LEM down on a deadly, boulder-strewn incline. Scott, recognizing the dark significance of the little peak, had named it Bennett Hill, after Floyd

Bennett, the NASA trajectory planner who had designed this dare-devil approach in the first place. The name Scott had chosen for his lunar module itself was equally apt: *Falcon,* he had decided to chris-ten the ship, in a nod to the acrobatic flying it would have to perform in order to get through this expedition intact.

The descent to the surface began not long after Scott and Irwin climbed into the lunar module and drifted away from the command module, when they at last instructed their computer to instruct their descent engine to begin firing. For an instant they felt nothing at all, then a low-register rumble shook the cockpit, while outside, an ut-terly silent column of flame erupted from the bottom of the ship. Al-most undetectably, Scott and Irwin noticed a faint sense of pressure as the LEM began to slow and the barest ghost of gravity appeared. On their instrument panel the numbers on the velocity indicator be-gan to fall and the trajectory readouts began to steepen. Nearby, the altitude gauge, which had been holding steady at 60 nautical miles, began to slip, too.

For more than seven minutes, the moonwalkers sailed silently along, riding the quiet fire issuing from their ship, as they soared over the Sea of Serenity, Mare Imbrium, and the occasional crater or wrinkly rill that here and there disfigured both. Finally, when the *Fal-con* had descended to the prescribed 7,000-foot altitude, its thrusters puffed to life and it swung itself slowly upright. Scott looked imme-diately out his left-hand window, Irwin looked out his right, and both men went wide-eyed.

Surrounding the ship wherever the astronauts looked—above them, below them, across the entire field of vision their forward-facing windows allowed them—stood the sheer, gray-white walls of the Apennine Mountains. For most of the last twenty-four hours the *Apollo 15* astronauts had been seeing the moon from a nosebleed-high five dozen miles that reduced even the grandest features on the lunar surface to little more than bumps. Now, Scott and Irwin were just over a mile above ground, flying through a mountain range whose loftiest peaks were nearly twice the altitude of the ship itself.

Scott looked out his window and took in the mammoth canyon walls that hulked all around him. Though the moment seemed to call for some acknowledgment, he decided to limit his comments to pretty much what he had been trained to say at a time like this—which was pretty much nothing at all. The final descent into the Hadley region would require utter concentration, and if there was any talk going on inside the cockpit at all, it would come from Irwin, whose responsibility would be to keep an eye on his instrument panel and read out the spacecraft's ever-changing speed, altitude, and angle of descent. Scott, in turn, would keep his gaze fixed out the window, working the ship's maneuvering stick in response to Irwin's call. Irwin let a few long moments elapse as the ship continued its steep descent, gave his gauges a fast scan, and announced his first reading.

"Five thousand feet," he said uninflectedly, "thirty-nine degrees."

Scott nodded to himself and tweaked his throttle higher. The pitch of the engine picked up a bit and vibrated through the cockpit. Irwin let nine more seconds elapse.

"Four thousand feet," he said, "forty degrees."

Scott nodded again, tweaked again, and listened as his engine climbed half an octave again. He kept his eyes fixed out the window and worked his stick by feel alone.

"Three thousand feet, fifty-two degrees," Irwin said.

With his throttle held steady, Scott squinted out over the landscape below, scanning for familiar features. In the distance he thought he could just make out the serpentine line of Hadley Rille and the dark well of Spur crater. Ten more seconds elapsed.

"Two thousand feet, forty-two degrees."

Scott now saw a few more craters freckling the landscape. All of them were familiar to him, just where the mapmakers said they would be. Nearby, he also spotted a patch of smooth, invitingly unbroken ground. He smiled slightly.

"Okay," he said, for the benefit of both Irwin and Mission Control, "I got a good spot."

"Good," Irwin answered. "Forty-two degrees."

Scott throttled his engine to slow his descent still further and the *Falcon* went into what briefly felt like a stationary hover. He engaged his array of sixteen thrusters and the ship edged forward toward the target spot on the surface. Over the next minute the ship descended to 1,000 feet, then 500, then just 100. Finally, sixty seconds before scheduled touchdown, it had slowed to a speed of just three feet per second and an altitude of just sixty feet, and the tongue of flame that was holding the *Falcon* up at last began to graze the ground. When it did, dust devils began to rise and churn, swirling wildly outside the astronauts' windows.

"Okay," Scott said evenly, "I've got some dust."

Some dust quickly turned into a lot of dust, and a lot of dust quickly became a whiteout. On the other side of his triangular porthole, Scott could now see nothing at all. Holding tight to his throttle, he eased the ship lower and lower, as the engine's vibration grew louder and louder. At last, when the LEM was just three feet above the surface, probes extending from the bottom of its foot pads made contact with the ground and a sea blue indicator lit up on the instrument panel.

"Contact!" Irwin called.

Scott immediately reached forward and hit his engine stop button. Instantly, the pillar of fire beneath the ship vanished and the *Falcon* free-fell the last thirty-six inches to the surface, hitting with a deadweight thud that caused every piece of equipment in the cockpit to rattle in its bracket.

"Bam!" Irwin called out.

Bam, Scott echoed in his head. Without a word he reached forward and threw a bristle of switches, disarming his engine and reconfiguring the controls of his now stationary ship. Finally, he drew a breath, keyed open his mike, and called out to his home planet 250,000 miles away.

"Okay, Houston," he said with satisfaction, "the *Falcon* is on the plain at Hadley." And so it was.

August 1, 1971, 10 A.M.

Dave Scott and Jim Irwin knew enough not to say too much when they found their $400 million rock. It was hard to determine exactly what the little lump of lunar rubble was worth, of course. But since it had cost $400 million to build and fly *Apollo 15,* and since this was precisely the rock they had traveled across the void of translunar space to find, $400 million seemed like a pretty fair figure. In any event, it didn't pay to talk too much about it.

It was Buzz Aldrin who had first learned the lesson of talking too much on a lunar geology expedition, and he learned it the hard way. Buzz had preceded Scott and Irwin here almost two years ago to the day, when he and Neil Armstrong put their *Apollo 11* LEM down in the Sea of Tranquillity, 423 miles to the southeast as the lunar crow flies. Like so many of the pilots who were chosen for the astronaut program before him, Aldrin got his education at West Point, studying history, science, and soldiering in pretty much equal measure. Like not so many others, he decided to continue his schooling after that, earning a Ph.D. from MIT in the brain-numbing business of astronautics and orbital physics. A man who could fly airplanes, master the sciences, and pile up degrees like so much kindling was just the kind of man NASA was looking for, and it was no surprise that when it came time to pick a crew for the first lunar landing, Dr. Buzz Aldrin earned himself a seat.

Buzz took enthusiastically to his astronaut training—no more so than when that training included geology. For nearly half a year before *Apollo 11* took off, Armstrong and Aldrin made regular trips into the California badlands to learn the spelunking skills they would need when they at last reached the surface of the moon. More than a few of the scientists who came along with them on these expeditions had serious questions about the value of the trips. Geology isn't a discipline you simply pick up over the course of a few excursions into the desert, they grumbled among themselves; it's a lifelong vocation that takes a lifetime of study.

The scientists were careful not to share these misgivings with the astronauts, of course, but Aldrin, no fool, was well aware of what was going on. Determined to prove the turf-protecting scientists wrong, he fell to his geology studies with an almost inexhaustible zeal, becoming a lay expert in the field within months. Later that summer, when he and Armstrong found themselves in the Sea of

Tranquillity—a field site the boys in the lab coats had never harbored any hope of visiting and never would—he figured he was carrying the torch for every scientist-astronaut who would ever follow him there. Almost immediately, he appeared to have dropped it.

Armstrong and Aldrin's most important job after they climbed down the LEM's ladder was to collect and bag a few quick samples of rocks in case an emergency required them to terminate their lunar stay early. This so-called contingency sample was the highest-priority exercise in the early phase of the moonwalk, and the astronauts were expected to get to it straightaway. Aldrin, who climbed down to the surface only minutes after Armstrong took his first historic step, needed almost no time at all to get his bearings before setting to work. Poking about the landing site in the immediate vicinity of the ship, he noticed a small, vaguely purplish rock, freckled with what looked like bits of crystal. Buzz had seen a lot of samples just like this one back on Earth, and he did not hesitate to announce his find now.

"Hey, Neil," he called out. "Didn't I say we might see some purple rocks?"

"Find one?" Armstrong asked.

"Yep. Very small. Sparkly. I would make a first guess at some sort of biotite, but we'll leave that to further analysis."

To most people watching at home, Aldrin's observations were innocuous ones. But to the geologists in Mission Control, they were an absolute howler. Biotite! On the moon? Had these silly flyboys learned *nothing*? Biotite contains hydroxyl and hydroxyl forms only in the presence of water, a substance that was utterly absent on the rock-and-powder moon. You might find a lot of interesting samples in a place as geologically rich as the Sea of Tranquillity, but you had about as much chance of finding biotite as you did of finding a jackrabbit.

Throughout the scientific back rooms at NASA, knowing smiles and rolled eyes were exchanged all around, but among the astronauts, Buzz's blunder was seen as no blunder at all. If there was one rule the scientists had stressed to the astronauts from the very first day of their training, it was that the most important job of a field geologist is to observe and describe. Describing isn't analyzing; describing isn't identifying. Describing is merely telling the scientists back home exactly what you're seeing, and doing so in the clearest terms possible. If you found a rock that looked like a tangerine, you didn't waste your breath describing its color, shape, or nubby-rough surface. You said it looked like a tangerine and moved on to the next sample. It was the astronaut's job to explain what a rock resembled; it was the geologist's job to explain what it was.

Buzz had followed that rule, and had done so at the expense of his very credibility. From that moment on, every other lunar astronaut made it a point to mind his words, lest he be the next one to give the boys in the back rooms reason to laugh. Now, a little over two days after arriving on the surface of the moon, the astronauts of *Apollo 15* had their own remarkable sample to describe and their own phrasing to consider.

At ten o'clock on Sunday morning, Scott and Irwin found themselves standing three miles away from their landing site at the foot of the towering Hadley Delta Mountain. Three miles was a lot of ground to cover anywhere, never mind on the moon, and on *Apollo*s *11, 12,* and *14,* trying to trudge that far from the LEM would have been out of the question. On *Apollo 15,* however, the crew had brought a little help along: a nine-foot-long, four-foot-wide, open-chassis rover in which they could scoot across the surface at speeds approaching nine miles per hour. Unstowing the collapsible car early yesterday, they had hopped aboard and spent the better part of the last two days tooling here and there, looking for likely sites to stop and prospect. Just a few minutes ago they had parked on the lip of Spur crater and climbed down from the rover to pick through the rocks that littered its rim. Almost immediately something caught Irwin's eye.

Like all of the other *Apollo* crews who had come out this way before, one of the first things that struck the *Apollo 15* astronauts when they arrived at the moon was its almost total lack of color. Here and there the prevailing gray might give way to white and the white to a pallid tan or even a very faint—and very rare—purplish. But if there was any real color at all—any of the voluptuous reds and yellows and blues of Earth—it was only what was imported aboard the machines that brought the crewmen here. Yesterday, however, as Irwin ran his eyes over a slope near the lander, he could swear he saw something green flashing back at him. Irwin pointed this out to Scott, who saw the colorful shimmer, too. But when the astronauts inspected the ground more closely, the reflectivity of the soil—its albedo—faded to the familiar gray. Now, one day later and three miles away, Irwin was being teased by green again. He toed the soil beneath his feet and shifted it around in the sun. The color didn't change. Warily, he called out to Scott.

"I think we kicked up some more green material here, Dave," he said.

"Sure it isn't that light gray albedo stuff?" Scott asked.

Irwin squinted through his helmet at the ground in front of him. "No, it looks green."

"I think it might be the contrast."

"No, no. I see white, I see a light green, and I see a brown."

Scott bounced over to the spot where his junior pilot was standing, scraped at the soil with his rock rake, and picked up a single stone in the teeth of the tool. Bringing it close to his visor, he peered at it and laughed.

"I've got to admit," he conceded, "it really looks green to me, too."

Scott rubbed at the surface of the stone. While the soil flaked away, the color didn't. He turned it this way and that and the sample continued to glint at him. What was doing the glinting, it appeared, was a freckling of tiny, shiny, glassy beads embedded in the matrix of the material. Scott was impressed despite himself.

If the commander was reading his geology right—and he was pretty sure he was—these beads were volcanic beads, and not just from any volcano, but from the explosively dramatic variety known as a fire fountain. In volcanically active places like Hawaii, percolating magma not far below the surface could sometimes accumulate in a single, high-pressure reservoir, causing the crust above it to spring a pinpoint leak. When it did, the magma below would hiss out in a fine, aerosolized mist, discharging in all directions and settling on the surrounding land in a hellish sprinkling. When the droplets cooled, they hardened into small glass beads, and if the beads contained the right elements—principally magnesium—they turned green. Now, Scott was holding a specimen in his hand suggesting that the same exotic process had taken place on the moon.

Scott stashed a few of the glittery rocks in a code-numbered sample bag, taking care to read off the bag's numeral for the Mission Control scientists who would need the geologic dog tag to help them identify the specimen when the crew got home. The two astronauts then turned to a different piece of business. Immediately after arriving at the site, Irwin had spotted another curious-looking rock—a pale one this time—about the size of his ungloved fist. The rock was conspicuously perched on a small pinnacle of dirt a few steps away—so conspicuously, in fact, that to a romantic the pinnacle might even look more like a pedestal, a means of deliberately distinguishing this solitary sample from all the other ones scattered around it. Irwin had pointed out the rock to Scott, and Scott had promised to get to it later. Now, that time had come.

"Okay," Scott said summarily, "let's go get that unusual one."

Scott and Irwin bounded a step or two over to the rock and Scott bent toward it. He inclined his head this way and that, checking to see how the sun played off the stone, and immediately, a pearly flash caught his eye.

"There's a little white corner to the thing," he mused.

Scott lifted the rock to his visor, brushed away a bit of dust, and as he did, both men reacted with a start. The sample, they could see, wasn't just partly white, but almost entirely white. More important, it was a *sparkling* white. Winking up at them through the billions of years of lunar dirt was a constellation of diamond-like crystals, far brighter than the pedestrian glass that had studded the last sample. Instinctively, Scott started to blow more of the dust away and then caught himself when his wind was stopped cold against the inside of his visor. Instead, he simply thumbed the dirt off the rock; the crystals, now reflecting more sun, shone brighter still.

"Oh, man!" Irwin said.

"Oh, boy," Scott answered.

"I got . . ."

"Look at that!"

"Look at that glint!"

Scott turned the rock slowly in his hands and, for a moment, he could manage only one other sound: "Ahhh."

This, he knew instantly, was it. Ancient anorthosite, made up almost completely of plagioclase—a scrap of the original crust of the original moon if ever there was one. Over the course of 4.5 billion years of lunar history, meteoric violence had regularly shaken the moon, gouging out continent-sized seas all over its surface—yet this bit of rubble had remained unchanged. Over the same period, nearby Earth hardened and hydrated and evolved into the planetary petri dish that supported millions of forms of life—and a quarter of a million miles away, this bit of crust remained unchanged. Over that same time, the very age of the very universe increased by half—going from ten billion years old to nearly fifteen billion years old—and while young stars switched on and old stars switched off and nebulae and novas burst into life or exploded out of existence, this bit of crust remained unchanged. This was a shred of the primal moon, cooked up in the days of the primal solar system, and Dave Scott, an Earthly animal who had been nothing but elements himself a preposterously brief thirty-nine years ago, was holding it in his hands— and, much more important, *understanding* what he was holding.

"Guess what we just found?" Scott said to everyone and to no one at all. "Guess what we just found?" In the background, Houston could hear Irwin laughing happily. "I think we found what we came for," Scott said.

"Crystalline rock, huh?" Irwin asked, for Houston's benefit.

"Yes, sir," Scott answered. "You better believe it."

"Yes, sir," came the voice of Joe Allen, the astronaut in Mission Control handling the air-to-ground channel.

The precise significance of the rock neither Scott nor Irwin would say, not with the lesson of Buzz and his biotite still fresh in their minds. But none of the hundreds of people in Houston needed to be told what crystalline rock meant.

"Look at the plagioclase in there," Scott muttered to Irwin. "Almost all plage."

"That really is a beauty," Irwin said.

"Bag it up," Allen called from his seat in Mission Control as he watched the two astronauts on the movie screen–sized video monitor in the giant auditorium. On the screen, the remarkable rock was little more than a pixellated smudge.

Scott pulled out a fresh sample bag, started to place the rock inside, and turned to Irwin. "Let me get some of that clod there," he said, pointing to the mound of dirt on which the stone was found. As soon as he said that, though, he reversed himself. "No, let's don't mix them." He took a quick look at the bag's serial number. "Make this bag, 196, a special bag."

"Yes, sir," Allen answered from Earth.

Scott and Irwin dropped the remarkable rock into bag 196, cinched the now-precious gunnysack up, and kept it close to them throughout the remainder of the day. Even before the men of *Apollo 15* returned to Earth, reporters covering their mission had already dubbed their treasured find the Genesis Rock.

January 1973

Nobody talked much about what would happen if there was a fire in the moon rock building at NASA's Manned Spacecraft Center in Houston. Talking about a fire meant you were thinking about a fire, and thinking about it was something you didn't especially want to do. There were a lot of buildings NASA could lose and easily replace on the campus of its giant Houston facility, but the moon rock building wasn't one of them.

The moon rock building was originally and less colloquially known as the Lunar Receiving Laboratory, or the LRL, and its name did a good job of describing what it did. Well before the return of *Apollo 11* with its first, precious 47.7-pound cache of moon rocks, NASA knew it would need a place to store and study the rocks—a place that could keep the ancient samples as clean and pristine as they had remained in the endless eons they lay on the surface of the moon. Since no such lab existed on the planet, NASA built one, equipping its rock facility with airtight sample rooms, airtight meeting rooms, and negative pressure common areas that would allow clean air to flow out without allowing contaminated air to flow in. The locks on the vaults that protected the samples would be impervious to even the nimblest criminal fingers; the foundations on which the building was built would withstand even the most robust hurricane-force winds; the building itself would be all but entirely fireproof.

NASA officials were happy with their state-of-the-art lab—most of the time. The problem was, it was always possible to conjecture a fire or other disaster the likes of which no one had ever conceived, and more than one agency administrator lay awake nights fretting about just such a calamity. In order to provide the agency a little accident insurance—and NASA chiefs a little peace of mind—a small sampling of the lunar specimens was shipped out to a fortified military base, a fully secure place whose exact address was never disclosed to the public. The rest of the rocks would remain forever at the moon lab—and it was here that they would slowly begin to give up their secrets.

The studies conducted in the Houston facility generally began well before there was anything there to study at all—when the moon rocks were still on the surface of the moon. Geologists chosen to analyze the samples a lunar crew brought home would spend as much of the flight as they could camped out in a back room in Mission Control. As the astronauts in the field went about the task of collecting their samples, the scientists at the space center would keep a constant eye on them via the video cameras the crew took with them to the moon. Each time an astronaut came upon a sample worth collecting, the geologists would note precisely where it was found and in precisely what condition. When the rock was picked up and stowed in its coded bag, the geologists would record the bag's number on an index card—one card per rock—and set it aside. During the *Apollo 11* mission, with its comparatively modest load of samples, the pile of cards was a short one. By the last lunar landing—

Apollo 17 in December 1972, with its 243.1 pounds—the stack grew dramatically.

While most of the stones the moonwalkers discovered were collected and catalogued this way, not all were. Now and again, the crewmen would stumble across a rock that looked so promising it deserved special treatment. Perhaps it was covered with a dusting of unusual volcanic soil; perhaps it glittered an improbable shade of green; perhaps it was a bit of primordial crust that had lain undisturbed for several billion years before the astronauts stumbled across it. Whatever it was that set the rock apart, the idea of simply dropping it in a gunnysack and tossing it into the lunar module did not seem like a good one. Once the door to the LEM was closed and its cockpit was filled with air, the atmosphere inside the ship—filled with everything from bacteria to food debris to salt to spores to aerosolized urine—would seep through the bag and contaminate whatever was inside. In general, geologists were prepared for this and taught themselves to recognize and ignore out-of-place organic chemistry on their inorganic specimens. For their most delicate samples, however, they didn't want to take any chances.

Stowed aboard the LEM was a small stack of metal boxes measuring thirty inches across and just twelve inches high. When the astronauts found a specimen that warranted special care, they would carry one of these boxes out onto the lunar surface and open it up. If the box contained any stray atmosphere, it would burp it out into the lunar void, leaving behind the same near-perfect vacuum found everywhere else on the moon. The bag containing the sample would then be placed inside the box and the lid would be shut. When the box was carried back inside the LEM and the spacecraft was pressurized, the five pounds per square inch of cabin atmosphere would bear down on the lid of the chest, squeezing it tightly shut and causing its lining—made of an alloy of soft, ductile metals—to form an airtight seal. Back on Earth, where sea level pressure is an even more oppressive 14.7 pounds per square inch, the seal would become stronger still. Only when the box had been transferred to a larger, enclosed case in the moon rock building—a case filled with inert nitrogen—would the lid at last be pried open and the unlovely treasure revealed.

Typically, when a sample of this kind was being unveiled, it wouldn't be unveiled alone. The box would be opened by a single technician who would manipulate the specimen by placing his hands into a pair of heavy rubber gloves built into the side of the nitrogen case. A glass window on top of the case afforded him a view of what he was doing, and a far larger window built into one wall of the sealed room in which he was working afforded a similar view to the knot of geologists who would gather in the corridor outside. When the sample was first extracted, it would sometimes be covered with a layer of lunar dust, preventing the geologists from getting a good look at it. The technician in the room would then obligingly clean the specimen with a few gentle breaths from a nitrogen gun; as the billions of years of accumulated dirt swirled away, the watching scientists would get their first view of the crystals or glass or spanglings of color that had first caught the moonwalkers' eyes.

Painstakingly, over the course of several years, one after another of the 838.2 pounds of samples brought back by *Apollos 11, 12, 14, 15, 16,* and *17* were lovingly displayed and inspected this way. Then, less lovingly but no less carefully, they were sliced into more than 80,000 samples and biopsied down to their last constituent grains. The story those grains told was an extraordinary one.

For all of the assumptions the scientists and pilots of NASA made about the moon before the *Apollo* missions were launched, a lot of what they believed was based on *just* assumptions. One of the most enduring questions that had to be resolved concerned not lunar rocks, but lunar soil. If meteors indeed played a major role in shaping the surface of the moon, it wouldn't only be *big* meteors that did the job. Far more numerous than the giant space rocks able to dig out an entire sea would be the smaller, dust-sized bits of debris that bombard the lunar surface in a sort of constant cosmic rain. Over the course of billions of years, this persistent sand-blasting would work the soil over pretty completely, reducing it to a fine-sifted powder that was the stuff of geology ground small. So numerous were the collisions of these micrometeorites and for so long did they occur, that it was possible that the upper lunar crust was covered not merely by a thin skin of powder, but by a vast ocean of it that could, in theory, swallow up any machine or man foolish enough to dip so much as a toe in it.

Before the *Ranger* missions flew, Vienna-born astronomer Thomas Gold of Cornell University was the most conspicuously vocal scholar of the lunar dust theory. Gold had counted craters and estimated impact forces, and had come to the grim conclusion that the moon's surface was smothered in a blanket of dust that might reach down thousands of feet. The first crew of astronauts who landed on the moon, Gold warned darkly, would in all likelihood be the last, sinking immediately out of sight and remaining forever entombed in the sarcophagus of their lunar module.

Gold's theory suffered a blow when the three successful *Ranger* spacecraft returned close-up images of a lunar terrain that was littered with rocks and boulders—not the kind of heavy debris the cameras were likely to have spotted on a world whose fluffy surface had so much give. Advocates of the dust theory were undeterred by these images, arguing that if some moon rocks were porous and light, and if the soil they lay on top of was compressed in spots, there was no reason the occasional bit of debris wouldn't remain on the surface.

Things got worse for the Gold dust theory when the five unmanned *Surveyors*—which surely weighed more than a piece of rubble—landed safely on the moon in 1966, 1967, and 1968, and showed no sign of sinking. Even then, however, Gold's followers remained unpersuaded. With no absolute proof that the dust blanket was not miles deep, the spacecraft's creators had constructed it of the lightest materials possible and given it a broad, splayed-legged shape that would allow it to support itself on a thick layer of lunar dust much the way a swamp bug supports itself on water. If the *Surveyors* failed to sink out of sight, it might have said less about the state of the moon than about the ingenious engineers who built the ships.

It wasn't until July 1969 that Neil Armstrong finally settled Gold's hash when he put his *Apollo 11* lunar module safely down on the Sea of Tranquillity and found that the footing below both his ship and his boots was as firm as it was on Earth. The dust he and other moonwalkers brought back to the LRL proved to be made up of an enormous variety of pulverized rocks, suggesting that the surface of the world had indeed been pulverized by just the kind of microcollisions Gold and others had hypothesized. The force of those collisions, however, had simply been less than what Gold had anticipated.

Deep dust layer or not, the discovery that the moon had indeed been subjected to a rain from space sufficient to grind its surface to powder was monumentally important, settling not just the question of how the lunar soil was formed, but—in all probability—how lunar craters were formed. After several centuries of debate, there was still no absolute agreement on the process of crater formation, with the majority of scientists concluding that meteors were responsible. The new soil findings—illustrating just how ravaged the lunar surface was—gave this theory a considerable boost. Nonetheless, there was still a vocal minority that insisted that, when it came to midsized craters at least, it was not meteors that were responsible, but volcanoes. The dissenters' arguments were not totally without merit.

Earth has its own fair share of craters, virtually all found at the top of volcanic mountains or on the surface of sprawling volcanic fields. The floors of most of these craters are usually comparatively smooth, with the exception of a solitary cone made of cinder and ash that accumulates in the center. Even relatively weak telescopes had long revealed similar structures at the center of lunar craters, and close-up photos returned by manned and unmanned spacecraft had confirmed that discovery. More tellingly, on Earth, large craters often do not form individually, but rather are surrounded by clusters or chains of smaller craters, caused by underground magma percolating to the surface in several spots at once. On the moon, almost all of the largest craters appeared to be surrounded by secondary craters, too, and the closer astronomers looked—first with telescopes, then with spacecraft—the more numerous the smaller pits became. Finally, and perhaps most persuasively, large lunar craters were frequently found to be circled by concentric, rippling rings, as well as by long, sunburst-like rays that extended from the central pit in all directions. Similar surface scars are often found around volcanic craters on

Earth, caused by moving lava undulating away from a volcano or fast-moving lava splattering explosively out of it.

If lunar craters posed unresolved questions before the *Apollo* astronauts brought home their samples, the moon's color and shine posed an even greater one. Astronomers studying the face of the moon had always been puzzled by the curious way it reflects sunlight. Aim a light source at any sphere—from a planet to a ball to a soap bubble—and the illumination will not bounce back evenly. Rather, the center of the sphere will reflect the most brightly, while the edges, which curve away from the object emitting the light, will fall partly into shadow. The moon, however, doesn't do that. When the great white platter of a full moon is hanging in the sky, it beams back sunlight almost the way a real platter would—with its rim and edges reflecting precisely as brightly as its center does. Rationally, astronomers knew that when they were looking at the moon they were looking at a sphere; from the evidence of their eyes, however, they might as well have been seeing a two-dimensional decal slapped against the ceiling of the sky.

Just as perplexing as lunar light was lunar gravity. On Earth, gravity is an almost perfectly consistent thing, with the great mass of the planet pulling equally hard in all spots at all times. Place a hundred-pound block on a scale in, say, Mexico City, and you can be certain it will register the same hundred pounds if you later weigh it in Ottawa, Bangkok, or Berlin. Not so on the moon. When the *Lunar Orbiter* probes were flying their missions, NASA controllers often noticed that as the ships moved through their circular flight paths they would occasionally dip or skid a bit, almost as if the lunar gravity had suddenly tugged on them a little bit harder and then let go with a silent twang. During the flight of *Apollo 16* in April 1972, these gravity flutters proved to be so strong that a tiny subsatellite the crew released into orbit to study cosmic radiation soon began to wobble and fail, crashing onto the surface just two weeks after its launch.

NASA, which knew a thing or two about gravity, realized that such anomalies had to be caused by mysterious concentrations of mass beneath the surface of the moon, and appropriately enough began calling the patches of high-density matter mascons. But naming a thing is not the same as explaining it, and whatever the scientists chose to call the formations, they couldn't begin to account for them.

It was only when geologists began chipping answers out of the rock samples the *Apollo* astronauts had fetched home that most of these mysteries at last got solved. *Apollo 15*'s Genesis Rock, while not among the first samples brought back from the moon, was among the most precious, and as scientists studied its twinkly innards, it yielded just the primal secrets they hoped it would.

One of the first things NASA looked for in analyzing the Genesis specimen was the presence of strontium, an alkaline metal thought to have been present in certain known concentrations during the early, gestational stages of the solar system. If the rock contained strontium in the same ratios, the scientists would know that it was indeed an artifact of the original moon, accreting directly out of the hot, cosmic goo that made up all the local worlds. Chiseling off a tiny fragment of the rock and melting it down, they immediately detected the chemical signature indicating that strontium was present. Measuring how *much* strontium was being given off, they found it matched the levels in the early solar system perfectly. The Earth and the other planets, astronomers had always known, were about 4.5 billion years old; now they knew the moon was, too.

Once the geologists had established the age of the Genesis sample, the next step was to compare it to the age of the rocks the *Apollo* astronauts had brought back from the moon's seas to determine if they were indeed created later in the lunar life span. While the strontium method was a reliable way to conduct this further dating, it wasn't the only way. Another, more precise method—particularly when you're dating lots of different samples that could be lots of different ages—is to look at its isotopes.

Assuming the seas were indeed created by mammoth meteor strikes, the energy released by the impact of the projectiles would not only have vaporized the incoming rock itself, but also liquefied a thin layer of ground beneath it. When land is shock-melted this way, it undergoes a curious atomic change. Radioactive isotopes in rocks and soil tend to decay at a certain, fixed rate. If the land gets hot enough fast enough, however, the isotopes cook down into an earlier, more primal state, causing the decay to start all over again. In effect, the isotopic clocks deep inside the stones become reset, causing the fused, melted land within the freshly dug pit to become atomically younger than the crust around it. If you find a shiny layer on the floor of a sea or crater, you can be pretty sure the pit was dug by an incoming meteor; if you read the isotopes frozen in the soil, you can calculate precisely when that meteor hit.

As *Apollo* astronauts picked about at the edges of seas, they found a wealth of shock-melted rocks beneath the lava that otherwise covered the basins. And as geologists trained their radioactive sniffers on the samples, a variety of ages revealed themselves. Samples from the Imbrium basin, adjacent to the crustal zone where the Genesis Rock was collected, appeared to be 3.84 billion years old, meaning that the crust from which Genesis was blasted had existed undisturbed for at least the first 660 million years of the moon's existence before the sea was created and the rock was deposited on its banks. The distant Sea of Nectar, in the moon's southern hemisphere, appeared to be older still, having been dug from the crust 3.92 billion years ago; the Sea of Serenity, hard up against Imbrium, was 3.87 billion years old; the Sea of Tranquillity was 3.84 billion; both the Ocean of Storms and the Inner Sea were a comparatively young 3.16 billion. After that, little other evidence of major impacts turned up until the Copernicus crater was blasted into existence 800 million years ago, and the Tycho crater was created a scant 110 million years back. For the first billion or so years of the moon's history it thus appeared that the cosmos had administered it a savage beating, pummeling the new world with all matter of loose geological change that had yet to accrete into worlds. After that, things quieted down considerably, allowing the injuries on the lunar surface to heal and the globe as a whole to remain relatively peaceful.

With the evidence of the isotopic clocks confirming that both the moon's seas and soil were formed by meteors, the scientific majority arguing that all lunar craters were created the same way closed in for the kill. Studies of nuclear weapon test sites on Earth revealed that the main crater dug out by the blast was often surrounded by a series of smaller craters, gouged from the surface by shrapnel and other debris that the exploding bomb unleashed. A similar scatter pattern on the moon would likely be a result of similar secondary impacts caused by meteor debris rather than by secondary lava vents caused by volcanoes. The rings that surrounded the big craters could be explained by meteors as well, and were probably the remains of topographic rippling that occurred when the projectile that dug out the crater caused the ground around the impact site to liquefy temporarily; the rays that projected from the center of the craters were similarly the scars of backsplashing surface material blasting away from the point of the collision. Even the apparent cinder cones at the centers of the pits were, the geologists concluded, not cinder cones at all, but rather rebounded crust that was driven up by the meteoric impact and then became frozen in place. With these arguments—and the evidence to back them up—the volcano advocates at last threw in the towel, accepting that apart from the occasional fire fountain, volcanism had simply not played much of a role in the formation of the moon.

When that debate was at last settled, the geologists were free to move on to other, less contentious matters. One thing that engaged their imagination almost immediately was the chemical composition of the shock-melted rock layer the astronauts scraped up from beneath the lava bed. This glassy material, they found, turned out to be extremely rich in iron and extremely poor in aluminum. Samples from the surface of the nearby highlands seemed, curiously, to be made up of just the opposite stuff—lots of aluminum and little iron. Superficially, this didn't make sense, since a moon that jelled from what was likely a uniform glob of hot matter should be a lot more homogenous than that. The geologists, however, had long suspected that the early moon was only homogenous for so long. As the molten mass that forms any rocky body cools, it tends to stratify, with heavy metals precipitating down and lighter ones staying buoyantly on top. Over time, the cooling, hardening moon would have become highly layered, and when a big projectile like a meteor hit, the heavy geological viscera from far below would be dug out and scattered across the surface of the sea. The more of the deep, iron-rich stuff the *Apollo* astronauts found in a basin, the more violent they would know the sea-forming collision had been.

This same multi-layered lunar anatomy helped explain another of the moon's mysteries: the location and distribution of its seas. Ever since the Soviet Union's *Lunik 3* beamed back the first images of the far side of the moon in 1959, cartographers had been wondering why all of the large lunar seas were on the side of the globe facing Earth. Certainly, the far side had its share of craters—some of them monumentally big—but they were nothing *more* than craters, while the big, hemisphere-dominating plains were entirely absent. The reason for the disparity, the geologists now figured, almost certainly involved gravity.

Though the gravitational attraction of Earth tugged on the entire body of the moon at all times, it would be likely to tug a little harder on its dense inner layers than on its light, outer ones. This would cause the lunar innards to shift slightly Earthward inside their crustal shell, and that, in turn, would cause the crust to be slightly thinner on the side of the moon facing Earth than on the opposing side. A meteor that hit the moon's visible hemisphere thus had a greater chance of punching through the crust and getting a good, sea-creating lava flow going than a meteor that hit the *in*visible hemisphere.

Once the geologists understood the formation and location of the seas, they, in turn, found it easier to understand the mysterious underground mascons that so destabilized satellites orbiting the moon. It had not escaped the notice of NASA's orbital mappers that the most dramatic mascon tugs generally occurred over the biggest seas. As lava accumulated at the site of an impact, they now figured, it would cool, contract, and grow denser; as it grew denser, it would sink. Even a slight increase in the density of the surface commensurately increases gravity. An exponentially greater increase leads to an exponentially greater pull. After oozing so much lava over so many millions of years, the moon's seas had grown surpassingly dense, and three or four billion years later, when tiny machines from Earth started flying overhead, the vast lunar plains pulled on them hard indeed.

Just as important to the *Apollo* scientists as ancient geologic violence on the moon was the possibility of current geologic violence—specifically moonquakes. No one knew how seismically active the moon was, but in order to learn, the geologists had equipped all of the *Apollo* landing crews with portable, radioactively powered seismographs that could be left behind on the moon to operate unattended for years. Between 1969 and 1972 six of these sensitive instruments were deployed on the lunar surface, forming a pair of seismic triangles that covered much of the northern hemisphere. The positioning of the instruments allowed geologists on Earth not only to detect moonquakes as they occurred, but to pinpoint exactly where in the moon's interior anatomy the disturbance originated.

To be sure, listening for quakes on a world 250,000 miles away was more complicated than simply switching on the seismographs and waiting for their needles to jump. In order for the instruments to work properly, scientists first had to calibrate them, making sure they weren't tuned so low that they missed the moon's softer geologic whispers or so high that they mistook a mere tectonic twitch for a globe-jolting spasm. The problem was, there was no reliable way to adjust the machines unless the scientists could somehow predict exactly when a moonquake would take place and exactly what its intensity would be, then listen to how the seismographs picked it up and set their volume levels accordingly. While the moon itself could not provide this kind of seismic predictability, NASA could.

After a lunar crew lifted off from the moon and docked with the orbiting command module, the LEM became immediately expendable. Once the astronauts cut the ship loose, all it would take was a small push from the lander's nearly spent thrusters to destabilize it and send it tumbling down to the surface. An impact by a projectile that big, moving that fast, would certainly wake up the seismographs, and since scientists knew the spacecraft's weight down to the last ounce and its speed down to the last foot per second, they could easily calculate how high the needles ought to jump. Watching how much they did jump and then tweaking the machines as needed, they could prepare their instruments to spot any authentic tremors that might occur later.

As it turned out, not many did. In the eight years the *Apollo* seismographs functioned, they picked up almost no quake activity at all on the old, cold moon. Once in a great while an incoming meteor would sting the skin of the world and, like a horse twitching off a fly, it would shudder a little in response. Just as infrequently, the moon would stir a bit from within, emit a tiny seismic burp, and then settle back into slumber. So small and infrequent were these moonquakes that even if all of them were combined and released at once on Earth, they would be almost impossible to detect. The moon, it appeared, may have been a hot, seething place once, but it had long since gone still and cadaverous.

This very quiescence, it turned out, was also at least partly responsible for one of the moon's final unsolved mysteries: its unusual reflectivity. As the *Apollo* astronauts closed in on the moon during their outbound journey, they could not help noticing the same uniform brightness that had so long mystified astronomers. Whether they looked straight into the center of the lunar bull's-eye or off toward its most remote horizon, the light of the sun bounced back at them with an unshadowed brilliance that never varied. It was only when the astronauts actually landed their LEM and began trudging across the surface that they discovered the reason.

In the dry, low-gravity environment of the moon, dust particles don't accumulate in the comparatively dense blankets they do on Earth. Instead, lunar powder kicked into the sky—generally as a result of meteorite impacts—tends to land softly and arrange itself haphazardly, accumulating in loose mounds like fallen leaves. Covering the surface from pole to pole and horizon to horizon, these lightly packed dust bits play funny tricks with sunlight. Even as the edges of the moon recede away from the brilliance of the sun, some of the dust particles, propped up at all manner of angles, point back toward it, reflecting the light and obscuring the shadows beneath them. Thus, rather than retreating into shadow, the lunar horizons remain brilliantly lit, causing the entire face of the moon to shine back into space like a single celestial headlamp.

Such fairy castle soil, of course, is a fragile thing. Wind, if there were wind, would easily blow it to nothing; rain, if there were rain, would reduce it to mud. The merest touch from a gloved hand or a booted toe—to say nothing of the landing pad of a lunar module—would flatten it to a dirt mat. The delicately stacked particles could thus be witnessed by astronauts and even photographed up close, but trying to move them intact from the spot where they lay, much less bring them back to Earth, would be forever out of the question.

The sleepy stillness that allowed the fairy castle soil to last for so long was, of course, a dramatic departure from the violence in which the moon was born. Though meteors tattooed the lunar surface for the first billion or so years of its existence, in all the eons since, the cosmos appeared to have pretty much forgotten about the moon, setting it aside as little more than a dead, paperweight world in a solar system filled with far more dynamic places. Cosmic corpse or not, however, the moon was undeniably there, and the only question left unanswered by all the elegant data being pulled from the rocks and read from the seismographs was *how* it came to be there. That had the scientists stumped.

One of the prevailing truths about the solar system's moons is that no matter how different they look, they are without exception relatively small. Jupiter, Saturn, Uranus, and Neptune, measuring between 30,000 miles and 88,000 miles in diameter, are circled by swarms of moons that, compared to the vast size of the planets themselves, are little more than orbiting fruit flies. Mars, one of the solar system's smallest planets, is attended by two surpassingly tiny moons, one of which is less than fourteen miles across, the other of which is less than nine.

Earth's satellite breaks this size rule—at least in relative terms. Measuring 2,155 miles in diameter, the moon is about the same size as the largest satellites that circle the largest planets; its own parent world, however, is barely 8,000 miles wide. Such a one-to-four size ratio between a moon and a planet is rare indeed, making the two bodies look less like a dominant world and a dependent satellite than a pair of strange sibling globes orbiting each other as cosmic coequals.

Also curious is the inclination of the moon's orbit. Unlike most large, natural satellites, which inscribe a more or less horizontal path around the equatorial waistline of their planets, Earth's moon appears to have been knocked slightly cockeyed, crossing 16 degrees above the equator at some points and 16 degrees below it at others. The orientation of the two worlds' poles is also a bit skewed. Earth is famously tilted at a near-drunken angle of more than 23 degrees, a list that accounts for the planet's wildly changeable seasons. The moon, on the other hand, is almost perfectly upright, tilted at just 1.5 degrees.

The existence of such a large moon at such an odd angle around Earth was never easy to explain. But even before the *Apollo* missions flew, a few lunar scientists were pretty sure they had the answer. Earth's moon, they figured, was not formed side by side with the planet it circles, but was instead taken captive by it. Racing alone through the outer solar system, the prehistoric moon was once a rogue world utterly untethered to any planet. On one of its mad passes through the planetary neighborhood, however, it happened to fly too close to Jupiter, allowing the massive gravity of the star-like world to flick it in toward the sun. Hurtling ahead on this suicide trajectory, the moon would surely have been consumed by the solar fires, but on its way toward the solar system's center it happened to pass by Earth, whizzing by just a quarter of a million miles away. Earthly gravity immediately grabbed the moon, yanking it off its straight, breakneck course and pulling it into a large, arcing one. Unsteadily at first, the moon began circling the planet. After a few wobbly laps, the new satellite stabilized itself, settling into a smooth, if forever-crooked orbit.

The theory was a good one, but only if you were willing to overlook a few critical flaws. First of all, a moon that had been flung sunward by a planet the size of Jupiter would be moving at a speed that was literally meteoric. Trying to stop a body so fast and massive would be a little like trying to catch a passing freight train with a fishing rod: You might get lucky and snag the thing with your hook, but the instant you did, the rod would be ripped from your hands. Earth was a powerful body all right, and the gravity field it emitted was impressive, but intercepting a target like the speeding moon was well beyond its powers.

Even if the moon were somehow moving slowly enough to slip into Earth's orbit, there was something about the newly discovered chemistry of the *Apollo* lunar rocks that also raised scientists' eyebrows. The three most abundant elements detected on the moon were oxygen, silicon, and aluminum, in that order. These elements in precisely the same order are also the most abundant ones found in Earth's crust. A free-floating world that formed deep in the cold ether of the distant solar system would not be likely to be made of the same elemental stuff as a body like Earth, which was cooked up millions of miles farther in, over the open fires of the nearby sun. The inclination of the moon may have suggested that it was a relatively new arrival in the Earthly neighborhood, but its chemistry and size suggested otherwise.

If Earth didn't capture the moon, it was always possible it gave birth to it. Early in the solar system's history, Earth, like the moon, was a molten body with a consistency less like rock than putty. Get a planetary mass soft enough and ratchet up its spin velocity high enough, and it just might fling off a globule or two. In the deep freeze of deep space, it wouldn't take much for one of those globules to harden into a moon.

This lunar fission theory had a lot of appeal—not least because it avoided the nasty gravitational problems of the capture theory. The trouble was, the sharp angle of the moon's orbit—the very angle that gave the capture theory whatever credibility it had—made the fission theory less than credible. A molten Earth that was throwing off debris would likely throw it off from its fastest-spinning point, and that was its equator. Matter flung from the equatorial plane would likely stay in the equatorial plane, not hop a random 16 degrees above and below it.

The only realistic alternative to both of these theories was that the moon neither budded from Earth nor was lassoed by it, but simply condensed from primal stellar matter along *with* Earth. If the planets were nothing more than leftover matter that failed to condense into the body of the sun, there was no reason that the moons couldn't similarly be leftover matter that failed to condense into planets.

This accretion theory had perhaps the most adherents of all, but it, too, had fatal flaws. As with the fission theory, the nature of the lunar orbit posed a problem, since planets and moons that formed from the same spinning cloud of matter would likely have the same angle and inclination. And as with the fission theory, once again lunar chemistry was a puzzle. While the moon is rich in the same oxygen, silicon, and aluminum so common on Earth, it has little or no volatile, easy-to-boil materials like water, sodium, hydrogen, and helium. Earth, by contrast, is fairly drenched in volatiles. If the moon hardened from the same raw cloud that created the planet, it ought to have virtually all of the same ingredients.

It wasn't until another twelve years had passed that lunar astronomers came up with a fourth—and, they hoped, final—theory, one that combined the best of the other three and left none of the same troublesome questions unanswered. When the solar system as a whole was just accreting, so the new scenario went, a protoplanet about the size of Mars was moving about the sun in an orbit similar to Earth's. Similar orbits, of course, can be deadly orbits, since if the trajectory of one world should drift by only a degree or two, it could directly cross the path of the other, with disastrous results. About 4.5 billion years ago that is precisely what happened, as the smaller of the two planets lost its way and cannonballed blindly into the flank of the proto-Earth. The collision was a slightly off-center one, gouging a massive divot from the larger world, tilting it slightly off center, and increasing its spin rate to one rotation every twenty-four hours.

While the young Earth was only wounded and dizzied by the blow, the smaller, incoming world was utterly destroyed. Much of its heavy, metallic mantle was incorporated into Earth's own mantle. The rest of it was liquefied, vaporized, and ejected into space, where it began to orbit Earth at an altitude of about 238,000 miles. The pulverized swatch of Earth that had been dug out by the impact drifted up and joined it there. For a time, this cloud of orbiting debris was nothing *but* a cloud. Quickly, however, it contracted into a ring, and just as quickly the ring contracted into a moon.

This new impact theory had a lot to recommend it. For one thing, it explained the similar oxygen, silicon, and aluminum content of the two worlds. Since a good portion of the new moon was made up of material from Earth, it was only natural that both bodies would contain some of the same elements. For another thing, it explained the absence of volatiles like water, sodium, hydrogen, and helium on the moon. The collision between the impactor world and Earth generated intense heat, more than enough to cause any materials with a low boiling point to evaporate entirely. On Earth, which largely survived the blow, plenty of these materials would have remained unaffected. On the moon, which was made up only of superheated collision debris, they would be completely depleted. The violence of the impact also explained the moon's curiously inclined orbit. Lunar material that was blasted from Earth, after all, would fly out a lot more explosively than lunar material that was gently spun from it, and it would be no surprise if the primal moon wound up outside the plane of its parent world's equator, moving through an orbit that perpetually took it first above and then below the Earthly midline.

That was the answer the moon scientists of the mid-1980s came up with—and that was the answer that largely stuck. But the moon scientists of the early 1970s—the ones who had first cracked open the rock boxes that the *Apollo* crews brought home—had not yet reasoned things through that far. The NASA administrators of that earlier era were not, to be sure, happy with these unresolved questions. The agency had spent more than $30 billion to send twenty-two manned and unmanned missions to the moon, and for that kind of money they expected some answers. The scientists who were charged with the job of coaxing the secrets out of the lunar rocks, however, were not so troubled. Over the course of three and a half years, nearly half a ton of the moon had been pried loose and carried back to Earth, and the stories those stones and dirt could tell would far outnumber the ones they couldn't. For the moment at least, most of the *Apollo* geologists seemed blithely untroubled by their inability to solve one lingering lunar riddle. Indeed, they often liked to tell one another with winks and grins, it was entirely possible the moon didn't exist at all. If it did, the theories they did have would explain it.

Part II

Far

3

The Grand Tour

Pasadena, Calif., Late December 1971

William Pickering had suspected for some time that the people in Washington weren't going to let him go to Neptune. The press had warned him; the NASA brass had warned him; even some of the junior members of his Jet Propulsion Laboratory staff—people who had no real business cautioning their boss about such matters—had occasionally presumed to warn him. Congress, however, had not yet said a word, so Pickering did not give the matter much thought. When the news at last came from Washington that Neptune was indeed out of the question, it thus came as something of a jolt.

Actually, it wasn't just Neptune Pickering had been thinking about visiting. Neptune was part of the itinerary, certainly, but it was only the last part. Well before Pickering's spacecraft ever got as far as the eighth planet, he had also planned to send it to visit the fifth, sixth, and seventh—Jupiter, Saturn, and Uranus—paying a call on the whole archipelago of large worlds from the mid-solar system out to the very end. Regardless of where Pickering intended his ship to go, however, according to the word that had just come from the East, it now looked as if it wouldn't be going anywhere at all.

To be sure, it wasn't that Pickering had had much cause to complain about the treatment he'd gotten from Washington over the years. Oh, there had been that nasty bit of business with the *Ranger* program back in 1964, when all of JPL's experimental moonships kept failing or falling long before they ever got where they were supposed to go and committees on Capitol Hill started calling for someone's head, and Pickering had had to serve up poor Jim Burke's and poor Cliff Cummings's just to appease the baying lawmakers until *Ranger 7* could at last fly the mission it was supposed to fly and effectively save the lab's bacon. But that was a long time ago, and it seemed unseemly to be thinking about that after all the places Congress had let Pickering fly his ships since. There had been the five *Lunar Orbiter* spacecraft JPL launched successfully to the moon in 1966 and 1967. There had been the seven *Surveyor* spacecraft sent to make soft lunar landings between 1966 and 1968. There had been *Mariner 2* and *Mariner 5,* which successfully flew by Venus in 1962 and 1967; *Mariner*s *4, 6, 7,* and *9,* which reconnoitered Mars between 1964 and 1971; and *Pioneer*s *7, 8,* and *9,* which traveled to the sun in 1966, 1967, and 1968. There was even talk of a *Mariner 10* next year that would fly past Venus, pick up a gravitational boost from the big Earth-sized world, and then fling itself on to little moon-sized Mercury—a single spacecraft that would visit two separate planets using no more hardware and no more manpower than it would usually take to visit just one.

What pleased Pickering especially when he thought about all these missions—and what, by rights, ought to please Congress, too—was how cheaply he had flown them. It had cost about $24 billion over the past decade for NASA to mount its manned space program, with the costs climbing especially fast in the last three years when the agency began launching its *Apollo* lunar missions, each one setting the government back a cool $400 million. From the beginning Washington had made it clear that the unmanned space program could expect barely a fraction of that kind of lavish spending, repeatedly reminding labs like JPL that if they were going to survive, they were going to have to do so on the table scraps and pan drippings *Apollo* left behind. Pickering had always taken pains not to complain about such comparative starvation rations, even making it a matter of pride to bring his projects in as far under budget as he could get them. Congress, however, had never seemed quite as impressed with Pickering's parsimony as Pickering himself, and no matter how tight the JPL director managed to pull his lab's fiscal belt, Congress always tried to make him pull it tighter still. Today, Pickering learned, Washington had tightened things up but good.

Pickering learned the bad news about Neptune the way he learned most other things that happened on Capitol Hill: through a twenty-something girl named Victoria Melikan. To anyone who worked in the Pasadena space community where Pickering made his living, a name like Victoria Melikan didn't mean much. If you did your space business in Washington, however, you got to know Victoria early and you got to know her well.

It was Pickering himself who had first spotted Victoria Melikan almost ten years earlier, when she was a University of Michigan undergrad picking up extra money working weekends and afternoons at the General Motors headquarters just outside Detroit. As pocket-money jobs went, Victoria's wasn't a bad one: Half her time was spent writing speeches for one of the company's executive vice presidents, the other half was spent working in the company's government sales department, helping to administer some of the lavish research and manufacturing deals GM had struck with Washington. Even by college girl standards, the income Victoria was pulling down wasn't much, but that could change dramatically: GM, she knew, rewarded its employees well, and if she could just stay aboard until she finished school and then join the payroll full-time, she might get a chance to share in some of the riches, too.

One day in 1961, shortly before graduation, Victoria learned that a William Pickering from one of the new space labs in California would be dropping by GM for a technology conference of some kind. Space work wasn't strictly part of Victoria's job, but making important visitors feel welcome was, and she was given the job of showing this Mr. Pickering—actually *Dr.* Pickering, she kept reminding herself—around the plant. Pickering himself didn't really care who escorted him during what was essentially a glad-handing tour, but when he met the impossibly self-possessed young woman who had been assigned to his care, he was impressed by her immediately.

He liked her calm, he liked her poise, he liked the way she shook his hand respectfully but not too gravely. By the end of his visit, he knew he wanted her for his lab.

"Miss Melikan," Pickering said as he was taking his leave, "what are you doing wasting your time with automobiles?"

"I like automobiles," she answered.

"Nothing wrong with them," Pickering said agreeably. "Me? I prefer spacecraft. The way I see it, anyone can build a machine that will take you down a road. But there aren't too many who can build ones that will take you to the stars." Pickering reached into his pocket and gave Victoria his card. "You call me if you ever get restless here."

A year later, Victoria graduated from college and was, as she had hoped, hired by General Motors full-time, working in a position not all that different from the one she had held when she was in college. Two years after that, in mid-1964, during the beginning of what looked like a promising climb up the GM trellis, she indeed, as Pickering had suggested, began feeling restless. Pulling Pickering's by now dog-eared card from the back of her desk drawer, she phoned his office. To her surprise, he remembered her; to her greater surprise, he agreed to see her.

"I might want to take you up on that offer you made me," she said with little preamble when she presented herself in Pickering's office.

"What do you make at GM?" Pickering asked.

Victoria told him.

"I can offer you half of that."

"GM gives me a bonus every year."

"No bonus."

"GM lets me have a new car every three thousand miles."

Pickering suppressed a laugh. "No car."

"GM gives me—" Victoria started to say, but Pickering raised a silencing hand.

"GM gives you what private industry can give you," he said. "JPL will never match it."

Victoria paused. "Let me give it a night's thought," she said.

That evening, she went home and, well before the night was over, told her father that she'd be leaving her much-treasured job at GM and going to work for a Dr. Pickering at the Jet Propulsion Laboratory in Pasadena. With her first modest paycheck she'd have to start saving up for an equally modest car.

Victoria's initial assignment at JPL was in the public affairs department, but Pickering quickly decided that given her experience with government contracts, where he really wanted her to go was Washington. All of the NASA facilities across the country made it a point to have at least one full-time employee on Capitol Hill to lobby for funding for the agency's current and future projects. None of the NASA representatives ever called what they were doing lobbying, of course, and in a place as coy about such things as Washington, nobody expected them to. What they were really doing, they said, was simply explaining their programs, clarifying their programs, buttonholing congressmen in corridors, offices, and committee rooms not so much to press them for their votes, but simply to keep them abreast of the imaginative things the engineers and flight planners in the NASA labs were up to every day. If what they were up to interested Congress enough to write the agency a bigger check this year than it did last, well, that was just fine, too.

In the first few years the NASA representatives were coming to Washington, they didn't have to work very hard at their job. For all the lawmakers' resistance to being lobbied, the fact was, most of them seemed to like space. They liked the big, sleek rockets that often got built in their very own states, providing jobs for their very own constituents; they liked the grainy pictures the rockets sent home as visual proof that the public works money had been well spent; they liked the buzz-cut astronauts who sometimes visited their districts or spoke at their high schools and always made it a point to shake the congressmen's hands and remind the crowd that if it weren't for the vision of elected officials there would be no space program at all. The NASA representatives came to Washington prepared to sell their programs, and on the whole, they found lawmakers who were enthusiastically willing to buy.

At least most of the time they were. But Victoria and the other lobbyists noticed something else about the congressmen. It wasn't evident in anything they said; it wasn't evident in how they voted. But it *was* evident in subtler things: in the menacing noises they'd make when a program slipped over budget; in the lusty way they'd piled on JPL after all of the disabled *Ranger*s died. It was even evident when a mission succeeded brilliantly.

In the spring of 1966, when *Surveyor 1* was preparing to make NASA's first controlled landing on the moon, setting itself down on a puff of rocket exhaust in the same hard-baked plains in which its *Ranger* brothers had so explosively crashed, national interest in the mission was keen. As with the *Ranger* flights, the press had been following the mile-by-mile progress of the ship from the moment it was launched; and as with the *Ranger* flights, on the day the spacecraft was scheduled to land, the auditoriums and media rooms of JPL were fairly overrun by visitors determined to be on hand for so historic a moment. Unlike the *Ranger* landings, however, which were observed mostly by reporters and JPL employees themselves, this one was attended by an entire House of Representatives subcommittee— the Subcommittee on Space Science and Applications.

A visit by so exalted a group was a first for JPL, and it was Victoria who was assigned the job of looking after the committee men while they were there, escorting them around the lab's giant campus and sitting with them in the Space Flight Operations Facility's glassed-in VIP area during *Surveyor*'s final approach to the moon. Attending the landing in the company of so many men who were in a position to help or hurt JPL would not be easy, and throughout the day Victoria would be concerning herself as much with the welfare of the spacecraft as with the mood of the lawmakers there to watch it fly. It was in the final hour before *Surveyor 1* touched down on the moon that all of the VIPs at last settled into their seats in the SFOF viewing gallery, and after they did Victoria found herself preoccupied with trying to read their moods, regularly glancing at them for almost any cue that would reveal almost anything about their states of mind. With each announcement the mission narrator made—singing out the spacecraft's dwindling altitude and trajectory figures through the public address system in the gallery—she made it a point to catch the eye of this or that congressman, offering each an all-is-well smile or an A-OK nod. She got little in return.

Finally, at the end of the hour, when *Surveyor 1* was just a few feet above the surface of the moon, the spacecraft went into a brief, planned hover and the calls from the mission narrator stopped. For long seconds, no one in the VIP gallery or on the SFOF floor spoke, as the ship throttled its engine down and attempted to lower itself into the ancient soil. Then, all at once, the public address system again broke the silence as word arrived from the moon that the maneuver, against all odds, was complete, and the ship was safe.

"*Surveyor* is down," the announcer called exultantly.

Immediately, the gallery broke into an ovation, as hands were shaken, backs were slapped, congratulations were exchanged. Victoria, casting her eyes back to her congressmen, saw with satisfaction that they were joining the general celebration, too. Suddenly, however, the mission narrator, who should have had nothing more to say to the deeply relieved crowd, called out again.

"*Surveyor* is up," he said flatly.

To the people in the gallery who knew something about the mission, this was not unexpected. A *Surveyor* ship was an extremely lightweight machine and even with its engine breaking its fall it would still hit the lunar surface hard enough to cause it to bounce once or twice. A mere bounce still counts as up for a spacecraft that should be down, and mission rules required the narrator to announce the ship's position, no matter what it might be. If this *Surveyor* performed the way the design specs said it should, it would take only a second or two for the energy of the landing to be fully spent and the spacecraft to settle back down in the soil. A second or so later, that indeed happened and the public address voice quickly made the announcement that *Surveyor* was once again safe on the surface of the moon.

In that same second, however, Victoria noticed something disturbing about her congressmen—something in the set of their mouths, in the slight nods they exchanged, in the way they leaned expectantly forward like a steeplechase crowd when the lead horse all at once pulls up lame. Something, ultimately, in all of those things taken together.

They'd shut us down, Victoria suddenly realized. *Give them any excuse and they'd shut us down.*

Five years later, in December 1971, when William Pickering was trying to build a marvelous space probe that would visit Jupiter, Saturn, Uranus, and Neptune all in a single mission, Congress began doing just that.

•   •   •

It wasn't Pickering who figured out how to fly a single spacecraft past four different planets, and he never pretended to take credit for it. Just who actually did work out the plan was always a matter of some dispute, but the granddaddy of the thing looked to many people at the Jet Propulsion Laboratory to be Mike Minovich. By any measure, Mike Minovich—a newcomer to JPL who first showed up at the lab in 1966—was too young to be an authentic granddaddy; indeed, he was too young even to be a fully credentialed JPL scientist. What he was, at least when he came through the door, was a graduate student, just one of an eager group of apprentice researchers on loan from UCLA or Caltech, putting in a semester or two at the lab in the hope of being asked to join full-time when his classes were done.

Or that's what everybody said Minovich was, anyway. The thing was, not a lot of people ever saw him to confirm this. Oh, they saw his work, all right; it was impossible not to see Minovich's work, stacked up the way it was in giant mounds of computer paper flowing from his desk to his floor and into the hallway outside his office. But they just never saw what he did to generate it all. Come in in the morning and Minovich had already punched out for the day. Leave in the evening and he had not yet punched back in. When Minovich did like to work was at night, generally *all* night, a twelve- or fourteen-hour stretch when the JPL campus was quiet, the distractions were few, and the lab's room-sized mainframe computer was not being used, giving a simple graduate student the only chance he might ever have to get his hands on the mammoth machine.

Minovich didn't talk to anyone else who worked the night shift about what he and the mainframe were trying to puzzle out, but if you stumbled across one of his hummocks of paper crowding the hallway it was immediately evident what he was up to. The pages were filled with trajectory calculations—arcs and angles and complicated thrust numbers that were pretty standard stuff for a JPL researcher. But unlike most trajectory profiles, which are designed to get a ship from one point, typically the Earth, to another, typically the moon or Mars or Venus, this one was aiming for all four of the giant planets in the outer solar system.

The idea of making such a one-ship, four-world trip had been kicking around JPL for months before Minovich started working out the arithmetic of the journey, ever since Jim Long and the other scientists in the future missions department had discovered that sometime in the late 1970s, Jupiter, Saturn, Uranus, and Neptune were going to arrange themselves in an orbital conga line that would not occur again for another 176 years. Start planning right away, and you could launch a spacecraft in 1977 to arrive at Jupiter by 1979 and reach the other three planets over the course of the following ten years. Dawdle too long and you wouldn't get your next chance to make the same 2.7-billion-mile trip until 2153.

Puddle-jumping from world to world when they were queued up this way would have advantages beyond mere navigational convenience. A tiny dust mote of a spacecraft approaching a massive body like a planet does not just sail blithely by without feeling some of its effects. Well before the ship makes its closest approach to the world, the long arm of the planet's gravity reaches out to grab it. If the advancing spacecraft is traveling slowly enough, the gravity grip cannot be broken and the little machine will simply crash into the unyielding surface of the world—or, in the case of gas giants like Jupiter, Saturn, Uranus, or Neptune, simply be swallowed up by the planet's huge swamp of an atmosphere.

If the spacecraft is moving at enough of a clip and gives the planet a wide enough berth, however, even the most powerful gravity can't hang on to it. Rather, the ship will simply arc partway around the rear side of the world, surf its gravity field back around to the front, and then continue on its way into space. This gravitational whipcracking has two effects on the spacecraft: It bends its trajectory this way or that, depending on the mass of the planet and the proximity of the approach, and it speeds the ship up. Plan your approach to a planet you hope to visit carefully enough and you could use its gravity both to point your ship to its next destination and to step on the gas and get it there faster.

To be sure, a Newtonian free ride like this is not *completely* free. Since energy can neither be created nor destroyed, but rather merely exchanged, a planet that speeds a spacecraft up must itself slow down a bit. A ship whipping around the far side of, say, Jupiter at 22,000 miles per hour would appear from behind the planet's eastern edge moving at a speed of close to 35,000 miles per hour. In exchange for the kick a planet this large adds to a ship's speed, the planet itself must lose a little velocity, decelerating by about a foot of motion every trillion years. Though the loss is real, it was one most astronomers figured the big planets could afford.

While describing the theory behind such slingshot trajectories was a simple matter, plotting the trajectories themselves was infinitely harder. It was this job Minovich the grad student assumed for himself, and as the semester wore on and the obscure apprentice slowly began revealing his work to the scientists who supervised him, it became clear that such a mission could indeed be flown. The real trick would be building a spacecraft that was able to fly it.

A ship being designed to travel to the deepest provinces of the outer solar system would face a lot of engineering challenges, but the biggest one would be power. Probes setting sail for a target in the inner solar system—Mercury, Venus, Mars, or the moon—were remarkably lightweight machines, in part because they didn't have to carry much energy-generating hardware along. Flying in such close proximity to the bright, white fires of the sun, the ships needed to do little more to power themselves up than unfurl a pair of solar panels and drink up all the juice they needed for however long they intended to fly. On those few occasions when they flew around the shadowed side of a planet or moon, a small on-board battery could pick up the slack temporarily.

In the outer solar system, things are different. A planet that sits twice as far from the sun as, say, Earth does, doesn't just receive half as much sunlight. Rather, owing to the three-dimensional geometry of light dispersal, it receives barely a quarter as much. Three times greater distance causes sunlight to be squared down by a factor of 9; ten times greater distance means a hundred times less light. The planet Neptune, which floats at the end of a gravitational tether nearly 2.8 billion miles from the sun, receives nine hundred times less light than Earth, which lies thirty times closer, making high noon at Neptune's equator about as bright as a moonlit Earthly night.

Flying through such a cosmic gloaming makes solar panels patently useless. The obvious alternative—an on-board battery system—is equally impractical, since carrying enough batteries to keep a ship alive through the twelve years it would take to fly to a target like Neptune would add uncountable pounds to the weight of the ship and uncountable dollars to its price tag. If the deep solar system spacecraft was going to fly, it would clearly need juice, but just how it was going to get it was, at the moment, a mystery.

Just as problematic as the spacecraft's power system was its communications system. The transmitters aboard the earliest *Mariner* ships to Mars and Venus—to say nothing of the still earlier *Ranger* ships to the moon—were simple affairs, sending out data at a mere eight bits per second, or about the same rate as a Morse code key. Even traveling at the speed of light, that trickle of information could take up to eleven minutes to reach Earth from Mars. Assembling a single picture from a data stream this slow and thin took more than eight hours and produced an image that was patchy at best. In the provinces of the outer solar system, where there were four large planets and potentially dozens of moons to photograph and where light-speed transmissions could take hours to reach home, such a feeble system was out of the question.

Even more troubling were the spacecraft cameras. Photographing a planet or moon in the deepest shadows of the deep solar system was a tricky business, made possible only if the shutter of the imaging system was kept open for up to a minute at a time. This was easy enough to do with a camera that was standing still. With a camera aboard a spacecraft moving at a clip of 35,000 miles per hour, however, any image that resulted from such a time exposure would be little more than a dark, cosmic smudge. The only way to prevent that from happening was to invent a picture-taking system that could somehow compensate for the motion of the spacecraft, allowing the ship to pivot as it passed a planet so that it could stay photographically fixed on its target without slowing down by as much as a single mile per hour.

Until these and other technical problems could be worked out, a grand tour of the outer solar system would remain little more than the febrile fantasies of a Mike Minovich or a Jim Long, men with lots of time on the mainframe computer, but no real responsibility for turning their paper cipherings into functioning hardware. William Pickering did not care much for febrile fantasies, preferring that his scientists either fly missions or not fly them, but not spend untold time and money merely contemplating flying them. In late 1968, two years after mission planners began speaking dreamily of the grand tour, Pickering called a meeting with Homer Joe Stewart, the head of the future missions department; Robert Parks, the head of the lab's planetary projects department; and Clarence Gates, a trajectory wise man who worked for Parks. When the three men were assembled in his office Pickering got directly to the point.

"What exactly do we have with this grand tour?" Pickering asked summarily. "Is it a real mission or isn't it?"

"Clarence has worked the numbers," Parks said, "and they suggest it is."

Pickering turned to Gates, who glanced down at a sheaf of papers in his lap. "The alignment is real," he said. "The planets should move into position around 1979."

"And stay that way for how long?"

"We'd have until about 1989 to get through the system."

"Do we know how to do that?" Pickering asked.

"The physics and navigation aren't a problem," Stewart said. "It's the hardware that we need to work on."

"And have we worked on it?"

"Not yet."

Pickering paused. "Then," he said, "I suggest you gentlemen get started."

Before the end of the year Pickering had contacted NASA headquarters and requested funds to help the lab figure out how to build a spacecraft capable of embarking on a twelve-year mission to the outer planets. Weeks later NASA responded, not only approving preliminary funding for the project, but also approving a brand-new name for it. The proposed flight to the outer solar system that had been previously known as the grand tour would from now on be known as the TOPS project, for Thermoelectric Outer Planet Spacecraft. Such acronymic aridness was not pretty, but for a federal agency like NASA, it was the surest possible sign that the project had just moved from a mere idea to a fully certified mission.

●   ●   ●

It took three years—until the fall of 1971—before the plans for the TOPS spacecraft were at last completed, and it was clear to anyone who saw them that if the thing ever flew it would be over Congress's dead body. The problem was not that the ship wasn't a good one; indeed, even to the untrained eye of the legislator, it was evident that the TOPS spacecraft was a dandy machine. The problem was what it would cost to build it, and the way the blueprints looked at the moment, it would cost plenty.

The knottiest problem the JPL engineers had been confronted with—the power problem—was the one they were able to solve most easily, and the solution they came up with was nuclear. Radioactive material gives off heat as it decays, and a lot of radioactive material gives off a lot of heat. The TOPS designers proposed to fuel their spacecraft with twenty-four pellets of plutonium 238 separated into clusters of eight and packed into three iridium spheres. The thermal energy radiated by all of this material could be channeled into a small, on-board generator that would then produce electricity, much the way turbines driven by nuclear-heated steam generate power on Earth. The spacecraft power system—which the designers called an

RTG, for radioisotope thermoelectric generator—would be an impressively compact machine, only seven feet long and roughly cylindrical in shape. What's more, it would be an efficient machine. As the plutonium in the generator decayed, it would lose only about 2 percent of its total energy per year. Barring an unexpected breakdown, this would give the portable power plant roughly half a century of life, more than enough for the mission the JPL planners envisioned.

Just as impressive as the RTG was the on-board computer the engineers designed for their ship. For a spacecraft that had to operate trouble-free for more than a decade, the traditional system of a prime data processor and a single backup processor would not do. Rather, the TOPS was equipped with no fewer than five redundant computers—three prime systems and two backups. Before a critical maneuver was attempted, the trio of main processors would consult one another and decide how and when it should best be executed. If one of the three disagreed, the other two main computers would investigate why. If it turned out that the dissenter had discovered a flaw in the plan, its proposed fix would be adopted. If the third computer was simply malfunctioning, it would be taken off-line and one of the two backups would be promoted. The JPL engineers ambitiously called their system the STAR computer, for Self-Test and Repair, and they were convinced the hardware would live up to that handle.

The communications system for the TOPS was souped up, too. The spacecraft was equipped with a twelve-foot parabolic antenna that was able to send information to Earth not in the trickle the *Mariner*s could produce, but in a comparative electromagnetic gush. Aimed at deep-space receiving stations in Australia, Spain, and the Mojave Desert, the transmitter could beam back information at a rate of up to 115,200 bits per second. Such a rich data stream would be able to carry a wealth of planetary information, including readings of electromagnetic fields, chemical spectra, cosmic rays, plasma emissions, and, of course, crisp color images. Taking those pictures would still be something of a challenge, since the engineers in the flight dynamics department still had not figured out quite how to pivot the ship so that its cameras could stay stationary long enough to do the work they were intended to do. But given the quality of the spacecraft the designers were presenting them with, few people at JPL doubted that this last bit of business could be figured out.

Picking a precise trajectory for the TOPS was a less complicated job than all this engineering and design work. Working with Mike Minovich's preliminary calculations, the flight planners had determined that in order to reach the outer planets when they were in their maximum alignment, the spacecraft would have to leave Earth during a narrow forty-day launch window that ran from mid-August to mid-September 1977. Once the ship got off the ground, there were more than ten thousand subtly different flight paths it could follow, depending upon the precise photographic angle mission managers wanted it to have as it approached any one planet. So flexible were all of the available trajectory options that the analysts concluded it might even be possible to include a fifth planet, Pluto, in the overall mission profile. Since Pluto would not lie in the same clean line as the other four worlds, no one spacecraft would be able to fly the comparative corkscrew trajectory necessary to visit all five. But if NASA was willing to build two ships in the TOPS class, the flight planners could target one to fly a relatively slow route past Jupiter, Saturn, Uranus, and finally Neptune, while another would travel first to Jupiter and then bypass all the other worlds, riding the big planet's gravitational slingshot straight out to Pluto.

Naturally, the cost for such a pair of Cadillac spacecraft would be considerable. Near as JPL budget makers could calculate, the TOPS ships would probably set NASA back at least $750 million. In a space agency that routinely spent $100 million or more for a single expendable booster, $750 million was arguably not all that much— not when it would produce two spacecraft and more than twelve years of planetary research. But that was $750 million before any unexpected hardware problems; $750 million before any unexpected design setbacks; $750 million before inflation and labor disputes and contract snags ratcheted the cost higher still. That kind of $750 million could easily turn into a billion or more and everybody in the space community knew it.

It was for that reason that Pickering and the rest of JPL figured that the TOPS program was probably dead on arrival in the halls of Congress. Certainly, it was always possible in situations like this to scale the project back before it ever got to Washington. But engineering programs weren't education or environmental or social services programs—programs that lent themselves to compromise and half-a-loaf concessions. Engineering programs had hard, empirical, non-negotiable answers, and when Pickering's designers presented him with those answers in the form of the blueprints for the TOPS ship, his scientific instincts told him that these were the plans he ought to submit to Congress—even if his political instincts told him Congress would kill the plans the moment they arrived.

It was shortly before the holiday recess in December 1971 that the Subcommittee on Space Science and Applications at last got around to deciding the matter, assembling to vote on the proposed NASA budget for 1972. For weeks before that, Victoria Melikan had been on Capitol Hill lobbying the congressmen hard for the JPL budget in general and the TOPS project in particular. She'd practiced her arguments about how much there was to learn from the five planets; memorized the figures showing how much more expensive it would be to fly five different spacecraft to five different worlds instead of making one majestic sweep past the whole group; remembered especially to remind the congressmen that the outer planets' alignment wouldn't occur again until the year 2153 and it just wasn't scientifically sensible—it wasn't even scientifically *moral*—not to seize such an opportunity when it presented itself. That last argument was her trump card, and in the weeks she was lobbying the legislators, she always made it a point to pull it out, enjoying how the congressmen's faces subtly changed as a debate that until now had been all fiscal and political suddenly turned vaguely existential. If any of Victoria's pitches was going to change congressional minds, that would be the likeliest one to do it. Shortly before the vote on the project, Minnesota representative Joe Karth, the chairman of the subcommittee, pulled her aside for a private moment to let her know if it had.

"This TOPS thing . . . ," he began hesitantly.

"Mm-hmm," Victoria said, smiling as agreeably as she could.

"We've given it a lot of thought."

Victoria was silent.

"And there's no way, just no way we can justify spending that kind of money."

"You know it's five planets—" Victoria began.

"—for the price of one mission," Karth finished. "I know."

"You know it would cost much more to visit them all separately."

"It costs enough as it is."

"You know this is an opportunity humanity isn't going to get for another hundred seventy-six years."

"Then humanity will have to worry about it, then."

Karth looked at Victoria, saw her struggling for something else to say, and decided to make it easier for her. "It's not going to happen," he said simply. "It's just not." He walked away without another word.

The formal vote on the TOPS plan came a few minutes later, and Victoria, as was her custom, sat in on the subcommittee session until the decision was officially rendered. Then she returned to NASA's Washington headquarters, phoned Pickering, and told him what had happened. As always, she was impressed by the seemingly unperturbed way he took bad news; as always, he remembered to thank her for her call; and, as always, she suspected that no sooner had he hung up the phone than he was busy formulating other plans.

At his desk in Pasadena, Pickering was formulating indeed.

•   •   •

The problem with trying to undertake a mission like the grand tour is that no matter how hard engineers work to design it and congressmen work to derail it, the planets themselves don't really care. Even as the Subcommittee on Space Science and Applications was taking its vote and Victoria was making her call to Pickering, the great worlds of the outer solar system were already beginning to ease their vast bulks into position, preparing for an alignment that was now just a few years away. Humanity could either find a way to be there for the event or not, but the planets—heedless of the doings of a tiny species on a tiny world deep within the inner solar system—were going to line up regardless. What Pickering and his designers had to do now was figure out if there was some other way to fly a mission to those worlds without relying on the budget-busting ship they had just designed. The answer, they suspected, might be in a far simpler spacecraft they had first built long ago.

For all of the different ships JPL engineers had invented in the thirteen years the lab had been in business, most were descended from a single progenitor machine: the *Ranger* moonship. With its sturdy armature, its simple solar power system, and its easily adjustable architecture on which any number of cameras and sensors could be hung, *Ranger* was the least expensive and most flexible spacecraft the lab had ever developed. The *Mariner* ships that flew to Mars and Venus were descended directly from the *Ranger*s, though modified for their longer missions and different targets. The still-to-be-built *Viking* ships—two-part, piggyback spacecraft that would both orbit and land on Mars sometime in the mid-1970s—were, in turn, descended from the *Mariner*s.

In designing the TOPS ship, JPL had done away with the *Ranger* bloodline altogether, seeking to develop an entirely new species of spacecraft from the inside out. Now, with that new species extinguished before it could even be birthed, Pickering instructed the engineers to return once again to the ship that had succeeded for them so many times before and see if they couldn't reconfigure it for deep-space flight.

That reconfiguration turned out to be relatively easy. Beginning with a basic *Ranger-Mariner* skeleton, the engineers stripped away its useless solar panels and replaced them with their newly designed nuclear power system. In order to prevent the more delicate ship from being damaged by the waste heat the system gave off, they mounted the generator on an eight-foot boom that reached away from the main body of the spacecraft like a bony arm. On another limb on the opposite side they mounted a cluster of off-the-shelf instruments, including a camera, an ultraviolet spectrometer, a plasma detector, and a photopolarimeter. In the center, they mounted the same prodigious communications dish they had designed for the

TOPS spacecraft, but in the heart of the ship they did away with the powerful, five-brained STAR computer and replaced it with a simpler, double-processor machine.

The spacecraft that resulted from all this workbench cobbling may have been less reliable than the magnificent TOPS, and it was certainly less lovely—nothing that the engineers would have designed if their hands hadn't been forced by Congress. But when the engineers stepped back and looked at their ship, they had to admit that lamentable as the machine may have appeared, it certainly looked as if it could fly. Just where it would fly was another matter entirely.

It was clear from the resounding repudiation TOPS received from Congress that even with a more cost-effective spacecraft, the government was not about to put itself on the hook for a five-planet mission that would consume manpower and resources for more than a dozen years. The answer for the flight planners was to start eliminating planets. Pluto was the first to go. A flash-frozen ball barely two-thirds the size of Earth's own moon, Pluto had been a questionable target from the start. With the thirteen years it would take to complete the trip to the distant world, it seemed like an out-of-reach target, too. Neptune was a far bigger world and a far richer one in terms of chemistry and overall scientific promise. But the problem was, Neptune also lay at a breathtaking distance from Earth. Indeed, so far away was the eighth planet that for twenty years out of every 250 or so, it actually became the ninth planet, as the 3.7-billion-mile egg-shaped orbit Pluto inscribes around the sun caused it to cross inside Neptune's 2.8-billion-mile circular one. That twenty-year period was approaching at about the same time the outer planets mission would have to be launched, meaning that a flight to Neptune would take only about a year less than a flight to Pluto—clearly too many years for the new, less sturdy ship the JPL team had designed.

That left Jupiter, Saturn, and Uranus. Traveling at 35,000 or so miles per hour, the new spacecraft would need less than two years to reach Jupiter, which lies about 400 million miles from Earth, and little more than three years to reach Saturn, which lies about 800 million miles away. Uranus, however, is more than twice as far as

Saturn, orbiting nearly 1.7 billion miles out. Trudging such a great way through the solar system would take at least a good eight years—or almost seven years longer than any other spacecraft had ever had to fly to reach its target planet before.

The only prudent answer then was to pass up Uranus as well and pare the grand tour down to a simple, two-planet hop to Jupiter and Saturn. The JPL planners still hoped to send two ships to fly the mission, launching them on essentially the same flight path to ensure that at least one of them would get where it was going if the other should fail. Nonetheless, even with a two-ship fleet to build and launch, JPL estimated it could bring the entire project in at a relatively pennywise $250 million, or just a third of the price tag of its TOPS predecessor—clearly a more appealing prospect to budget-conscious legislators. In mid-1972, NASA submitted this scaled-down plan to Congress and this time Congress quickly approved it. In July the new project—which Congress began referring to by the clumsy moniker *Mariner* Jupiter-Saturn 1977, or MJS77—officially got under way on JPL's Pasadena campus.

At first, work on the MJS77 program began like work on any other JPL program. The engineers were given the spacecraft specs, the flight planners were given the planetary targets, and both went about the job of preparing to fly the mission NASA told them to fly. But at least a few of the designers and planners were contemplating something else. Sending the new spacecraft to Jupiter and Saturn was a fine shakedown cruise. But suppose the ships were still operating after that? Suppose they finished their work at Jupiter, finished their work at Saturn, and then found themselves fit and well, poised on the gravitational precipice between Saturn and the true cosmic deep? To be sure, that was a big supposition, what with the makeshift design of the spacecraft that would be flying the mission. But if by some chance the machines *were* still functioning after their 800-million-mile journey, did it make sense—any sense—simply to shut them down? It was one thing to refuse to spend the money to build a spacecraft capable of making the long-haul journey to Uranus, Neptune, and Pluto. It was another thing entirely to build a cheaper spacecraft that just happened to be able to fly farther than the spec sheets said, and then not push it as far as it could go.

Congress, clearly, had already said no to such a plan; but in the space game, there were noes and there were noes. And three years after launch, when the new ships were deep in space and had completed their prime mission, it was always possible that just enough cajoling and just enough wheedling could persuade Congress to change the no to a yes. The key for now was to shut up about any mission but the formally approved MJS77, stay within the $250 million budget, and be exquisitely careful about how the new spacecraft were designed. Nothing should be built into either ship that had an absolute life span of only three or so years. The two machines that reached Saturn around 1980 had to be as fit as possible for another eight or nine years in space so that if the green light was given for the planets beyond, they could simply point their prows and go.

Certainly, few of the engineers at JPL would speak openly about so seditious an idea. But in other, subtler ways the word got out. Bud Schurmeier, the engineer who had assumed the reins of the *Ranger* program when Jim Burke was sacked, had been named manager of MJS77, too. John Casani, another *Ranger* veteran who had been working on other projects at JPL, was named his lieutenant. Coming to work for Schurmeier meant Casani would have to change offices, and changing offices meant he would have to change his telephone extension. The day he moved in, he called the JPL operator and made a special request.

"I'm new to the MJS77 office," he said. "What's the chance of getting the number 6578 assigned to me?"

The operator flipped through her phone log. "It's pretty good," she said at length. "No one else is using it."

"Let me have it then," Casani instructed.

"It has to be 6578?" the operator asked as she made the notation.

"Precisely 6578."

Later that day, with his phone operating, Casani dropped by Schurmeier's office and handed him a slip of paper.

"New phone number," he said to his boss. "Like it?"

"It's a number," Schurmeier said with a shrug.

"Not just a number. Look at the paper and then look at your phone," Casani instructed. Schurmeier complied. "6578 is MJSU," Casani said with a broad smile. "We're going to Uranus, Bud."

Schurmeier dropped his head in his hands. "Jupiter and Saturn, John. I want you thinking only about Jupiter and Saturn."

"I will," Casani said cheerily. "For now."

•   •   •

Enthusiastic as the JPL engineers were about the idea of sending their spacecraft to as many planets as possible, they did not much care what the ships actually did when they got there. If the planetary scientists wanted to take pictures the engineers would build them a camera. If they wanted to study chemistry or magnetism they'd build them a spectrometer or a magnetometer. The engineers were in the business of designing hardware and assembling the ships; after that, it was up to the scientists to put them to use. The scientists themselves had very definite feelings about the matter and very specific plans for the spacecraft. And while they were intent on visiting and studying the outer planets, what they wanted to study especially closely were those planets' moons.

It had been a long time since scientists began to appreciate the curious appeal of the solar system's moons—more than 350 years, in fact. Unlike many of the pivotal moments in the near-timeless field of astronomy, the study of the moons could be traced to an extremely specific date: Thursday, January 7, 1610, at about one o'clock in the morning.

It was cold and clear in Padua, Italy, that night, one of many punishingly cold nights the city had experienced that winter. Cold was not so good if your work was stargazing, but clear was very good indeed. When the night broke cloudless and sparkling, Galileo Galilei, a tenured professor at the University of Padua, who had made it his business to do a lot of stargazing lately, carried his little two-foot-long telescope outside for as many hours of studying the sky as the frigid temperatures would allow.

Galileo's telescope was a fine one—lightweight, portable, and yet capable of magnifying objects a remarkable thirty times. To be sure, it was not the first telescope in the known world; the first telescope— or at least the first Galileo had heard of—had been developed the year before by a Dutch optician, Hans Lippershey. Living in faraway Padua, Galileo had never had a chance to examine Lippershey's instrument himself, but after reading a detailed account of it in a scientific journal, he was pretty sure he could design a better one. Before the year was out, he had indeed built a modest scope with a nine-power magnification that rivaled Lippershey's own.

When Galileo completed the instrument, he knew that this was not an achievement he should keep to himself and quickly invited the members of the Venetian Senate to come see what he'd built. So dazzled were the lawmakers when Galileo led them up a Paduan tower and showed them the swollen, swimming images of the local countryside that appeared in his lens that they quickly doubled his salary and made his provisional tenure at the local university a permanent one. With the time and wealth such good fortune brings, Galileo was able to devote himself to building his improved, thirty-power instrument. As he grew more skilled at using the new scope, he'd begun putting it to work to study not ships at sea and remote hills, but the bodies of the distant cosmos themselves. The cloudless night of January 7 struck Galileo as just the right opportunity for those kinds of observations.

As this evening's work began, Galileo first turned his attention to the planet Jupiter. The fifth of the six known planets had always been a seductive target for stargazers, mostly because of its great size and brilliance. On other nights, Galileo had pointed the maw of his telescope toward the remote world, framed its fuzzy, vaguely striated face in his narrow field of vision, and followed the planet as it swam through the stationary stars that surrounded it, wandering across the firmament from one evening to the next, even as the stars themselves hung fixed in the sky.

Most of those stars were of little interest to Galileo—at least on a night when he had set his sights on a planet. This evening, however, as he brought Jupiter into focus in his eyepiece, three unusual stellar pinpoints caught his eye. Unlike the other stars that surrounded giant Jupiter in an utterly random spangle, these three seemed almost to be clinging to the planet—two lying just to the east of it and one just to the west. More curiously, they were not located just anywhere in the immediate vicinity of the world, but in what appeared to be a straight line, queued up almost precisely along the same latitude as the planet's equator. Though the soupy Earthly atmosphere and Galileo's own tired eyes made it impossible to be certain, it also appeared that the three points were not all the same size, with the western- and easternmost ones appearing slightly larger than the one between them.

The queer new stars intrigued Galileo, but only mildly. Even amidst the most chaotic disorder there could sometimes be little pockets of order, and if the ancient constellations proved anything, it was that, with an infinite number of random stars decorating the sky, at least a few would occasionally position themselves in not so random ways.

The next night, however, when Galileo once again turned his telescope to the sky, he noticed a different state of stellar affairs. The three little stars were still visible, still lined up along the same plane as Jupiter's equator, but tonight, all of them had somehow migrated to the western side of the planet. This kind of motion was beyond the powers of any known star and that left Galileo feeling uneasy. Generations of astronomers before him had painstakingly calculated the speed at which all of the planets change their position in the sky from a particular hour one night to the same hour the next night. From what Galileo was seeing now, their calculations of Jupiter's speed were apparently wrong, with the planet moving far faster than once thought—so fast, in fact, that in a span of just twenty-four hours, it had left all three stars in its western wake. By tomorrow night, Galileo figured, the three little points should be even further behind.

But the little points weren't. During his next observation Galileo noticed that one of his mysterious stars had inexplicably disappeared altogether, while the remaining two had returned to Jupiter's eastern side. The night after that, one of the two stars had moved to the west; the following night, a third star reappeared on the east. The

Io   Springing volcanic leaks all over its scalded surface, Io may be the most explosively active world in the solar system. Gravitational pressure from Jupiter and its other satellites causes rhythmic pulsing deep within the moon, which in turn leads to volcanoes. This picture was taken by the *Galileo* spacecraft at a distance of 302,000 miles.

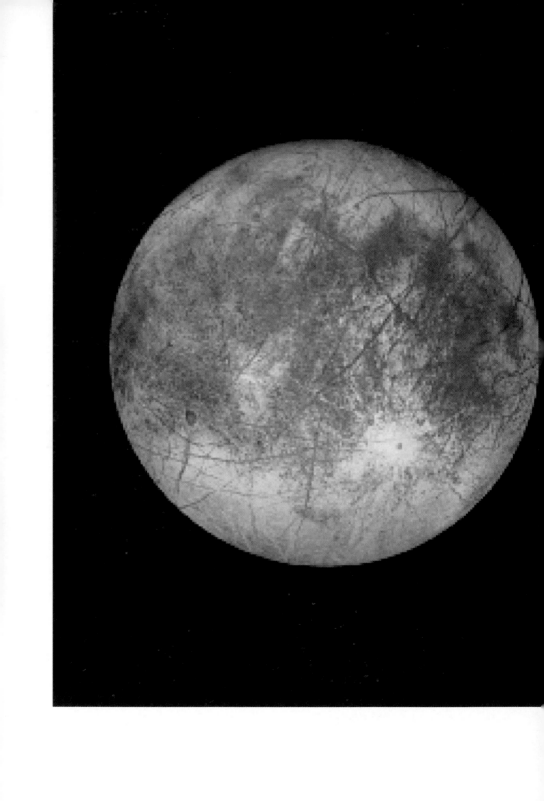

EUROPA   Europa's brittle, eggshell surface is made entirely of water ice. The absence of craters suggests that the ice regularly melts; the webwork of visible cracks suggests that something is surging not far below. Taken together, all of this points to a rich, briny, global ocean lying just beneath the surface—an ocean that could easily support Earth-like life. This picture was taken by *Galileo* from a distance of 417,900 miles. (The image at left is Europa's natural color; below, enhanced color highlights surface differences in the moon.)

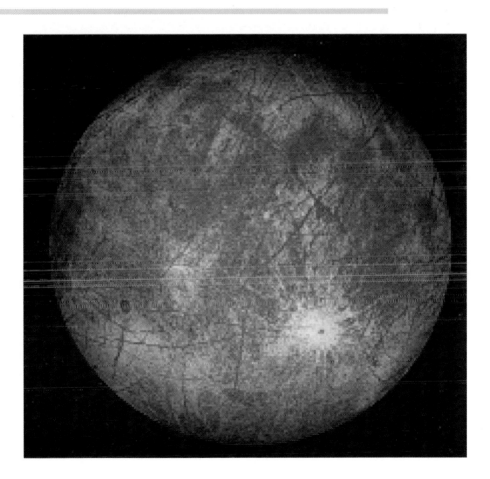

GANYMEDE   The biggest moon in the solar system, Ganymede has a diameter of 3,262 miles, making it larger even than some planets. A rice-pudding world composed of rocks and ice, Ganymede is covered by a crust made up of ancient, cratered terrain and smoother plains where the topography was refreshed by upwelling slush. This picture was taken by *Voyager 2* from a distance of 744,000 miles.

CALLISTO   The most distant—and the coldest—of Jupiter's four large moons, Callisto has remained virtually unchanged for the better part of 4.5 billion years. Its frozen face still carries the scars of nearly every meteorite that ever hit it, though over time, the contours of those craters have softened a bit as the surface ice has slowly slumped. This picture was taken by *Voyager 2* at a distance of 1.4 million miles.

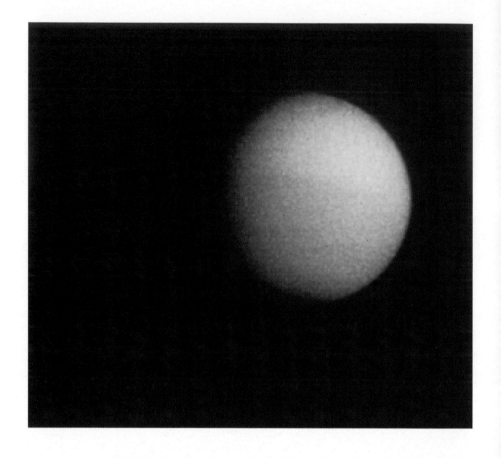

TITAN   Saturn's largest moon, Titan is one of the most chemically rich places in the solar system. Drenched in ethane, methane, and other organics, it is thought to be a dead ringer for the prebiological Earth—if the prebiological Earth were somehow transported 800 million miles away and flash-frozen in deep space. *Voyager 1* flew by Titan at a distance of just 4,000 miles, but the moon's dense atmosphere prevented the spacecraft's cameras from seeing a thing.

MIRANDA   Flying by at a distance of 19,000 miles, *Voyager 2* photographed the fractured face of Uranus's moon Miranda. Gravitational turbulence from surrounding bodies may have shattered Miranda to shrapnel. The accumulated gravity of all those fragments, however, allowed the moon to reassemble itself into whatever shape best approximated a sphere.

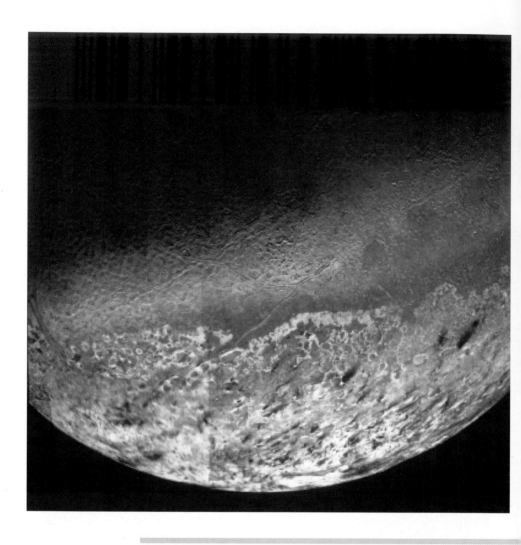

TRITON   At −391°F, Neptune's large moon Triton is the coldest place in the solar system. Its surface is made up of a permafrost of nitrogen, carbon dioxide, carbon monoxide, and methane—materials that would normally vaporize at the barest breath of solar warmth. Flying by just 15,500 miles above the moon, *Voyager 2* nonetheless noticed dark carbon geysers occasionally blasting through the ice, as the faint flickers of sunlight that reach Triton heat up subsurface materials and cause them to expand.

stars continued changing position and numbers until finally, on the ninth night of his observations, Galileo saw no fewer than four of the mysterious points, all in close proximity to Jupiter, all in a tidy rank along its equatorial waistline. Now, suddenly and simply, the astronomer understood what he was seeing.

These seeming stars weren't part of the stellar background canvas at all. They were wards of Jupiter, moons of Jupiter, as inextricably tied to the planet as if they had been *lashed* to Jupiter. The great, distant world did not hurry alone through the nighttime skies like Saturn, Mars, Venus, and speedy little Mercury. Rather, like Earth, it was forever accompanied on its journey; while Earth was circled by just one companion body, however, Jupiter was surrounded by four. When that evening began, Galileo—and, by scientific extension, humanity as a whole—knew of but one moon in the vast solar system. When that evening ended, the number had suddenly grown to five.

Coming up with names for the clutch of new moons was not a simple matter, not like naming the planets had been, anyway. It was the ancients—the Romans, the Greeks, the civilizations that came even earlier—who discovered the planets. In those less cynical, less clinical times, grand celestial bodies needed grand, heavenly names, so the planets were given the names of the gods.

But the first century was not the seventeenth, and in the era of science and finance, there were other, less exalted things to consider. It was not an accident that even before designing his telescope, Galileo had already distinguished himself as one of the great academicians of his land, having risen to his provisional and then permanent professorships. Such academic positions required influence, influence required powerful supporters, and Galileo knew how to attract both. Gifted in language, Galileo enjoyed writing, and did so with cunning versatility. When he was composing an essay for popular consumption, he generally wrote in the same slouchy vernacular he knew his readers used. When he was writing for other academics or potential benefactors, however, he would drop such studied plainspokenness in favor of the florid and floral Latin that was his natural tongue. If he had a controversial idea to introduce, an idea that was sure to inflame scholars or government officials, he would trick out his writing in even more decorative language than usual, pick a few of his likeliest critics and then—with gratitude and blandishments— dedicate the work to them. Their eminences generally appreciated the homage and often as not rewarded Galileo with their support and their applause.

It was no surprise, then, that when Galileo discovered his four new moons he should take the occasion to honor a group of people far more worthy of acclaim than he—a mere salaried professor— could ever hope to be. With a humbleness that surprised none of his peers, Galileo announced that the new worlds he had just found would henceforth be known as the Medicean Stars.

The Medicis, of course, for whom the moons were self-evidently named, were members of a powerful family of financiers and business leaders who had been ruling Florence and Tuscany since 1434. It was the Medicis who had long determined how local land was held, how wealth was distributed, how industries were run, and even how the occasional war was fought. And it was the Medicis, not incidentally, who determined who should fill the chairs and win the tenures at the country's great universities. In Padua, where Galileo lived, it was Cosimo de Medici, the grand duke of Tuscany, who handled such affairs, and when Galileo published his first book, *The Starry Messenger,* in 1610, making public the discovery of his Jovian moons, he did not shrink from explaining how he had chosen the names for the new bodies.

". . . the maker of the stars himself has seemed by clear indications to direct that I assign these new planets Your Highness' famous name in preference to all others," he wrote. "For just as these stars, like children worthy of their sire, never leave the side of Jupiter by any appreciable distance, so as, indeed, who does not know clemency, kindness of hearth, gentleness of manner, splendor of royal blood, nobility in public affairs, and excellence of authority and rule have all fixed their abode and habitation in Your Highness." Cosimo de Medici himself saw little to quarrel with in this, and months after the new book appeared, decided that a scholar of Galileo's stature was wasting his considerable skills in an academic backwater like Padua.

From now on, Medici declared, the gifted astronomer would fill the chair of first professor of philosophy and mathematics at the University of Pisa.

Simply because a celestial name served the career of Galileo Galilei, however, did not mean it served the interests or satisfied the tastes of astronomers as a whole. Simon Marius, a German stargazer and righteous rival of the favor-seeking Galileo, was more dissatisfied than most. In 1606, Marius was studying at the University of Padua when he invented—or believed he invented or pretended he invented, depending on just who at the university was telling the story—a marvelous little calculating instrument called the proportional compass. No matter how Marius came by the instrument, he quickly went public with it. No sooner did he do so, however, than an outraged Galileo rose up to challenge him, insisting that he had dreamed up just such a device nine years earlier and that Marius had obviously filched the idea. Wielding far more influence at the university than Marius ever could, Galileo quickly had the insolent young student expelled on a charge of plagiarism and sent packing back to Germany.

Three years later, when Galileo began popularizing the use of the telescope, Marius, learning of the development, claimed that he himself first got a look at a simple telescope in Frankfurt in 1608 and that there was nothing new in Galileo's work. When Galileo announced that he had discovered his four Jovian moons in January 1610, Marius insisted that he had spotted them, too, and had done so in December 1609.

But the grim Marius was no match for the glad-handing Galileo, and it was Galileo who earned universal credit for finding the new satellites. Marius, however, still had an argument to make. Regardless of who discovered the bodies, was it right for the self-interested Galileo to determine what they would be called? The pandering evident in naming such magnificent worlds after a mere Tuscan industrialist demeaned the very moons themselves, Marius argued. He recommended something else entirely. Since the world around which the moons orbited was named in honor of the Greek's Zeus and the

Roman's Jupiter, its retinue of moons should justly be named after members of Jupiter's own court. The innermost satellite, he insisted, should be named Io, after one of Zeus's most beloved maidens; the second would be Europa, a Phoenician princess abducted by Zeus and later the mother of three of his children; the third would be Ganymede, a Trojan boy whom Zeus made cupbearer to the gods; the fourth and last would be Callisto, a nymph loved by Zeus.

Marius knew instinctively that astronomers would take to these names—certainly more readily than they had taken to Galileo's—and, as it turned out, he was right. No sooner had he floated his idea than scientists around Europe began writing, speaking, and teaching about the marvelous new worlds Io, Europa, Ganymede, and Callisto. With each such reference, the gulf between Galileo's lofty position in the scientific world and Marius's far more modest one closed a little bit. Galileo, after all, may have birthed the Jovian moons, but it was the lesser Marius who determined how they'd forever be known.

•   •   •

Over the years that followed Galileo's discovery, the business of spotting and naming moons picked up momentum. Before the seventeenth century was out, Holland's Christiaan Huygens found a vast, yellow-orange moon circling the already gaudily ringed planet Saturn. Huygens named his new moon Titan, after a family of giants in Greek mythology who once sought to rule the heavens but were overthrown by Zeus. Giovanni Cassini—a countryman of Galileo—followed with the discovery of four other Saturnian moons, which he named Iapetus, Rhea, Dione, and Tethys, after other mythical Greek figures. In the eighteenth century, the British astronomer William Herschel discovered the planet Uranus and quickly found Oberon and Titania—which he named after characters created by his beloved Shakespeare—circling it. He then found Enceladus and Mimas, two additional moons orbiting nearby Saturn. In the nineteenth century, other astronomers found the planet Neptune and its big moon Triton. After that came Hyperion, yet another moon for Saturn, and Ariel and Umbriel, two more for Uranus.

There was not much beyond the obvious that the astronomers could determine about any of these new satellites from the fantastic distance at which they were being observed. The worlds were obviously rather big, since how otherwise could they be seen? They were obviously round like Earth and the other planets, since what other shape could they be? More than that, the scientists couldn't say. So taken were they with the very fact of the satellites' discovery, however, that many of them didn't even *care* to say.

But even as the family of planets and moons was growing so steadily, something about the solar system puzzled astronomers— something that appeared to be missing from it entirely. In 1766, Johann Titius, a Prussian professor of astronomy, physics, and biology, had happened upon a troubling little equation. For no reason that he could determine, Titius was fooling about with numbers one day and found himself writing down a string of utterly meaningless figures— 0, 3, 6, 12, 24, 48, and 96—connected only by the tidy way the numbers doubled as they went along. Titius stared at the figures, then stared at them some more, and something faintly, faintly familiar seemed to tease at him—or something that *would* be familiar if he just did something to the numbers. Without knowing why, he realized that what he had to do was add to each one the numeral 4, the smallest whole number whose nonfractional square root is something other than the number itself. This changed Titius's pointless string of figures to a still pointless—and now far less tidy—4, 7, 10, 16, 28, 52, and 100. Finally, though again he did not know the reason, he divided those numbers by 10. The string of ugly fractionals he had left—.4, .7, 1, 1.6, 2.8, 5.2, and 10—had no meaning to him at all.

Or at first they didn't. But as Titius looked at the numbers, he all at once realized that they weren't just an arbitrary string of digits at all, but a ratio. And they weren't just any ratio; they were a planetary ratio. If the third number, the 1, were assigned to Earth, the first number would become Mercury, which Titius and other astronomers had calculated was .4 times as far from the sun as Earth is. The second number was Venus, which was .7 times Earth's distance. The fourth number was Mars, which was 1.6 times as far as Earth; later came Jupiter, at 5.2 times and Saturn at 10 times.

What struck Titius most wasn't just the sweet neatness of the astronomical coincidence; rather, what struck him was that the figures made gravitational sense. As the planets were coalescing out of the great cosmic windstorm that gave birth to the solar system, they couldn't pop up just anywhere. Form too close to a neighboring planet and the gravity of the two worlds would cause them to take hold of one another, drawing them together into a single megaworld. In order to survive as a discrete body, each fetal planet had to keep its distance from the next, with the precise distance being determined principally by the size of the planets on either side of it. In the case of the solar system as it was known so far, the balance wheels of gravity caused the planets to fall into orbital slots that precisely corresponded to Titius's seemingly random string of numbers.

But the equation had a flaw. In Titius's numeric sequence, there was one orphan number, his 2.8, that had no planet associated with it. If the formula was to be believed, somewhere in space, about 260 million miles from the sun—gravitationally centered between Mars and Jupiter—a world as real as any other by rights ought to be orbiting. But when centuries of astronomers looked in the direction of that vast stretch of cosmic real estate, empty space was all they ever saw. Things grew even more perplexing several years later when William Herschel turned his telescope toward the deep sky and found Uranus, just over nineteen times as far from the sun as Earth is—or exactly where the gravitational formula said it should be. Titius's orderly equation had now gotten more orderly still at its far end while remaining stubbornly corrupt in the middle.

That was how the equation stayed for the next thirty-five years, the mystery outliving even Titius himself. It was only on January 1, 1801—the first night of a new year in a new century—that the puzzle was solved. On that evening, Giuseppe Piazzi, a Theatine monk and a professor of mathematics at the Academy of Palermo, was conducting some telescopic observations of his own when he discovered a small, nearly insectile flicker somewhere between Mars and

Jupiter. Piazzi knew he couldn't be looking at a planet, since planets were great, hulking spherical things made out of gas or rock. What he was looking at could be no bigger than 125 or so miles across, a scrap of nothing that, given its small size, might not even be properly round.

If the body wasn't a planet, the only thing it could be was a comet. But when Piazzi tried to track its path, he found it didn't move like a comet at all. Comets typically careen through the solar system at a drunkenly inclined angle, cutting heedlessly across the flat orbital plane as they dive in toward the sun. Piazzi's putative comet moved in a much more orderly way than that, plodding in what appeared to be a tidy solar circle that was somewhat wider and slower than Mars's orbit and somewhat smaller and faster than Jupiter's.

Piazzi wasted little time going public with his mysterious new body, trusting that the rest of the astronomical brotherhood would help make sense of its meaning. The rest of the astronomical brotherhood quickly did just that. Though Piazzi's rock might be too small to be a planet, the other scientists concluded, the fact that it was orbiting the sun precisely where Titius's equation said a big body should be suggested that it might well be one of the bones of a planet. Long ago, it now seemed possible, a protoplanet might have begun forming in the Mars-Jupiter void, only to be shattered by a collision or some other cataclysm. Tumbling through so gravitationally complex a region, the rocky detritus that resulted might never have been able to reassemble itself into a world, and instead would simply orbit the sun as a permanent debris stream—a former planet, a mortal planet, reduced by violence to nothing but rubble.

That was the theory anyway, but before it could be confirmed, additional bits of the lost world had to be found. Over the next six years several of them were, as astronomers turned up more of these "asteroids"—or star-like bodies—in the Mars-Jupiter gap, none more than a few dozen miles across. The collected rubble still didn't amount to much, certainly not enough to form a full-sized planet, but if a few big bits of rock were turning up, hundreds, perhaps thousands, of smaller ones were almost certainly there, too.

Just how many asteroids there were and just how big a planet they once made up was impossible to say. Indeed, even if observers on Earth could somehow count every rock and pebble floating in the Mars-Jupiter rubble stream, they would still have a hard time making such a calculation, mostly because it eventually became clear that bodies orbiting in the asteroid belt didn't have to *stay* in the asteroid belt. Late one night in August 1877, a full human lifetime after Piazzi spotted the first asteroid, Asaph Hall, an American astronomer working at the United States Naval Observatory in Washington, D.C., was observing the planet Mars with the observatory's stout, twenty-six-inch-diameter telescope. Hall had picked a bad night for skygazing, with the air hanging thick and stewy in the former swampland that abutted the Potomac. Nonetheless, the sky was sufficiently free of clouds that when the steamy atmosphere occasionally held still, he was able to get an unusually sharp view of the planet, which was at that time an uncommonly close 53 million miles from Earth.

Bringing his instrument into focus, Hall began studying the familiar Martian disk with its ruddy color and its darker dapplings, when all at once he was brought up short by something not so familiar. Hovering directly by the world were two vanishingly small objects that appeared to be orbiting it. It was impossible to determine the size of the tiny satellites, but they couldn't be more than ten or twenty miles across, small enough not to be true spherical moons at all, but mere jagged boulders. Shards that small probably had not simply formed alongside their parent world the way the big, properly round moons had. Rather, they were almost certainly asteroids, refugee rocks from the nearby rubble belt that had somehow been knocked into free space by a collision of some kind and then been grabbed up and reeled in by the gravity of Mars. Once part of a true planet, the twin rocks had long ago been broken down to mere asteroids when the world they helped make up had fallen apart. Now, thanks to a few gravitational ricochets and a lucky catch by Mars, they had been upgraded again, orbiting a new planet and earning the status of fully certified moons. Hall named his tiny satellites Phobos and Deimos—or fear and panic—the mythical sons of Mars.

Hall's moons may have been modest things, but their meaning was enormous. If the silhouettes of Phobos and Deimos were so different from those of the other moons that circled the larger planets, their chemistry and history were likely to be different as well. And if huge moons and tiny moons could exist in the same solar system, who knew how many other species of satellites lay between them? Over the next century astronomers began to find out, as other big moons, other tiny moons, and litters of modest midsized moons tumbled through the telescopes. By the second half of the twentieth century, the solar system's satellite census had risen to a robust thirty-four, with one moon circling Earth, two circling Mars, thirteen orbiting Jupiter, eleven orbiting Saturn, five orbiting Uranus, and two orbiting distant Neptune. In the summer of 1977, a hundred years nearly to the day after the discovery of the Martian moons, a team of scientists in a scabby patch of desert in the southern half of California were preparing to launch a pair of spacecraft to visit most of those curious satellites and discover untold others that may lie among them.

4

Fire and Ice

A lot had changed around the Jet Propulsion Laboratory in the five years since the MJS77 program had gotten under way. For one thing the MJS77 program didn't exist anymore. Oh, the spacecraft JPL hoped to launch to the deep solar system were still being built all right, and they were still being launched to Jupiter, Saturn, and who knew where else. But nobody had called the program by the name MJS77 in a long, long time.

Almost from the moment the *Mariner* Jupiter-Saturn 1977 mission was given the congressionally mandated moniker it was, most people around JPL figured it would have to change. First of all, the name was a mouthful. JPL mission planners had long made it a point to give their ships only clean, uncluttered, single-word names—*Explorer, Ranger, Mariner, Viking*—names as simple and stripped-to-the-chassis as the vehicles themselves. This one, however, was a polysyllabic mess.

For another thing, the name revealed too much. While most of JPL's projects had to be approved by administrators and legislators, it was generally the engineers themselves who called all the technical shots, choosing the places they wanted to fly and designing the spacecraft that would take them there. MJS77 had been a different story. From the outset, lawmakers in Washington had had their fingers in this particular mission's pie, using their budget-writing power to pick its destinations, define its objectives, and even dictate the

kind of hardware that would make the trip. Give a group of legislators a chance to plan a project and they'll almost certainly want to
name it. Give them a chance to name it and they'll almost certainly
bungle the job, jamming as much information as they can into the title to prove that the money they're allocating will be well spent. A
name like *Mariner* Jupiter-Saturn 1977 might have told you the
model of the spacecraft, the places it was going, and the year it was
scheduled to leave, but it also told you how the ship came to be—
and in this case it clearly came to be in Washington, just another
public works project brokered in the halls of Congress and approved
in its committee rooms. That, the planetary explorers of JPL figured,
was not the way taxpayers liked to think of their spaceships.

Shortly after the mission was approved, John Casani, the deputy
project manager, held a competition among the JPL staff to see if it
wasn't possible to choose a less unlovely handle for the new spacecraft. To his disappointment, however, the engineers came up empty.
Nomad was a popular choice, but nobody especially cared for the
vague whiff of aimlessness the name carried. *Pilgrim* was another fa-
vorite, but it too had a meandering feel about it. What Casani and
his senior project managers really wanted to call the ship was *Voyager*, but they wondered whether they dared.

Several months before the grand tour was originally proposed,
NASA and JPL had been considering another, equally ambitious mission to nearby Mars. Under that plan, the engineers would stack a
pair of interplanetary spacecraft piggyback-style inside a Saturn-
class booster and launch them into space together, targeting them to
explore entirely different areas of the Martian surface simultaneously. The project, which had tentatively been dubbed *Voyager*, had
an attractive simplicity about it, but ultimately it proved to be *too*
simple. Launching two spacecraft atop just one rocket meant that if
that rocket should blow during launch, both ships would be lost instead of just one. For a space agency that lived and died by the credo
of hardware redundancy, that kind of twin-ship disaster was unacceptable. Eventually, nervous NASA officials scrapped the *Voyager*
project, and when they did, they scrapped the name along with it.

Now, there was another mission getting ready to fly—another

mission in need of a swashbuckling name—and the orphaned *Voyager* seemed like a natural. The JPL bosses, however, who ordinarily were not a suspicious lot, were uncharacteristically skittish about giving their ships a borrowed name from a junked spacecraft that hadn't even been able to get itself off the drafting board and out to the launch pad, much less catapult itself into space. Every time the *Voyager* name came up for consideration, the members of the MJS77 team would fall into an awkward silence, knowing *Voyager* was a perfectly good choice, but knowing, too, that it was a mystically marked one. After a few awkward moments, somebody would come up with a reason to table the decision and the scientists would move on to other matters.

Eventually, however, with the project approved and the space agency press machine preparing for a big publicity push, the NASA bosses began insisting that the project managers pick *something* and be done with it. With nothing else remotely appealing—and none of the JPL staffers willing to admit why they were refusing to use the perfectly suitable *Voyager*—the engineers threw up their hands. There would be two ships built and launched on this outrageous mission to the outer planets, and for better or worse, they would henceforth be known as *Voyager 1* and *Voyager 2*. If fate was going to punish that choice, that was simply a risk the engineers would have to take.

Undergoing an even more important change than the name of the ships was the trajectory they would fly. Of all the planets and moons the twin spacecraft would be sent to visit, the one that intrigued the scientists the most was Titan, the largest satellite of the planet Saturn. At 3,193 miles in diameter, Titan is the second biggest moon in the solar system—after Jupiter's Ganymede—and is, in fact, bigger than two fully certified planets: Mercury and Pluto. What's more, Titan was thought to be an extremely chemically complex moon, dense with ethane, methane, and other hydrocarbons essential to life. A world this big with this kind of biological potential clearly warranted a good, close look. The problem was, actually getting to Titan was not going to be an easy matter.

In order to make the most thorough study of the moon possible, it wouldn't suffice for the *Voyagers* merely to fly by it on a horizontal trajectory that would take them past Titan's equator. Rather, it was necessary to make a steep, ascending approach, climbing vertically alongside the world in a trajectory that would allow a spacecraft to soar over both Titan's south and north hemispheres, while at the same time getting a glimpse of at least parts of its eastern and western halves as the moon rotated obligingly beneath it.

A *Voyager* attempting to make such a sharp south-to-north turn over a moon could not rely on its comparatively feeble engines alone. Instead, it would have to fly below the south pole of Titan and allow the gravity of the moon to grab it and flick it northward, helpfully tossing it along just the trajectory the JPL planners wanted it to follow. As powerful as Titan's gravitational throw would be, it would be impossible for the spacecraft to reverse course after its flyby of the moon was over and resume its horizontal trajectory past the planets. Instead, it would simply soar up and out of the solar system, sailing off forever into empty space. For a ship that was only going to visit Jupiter and Saturn, this would not be such a bad thing. For a ship whose builders were still harboring secret dreams of continuing on to Uranus and Neptune, however, this meant trouble.

For this reason, the JPL planners decided to send *Voyagers 1* and *2* on two different trajectories. *Voyager 1*, it was agreed, would be launched on a relatively speedy trajectory that would take it to Jupiter and Saturn, with special emphasis on the close Titan flyby, after which it would conclude its mission. *Voyager 2* would similarly be targeted for both Jupiter and Saturn, but would fly a slower route that would carry it past both worlds in the flat, allowing it to maintain a horizontal trajectory that would give it a far poorer look at Titan but at least a fighting chance at Uranus and Neptune. If *Voyager 1* for some reason failed in its Titan flyby, *Voyager 2* would be retargeted en route to make a second attempt at the giant moon. Such a course change would, of course, send *Voyager 2* out of the solar system, too, ruling out a visit to any worlds beyond Saturn. But so alluring was the biologic promise of Titan that the JPL scientists figured it was worth the risk.

Even as the *Voyager* mission's flight plan was changing, its trailblazing status was, too. Ever since the grand tour was proposed in 1966, JPL had been boasting that its ships would not only explore the outer solar system, but that they would be the *first* to do so, flying out to Jupiter and Saturn and perhaps Uranus and Neptune before either the Russians or any other NASA lab had even made it beyond Mars. But no sooner was the mission approved and the funding in place than two other spacecraft beat JPL to the punch.

In March 1972 and April 1973—years before the *Voyagers* were even scheduled to leave the ground—the aptly named *Pioneer 10* and *Pioneer 11* blasted off the pad in Cape Canaveral and into space toward their own planned rendezvous with Jupiter and Saturn. The ships were not designed and built by one of the grand NASA facilities in Houston or Florida or Pasadena, but rather, by the comparatively obscure Ames Research Center in Moffett Field, California. Ames, like JPL, was an old facility, having been in operation for years before joining the NASA family in 1958. Unlike JPL, however, Ames did not begin its life as an academic institution, but rather as a military one, serving as a research base and a training ground where new airplanes were developed and tested.

With its more modest scientific credentials, Ames was never considered one of NASA's frontline labs and was rarely given the glamorous mission-planning and spacecraft-building jobs JPL was. Instead, the Ames scientists were assigned the lower-profile job of conducting basic research and testing basic hardware, helping to support the more celebrated NASA facilities that were busy doing their more celebrated work. It thus came as something of a surprise in the early 1970s when Ames announced that it was planning to strike out on its own, building a pair of brand-new spacecraft and flying them to the same two planets that were the *Voyager* mission's prime targets.

The ambition—to say nothing of the presumptuousness—of the project took the space community by surprise. For all its derring-do, however, the *Pioneer* project was a relatively simple one, mostly because the spacecraft themselves were simple, too. The ships Ames was building were spindly little things, weighing barely 570 pounds apiece. Lightweight ships could carry only lightweight hardware, and the *Pioneer*s would not be carrying much at all—a camera, a few Geiger counter–like sensors, a radio, and a computer. A putt-putt spacecraft like this could be built on the cheap, and when the Ames researchers presented their plan to NASA and Congress, the administrators and lawmakers quickly approved it. If the *Pioneer*s failed, there'd be little in the way of resources lost; if by chance they succeeded, much would be gained scientifically. Meanwhile, JPL could continue working on its own, more ambitious project scheduled for later in the decade.

Though most of the engineers at JPL wouldn't admit it, they were only too happy to let Ames be the first to dip a toe into deep space. Traveling to the outer solar system meant traveling through the asteroid belt, a shooting gallery of flying rocks that could easily pulverize any ship that tried to pass through it. No spacecraft had yet attempted the tricky piloting that would be necessary to negotiate this region, and if the Ames planners were volunteering to run a little reconnaissance, the JPL planners would not argue.

Just as menacing as the environment between Mars and Jupiter was the environment in the immediate vicinity of Jupiter itself. Astronomers had always suspected that a world as big as Jupiter emitted a powerful—even lethal—radiation field, one that could poison a spacecraft's instruments before it could take so much as a single reading. No one knew exactly how toxic the radiation field was, but once again, if Ames was willing to be the canary going down into that particular coal mine, JPL would not stand in its way.

When *Pioneer*s 10 and 11, against all predictions, were successfully launched, and then, against all odds, actually arrived at Jupiter in 1973 and 1974, the news they sent back was both good and bad. The asteroid belt, it appeared, was a navigable place. When the ships reached that region of the solar system, they found that the population of rocks was even greater than expected, but the stretch of space in which they moved was so vast that an object as small as a spacecraft could pick its way through with little danger.

If the asteroid belt was less perilous than feared, however, the radiation field around Jupiter was far more so. Even when the two ships were nearly four million miles from the giant planet, their

Geiger counters began to twitch off the high end of the scale. The closer they got, the more punishment they absorbed, until the detectors themselves, as well as several other minor instruments aboard the spacecraft, became saturated and failed. At the ships' closest approach to Jupiter they were absorbing a radiation dose of about 450,000 rads, or about a thousand times the 450-rad level that is usually considered fatal for a human being.

After the ships' sizzling encounters with Jupiter, *Pioneer 10* was steered out of the solar system altogether, and *Pioneer 11*—or what was left of it—was sent on a lazy trajectory to Saturn that would allow it to reach the planet by sometime in late 1979. Ames, abiding by the compulsory esprit de corps NASA expected of its labs, shared all of its Jupiter findings with the scientists at JPL, who thanked their fellow engineers for their data, and then promptly put the information to use, adjusting their asteroid-belt flight path to follow *Pioneer*'s own, and reinforcing their spacecraft with radiation shielding that would prevent the ships from growing sick the way Ames's had.

The most significant change of all around JPL in the last few years, however, concerned not the spacecraft the lab built, but the people who built them. In 1976, William Pickering at last retired from the Pasadena lab he helped establish, deciding to go to work in the oil industry and, for the first time, earn the kind of private sector money his long-ago Caltech education entitled him to make. Stepping down at the same time was Bud Schurmeier. Chosen to head up the MJS77 flights almost the moment they were approved in 1972, Schurmeier had been working steadily on the project ever since. Half a decade was generally considered more than enough time for any one manager to oversee any one mission, and not long before the first scheduled launch of the first *Voyager*, Schurmeier was rotated out of the manager's spot, assigned to other JPL projects, and replaced by his assistant, John Casani. The cautious and conservative Schurmeier had been just the right man to steer the *Voyager* project from its planning stage through its development stage and up to the moment of launch. Casani, who from the very beginning had wanted to take the whip hand to the *Voyager*s and gallop them as deep into the solar system as their off-the-shelf parts could take them, would be just the right man to send them to the planets.

On August 20, 1977, at 8:56:01 A.M. Cape Canaveral time, the first *Voyager* interplanetary probe left its launch pad on the east coast of Florida and soared into space, carried atop a 159-foot Titan-Centaur booster. On September 5, 1977, at 10:29:45 A.M., the second Voyager followed. It took only minutes for the ships to reach Earth orbit and only an hour more for them to leave that safe, circular path and head out toward the stars. The journeys ahead of them would last years.

•   •   •

Scientists didn't know much about the moons of Jupiter in the months before the pair of *Voyager* spacecraft arrived at the planet. They knew their names, they knew their approximate size, and they knew—more or less—where to find them. For researchers hoping to take the census of the entire solar system, that wasn't much, but such fragmentary data at least gave them a place to begin their studies.

Io, Europa, Ganymede, and Callisto—the four moons Galileo had discovered 367 years before the *Voyager* launches—were by far the largest of the Jovian satellites. Measuring from 1,945 to 3,262 miles in diameter, they stay relatively close to the planet, flying in circular orbits that range from a low of 261,000 miles for Io to a high of 1.2 million miles for Callisto.

Far closer to Jupiter is tiny Amalthea. An irregularly shaped moon discovered in 1892, Amalthea measures only 117 miles across and flies at a cloudtop-grazing altitude of just 112,000 miles. Smaller and much more distant is an additional cluster of four tiny moons—Leda, Himalia, Lysithea, and Elara—discovered by different astronomers at different times throughout the twentieth century. Orbiting in reasonably close proximity to one another about 7 million miles above the surface, the moons measure from just 10 miles in width to a somewhat more substantial 115 miles. Farther still—some 15 million miles above the Jovian surface—is another four-moon cluster, Ananke, Carme, Pasiphae, and Sinope, also discovered one at a time throughout the twentieth century and also orbiting relatively close to one another. These four moons are even tinier than the lower cluster—measuring between 19 and 31 miles across—and circle the planet in a highly eccentric orbit, inclined about 105 degrees from Jupiter's equator. In addition to their sharp angle, the orbits of Ananke, Carme, Pasiphae, and Sinope are backward, following a clockwise path around the planet, unlike virtually all of the other bodies in the solar system, which orbit counterclockwise. How many other undiscovered moons there might be swarming among Jupiter's known litter of thirteen was impossible to say.

For *Voyager* scientists hoping to explore the Jovian system, the answers to these kinds of questions would come from the outside in. Approaching the planet and its moons would be like flying into a great metropolis. The spacecraft would pass the planetary provinces first, the deep-space wilderness where little was evident of Jupiter but the lines of magnetism and the waves of charged particles it gave off—the smokestack exhaust of a world still too far away to see clearly. Next would come the fiefs and duchies of the smaller outlying moons, then the hulking Galilean moons, and last the great body of the planet itself.

During the first week of 1979, close to a year and a half after the *Voyager*s had been launched, *Voyager 1* drew within 40 million miles of Jupiter, a distance at which the planet at last swam milkily into view. On January 4 the signal went up to the ship ordering it to open its long-shuttered eyes, switch on a pair of high-resolution video lenses—one built for close-ups, one built for panoramic shots—and beam back to Earth an image of what it was seeing. At first, what it saw was not much at all—a vaguely orange saccharin tablet of a world in which cloud patterns were visible, but just faintly; big moons were visible, but fainter still; and small moons were not discernible at all.

At just 40 million miles from the planet, however, *Voyager 1* needed to travel only another 20 million miles—a tiny fraction of the nearly 400 million miles it had covered so far—to halve its distance and double its resolution; at 20 million miles it had to travel just 10 million to halve it again; at 10 million miles another 5 million would do it. In steadily dwindling increments, therefore, the ship's eyesight grew better and better, and the improvements came faster and faster.

Even before the quality of the pictures the spacecraft could return had sharpened appreciably, the number it could return was already impressive. With its high-performance cameras gathering even the faintest shimmer of Jovian light and its cannon of an antenna blasting data back home, the ship was capable of taking and transmitting one complete picture every forty-eight seconds. This meant a lot of pictures indeed in the four months Voyager 1 would spend in camera range of the Jupiter system—two months shooting straight on as it approached the planet and two months looking back over its shoulder as it passed.

As the pictures streamed in during those initial two months, the atmosphere changed palpably at JPL. The conference rooms and other common areas across the sprawling lab had long ago been fitted with closed-circuit televisions hung from the ceilings in heavy metal brackets. Most of the year these monitors carried nothing of much interest—time and weather readouts usually; perhaps a status report of an upcoming launch. Now, however, the screens fairly shimmered. From early in the morning to deep in the night, ever bigger, ever crisper pictures of Jupiter and its marble bag of multi-colored moons flashed and flashed and flashed again. In hallways and reception areas, knots of scientists and visitors would gather to watch as the newest images resolved themselves on the monitors. In the employee cafeteria, lines slowed and food service stopped as screens that only moments before had been sizzling with static suddenly began to flicker to life with the latest image of the giant planet 400 million miles away.

Though the Voyagers were designed to send home a wealth of other scientific data from the planets they visited, it was, not surprisingly, these cosmic portraits that caused the greatest sensation around JPL. And it was, also unsurprisingly, the technicians from the imaging team—the people who received the incoming signals from the ships and helped transform them from a mere electromagnetic data stream to crisp and comprehensible pictures—who earned the most applause from their Pasadena colleagues. Just as interested in the pictures that were coming back from the Voyagers, however, was another group of JPL investigators: the obscure and uncelebrated members of the *Voyager* navigation team.

There were few jobs in the entire JPL organization that were seen as less glamorous than the ones held by the scientists on the navigation team, who spent their days studying sky maps, calculating trajectory arcs, and scrutinizing hundreds, often thousands, of numbingly similar pictures the spacecraft took of star fields, in an attempt to confirm that the ships were pointed the right way. The environment in which this grindingly tedious work was done was spartan at best. With the exception of a few section chiefs and overseers, nearly all of the *Voyager* navigators were assigned to one end of the second floor of JPL's Building 264, a warren of open offices and cubicles tossed together in an overloud, overlit, utterly public jumble.

When the *Voyager* images were returning to Earth, the buzz in this noisy corner of the lab often grew overwhelming. Daily—sometimes it seemed hourly—the grand investigators from the grand imaging team would arrive with another stack of pictures for another busy navigator, wondering if it might be possible to chart this or that course or this or that coordinate from the sleet of static that flecked most of the pictures.

One day during the Jupiter phase of the *Voyager* missions, Steve Synnott, a member of the navigation team, was laboring over some images when he was approached with a slightly different request by Dave Jewitt and Ed Danielson, two Caltech researchers who were serving as consultants on the *Voyager* project. Earlier in the day Jewitt and Danielson had been studying a badly overexposed *Voyager* image and had noticed something strange. When a picture was overexposed, it meant that the camera shutter had been left open too long, causing things to run and blur. In the background of this particular picture, things had blurred indeed, so much so that the pinpoint-like stars the camera caught in its frame had all turned into tracer-like streaks. This in itself was not unexpected, but there was a detail in the picture that the imaging scientists didn't anticipate: One of the streaks, it appeared, was moving at an angle that was completely different from the rest of the star field. Since all stars remain fixed relative to a moving spacecraft, they should distort in the same direction. The fact that one was diverging from the others meant it couldn't be a star at all.

What Jewitt and Danielson wanted to know was what it was. The least exciting possibility was that the streak was nothing truly meaningful—a stray asteroid moving through space, perhaps, or an imaging glitch in the JPL computers. In the alternative, however, it might be something decidedly more important: a tiny, new, 10-mile-wide moon to add to Jupiter's already considerable brood.

If the object was indeed a moon, it ought to be in a calculable orbit, but there wasn't much in the image that would allow even a seasoned astronomer to determine what that orbit could be. The entire streak covered only about 40 pixels on the 800-pixel screen, and the entire screen represented only about 5 percent of the theorized moon's theorized orbit. Would it be possible, Jewitt and Danielson wondered, for Synnott to extrapolate the angle, altitude, and shape of the satellite's path around the planet from just that scrap of data and determine if it was in fact a satellite?

Synnott acknowledged that yes, such calculations might indeed be possible and yes he'd be willing to try. The job, he knew, would be a thankless one, since even if he did confirm Jewitt and Danielson's moon, the Caltech researchers would be the ones credited with its discovery. Nonetheless, he took on the assignment, deciding to tackle it only on weekends, when it wouldn't interfere with his other, more pressing work.

One Sunday evening not long after he first received the picture, Synnott was sitting at his desk in the uncharacteristically quiet navigation department, bent over a stack of Jupiter pictures, looking for more evidence of Jewitt and Danielson's maybe-moon. Giving one particularly vivid image a quick scan and preparing to slide it off his still-to-be-studied pile, he noticed something that brought him up short. There, at the edge of the image, small enough to be overlooked if you weren't trained in the business of *not* overlooking such things, was a pair of small black specks, ones that were nowhere near the likely location of the fleck he was looking for. The specks were situated directly over the central girth of Jupiter, directly over its equator, in fact, in a spot where the swirly orange-red mottling of the planet's atmosphere gave way to a cleaner all-orange band, making it far easier to pick out an object that might be flying overhead. Synnott snatched up another image and the dots were there again, still in lockstep with one another, but now in a different position relative to the cloud formations. A third picture revealed the same thing, but this time the objects had moved further still.

Synnott gulped, gaped, and had a sudden impulse to stand up—which he did for no reason at all. He then sat down again and smiled broadly. What he was seeing, he knew, was not an image of two bodies at all, but one: a Jovian moon sailing over the planet and casting its shadow on the cloudtops beneath it. That moon and that shadow were moving fast, he could tell, tearing along in an orbit that was far brisker than the leisurely flow of the ever-changing atmosphere below. What's more, the body and its shadow were relatively big, anywhere from 60 to 120 miles across, far bigger than the 10-or-so-mile body Jewitt and Danielson believed they had found, and located in an entirely different patch of Jovian sky. If this was indeed a moon, it might well be a brand-new one, and if it was a new moon, it belonged to Synnott alone.

Before Synnott could claim the discovery, he had to make sure that the satellite he had just spotted was not one of Jupiter's thirteen existing ones. If it was, it would most likely be the little, low-flying Amalthea. This innermost of Jupiter's moons, discovered nearly ninety years earlier by astronomer Edward Barnard, was more or less the same size as Synnott's theorized moon and orbited, as nearly as Synnott could tell, at more or less the same altitude.

For the next two weeks, Synnott continued to find his moon in picture after picture returned by the *Voyager*s, calculating its size and orbit more and more precisely. As he did, he became more and more elated. This was not the 117-mile-wide Amalthea, sailing along in its 112,000-mile-high orbit; and it certainly wasn't the little pebble of a moon Jewitt and Danielson thought they had seen. This was a body sixty miles wide, sailing through an orbit roughly 138,000 miles up. This was going to be Synnott's moon after all.

Within the week Synnott had amassed all of the pictures and all of the data he needed to make his claim, and had set them aside, waiting for the next time Ed Stone, the chief project scientist for the *Voyager* program, made one of his frequent walking tours of the *Voyager* bullpens. When he did, Synnott flagged him over to his desk and handed him a typewritten draft of a one-paragraph letter he was about to send to the International Astronomical Union in Cambridge, Massachusetts. The IAU, as anyone in the astronomical community knew, was the group you contacted if you had a new heavenly body to announce, and a one-paragraph letter was how you announced it. While astronomers everywhere were aware of what such a letter should look like, few of them ever got the chance to write one themselves. Stone read Synnott's single paragraph wordlessly and then looked up at him with just a ghost of a smile.

"Do you know its orbital period?" Stone asked.

"About eighteen hours," Synnott answered, handing Stone a sheet of figures.

"Its size?"

"About sixty miles," he said, handing over a small stack of especially sharp photos.

"Altitude?"

"138,000."

Stone glanced at the letter again and gave the pictures and the page of calculations a practiced scan. After a long moment he looked up and allowed the ghost of a smile to bloom into a real one.

"Well," he said with a good-natured shrug, "it looks like you've found yourself a moon."

Later that day Synnott sent off his letter, and a short while later the IAU responded, congratulating him on his discovery and sending along a list of possible names for his moon. Synnott studied the list and, at length, chose the name Thebe, after an ancient mythical nymph. Before long he found that he had cause to consult the same list again. Wading through more *Voyager* images, he found another, more modest, 25-mile moon orbiting 79,360 miles below his newly christened Thebe. This newer, smaller satellite he chose to name Metis, after the first wife of Jupiter. Shortly after that, he at last confirmed the existence of Jewitt and Danielson's moon, a 12-mile pebble of a thing orbiting just 79,967 miles up. The two Caltech scientists named their single satellite Adrastea, after the daughter of Jupiter.

The moon total of the largest known planet in the cosmos had now risen to sixteen, and not long after, the two *Voyager* spacecraft that had discovered the new worlds glided out of the Jovian system forever. Before the ships were gone, however, Linda Morabito, one of Synnott's colleagues in the navigation department, made one final and explosive discovery. Sitting down with a stack of overexposed images like the ones Synnott had been studying, she noticed a huge and curious bulge rising over the horizon of the big moon Io. The bulge could not be a mountain, she knew—it was just too big for that. Nor could it be another moon hiding behind Io; a body that large would long since have been spotted from Earth. The bulge, she knew, could only be a volcano, a mammoth surface eruption sending a vast cloud of exhaust high into space. Other pictures revealed other bulges indicating other eruptions elsewhere on the surface—as many as nine at any one time. Io, a world that had long been thought to be inert as a stone, was, in fact, fairly exploding with volcanoes, the only body in the solar system other than Earth known to be geothermally active. Morabito might not have spotted a moon of her own, but as scientific discoveries went, hers was even more significant than anything even Steve Synnott had achieved. Learning that the Jovian system was a crowded one was one thing; learning that it was a geologically living one was something else entirely. Closer study of the *Voyager* images would reveal just how alive the Jupiter family as a whole might be.

•   •   •

When the *Voyager* scientists at last had the time to analyze their portfolio of Jupiter pictures, one of the first things that caught their attention was neither the planet itself nor any of its moons, but another, related formation: a fine set of rings. The seventeenth-century astronomers who first turned their telescopes toward Jupiter's sister world, Saturn, were struck as much by the size and color of the planet itself as by its vast ring system. The bands were broad, bright, ostentatious things, more than twice as wide as the diameter of Saturn proper. Indeed, so vast and opaque were the rings that they actually appeared to obliterate the very stars that twinkled behind them, clearing a blacked-out zone ahead of and behind Saturn as the hoop-skirted world moved from one place in the sky to another. Had Saturn's rings been more tenuous things—straggly formations that were closer to mere strands than bands—they might have been utterly invisible from Earth; when they passed in front of the stars, however, they still would have blotted at least some of them out. When twentieth-century astronomers turned their attention toward Uranus and Neptune, they noticed precisely this kind of star-snuffing—which they called occultation—and concluded that the supposedly naked worlds were probably decorated by extremely faint rings of their own. No matter how powerful the telescopes, however, nobody had found a trace of even the faintest bands around Jupiter.

Voyagers 1 and 2 changed all that. As the spacecraft closed in on the planet, the relatively bright sunlight pouring in from over the ships' shoulders illuminated the world and its moons but bleached out anything smaller and fainter. When the spacecraft moved past Jupiter, however—placing themselves in a position that allowed the planet to blot out the sun—and then looked back at where they had just been, the world looked entirely different. Surrounding the now-shadowed sphere was an eerie, backlit nimbus that shrouded the planet like a luminous cloud. And within that cloud was a distinct set of nested rings.

Unlike Saturn's rings, a dense river of rubble and ice, Jupiter's rings were hopelessly tenuous, little thicker than the airborne dust that appears in the shaft of light pouring through an uncurtained window. When that dust forms itself into a circle and begins to orbit a world with a 279,000-mile waistline, however, even so faint a formation can look dramatic indeed. The *Voyagers* discovered at least three—and perhaps four—distinct rings around Jupiter, the biggest orbiting more than 35,000 miles above the planet, the smallest practically grazing its cloudtops.

Astronomers could not be certain what was responsible for Jupiter's bands—and all other planetary rings for that matter—but for more than a century they had had a pretty good idea. In 1850,

French astronomer Edouard Roche marshaled what knowledge he had of gravity, orbital dynamics, and lunar distances, and devised a formula for moon formation that became known, unsurprisingly, as the Roche limit. Assuming a planet and a moon have the exact same densities, Roche calculated that the moon could never safely move any closer to the planet than a distance equal to about 2.4 times the planet's radius. Slip too far inside the limit and the planet would tear the moon apart and drag the resulting debris down to the surface. Slip less deeply inside the limit—just barely cross the gravitational line—and the planet would still pulverize the moon, but the rubble that remained would be able to stay aloft, dispersing around the planet and transforming itself into a ring. As the relative densities of the planet and moon changed, the Roche limit would move in or out accordingly, but the satellite-shattering principle behind it would remain the same. In the case of Jupiter, the tenuous, planet-hugging nature of the newly discovered rings suggested that they had been formed by a small, low-flying moon, one that was no different from all of the other small Jovian satellites except that it had allowed itself to be grabbed too tightly by Jupiter's gravity fist and had paid a dreadful price.

"The stronger a moon is, the closer it has to be for the Roche limit to break it apart," said Rich Terrile, a planetary scientist at JPL. "Apply enough gravitational pressure and you can overcome the strength of any material. A moon inside the Roche limit experiences just that kind of pressure, and once it breaks apart, it all at once finds itself living in a world in which it can't put itself back together. Planetary rings are sometimes thought of as nothing more than the grave markers of moons."

Simply because one of Jupiter's small satellites was murdered by its parent planet, however, did not mean the others couldn't somehow survive, and plenty of them did. Of all of the undersized satellites, it was the innermost group of four—Metis, Adrastea, Amalthea, and Thebe—that intrigued scientists the most, particularly when they tried to figure out how the tiny satellites came to be there at all. The best guess for most of the astronomers was that they came from the asteroid belt between Mars and Jupiter. The more time scientists spent studying the giant debris field, the more they came to appreciate just how mobile the rocks that populate it could be. In addition to Phobos and Deimos, the two small Martian moons that appeared to have migrated out of the asteroid belt, there were at least two other clusters of belt rubble that seemed to have drifted away from the main stream. Those asteroids—which astronomers had dubbed the Trojan asteroids—had not gone into orbit around a planet the way Phobos and Deimos had, but into orbit along *with* it, taking up positions precisely 60 degrees in front of Jupiter and 60 degrees behind it and constantly maintaining those distances as the giant world inscribes its own orbit around the sun.

If both the Trojan asteroids and the Martian moons had come from the Mars-Jupiter debris belt, there was no reason that Metis, Adrastea, Amalthea, and Thebe—with their asteroid-like size and shape—wouldn't have originated there as well. If that was indeed their source, they could be carrying with them some revealing cosmic chemistry.

Meteorites falling to Earth are themselves often assumed to have originated in the asteroid belt, if only because the relative proximity of such a huge reservoir of free-floating debris makes it the statistically likeliest source. Analyzing the hot rocks when they do land, geologists have found that they fall into one of two chemical categories. The first group is far and away the more primitive one, made of some of the most ancient cosmic stuff imaginable. Superficially composed of iron and silicates, the samples also contain microscopic diamonds and tiny beads of a simple, primitive species of carbon. Such a composition is consistent not with the commonly accepted chemistry of planets, but with the commonly accepted chemistry of stars, and suggests that the meteors were cooked up long ago in some primal stellar oven—almost certainly the solar system's own sun. Indeed, so rich are the rocks in carbon and other primitive materials that they give off vapors of the stuff, tiny breaths of highly volatile compounds that even after billions of years still evaporate from the meteorite in pungent wisps. Astronomers with such an ancient sample in hand often amuse one another by dropping the rock into a jar, screwing on the lid, and coming back a few hours later to take a whiff of what's inside. The vaguely solvent-like scent, they remind one another soberly, is the oldest smell in the universe.

Ancient as such artifacts are, they are surprisingly easy to destroy. One way to annihilate them is to cook them down into a planet. As the gas, dust, and rubble that make up a fetal world begin to accrete, the gravitational power of the growing mass dissipates as heat, causing the world as a whole to reduce itself to a soft, hot, lava-like ball, much like early Earth and its early moon.

When bodies melt down this way, they undergo a dramatic transformation. The fragrant volatiles that so entertain contemporary scientists immediately boil away, unable to tolerate temperatures of even a few hundred degrees, never mind thousands. More important, the anatomy of the world itself begins to change, separating into layers as the heavier metals in the rocks precipitate toward the center of the body and the lighter ones remain on top. If a mass like this somehow fragmented back down into asteroids, each individual fragment would no longer be made of the mix of fundamental materials that characterize older, unmelted samples, but rather, of a single heavy element like nickel or iron, depending upon where in the planet's interior the rock originated. The second type of meteorites geologists recover generally have this type of homogenous composition. Since both types of meteorites routinely fall to Earth, it is likely that both types populate the asteroid belt. And if the rocks in the belt are made of this kind of material, rocks that wander out and become moons, like Metis, Adrastea, Amalthea, and Thebe, should, too.

That was the theory, anyway. But while the *Voyagers* could photograph the four inner satellites, the small size of all of the bodies and the fleeting look the spacecraft got at them made it impossible for the ships' chemical and light sensors to determine their makeup conclusively. The scientists could thus make all the inferences about the innards of Metis, Adrastea, Amalthea, and Thebe they wanted, but for now at least, they were powerless to confirm them. Mysterious as the guts of the inner four moons remained, however, the *Voyager* cameras were able to learn a lot about their surfaces.

Sixty-mile-wide Thebe, the highest flying of the four inner satellites, appears to have a lot in common with Earth's own 2,100-mile moon, at least when it comes to its history of meteor bombardment. One side of Thebe is scarred by three or four large impact craters, all nearly big enough to have fractured the moon entirely. Worse for the fragile satellite, the side that suffered the bombardment is the leading side—the one that faces forward as the moon moves through its orbit—meaning that Thebe took all of its biggest blows head-on. Had Thebe been able to rotate freely, the wounds it suffered would have been distributed more evenly across its surface. Like Earth's moon, however, the little Jovian satellite keeps one of its hemispheres pointed toward Jupiter at all times, locked in an orbital grip from which it can never turn away. Indeed, so tight is the gravitational hold Jupiter has on Thebe that the side of the moon that faces the planet actually bulges slightly. This small protrusion causes the mass on that side to increase, which in turn causes Jupiter's gravity to hold it even tighter, making it even harder for the moon to turn its face.

Metis and Adrastea, much smaller than Thebe, do not appear to have suffered the same kind of bombardment their larger sibling has, their modest size evidently allowing them simply to avoid many of the blows they otherwise would have absorbed. But while the two moons are not so remarkable for how they look, they *are* remarkable for how they behave. Lying even closer to Jupiter than Thebe does, Metis and Adrastea are deeper inside the planet's Roche limit, meaning that by rights they ought not to be there at all, having long since been crushed down into ring rubble. The fact that the moons do exist suggested to the *Voyager* scientists that the Roche limit rule has its exceptions and that its powerful gravitational hammer can be survived, provided that a moon orbiting a planet is small enough and sturdy enough. In the same way a large, creaking, multi-floor home will collapse in the face of an earthquake while a small stone cottage may emerge undamaged, so too will a relatively large moon—with all its flexing and churning and structural straining—suffer worse Roche damage than a static, asteroid-like rock.

But the mere act of surviving the worst the Roche limit has to dish out does not mean Metis and Adrastea are entirely unaffected by it. The moons' positions so close to Jupiter place them in direct proximity to the planet's rings—the presumed remains of a moon that hasn't fared so well gravitationally—and that has a curious effect of its own. Given the dusty nature of ring particles, any bands they form should be sloppy things, swirling at their centers and disintegrating at their edges before dissipating into space altogether. The bands around Jupiter, however, appear to be surprisingly sharp—tidy circles that maintain a relatively constant thickness across their entire width until they reach their outer rims, where they stop as cleanly as if they had been cut away by a blade. It is just beyond those rims that Metis and Adrastea are found.

Clearly, it appears, the two small moons act like a pair of orbital shepherds, moving constantly at the perimeter of the rings and rounding up stray particles that manage to escape. If Metis and Adrastea weren't there, the rings would probably still exist, but they would almost certainly be much messier, much more tenuous things, far less lovely for the loss of their two tiny attendants.

The last of the innermost moons, Amalthea, appears to be a far more passive player in the Jovian system, one that has not done much to affect the bodies around it, but has clearly been affected by them. Lying between Adrastea and Thebe, Amalthea is smaller than the Earthly island of Sicily and just as irregularly shaped, yet includes topographic formations seen nowhere on the distant Earth. *Voyager* scientists studying images of Amalthea spotted a sixty-two-mile-wide, five-mile-deep crater, which they dubbed Pan, and a second crater that wasn't quite as wide, but plunged twice as deep. Nearby stand two looming mountains, Mons Lyctas and Mons Ida, that appear to have been dug out and thrown up by the meteorite hits that created the vast pits. Soaring 12.4 miles above the crater basins, the mountains are more than twice the height of Mount Everest, a fact that is all the more impressive since the moon itself is more than sixty-eight times smaller than Earth. On a body this small, a mountain this big doesn't serve merely as topographic decoration, but as a sort of gravitational keel, helping to position the moon so that, as with Thebe, one face of the jagged world remains forever turned toward Jupiter.

More striking than the size of Mons Lyctas and Mons Ida is the color. Portions of the mountains' flanks—as well as the valleys and plains that lie below them—appear to be covered in an unmistakable dusting of ruddy red. Unlike Mars, the only other solid red body in the solar system, Amalthea doesn't appear to be uniformly colored. Rather, its surface is blotchy and uneven, almost as if the red did not originate on the moon, but instead was deposited from somewhere else. The *Voyager* astronomers believed they knew where that somewhere else was.

Barely 149,000 miles above the orbit of little Amalthea is the orbit of giant Io. On the scale of the Jovian system, 149,000 miles is just around the block, and for Amalthea that kind of proximity has real meaning. Linda Morabito and the other JPL scientists discovered a wealth of active volcanism on Io, and with the moon's low gravity, the red sulfur and silicate smoke emitted by the massive vents in the surface was unlikely to settle back down onto the moon. Instead, the emissions would simply be blown into Io's sky and permanently expelled into space. Chugging along downwind of Io, Amalthea would be unable to avoid the bigger moon's exhaust and—like a tugboat following a giant steamer—would soon be covered with the stuff.

"Io's volcanoes drive material up to enormous altitudes," said Torrence Johnson, another JPL planetary scientist. "When you take away atmosphere, the vent acts like a rocket nozzle, blasting the material much higher than it can go on Earth, and allowing it to settle on anything it might find in space."

The relationship between Io and Amalthea was far and away the most dramatic discovery the *Voyager* scientists made about the four inner Jovian satellites. The fact that they were able to learn even this much about moons as small and photography-defying as Amalthea, Thebe, Metis, and Adrastea surprised even them. When the *Voyager*s peered farther out into the Jovian system, toward the spots where the tiny, far more remote moons Leda, Himalia, Lysithea, Elara, Ananke, Carme, Pasiphae, and Sinope orbit, the scientific yield was, predictably, far less. Such distant, pebbly moons were completely beyond the reach of the *Voyager*s' sensing instruments, and even the spacecraft's cameras, built for just this kind of long-range snooping, were unable to return images that looked like much more than cosmic flyspecks. Nonetheless, the position and grouping of the specks was clear, with the eight moons clustered together in their characteristic flocks of four, and that at least told the scientists something.

When rubble gathers together so high above a world as these moons do, it probably did not simply accrete there, congealing out of the same cosmic raw material that formed the planets and the big moons. Rather, the rocks are likely the remains of deep-space asteroids, bodies that flew in from the cosmos beyond Jupiter, got nabbed by the planet's gravity, and broke into pieces as they slowed down and dropped into orbit. The highly eccentric paths of all of the moons in the two clusters seemed to confirm this caught-on-the-fly scenario: Leda, Himalia, Lysithea, and Elara circle the planet at a 28-degree tilt, an orbital angle that is among the sharpest in the solar system. Ananke, Carme, Pasiphae, and Sinope, with their 105-degree inclination and their backward orbit, are more dizzying still. The composition of all eight satellites is probably similar to that of the more primitive, star-baked asteroids that fall periodically to Earth since the moons would have originated in the distant reaches of the solar system and thus would not likely be forged inside a planet that later fractured. Once again, however, given the tiny size of the worlds and the distance at which the spacecraft flew by, the JPL team could speculate about these matters, but *only* speculate.

The *Voyager* spacecraft's scientific haul was much greater when the ships turned their cameras—and the investigators turned their attention—to the four large Galilean moons. Of all the data Earthbound scientists wait for when a spacecraft is flying by a large extraterrestrial body, there is perhaps none they wait for more anxiously than a reading of the body's density. Density is the first important indicator of the world's composition, which itself helps determine chemistry, and from chemistry, everything else flows. Unlike radiation or temperature or magnetism readings—all of which can be read only with sophisticated on-board scanners and sniffers—density measurements can be taken with little more than the spacecraft's own navigation system. Fly by a world with a gravity field powerful enough to pull hard on your ship, and you can be pretty sure you've flown by a dense world, one rich in metals and rocks and other heavy elements. Fly by a world that barely flicks your ship, and you've flown by one made of far fluffier stuff. Analyze those gravitational readings carefully enough, add a few magnetometer and spectrometer readings, and you can infer the anatomy of a large planet or moon as accurately as if you were down on its surface drilling.

When the *Voyager*s flew by the outermost of the four large moons Galileo had discovered—Callisto—the scientists were surprised to find that the ships seemed relatively unperturbed by the encounters. Though the moon measures 2,976 miles across, about the same size as the planet Mercury, it appeared to have only a third of Mercury's mass. While Mercury itself had long been known to be a sort of cosmic ball bearing—made up largely of iron—Callisto appeared to be only 60 percent metal and rock and the rest ordinary water ice. Moreover, unlike Earth and its own moon, whose interiors are stratified into mantles, crusts, and cores, Callisto appears to be a relatively homogenous place, with its rock and ice stirred into a well-blended mass that changes little until the very center of the world, where a small, rocky core might or might not lie buried.

For a moon with so lightweight a makeup, Callisto has survived a lot over the eons. The *Voyager* cameras revealed almost no patch of the Callistan surface that wasn't covered by craters. There were sea-sized craters and divot-sized craters, craters overlapping craters and craters within craters. There were even craters that formed in long chains—one stretching nearly four hundred miles—apparently created when an incoming asteroid was ripped apart by the gravity of nearby Jupiter, causing it to strike the surface of Callisto in a sharply angled, machine gun–like tattoo of small projectiles. A crust in which so many scars have been so perfectly preserved is a decidedly ancient surface, and the *Voyager* scientists concluded that the Callistan skin probably dates back to the creation of the very solar system, 4.5 billion years ago.

But while Callisto's craters are old, they are apparently not eternal. As the *Voyager*s dipped low over the moon, the cameras noticed some curious features. Unlike the largest craters on Earth's moon, which are surrounded by deep basin rings, the rings that surround the largest Callistan craters appear to be shallow, almost undetectable. While large craters on Earth's moon project bright rays from their centers—some of which can be easily seen from Earth itself—the Callistan rays are almost invisibly faint. Most dramatically, while the biggest basins on Earth's moon are circled by mountains driven up by the force of the incoming meteorite, there is no such uplift surrounding the basins on Callisto; the craters were dug out of flat plains, and plains the surrounding area remained, no matter how much surface material was thrown about by the impact.

To the geologists, the answer was obvious. The only sure way to preserve craters is to carve them out of something hard, like metal or rock. If you carve them, instead, out of something as ductile as ice, it won't be long before the surface begins to slip and slump, slowly softening the contours of even the deepest impact scars and erasing their details entirely. On a moon like Callisto, which was more than four-tenths ice, this surface relaxation would be going on constantly, and as it did, the face of the entire globe would be steadily reshaped.

Notwithstanding such constant crustal creep, the prevailing impression the scientists got of Callisto was of a static, almost inert world. With the moon's low density and relatively modest mass, it was simply not able to generate enough gravitational heat to stir up its innards in any significant way. But what if such a wet, chemically rich world managed to get just a little bigger and gravitationally hotter? What would happen if it could roil and warm itself in some meaningful way? The scientists found out when the *Voyager*s passed Ganymede.

Just what distinguishes a planet from a moon had never been scientifically certain before the *Voyager* spacecraft ventured to the outer solar system, but one obvious distinction was size. Planets, after all, could be mammoth things, with a world like Jupiter measuring more than 88,000 miles in diameter—or eleven times the size of Earth. Moons, by comparison, could be little more than cosmic pebbles, with some measuring just a handful of miles from one end to the other. As moons get larger, they become more and more planet-like, with a few—like Callisto or even Earth's own moon—pushing the very species line that separates a parent world from its satellite. Ganymede, with its 3,262-mile diameter, crossed that line entirely. If the gravitational cord that connects Ganymede to Jupiter were somehow cut and the moon were allowed to sail into orbit around the sun, it would not only readily join the family of planets, but also qualify as one of its more complex ones.

Though Ganymede, like Callisto, appears to be a relatively low-density rock-and-ice world, its sheer bulk is enough to give it a far more elaborate anatomy than its simple sister. Ganymede's original gravitational heat allowed it to stratify into no fewer than four distinct layers: a small metallic inner core, a larger rocky outer core, a mantle of relatively warm ice, and a crust of harder, more brittle ice. A world with so differentiated a structure and a pure metal center can generate intense magnetism, and when the *Voyagers* switched their magnetometers on, they found that Ganymede—like Earth, but unlike Callisto—is indeed surrounded by a complex magnetic field.

The kind of subsurface fever that allowed Ganymede to develop so complicated an interior played a significant role in shaping the exterior of the globe, too. The surface of the moon appears to be a combination of two distinct topographies: dark, ancient stretches of meteorite-blasted land, and lighter, younger expanses of refreshed terrain, where extensive cracks and channels have allowed subterranean flows of some kind to well up and cover the worst of the world's impact craters. On a body as icy as Ganymede, it was a good bet that the upwellings weren't lava—not with the moon's rocky layer buried as deeply as it is beneath the icy crust and mantle. Instead, the *Voyager* scientists were betting on water. As the once slushy mantle of Ganymede began to freeze earlier in the moon's life, it would behave like freezing water anywhere and expand. This would cause the overlying crust to stretch and fracture, allowing the slush to creep out and at least partially smooth the surface of the moon. The areas in which craters survived were those that had managed to hold back this warmer, upwelling flow. The areas that looked smoother had at one time been inundated.

If Ganymede was warm once, however, it clearly was no more. After its gravitational energy dissipated, the moon had only the heat of the sun to keep it from freezing over altogether. Half a billion miles from the center of the solar system, however, is a frightful place to hope for sunshine, and while Ganymede might have been forged in a furnace, it was now suspended in an icebox, its dynamic crust and mantle having long since gone still and cold.

The irony for poor, frozen Ganymede is that despite its cryogenic state, it is not located all that far from a source of world-warming energy. Just over 660,000 miles away lies the vast bulk of Jupiter, with its sizzling energy fields and its bone-crushing gravity. Just as Ganymede's bulk made it a slightly more dynamic world than light-weight Callisto, so, too, could a moon that lay closer to Jupiter be far more dynamic than Ganymede itself. Io and Europa fit that description.

When Linda Morabito spotted the volcanic plume streaming out of a crustal vent in the orange-yellow surface of Io, she became the first person ever to spot a volcano on the little Galilean moon, but not the first person ever to consider one. Only a few weeks before Morabito's discovery, a paper began circulating in the astronomical community theorizing that just such volcanic unrest might exist on Io. The evidence was not all that strong—nobody had actually seen volcanic exhaust yet—but it was nonetheless tantalizing. Both Earth-based telescopes and the *Pioneer* spacecraft had noticed a curious doughnut-shaped formation around Io, a cloud of particles that surround the moon in an all but invisible haze. There were only two possible sources for such a cosmic cloud: Jupiter, which could be sputtering off some kind of material that migrated into orbit around the moon; or Io itself, which could be outgassing volcanic exhaust that dispersed into nearby space. Most of the scientists who read the paper were betting on Io—and with good reason.

As the innermost of the four Galilean moons—orbiting just 261,000 miles above the planet, or roughly the same altitude at which Earth's own moon orbits—Io was about as deep inside the planet's gravity well as it was possible for a large moon to be and still remain intact. This kind of proximity has a profound effect on the satellite. While Earth's moon needs a full month to make one lazy circle around the planet's 25,000-mile circumference, Io is whipped around Jupiter's far larger, 279,000-mile equator in just 1.8 days. Io, like Earth's moon, is orbitally locked on Jupiter, with the gravitational hands of the planet keeping one face of the moon turned toward it at all times. In the Earth-moon system, this intimate gravity dance affects the very shape of the bodies themselves. The constant tug of the moon's gravity raises oceanic tides on Earth, causing the waters to recede and advance depending on which side of the planet is facing the always present moon at which point in the day. In return, the constant tug of Earth's gravity raises solid-body tides on the moon, temporary bulges in the lunar crust that subtly distort the overall shape of the globe. In the Jupiter-Io system, where there is a far greater disparity in the size of the bodies, things are all one way; Io has little gravitational effect on Jupiter, but Jupiter causes a solid-body tidal bulge on Io that rises more than three hundred feet high. What's more, since Io's orbit is slightly elliptical, the bulge tends to rise and fall, growing bigger when the moon is making its closest approach to the planet and smaller when it's at its farthest.

Such steady, wrenching changes in Io's shape would itself impose a lot of strain on the moon, but it is not Jupiter alone that has its gravitational way with Io. As the closest and speediest of the Galilean satellites, Io makes two complete orbits of Jupiter for every one its nearest sister, Europa, makes, and more than four for every one Ganymede makes. Each time Io passes one of these relative slow-poke moons, the neighboring world gives its orbit a quick gravitational pluck. As it does, Io strains briefly away from Jupiter, developing a tiny tidal bulge on its opposite hemisphere. The bulge quickly subsides, but no sooner does it vanish than another pass by another moon causes another one to rise. This much stretching and relaxing of the crust and mantle creates enormous subsurface friction, ultimately causing the moon's metabolism to speed up and its temperature to rise. As it does, volcanoes ought, in principle, to form.

That was the theory in any case before *Voyager* arrived at Jupiter, and when it did, Morabito discovered that the theory was explosively true. In the months *Voyager*s 1 and 2 were barnstorming the world, volcanoes were discovered all over the face of Io. In just that time, whole new volcanic mountains rose up on the moon, whole new volcanic calderas opened, and whole stretches of Io's landscape were resurfaced practically overnight—by geological standards at least.

"The geology on Io changes on the scale of human lifetimes, on the scale of weeks or months," said Terrile. "If you lived on Io, you wouldn't have much need for weather reports, but you'd have an enormous need for geology reports: mountain building in the east, lava flows in the west. It happens that fast."

The eruptions that cause all these topographic changes are not only remarkably frequent but remarkably violent. The lava that the Io volcanoes emit gushes out at speeds of nearly 3,000 feet per second, ejecting more than 10,000 tons of material in a single blast. The moon's relatively thin atmosphere, which was found to include traces of oxygen and sodium, is rich in aerosolized sulfur, virtually all of it outgased by the explosions. Located so far from the incubator of the sun, the surface of most of Io rarely rises above −230 degrees; in the vicinity of the volcanoes, however, the temperature was measured at 60 degrees.

In addition to acting as a natural volcanic incubator, Io turned out to be a natural dynamo. Orbiting as close to Jupiter as it does, the moon lies deep within the planet's magnetic field. Flying so fast through such powerful lines of force, Io develops a current of nearly one trillion watts that crackle constantly across its diameter and leave a charged trail behind it as it soars through near-Jovian space. A moon that had once been thought to be little more than an inert sphere turned out to be not only thermally alive but also electrically alive, one of the most dynamic bodies anywhere in the solar system.

For all this violent energy, what Io lacked was some sign that it was organically alive, too. Unlike Ganymede and Callisto, this innermost of the Galilean moons is without any discernible concentrations of water, since whatever moisture it might once have had would have long since evaporated as a result of Io's tremendous internal heat. A world without water, of course, can never support known life. If it were possible to contrive a slightly different moon, however, one that could get itself even half as warm as Io and keep itself even half as wet as Ganymede and Callisto, that would be a moon to be biologically reckoned with. Such a moon, in the form of Europa, appeared to exist right next door.

When *Voyager 2* returned the first close-up images of the frozen crust of Europa, the JPL scientists knew they were looking at something remarkable. Located in the same punishing part of space where incoming projectiles had once done so much damage to the surfaces of Callisto and Ganymede, Europa should, by rights, have been covered with craters itself. *Voyager*'s first shimmery images of the moon, however, revealed nothing of the kind. Europa's crust, it appeared, was all but completely pristine; here and there might be a divot or two from a relatively recent meteorite hit, but apart from that, the battlefield look of the badly ravaged Ganymede and Callisto was utterly absent. What did mar the surface of Europa were cracks—hundreds, even thousands of them, forming a fine, spidery webwork across the face of the moon. It was the absence of craters that first caught the scientists' attention, but it was the presence of the cracks that held it fast. In both cases, they figured, they knew what was responsible.

As nearly as the *Voyager*s could tell, Europa is a relatively simple world. The smallest of the Galilean moons—just 1,945 miles across—it appears to be made up of almost nothing but core, a hard sphere of silicate rock nearly 1,800 miles wide. Surrounding this is a thin mantle of water ice and surrounding that is an even thinner water-ice crust. If the crust was uncratered, it must have periodically melted, and the *Voyager* scientists suspected that tidal pulsing was doing it. Though Ganymede and Callisto are too far from Jupiter to experience a significant amount of the kind of steady gravitational squeezing Io undergoes, and Io is so close it experiences too much, the roulette slot into which Europa fell as the moons were accreting around Jupiter was another matter. Located 415,000 miles from the planet, Europa was able to stay warm but not hot, gravitationally massaged but not gravitationally crushed. When an ice world is gently stirred up this way, it won't be long before it turns into a water world.

On Europa, that appeared to be precisely what was happening. As the moon's underground ice shifts and churns, it soon becomes slush, and the slush soon becomes water. The sugar-shell crust above it, in turn, continually thaws and refreezes, erasing any craters that once freckled it. And as the freshly refrozen crust grows stiffer and more brittle, it also repeatedly cracks, leading to the filigree of fractures that cover the world.

When the *Voyager* scientists peered into those cracks, they got another surprise. Most of the fractures were unexpectedly dark, suggesting both that water beneath the surface had welled up to fill them, and that that water was remarkably dirty. Given the prevailing chemistry of the Jovian system, JPL scientists were pretty certain that the contaminants that had muddied the water included salts, carbons, and various minerals, as well as nitrogen and dissolved oxygen and hydrogen, precisely the same organic materials that give lake water its murky appearance on Earth. Fill a warm, bathwater ocean with those kinds of ingredients, let constant tidal pumping keep the chemicals moving, and give everything 4.5 billion years or so to mix and marinate, and there was every reason to believe that life—ordinary, unremarkable, Earth-like life—would emerge.

"What we're finding with terrestrial organisms," said Terrile, "is that life has an extraordinarily high tolerance for conditions that we once thought were intolerable. There is almost no environment on Earth too punishing for some kind of biology to take hold as long as there's liquid water present. If there's water on Europa, too, then there are life forms on Earth right now with which we could probably contaminate that world's oceans. Whether there is any indigenous life on Europa already is the million-dollar question."

For the *Voyager* scientists, determining if a Europan ocean indeed exists beneath the ice, how warm and deep it is, and whether it is in fact home to its own native life-forms was, of course, pure conjecture. Answering those questions would require a closer look at the world, and a closer look at the world would take another spacecraft. For their part, the *Voyager*s were done with Jupiter. In the late summer of 1979, the second of the two ships flew beyond the last of the distant moons and vanished from the Jovian system forever, leaving the planet, its rings, and its flock of satellites far behind. Far ahead lay the other three giant worlds of the outer solar system—and separating those worlds lay hundreds of millions of miles the spacecraft would have to travel to reach them.

5

Cosmic Hoopskirt

Autumn 1980

When the engineers overseeing the *Voyager* mission arrived in the Space Flight Operations Facility late in the evening on November 12, 1980, they were not, by any measure, a relaxed group. Certainly, the walk over to the SFOF was pleasant enough. Getting to the big, boxy building on the west end of the JPL campus took you up the hill of Explorer Road, down the small path of Ranger Road, and through the slightly twisty stretch of Deep Space Network Lane, depositing you directly in front of the big, glassy entrance of the SFOF. The campus, as always, was largely empty by this time of night, and with the late autumn sun long gone, the streetlights on the footpaths had all blinked on, giving the grounds less the look of a large lab than a small village.

It was a pretty enough place to be all right, and on any other evening the *Voyager* engineers might have been able to enjoy it. On this particular evening, however, they had something else on their minds. What was troubling them mostly was the very real likelihood that the *Voyager 1* spacecraft—the spacecraft they were coming to the SFOF to fly tonight—was not going to live through the evening.

The engineers had gone through a lot to get to this night on which their spacecraft might well die, and the idea of losing the ship now was not an appealing one. It had been more than twenty months since *Voyager 1* had soared past Jupiter and its thirteen—actually, now sixteen—known moons, and the spacecraft had been in a flat-out sprint ever since. Clipping along at roughly 35,000 miles per hour, the ship had put more than 400 million additional miles on its odometer—or a tidy 20 million miles every month. Such a long-haul journey was just about sufficient to close the gap between the fifth planet from the sun and the sixth one, and before the evening was out, *Voyager 1* would indeed enter the neighborhood of Saturn, beginning its thirty-five-hour and four-minute trip past the planet, around its rings, and by at least seven of its eleven known moons, before flying out of the solar system altogether and vanishing into interstellar space. That was the plan anyhow, but standing between the spacecraft and this ambitious itinerary was the giant Saturnian moon Titan, and if the trajectory numbers streaming back to Pasadena were correct, the ship could easily be on a heading that would take it not past the vast moon, but directly into its rocky surface.

In the three-plus years *Voyager 1* had been in space, its flight plan had not changed significantly. As the ship flying point in this two-ship expedition, the spacecraft would be attempting the closest possible flyby of Titan, sweeping beneath its south pole, using the massive body's gravity to make a sharp turn north, and closely barnstorming the moon before it soared out of the solar system altogether. As long as *Voyager 1* succeeded, *Voyager 2* would be permitted to continue flying its horizontal course past Saturn and, possibly out to Uranus and Neptune. If *Voyager 1* failed, *Voyager 2* would be retargeted for the same acrobatic Titan flight.

In order to give *Voyager 1* any chance of pulling off the navigational stunt it would be attempting tonight, the flight dynamics engineers had charted a hair-raising route for it. According to the scientists' ciphering, if the ship was going to make the most of Titan's gravity field, it would have to fly just 4,023 miles above the bright orange cloud decks of the moon, burning its engine all the while to keep the ship's speed high and its direction true. Four thousand miles is a considerable distance by terrestrial standards, but terrestrial standards were not what was being applied this evening. When a spacecraft has already covered nearly 800 million miles on its looping trajectory from Florida to Saturn, a 4,000-mile miss distance does not amount to much. When the maneuver being attempted requires Earthbound engineers to transmit the spacecraft fly-by-wire signals that, even at the speed of light, take seventy-one minutes to reach the ship, the same 4,000-mile margin of error vanishes entirely. But 4,000 or so miles was all the *Voyager 1* engineers had to play with at the moment, and on this particular evening it would be Ray Heacock's job to see to it that that was enough.

Ray Heacock was, if anyone was still counting, the fifth project manager of the now nine-year-old *Voyager* program. First, there had been Bud Schurmeier, who oversaw things when the project was in its development phase; then came John Casani, who ran the show up through the launch; next was Robert Parks, who was in charge during the cruise to Jupiter; after that was Peter Lyman who took over from Parks for *Voyager 1*'s pass through the Jovian system. Now it was Heacock who had taken over the reins, and so far he had done a good job, commanding *Voyager 2* during its own Jupiter encounter and nursing both ships along as they made their transit to Saturn. At last, the first of those two spacecraft had arrived, and Heacock dearly did not want to become known as the only one of the five managers to wreck his ship on the shoals of his target world before he had a chance to collect so much as a single shred of data.

To hear Heacock tell it—when he cared to be completely honest, that is—there was no real reason he had to be in this position at all this evening, mostly because there was no real reason any spacecraft should be cozying up as close to Titan as this one was going to. It wasn't that Heacock didn't appreciate the appeal of the big, cloudy moon. Titan's sheer size, plus its rich hydrogen-oxygen-nitrogen chemistry, made it an irresistible target. The problem was, no matter how tempting Titan seemed, there were still dozens of other moons out there, too, and no matter how Heacock worked the arithmetic, he had never seen the sense in risking one spacecraft—and possibly two—studying just a single one of the solar system's satellites circling just a single one of its worlds.

Recently, a small but vocal group of dissenters at JPL had begun seeing things this way, too, and had begun arguing that perhaps it was time to consider removing a close flyby of Titan from the *Voyager* itinerary altogether. Heacock was part of this group and was not shy about making the case against Titan whenever the topic came up for discussion. The problem was, in the twenty months the ships had been traveling to Saturn, a larger, more vocal group of Titan advocates had taken precisely the opposite position, working hard to make sure the big orange moon stayed on the agenda. The most conspicuous of these pro-Titan spokesmen was astronomer and scientific pitchman Carl Sagan, and around JPL the accepted wisdom had always been that if you found yourself in a public debate over two competing scientific positions, Sagan was not the man you wanted at the opposite lectern. Equal parts Cornell University professor and television personality, Sagan had been serving as a consultant to the *Voyager* team for years. Whenever the conversation turned to Titan, he made it his business to champion the moon, touting it in that damnably seductive voice of his as "a uniquely interesting place," one where there was "undoubtedly organic chemistry going on." When Carl Sagan started going on about organic chemistry and the best you had to counter with was a lot of fretting and hand-wringing about 4,000-mile miss distances, there was little question whose argument was going to prevail.

The result was, on this November evening near the end of 1980, *Voyager 1* was indeed preparing to surf its way through near-Titanian space, and it would be up to a reluctant Ray Heacock to pull off the fancy flying that would make the encounter happen. For all his misgivings, Heacock knew he had at least a few things on his side—not the least being that his close approach to the moon would be over the friendly south pole.

Titan, as the *Voyager* team knew, flies a slightly inclined path around Saturn, one that takes it a bit above and a bit below the planet's equator—and thus a bit above and below its rings. Trying to fly a spacecraft across the Saturnian ring plane was not something the JPL flight planners had ever tried and not something they especially *cared* to try. While the rings were made up primarily of empty space, that empty space was still aswarm with tens of thousands of small and not so small particles. A mere grain of sand that collided with an object moving at the 35,000-mile-per-hour speed at which a *Voyager* traveled could pack the same energetic wallop as a bowling ball traveling at 120 miles per hour. A bigger bit of debris would hit harder still.

Happily for the scientists, Titan was, at the moment, orbiting slightly below the rings. This meant that as long as the spacecraft flew beneath the south pole of the moon before turning north, it would not even encounter the rings until it had already passed Titan completely. If there were any stray particles in the path of the *Voyager,* they would thus not have the opportunity to do any damage until after the spacecraft had done its work, taken its pictures, and beamed the data home to Earth. If the ship met an untimely end after that, it would be an end the scientists could accept.

Also boosting Heacock's confidence was that, hard as the kind of deep-space flying he'd be attempting tonight was, it was not completely without precedent. Just such a stunt had been pulled off nearly ten years earlier when William Pickering's *Mariner 9* was in orbit around Mars, and the JPL folklorists had been talking about the maneuver ever since.

The mission profile for *Mariner 9* had been an ambitious one, calling for the spacecraft to enter orbit around Mars—the first ship ever to do so—and spend eleven months circling the planet and sending home more than seven thousand pictures of its surface. Pickering's scientists had approved of this flight plan when it was first proposed, but secretly, most of them had wanted even more. As long as *Mariner 9* was out in Martian space anyway, they had wondered, why not up the exploratory ante and also send the spacecraft on a detour route that would take it by Phobos, the larger of the planet's twin moons? As tiny as Phobos is—just over thirteen miles wide—getting a good picture of it would require a breathtakingly close approach. That kind of maneuver was not something that had ever been contemplated, and for that reason as much as any other, the scientists had not initially raised the possibility of attempting it now. It was only when the ship was actually in Martian orbit and functioning perfectly that they drew themselves up and asked Pickering if it might be possible to give it a try.

Pickering, who was in no mood to risk the health of his spacecraft on an orbital thrill ride, was at first inclined to say no. The more his scientists made their case, however, the more he realized it was a good one. *Mariner 9* was there to study as much as it could about the Martian system, and the moons were a critical part of it. The only legitimate reason to resist the idea of steering his spacecraft past Phobos would be if he didn't have confidence that his flight controllers could manage the maneuver; and if his flight controllers couldn't manage such a maneuver, they shouldn't be flight controllers in the first place. To his own surprise as much as anyone else's, Pickering okayed the plan, dispatching one of his scientists to the SFOF to pass the order on to the men who would have to execute it. The scientist went to the control center as instructed, tracked down the chief navigation engineer, and begin explaining what he needed.

"You know," he said, as nonchalantly as if the idea had just occurred to him, "we've been thinking upstairs that we'd really like to get some close-up photos of Phobos."

"Okay," the engineer said absently.

"You can do that?"

"We can do that."

"How close can you go?"

"How close do you want to go?"

"As close as possible, I guess," the scientist said.

The engineer considered that for a moment. "How's twenty miles?"

The scientist's eyes widened. "Twenty miles?" he said with a slight laugh. "Sure, twenty miles is good." After a pause he added warily: "Just don't hit it."

The engineer looked back at him fixedly. "We're not going to hit it," he said and turned on his heel.

Before the day was out the men in the SFOF had figured out a way to fly precisely the maneuver Pickering wanted. Before the week was out the spacecraft had flawlessly flown it, sending the JPL scientists the best pictures ever taken of the little Martian moon. That, in any event, was the way the story was told, and that was the way it had *been* told since 1971. Now, on a cool night in 1980, Ray Heacock was going to have to live up to it.

When Heacock arrived in the SFOF to prepare for the Titan encounter, he quickly queried his controllers to be certain that both the ship and the men charged with caring for it were ready for the maneuver. When he had satisfied himself that they were, he instructed his flight dynamics officers to proceed, then listened from his console as they chatted among themselves, reading off the codes that would direct the ship to rotate to the appropriate attitude and fire its engine for the planned flyby. From there an engineer in a back room tap-tapped the command into a navigation computer, and from there the command was fired into space.

What happened after that was essentially nothing at all. Not only would it take the signal from Pasadena seventy-one minutes to reach the distant *Voyager* probe, it would also take another seventy-one minutes for the confirmation to come back that the maneuver had been executed. This 142-minute stretch far exceeded the length of time scheduled for the entire Titanian flyby, meaning that if the maneuver was a failure, the transmission *Voyager 1* sent back home would likely be a posthumous one, a beyond-the-grave shout that would not reach Earth until an hour after the spacecraft had already tumbled in toward Titan and annihilated itself on the surface of the moon.

Heacock spent the two hours and twenty-two minutes of waiting time uneasily pacing from his console in the SFOF to his office down the hall and back again, occasionally collaring a flight controller to ask him how he thought the maneuver was progressing. Each time the controller would answer with a no-problem nod that Heacock knew was either pure bravado or real confidence, and probably a little bit of both. Finally, as the 142nd minute of light-time concluded, Heacock resumed his seat, donned his headset, and waited, like the other men in the SFOF, for some word from space. After a few seconds a stream of numbers began to flicker onto the trajectory consoles in the middle of the room. The technicians watched as the data appeared, and after a moment one of them spoke evenly into his mouthpiece.

"A signal has been received," he said.

Heacock leaned forward, pressed his headset closer to his ears, and said nothing, waiting for the remainder of the call. Across the room most of his controllers mirrored him. A bare second or so later, the trajectory engineer called out again.

"Maneuver executed," he announced. "Titan encounter achieved."

Throughout the SFOF a cheer went up, applause broke out, and the controllers, who had been bent so worriedly over their screens, leaped up for a round of handshakes. Heacock, at his project manager's console, stood, too, and allowed himself a brief celebratory moment—but *only* a moment. *Voyager 1*'s near-Titan approach was not planned to be a long one. Arcing over the moon's south pole and up toward the north at a speed of 9.7 miles per second, the ship would pass Titan's equator in less than two and a half minutes and complete its reconnaissance of the entire moon in less than five. From there it would fly free in the Saturnian system, passing six of Saturn's other moons—Mimas, Enceladus, Tethys, Dione, Rhea, and Hyperion—as well as the giant ringed planet itself. In just over a day it would leave the Saturn neighborhood altogether.

This meant that the imaging team had to act fast. Even before *Voyager 1* began its Titan maneuver, the spacecraft had switched on its cameras and started beaming home pictures of the moon it was so closely barnstorming. Well before those signals had traveled the 800-million-plus miles back to Earth, the imaging team scientists had already gathered in their modest meeting room in the southeast corner of Building 264 to watch as the Titan portraits flickered to life on their multiple monitors. These first images, the scientists knew, would not be much—unenhanced black-and-white snapshots that would initially show only the moon's brightest features. Analysis of Titan's chemical spectra would allow the scientists to add inferred color to the image; subsequent refinement of the data stream would help them sharpen the picture and bring it into crisper relief. Nonetheless, for scientists trained to read interplantary pictures as a physician reads an x-ray, even the murkiest, milkiest Titan image could carry a wealth of information.

Just after the signal confirming the Titan flyby was received in Pasadena, word also came back that the first of the images had reached the deep-space antennas in the California desert, and was being relayed to JPL's massive mainframe computers, and from there to the viewing room monitors. For long minutes after the transmission arrived on Earth, those screens showed nothing but static. The scientists seated around the conference table stared fixedly at the glass, leaning expectantly forward when the static would flicker promisingly and then sitting back again when the flickering would stop.

Finally, after three or four such false alarms, the monitors began to blink and clear, and a faint ghost of a circle appeared. The circle faded, then reappeared, and at last sharpened itself into a stark gray-white ball. For a full five seconds, the scientists around the table stared at the image, and then, as one, they slumped.

The ghostly circle—the circle that was Saturn's Titan, the moon the scientists had waited more than three years to see—looked like nothing at all. There was not a shadow of a surface feature visible anywhere on it, not a hint of a cloud formation roiling its atmosphere. There was only the globe itself, an egg-smooth, egg-blank sphere, shrouded in an atmosphere that was completely opaque. For 325 years, astronomers on Earth had been peering at the moon from a distance of 800 million miles and wondering about its makeup. Now they were looking at it from an eyelash distance of just 4,000 miles and they were still left wondering.

"Nothing," somebody in the room muttered.

"A washout," somebody else responded.

None of the imaging scientists was prepared to say much more than that. There were plenty of pictures yet to stream in, and it was always possible that some of these would reveal something useful—a break in the clouds, perhaps, a storm in the atmosphere. Even without that, the scientists could still try to stretch and squeeze and brighten the data they did have, wringing at least a little more information out of all of the images. Intuitively, however, they knew that a little was all they were ever going to get. The moon that was perhaps the most promising world in the solar system save Earth itself was photographically blacked out by the very gaseous shroud that helped make it so chemically rich in the first place.

The scientists spent the better part of the next hour watching the blank, billiard-ball images of Titan arrive, and as each did, word seeped deeper and deeper into the JPL community that the grand visit to the grand moon had been an utter bust. Such a development should have hit the scientists hard, and indeed, the mood on the campus was grim. But there was cause to feel not so grim, too. First of all, there were still at least ten other Saturnian moons for *Voyager 1* to see, and before the day was out, images of those worlds should be in hand as well. More important, there was always *Voyager 2*. The impenetrability of the Titanian atmosphere meant that there had really been no reason for *Voyager 1* to visit the moon in the first place. And if there was no reason for *Voyager 1* to have gone to Titan, there was certainly no reason for *Voyager 2* to follow it there. The anti-Titan camp, it appeared, was going to carry the day after all. The second of JPL's magnificent deep-space ships was going to be able to remain safely in the plane of the solar system, where it could visit not just the Jovian system, not just the Saturnian system, but the Uranian and Neptunian systems that lay beyond.

Provided, of course, it was able to make it through its own Saturnian encounter alive.

August 26, 1981

Voyager 1's visit to Saturn was nowhere near the total loss it had started out to be. Just over sixteen hours after its fruitless flyby of Titan, the spacecraft passed within 258,000 miles of the midsized moon Tethys. An hour and twenty-nine minutes later, it came within 77,000 miles of Saturn itself. An hour and fifty-seven minutes after that it came within 55,000 miles of Mimas. That was followed by a 125,000-mile flight over Enceladus; a 100,000-mile approach to Dione; a close, 46,000-mile pass by Rhea; and, finally, a distant, 546,000-mile glimpse of Hyperion before the spacecraft vanished into space forever.

Nine months later, in August 1981, *Voyager 2* at last reached Saturn as well. Flying horizontally through the system, it first passed Titan at a great and indifferent distance of 413,000 miles. Next it flew by Dione at 311,000 miles, then Mimas at 192,000 miles, then Saturn itself at just 62,500 miles. Finally, it buzzed both Enceladus and Tethys at a distance of barely 54,000 miles and 58,000 miles, respectively, before sailing out of the Saturnian neighborhood just twenty-one hours and five minutes after it arrived, taking a last parting look at Rhea at a distance of 400,000 miles as it went.

Of all of the moons the ships saw after the disappointing Titan encounter, it was Enceladus and Tethys that most intrigued the JPL scientists. Measuring just 310 miles across, Enceladus is the second innermost of Saturn's moons and, as such, was thought to be gravitationally pumped by outlying Tethys and Dione, much the way Jupiter's Io and Europa are squeezed by the other Galilean satellites. The juxtaposition of the three moons alone suggested that this kind of gravitational dynamic might be under way, and the appearance of Enceladus—at least from the distant look *Voyager 1* got of it— seemed to confirm it. The moon seemed to be covered with a snowy surface material so bright it reflects nearly 100 percent of the sunlight that strikes it. Back on Earth, a full moon on a bright night reflects only 12 percent. This kind of ground cover would probably not lie exposed on the surface of Enceladus indefinitely without growing damaged and dimmer. The fact that it was as bright as it was suggested that some kind of tidal or volcanic force was causing upwellings or eruptions that were regularly refreshing the crust.

Neighboring Tethys was significant for just the opposite reason: the ancient, tortured appearance of its surface. *Voyager 1* discovered that Tethys, measuring 657 miles across, is scarred by a large crater and an associated crack that stretch nearly three-quarters of the way around its middle. The size of the scar suggested to scientists that Tethys was wounded almost mortally by the projectile that struck it. A slightly harder hit by a slightly bigger rock and the moon would have split open like a soft melon. Studying Tethys might provide clues to the internal stability of rocky worlds and how they are born and die. But while *Voyager 1* was able to confirm that the crack in the moon existed, the ship's quarter-million-mile flyby prevented it from determining much more.

For these reasons, getting good pictures of both moons was a

Voyager 2 mission priority, and in the additional months it took the ship to reach Saturn, the trajectory engineers worked hard writing the lines of programming code that would allow it to barnstorm the worlds. It was at the twenty-hour mark in *Voyager 2*'s twenty-one-hour visit to Saturn that it was scheduled to make those Enceladus and Tethys flybys, and this time it was Esker Davis who was in charge. Davis was the sixth *Voyager* project manager, taking over from Ray Heacock in early 1981. Unlike Heacock, Davis would not have to worry about sending his ship through any acrobatic flips as it sailed through the Saturnian system. Instead, he would simply have to point the bow of his spacecraft straight ahead, steer the ship this way and that as he executed the navigational programs his engineers had developed for him, and prepare to collect the album of images the ship would send home.

Late in the day on August 26, the first of those pictures—from Titan, Dione, and Mimas—began to stream in, and by any measure they were dandies: clear, colorful, full of detail, everything the inscrutable Titan pictures had failed to be. So pleased was Davis with these early results that he promptly released a few of the preliminary images to the media. So pleased were the reporters with what they saw that they asked Davis to hold a late night briefing to describe what the pictures revealed so far. Davis readily agreed, summoning the reporters to a large conference room in Building 264 just before midnight. Holding up picture after picture, he pointed out some of the salient features in this one or that, begged the writers' pardon for not having had a chance to analyze them better, and promised to call another press conference after the sun came up, when there would be more science to report and more pictures to see, principally from the much-anticipated Enceladus and Tethys flybys. As he spoke, the reporters could listen to a public address speaker mounted on the wall behind him, crackling with round-the-clock chatter from the SFOF, where the laconic exchanges of the flight controllers provided the best possible evidence that the mission was proceeding well.

After just a few minutes Davis finished his presentation and the reporters left happy, making it a point to congratulate the new project director on his accomplishment so far and shake his hand on the way out. As the last of them was leaving, Davis, now alone, took it upon himself to shut off the lights in the conference room and close and lock the door. Just as he was doing that, however, he paused for a moment to listen to the SFOF speaker and heard something disturbing.

"Ace, this is Bus," an imaging technician called out to the flight director. "We have a problem."

Davis, who had spent plenty of hours himself in the SFOF, knew that there were big problems and small problems in any mission control room and that most of the time it was not the words of a flight controller that distinguished one from the other but the tone. The tone this time told him there was something seriously wrong, indeed. Slamming the door to the conference room, he ran across the campus to the SFOF, and as soon as he arrived saw what the problem was. The first pictures of Tethys had come in as scheduled, and they were just as crisp and rich as all of the others that had arrived that day had been. Good as the pictures were, however, they were also badly off center. In the first image Davis looked at, the moon was flying out of the right side of the frame; in the next one, it was flying off to the left; in the ones that followed, it was soaring out of the top or barely peeking in from the bottom. Davis quizzed the SFOF technicians and learned, to his distress, that the problem was not originating on the ground. A glitch in the guidance computers or imaging hardware at JPL itself could be fixed with little more than a kick. But the ground-based systems were working fine. The breakdown, it was evident, had almost certainly occurred in the body of *Voyager 2,* and from what Davis could tell, it was the ship's sensitive scan platform that was to blame.

In order to give the *Voyager*s the widest field of vision possible, the cameras on its instrument boom were mounted on a motorized platform that spun left and right, allowing the ship to aim its eyes without having to power up its thrusters and shift its entire body. Useful as an assembly like this was, it could also be maddeningly temperamental. Early in the *Voyager* missions, flight engineers had sent *Voyager 1* a signal instructing it to flex its scan platform for a routine check. The platform started to comply and then came to a stop with a grind the engineers could practically hear. The signal went out again, and the platform balked again. With the spacecraft utterly beyond any hands-on repairs, engineers hoping to fix it had to rely on the next best thing: a full-sized, working model of a *Voyager* that was kept in a clean room on the JPL campus. Put this understudy spacecraft through the same paces the real one had just gone through, and you could often re-create any breakdown that had occurred in space and determine what was behind it. In this case, the engineers re-created the problem easily and quickly found its probable cause.

Most of the mechanical assemblies on the *Voyager*s were insulated by multiple blankets of Teflon, a material that did a good job of protecting the spacecraft's instruments, but also had an occasional tendency to flake. In most cases, flaking Teflon was not a problem since any stray bits would simply drift harmlessly into space. When Teflon flaked near the scan platform, however, there was always the chance that a few bits could drift up into the motor assembly and cause the entire apparatus to seize up. That was what appeared to have happened to *Voyager 1,* and if that was indeed the case, the solution ought to be wonderfully simple: Just send the ship a series of commands ordering it to keep trying to turn, no matter what, and eventually the teeth of the gears in which the Teflon was caught should chew the contaminating flakes up completely. The SFOF controllers sent up the instruction, the platform did what it was told, and the Teflon, as predicted, crumbled away. Now, years later, *Voyager 2*'s scan platform seemed to have gotten itself hung up the same way.

Before concluding that that was indeed the problem, however, Davis wanted to be sure. Shortly before 1 A.M. he called his engineers back into the conference room in which he had just held his celebratory press conference and instructed them to bring him all of the performance readouts they could from both *Voyager 2*'s scan platform and the platform on the stand-in model in the clean room. The engineers complied and Davis immediately began comparing the data from both—placing the printouts side by side and running his finger down columns of corresponding numbers and along the peaks and troughs of corresponding graphs. As he did, he concluded that the profiles of the two just weren't similar enough—the numbers simply didn't line up as they ought to, the graphs simply didn't match. The way Davis read the readouts, the cause of *Voyager 2*'s off-center images was not Teflon at all; instead, it appeared to be an overheated actuator, a drive-train shaft that had lost its lubrication and was becoming jammed as it attempted to turn in its sleeve. Try to force the platform on *Voyager 2* the way the engineers had on *Voyager 1*, and he would likely rip the innards right out of the system.

The answer, Davis decided, would be to do almost nothing at all. Instead of forcing the platform, he would simply instruct his engineers to nudge it a bit, coaxing it slowly this way and that over the course of the upcoming days and weeks. In time, changing temperatures and migrating lubrication ought to help things loosen up a little, allowing the platform to spin free again. Davis announced this decision as soon as he reached it, and only hours after the crisis began the relieved project manager was able to send most of his engineers home for a little sleep. Returning to his office, Davis was preparing to collect his belongings and punch out for the day himself when his door flew open and a scientist from the imaging team burst in.

"What do you mean you refuse to command this platform?" the scientist half shouted.

Davis jumped a little and then forced himself to answer levelly.

"I'm not going to send any commands that are going to rip out the gears," he said.

"But we need Saturn pictures!"

"We've *got* Saturn pictures—20,000 from *Voyager 1* and 20,000 from *Voyager 2*. We probably have enough."

"But the press is still out there. We've got to send those commands."

Davis nodded no.

"Move that platform!" the scientists barked.

Davis nodded no. "We're not going to do that," he said to the obviously fatigued man. "We're just not."

"We'll see," the imaging scientist said. "I'll call headquarters." Turning on his heels, he stalked out of the office and slammed the door loudly behind him.

Davis stared at the door as it shook in its frame, rubbed his eyes, and stifled a yawn, idly trying to predict how long it would take before the scientist would regret his outburst. As it turned out, it didn't take long. Before the next day was out, the scientist—who by now had gotten a few hours' rest—reappeared in Davis's office with a mumbled apology and a promise that such a breach in protocol would never happen again. Davis told him to put it out of his head and dismissed the matter with a wave. Over the course of the next few weeks the engineers determined that *Voyager 2*'s problem was indeed not caused by Teflon at all, but by the very actuator problem Davis had suspected. Over the course of the months after that, the motor began to limber up. Well before it did, the JPL planetologists had already begun analyzing the pictures—both well-centered and off-center—that *Voyagers* 1 and 2 had returned. Temperamental scan platforms or not, it was clear the ships had gotten what they came for.

•   •   •

Like centuries of astronomers who had observed Saturn before, the JPL scientists had a lot to look at when they studied the planet, but what claimed their attention first was its most conspicuous feature: its rings. The long-wondered-about ring system had never been terribly easy to study in much detail from Earth, but as nearly as astronomers could tell, it was made up of seven discrete bands, each of which appeared to be composed of a relatively homogenous river of rubble. The *Voyagers* found out that the structure is a good deal more complex than that. All of the bands in the Saturnian rings, it appeared, are made up of thousands of far finer strands, each just a fraction of the width of the bands as a whole and all of them nearly impossible to see from any appreciable distance. Up close, however, where the *Voyager*s flew, the ring system had the look of a grooved LP record, down to the distinct gaps that separate song from song or, in the case of Saturn, band from band.

Of the seven bands, the one closest to the planet is a precariously low-flying one, orbiting just 4,100 miles above the Saturnian cloud-tops. The band farthest away begins 74,200 miles up and reaches about 260,000 miles into space, or about the same as the distance between Earth and the moon. Despite their enormity, Saturn's rings, like Jupiter's, appeared to be made of extraordinarily tenuous stuff— ordinary clumps of water ice, a few of them with bits of rock at their center, but most of them as pristine as snow. While the largest of these frozen masses might be the size of big boulders, most of them are far smaller, measuring anywhere from a foot or two down to mere fractions of an inch. Viewed edge-on, Saturn's entire sparkly ring system is barely a mile thick, or sharper than a knife blade on its own massive scale.

Though the rings are almost certainly the remains of a lost Saturnian moon, it was clear that it hadn't been much of a moon. If the material that makes up all seven rings were scooped up, balled together, and tamped down into a single body, the re-created satellite would have a diameter of less than 120 miles. Such a modest body would have been easy to destroy with even a relatively small meteor hit, and most of the *Voyager* scientists agreed that that was probably just the disaster that struck the former satellite.

What impressed the scientists more than the fact that such a cosmic calamity befell the small moon was when it befell it. A ring formation this gauzy circling a planet this big would not be likely to have an especially long life span. No sooner would the moon be blasted apart and its remains dispersed around Saturn than at least some of the material would migrate inside the Roche limit. There, the planetary gravity would begin eroding the rings from the inside out, pulling more and more of the particles out of orbit and down to the surface, until the once broad bands crumbled to nothing. The fact that the rings exist today suggests that they are relatively new formations, existing for tens or hundreds of millions of years instead of the 4.5 billion years the solar system as a whole has been around. Had human beings emerged just a little earlier or a little later in celestial history, they would likely have seen no rings at all, coming to know a Saturn entirely denuded of the very feature that has so come to define it.

Short as the lives of the rings will be, they are lives that are being spent dynamically. Much of that dynamism, the *Voyager* team found, is caused by the planet's moons.

Not long before the *Voyager*s reached Saturn, astronomers working with ground-based telescopes were already adding to the planet's family of 11 known satellites, discovering tiny 19-mile Telesto and 16-mile Calypso orbiting 183,000 miles above the planet, and 20-mile-wide Helene orbiting 233,000 miles up. The *Voyager*s then spotted a cluster of three more moons: 19-mile Atlas, 62-mile Prometheus, and 56-mile Pandora, all circling the planet in close proximity to each other, between 85,000 and 88,000 miles up. All of the new moons were unlovely things—jagged, irregularly shaped objects that looked less like asteroids than moons at all. Nonetheless, all were undeniably part of the Saturnian family, raising the planet's total of satellite offspring to a whopping seventeen.

So much moon traffic moving through so much ring traffic made some kind of gravitational interplay inevitable, and the *Voyager* astronomers found plenty of it. Prometheus, like the rings themselves, was found to be a clean, porous body made principally of water ice, with a little studding of rock to give it some heft. Its small size and low gravity have helped it dodge most incoming meteors over the course of its long, long life, and Prometheus is thus relatively free of craters. As the little moon sails through its orbits, it appears to stay in close proximity to the inner edge of Saturn's narrow, 310-mile-wide F-ring—the fifth and one of the most tenuous of the planet's seven bands. Just on the opposite side of the F-ring, flying equally close to the outer edge of the fragile band, is Pandora, a body only slightly smaller and slightly more cratered than its sister. Plying orbital paths separated by just 1,457 miles, Prometheus and Pandora move at such similar speeds that they remain in near-lockstep, tightly straddling the F-ring as they move round and round the world. And just as Metis and Adrastea cooperate to keep part of Jupiter's ring formation tidy, so do Prometheus and Pandora continually groom Saturn's F-ring, keeping the zone of space on either side of it debris-free and the ring itself sharp-edged and neat.

A little closer to the planet, where the far broader, 9,000-mile-wide A-ring orbits, tiny Atlas performs similar sweep-up work. With less than half the size and gravity of Prometheus and Pandora—not to mention a much bigger ring mass to deal with—Atlas can do nowhere near the thorough job its two sister moons can. Nonetheless, it makes its gravitational presence felt. Patrolling the outer border of the sprawling A-ring, it manages to keep that edge of the band at least as clean as the F-ring's edges are—and certainly far cleaner than the A-ring's distant, unpoliced inner edge is.

Also doing its share of ring-cleaning work is the comparatively large, 243-mile-wide Mimas. Another relatively lightweight ice world, Mimas by rights should not exist at all. *Voyager* scientists found that this moon, like its sister Tethys, is scarred by an impact crater so big it nearly tore the moon apart. In the case of Mimas, the crater measures eighty miles across—or nearly a third of the diameter of the moon itself—and radiates stress cracks as far as its opposite hemisphere. The central peak of the crater—the cone of material that rebounds at the heart of an impact site—is nearly as tall as Mount Everest, climbing nearly four miles into Mimas's sky. On a world the size of Earth, this would be the equivalent of a mountain rising more than 131 miles above sea level.

Nonetheless, sturdy Mimas survived the hit and thus lived to help shape the appearance of the Saturnian system as a whole. Orbiting at the inner edge of the outermost ring—the vast, 186,000-mile-wide E-ring—Mimas seems to be associated with a huge, 5,000-mile-wide gap in the rings first spotted by Giovanni Cassini in 1675 and named, appropriately enough, the Cassini gap. The moon is certainly not large enough to have swept all of the material out of a stretch of space so vast, and some other body must have played at least some role in creating the Cassini gap. Mimas, however, did clear out much of it, helping not just to smooth the rings, but in this case to obliterate an entire stretch of them.

Slightly larger than Mimas—and infinitely more dynamic—is neighboring Enceladus. The bright, white surface of the 310-mile moon that *Voyager 1* photographed led astronomers to conclude that some kind of subsurface slurry was regularly resurfacing the world, and *Voyager 2* seemed to confirm that. Detailed images captured by the spacecraft as it buzzed the moon in its low, 54,000-mile approach revealed vast stretches of repaved terrain where new frozen flows appeared to have obliterated valleys, filled up basins, and in some cases cut craters cleanly in half, obscuring one side of them while turning the other into a sort of open-ended, topographic horseshoe.

Such slushy upwellings—cryovolcanism, as the *Voyager* scientists called the phenomenon—are probably due to the same tidal pumping that heats up Jupiter's Io, causing its more traditional silicate volcanism. While Io is squeezed by Jupiter itself and by nearby Europa and Ganymede, Enceladus is probably pumped by its sisters Tethys and Dione, both of which are more than double its size and orbit nearby, allowing them to pack a considerable gravitational wallop. So powerful is the constant pressure Tethys and Dione apply to Enceladus that the crystalline volcanoes that erupt across the smaller moon's face affect more than the moon itself. Orbiting 147,000 miles above Saturn, Enceladus is located deep inside the planet's huge E-ring. As *Voyager* scientists were studying the distribution of material in the ring, they noticed a curious thickening in the particle band that corresponds precisely to the orbital path Enceladus inscribes. The only sensible explanation was that the material comes directly *from* Enceladus, a sort of snowy, volcanic exhaust that the moon leaves in its wake as it sails around the planet, much the way a steamship trails a thick cloud of smokestack soot as it moves across an ocean.

"The peak of the E-ring in terms of density is right about the orbit of Enceladus," said Terrile, "so it's always been very suspicious. To all appearances, the moon doesn't merely shape the ring the way shepherd moons do. Instead, it actually *feeds* the E-ring."

To the *Voyager* scientists, it was no surprise that Saturn's satellites could play such a considerable role in determining the look of the planet's bands. The material that makes up the formations, after all, is the most fluffily insubstantial of stuff. Bring a massive, chugging moon into their vicinity and it would be hard-pressed *not* to distort things in some way. Far less expected than the impact the moons have on the bands is the impact they have on one another. Nowhere was this more evident than when the *Voyager* scientists turned their cameras toward Iapetus.

When Giovanni Cassini discovered the 905-mile-wide moon in 1671, he had no idea exactly what he was seeing. On the whole, a satellite circling a planet ought to be visible for very predictable portions of its orbit. When it is on the eastern side of its parent world, the moon ought to appear as a dot in telescopes as it moves counterclockwise away from the viewer. For a fixed period, the moon should then vanish as it passes behind the planet, eventually reappearing on the western side as it continues its counterclockwise path, this time moving toward the telescope. When Cassini studied Iapetus, however, he noticed something strange. No matter how many times he observed the moon, he was able to see it only when it was on the eastern side of Saturn, orbiting away from him. When it reappeared around the western side—or when it should have reappeared, at least—he saw nothing at all. Only if he was patient, waiting another thirty-five days for the moon to complete its half revolution and cross back over to the eastern side, was it once again visible.

Generations of astronomers who observed Iapetus after Cassini noticed the same phenomenon, though none of them could say with certainty what was causing the moon to flicker on and off this way. When the *Voyager*s flew by and looked up close, they provided the answer. Iapetus, the spacecraft's cameras discovered, is an almost perfectly two-toned world, with its forward hemisphere colored a deep, tarry black and its hindquarters colored a bright, snowy white. So dark is the prow half of the planet that it reflects as little as 3 percent of the light that strikes it—little brighter than asphalt. So bright is the rump half that it reflects a full 60 percent, or five times more than Earth's own moon. Since Iapetus is in a fixed orbit, keeping one hemisphere facing the planet at all times, its bright, stern end is always pointed toward Earth when it is on the east side of the planet, making it easily visible; its dark bow end is always exposed when the moon is on the west side, causing it, effectively, to vanish.

Without having a sample of the dark material on the front half of the moon, there was no way for the *Voyager* scientists to know precisely what it is made of, though its sooty color suggests that it is some kind of organic hydrocarbon. Where it comes from is more of a puzzle. Given the huge population of the Saturnian system and the gravitational turbulence that implies, it's always possible that Iapetus is tidally pumped by other worlds and simply releases the material from deep within itself. With no big moons within arm's reach of the world, however, such a volcanic process seems unlikely. That means the dusty stuff came from somewhere outside, and the *Voyager* team could guess the place.

Orbiting Saturn at a distance of more than two million miles, Iapetus is the second most remote of the planet's satellites. The one that earns the distinction of being the most remote is 136-mile-wide Phoebe, circling Saturn more than eight million miles high. Phoebe, unlike the other Saturnian moons, orbits the planet in a retrograde direction—clockwise instead of the customary counterclockwise. Also unlike the other moons, it does not orbit in the general vicinity of Saturn's equator, but rather flies at a sharply inclined angle of more than 175 degrees. For this reason, the *Voyager* astronomers concluded that Phoebe is not an indigenous satellite that accreted around Saturn, but rather an immigrant moon, a captured asteroid that was speeding through space until it became snared by Saturn's gravity.

If so, Phoebe, like other free-flying asteroids, ought to be made up of extremely primitive carbon compounds, and when the *Voyager* cameras caught a glimpse of the rocky world, they found that it is indeed colored the same telltale black as the asteroid bits that occasionally fall to Earth. This led the JPL scientists to concoct a complex scenario. If a wildly flying moon like Phoebe were to get whacked by another, smaller projectile—a collision that was by no means unlikely in the solar system's early, more violent days—a substantial chunk of its surface would get pulverized and blown into space. If this Phoebe-powder were then to spiral in through the Saturnian system—something that was also quite probable—it would settle on the first big object that got in its way, and that would likely be Iapetus, passing face first through the dark cloud the outer moon had given off. Every time Iapetus flew through the spot in space where Phoebe's debris hung, it would pick up more and more soot, growing darker and darker with each orbit. Ultimately, the leading half of the moon would have grown utterly black. From the point of view of Earthly observers, the spherical world would have been mysteriously reduced to a half sphere, one that both existed and didn't exist, depending upon its position at any given moment.

"For the most part the Saturnian system is clean as a whistle," said Torrence Johnson, "filled with moons made largely of ice. Then in the midst of it you have this half-blackened world, dirtied by processes that may take place on a different moon entirely."

Even more closely connected than the Phoebe-Iapetus pair is the Janus-Epimetheus pair. First spotted by Earthbound observers in 1966, Janus and Epimetheus, measuring 118 miles and 74 miles across, respectively, are two small, irregular moons orbiting about 94,000 miles high. When the trajectories of the two satellites were first analyzed, however, they were found to lie so close to one another that scientists studying them often grew confused, never quite sure which of the two moons they were looking at, or even if they were looking at two at all. The mystery stood for a decade and a half until the *Voyager* probes flew past the worlds and cleared up the confusion.

According to the images the spacecraft returned, Janus and Epimetheus typically orbit within thirty-one miles of one another, with one flying slightly lower—and thus moving slightly faster—than the other. Every four years whichever moon is on the inner, quicker track overtakes the one on the outer, slower one, causing the twin bodies to exchange a bit of momentum; this causes them suddenly to switch positions, with the lower moon boosting itself to the higher orbit and the higher one dropping down to the lower spot. Locked in this new clinch, they continue to waltz around the planet until four more years—or 2,100 more revolutions—have elapsed, when just enough new momentum will have built up to allow the moons to switch once again. The odd pirouette has probably gone on since the moons formed and will probably continue for as long as they survive.

A similar orbital fandango is danced by a trio of other Saturnian moons, Telesto, Calypso, and Tethys. Of the three, 657-mile Tethys—with its icy surface and globe-spanning crack—is far and away the largest and most complex, easily dwarfing the rocky, asteroid-sized Telesto and Calypso. Orbitally, however, the moons are co-equals. Flying at identical altitudes 182,689 miles above Saturn, the bodies form a neat, three-member procession, with Telesto walking point, Tethys flying behind it, and Calypso following them both as a sort of cosmic trace horse. The route the three moons follow around Saturn is reminiscent of the far longer march Jupiter makes around the sun accompanied by the two clusters of Trojan asteroids that precede and follow it. In both cases, the leading and trailing bodies are separated from the larger central body by precisely 60 degrees, the closest approach they can evidently make to one another without creating orbital instability.

Precisely 51,299 miles above Telesto, Calypso, and Tethys, another pair of moons, Helene and Dione, march in an orbital line of their own, also a precise 60 degrees apart. Helene, like Telesto and Calypso, is a mere jagged scrap of a world; Dione, like Tethys, is a bigger, rounder, icier one. Despite this disparity in size and gravitational power, the two moons are apparently able to maintain stable orbital positions, even without a third world like Calypso acting as a gravitational tailfin.

Of the remaining two moons in the Saturnian system, one of them—Rhea—held few surprises. Located one orbital slot above Dione and Helene, the 949-mile-wide moon turned out to have roughly the same snowball makeup as Dione, Tethys, and Iapetus. In its relatively isolated orbital perch, however, it is neither gravitationally linked to any other moons nor tidally pumped by any of them—meaning that it generally remains both cold and quiet.

The final remaining satellite, Hyperion, is another matter. One of the solar system's many irregular moons, Hyperion measures nearly 180 miles across. At that size and with that shape, the satellite is clearly pushing the limits of moon-forming physics. The larger a body gets, the likelier it becomes that the force of its accumulated gravity will cause it to collapse on itself, making its highest peaks crumble, its deepest valleys fill, and the moon as a whole assume at least an approximate spherical shape. Hyperion is almost big enough to round out this way—and indeed, it almost certainly would have if it had managed to accumulate just a bit more mass. Instead, it has remained forever imperfect, its slightly-too-small size keeping it permanently poised at the brink of the sphere it clearly wants to be, but never actually letting it make the transformation.

Hyperion's irregular shape is matched by its irregular rotation. Unlike most other satellites, which spin on their axes at a fixed and predictable rate, Hyperion is in a permanent state of rotational chaos, somersaulting through its orbits in a constant tumble that causes it to point toward a different spot in space at any given moment. At least part of the reason for this erratic motion is Hyperion's violent past. The little moon appears to be heavily cratered, with one impact scar stretching seventy-five miles across its face—nearly half the length of Hyperion itself. The moon's dense, rocky interior—far denser and far rockier than that of most other Saturnian moons—probably prevented it from being destroyed by this bombardment, but it could not prevent it from being knocked off its pins, utterly losing its bearings and never quite regaining them.

A moon this erratic, of course, is not the rule for Saturn. With the exceptions of Hyperion and high-flying Phoebe, things generally remain orderly in the Saturnian system, with clean snowball moons sailing through light, fluffy rings, all of them maintaining a sense of near-perfect gravitational order. Most of the seventeen moons have some of the essential ingredients of simple biology—carbon, water, perhaps some free oxygen and hydrogen—but without sufficient gravitational turbulence to keep them terribly hot, they are unlikely to make much of these raw materials. It was only Titan—with its rich, opaque haze of chemistry—that stood any chance of achieving something biologically extraordinary, and for now, at least, Titan would remain a mystery.

On August 26, 1981, about 4.5 billion years after Titan and its sister moons were formed and about four years after two tiny rocket ships left the ground on a peninsula of land on faraway Earth, *Voyager 2* left the Saturnian system behind, heading in the general direction of Uranus, the next world down the planetary line. It had been exactly two hundred years since William Herschel discovered the Uranian system, and in all that time it had eluded detailed study. In just four more years it might at last be seen up close.

The Edge

Most of the people at the Jet Propulsion Laboratory had heard all they cared to hear about the seven astronauts aboard the space shuttle *Challenger*. The press could write about them all they wanted, the public could read about them all they wanted, but the folks at JPL were just about fed up.

It had been almost five years since NASA's celebrated space shuttle had started flying—the maiden ship going up in April 1981, just four months before JPL's own *Voyager 2* probe made its historic flyby of Saturn. At the time the huge new space plane had been nothing short of a sensation, taking off like a rocket, landing like a jet, and shedding almost no expendable parts along the way. But sensations fade and years pass, and what was novel once becomes not so novel anymore. By January 28, 1986, NASA was ready to launch yet another shuttle—the twenty-fifth such mission, according to the space agency's reckoning—and by all rights nobody should really have cared.

The fact that so many people did care this time around had nothing to do with the work the ship was being sent up to do—though the workload on this flight was considerable. Rather, it had to do with whom NASA was sending up to do it.

Ever since the first shuttle flew, NASA had been boasting that the era of high-risk spacecraft had come to an end. The shuttle would be so safe and reliable a spacecraft—a space*liner,* really—that it would no longer even be necessary to send only pilots aboard. While at least two skilled stick-and-rudder men would always have to fill the shuttle's two main cockpit seats, the other five seats in the ship could now be occupied by almost anybody. Over the years the space agency had made good on this promise, filling out the shuttle's flight manifests with biologists, physicians, and even politicians, all civilians who never would have remotely qualified for a mission before. Now, after two dozen shuttle missions, NASA was set to do it again. Included in the five-man, two-woman crew of the space shuttle *Challenger*'s current mission was a previously unknown New England schoolteacher, a soon-to-be space traveler who had spent the last year training for a mission that once again would help the space agency prove that in the modern era it wasn't necessary to *be* an astronaut in order to fly like one. For this reason as much as any other, the upcoming *Challenger* flight had generated immense publicity, and for the scientists at JPL, who could always use a little publicity of their own, it had generated it at the worst possible time.

In 1981, long before the *Challenger* crew had been chosen—indeed, long before all but the earliest shuttle flights had even flown—NASA had made a quiet decision. Somewhere in space, a billion or so miles from Earth, the *Voyager 2* probe had just passed the planet Saturn and headed off in the direction of Uranus. By rights, *Voyager 2* had no business going anywhere near Uranus since NASA had made it clear from the moment the spacecraft left the ground that it was not budgeted for any mission beyond Saturn. But now, with the ship having completed its visit to the big ringed world, and with its systems operating so well you would almost think, if you didn't know better, that someone had designed them for more than their stipulated three years of life, it didn't seem scientifically sensible—or even scientifically *moral*—not to let it fly on and see where it could go. For years JPL engineers had been making the case for just this kind of extended mission, but the NASA administrators had always said no. In an agency as big as NASA, however, minds change and policies change, and by 1981, when the *Voyager* ships had already returned a dazzling portfolio of images from Jupiter and Saturn, it didn't take long for that no to change to a yes.

On January 24, 1986, just four days before *Challenger*'s intended launch, the *Voyager 2* spacecraft was scheduled to reward the agency's faith in its abilities, completing its fifty-three-month, 916-million-mile journey from Saturn to Uranus, flying directly through the heart of the densely populated Uranian system and out the other side. The pictures the ship sent home from this encounter promised to be dazzling, but in order to take them, the sometimes balky *Voyager 2* would have to do some fancy flying.

Long ago ground-based scientists had discovered that Uranus is a planet like none other in the solar system. While the other eight planets all orbit more or less upright, Uranus is tipped loopily on its side, completely toppled so that it rotates horizontally, with one pole pointing back toward the sun and the other pointing off into space. The planet probably assumed this drunken position as a result of a collision long ago with a projectile at least as big as Earth. Whatever tipped the world over, such an odd posture meant no end of navigational headaches for spacecraft trajectory planners. Stars flickering on and off in the vicinity of the planet indicated that Uranus probably had at least nine faint rings, all of which were also turned at a sharp angle to the horizontal. Also circling the planet were five comparatively large moons—Miranda, Ariel, Umbriel, Titania, and Oberon, ranging from 292 to 978 miles in diameter—that orbited more or less vertically as well. How many other smaller moons there might be, flying at how many other vertiginous angles, was impossible to say.

Tiptoeing through this planetary train wreck, and doing so in a way that would give *Voyager 2* any hope of flying on to Neptune afterwards, would require some creative navigation—and some necessary compromises. The way the mission overseers figured things, if they were going to get out of the Uranian system alive, they would be able to make a close flyby of only the smallest of the known Uranian moons—Miranda—using its comparatively gentle gravity field to nudge the ship along. This trajectory would help the ship keep a safe distance from both the planet itself and the other satellites—a prudent precaution, but one that would prevent it from taking anything better than arm's-length pictures. Nonetheless, arm's length was still breathtakingly closer than any human or robotic eye had ever gotten to Uranus before, and the JPL scientists were willing to settle for almost any look their spacecraft could give them.

Preparing for this high-wire encounter with Uranus had been a long process. In almost any deep-space mission, cruise time—the time it takes a ship to travel to its intended destination—could be dead time, with nothing for either the spacecraft or flight controllers to do but wait for the ship to get where it was going. Between Saturn and Uranus this four-year-plus period of idleness was longer and deader than most—so dead, in fact, that *Voyager 2* could essentially be put to sleep. Not long after the spacecraft passed beyond the harbor lights of the Saturnian system, its cameras were shut off and most of its instruments were powered down. Now and then a few sleepy sensors might be prodded to life to take this or that measurement of this or that particle field. But for the most part, the ship was kept electronically anesthetized, programmed only to send home a regular readout of its own vital signs and to sound an alert if one of those signs began to flicker or fail.

For this kind of graveyard-shift work, a team of trained scientists was not necessary, and at the same time *Voyager 2* was being unplugged, most of the people responsible for caring for the ship were reassigned to other missions. Shortly before the spacecraft at last reached Uranus, those on-loan scientists were recalled and put back to work preparing for the encounter. Once they were, however, they found to their distress that even as *Voyager 2*'s cameras were pointing toward the steadily growing blue-green world 1.7 billion miles away, the media's cameras were stubbornly pointed toward the launch pad at Cape Canaveral, where the *Challenger* crew was preparing for a decidedly less ambitious mission two hundred miles into space.

Nonetheless, as the January 24 encounter date approached, the members of the *Voyager* team worked hard to publicize the flyby to any member of the media who cared to pay attention to it. By now, Esker Davis had stepped aside as manager of the *Voyager* project and had been replaced by Richard Laeser, a JPL scientist who had worked on a number of planetary projects before and had come to appreciate that one of the first jobs of any project manager was not just to fly a mission successfully but to promote it successfully, selling both the science and the spacecraft to a public that didn't always want to buy. When it came to *Voyager*, he knew, the project managers had always promoted well.

As the first *Voyager* was approaching Jupiter in 1979, reporters from all over the country clamored to be on hand in Von Karman Auditorium for the encounter. While the crowd shaped up to be a good one, however, it was hardly standing room only. In order to fill things out and avoid the embarrassing spectacle of a half-empty Von Karman on national TV, the public affairs office decided to loosen its media standards a bit, granting credentials to reporters from science fiction magazines and other fringe publications who never would have earned a press pass before—the thinking being that a flaky journalist and a mainstream journalist would look pretty much the same to anyone watching the press conference at home on TV.

By the time the Saturn encounter arrived, things had changed dramatically. The Jupiter pictures *Voyagers 1* and *2* beamed home had created such a stir in the media that the requests for Von Karman press credentials soared. The public affairs office once again had the luxury of picking and choosing who would attend the event, and this time the science fiction writers would be excluded. When the Uranus flyby approached more than four years later, competing so futilely with the *Challenger* launch, the demand for access predictably dipped. Regardless, Laeser decided to make the most of the event, scheduling as big and ceremonial a press conference as all of the project managers who preceded him had, and indeed making his bigger still, flinging open the doors not only to Von Karman, but to a smaller press room adjacent to the main hall. Reporters of all stripes would be invited to attend, and Laeser, eschewing his customary seat in the control center, would be there, too, circulating through the two rooms, shaking the hands of his hard-core media supporters, and making sure they knew how grateful his lab was for their ongoing attention.

On the morning of the Uranus flyby, the turnout was a bit better than Laeser had expected—and the show the reporters got was more spectacular. *Voyager 2*'s initial encounter was its close approach to Miranda, a maneuver that would take it to within 19,000 miles of the moon. The rendezvous would occur early in the day, but the signals, traveling 1.7 billion miles from the Uranian system, would need over two and a half hours to arrive. After that they would need another few minutes to be assembled, and only after that would the crowd in both press rooms see the sights that *Voyager* had seen. When the spacecraft's pictures at last did appear on the JPL TV screens, those sights moved the Von Karman crowd to speechlessness.

Glowing on the multiple monitors in the press rooms was the familiar circle of an otherworldly world, filled with one of the most unfamiliar landscapes the reporters had ever seen. The surface of Miranda was a fractured, fragmented, jigsawed mess, a smashed-looking place in which a smooth plain lay hard up against a jagged scarp which itself lay hard up against a capsized cliff or a deep valley. It was almost as if Miranda had somehow been utterly shattered, broken into countless ragged bits, and then had reassembled itself into any pattern that best approximated a sphere.

For a long half minute after the images appeared, the normally voluble reporters simply stared; then, at last, a cheer went up. Laeser, standing in the midst of this, was quickly set upon—his hand shaken, his back slapped, his attention pulled this way and that as the reporters shouted a volley of questions. The scene was repeated again and again throughout the morning, as the speeding *Voyager 2* sent back snapshot after snapshot of the tortured Miranda. After that, the ship proceeded through the Uranian system, returning equally sharp images of other moons, twinkly rings, and at last, Uranus itself, before finally flying away from the planet and heading in the general direction of Neptune. For this one morning, in this one room, Laeser could convince himself that there was no *Challenger* poised on its launch pad in Florida, turning so many media heads as it readied itself for its launch later in the week. Today there was only *Voyager* and Uranus and the clutch of reporters who came to see what the ship and planet had to tell them.

Now, four days later, at 8:30 A.M. on the morning of January 28, a more sober public relations reality had returned. The *Voyager* scientists had had a good ninety-six hours to study the batch of pictures the spacecraft returned and were ready to discuss their preliminary findings with the press. Such a follow-up briefing, Laeser and the others knew, was where the science in a mission like this started getting done, where the geology and planetology work the spacecraft had been built and launched for in the first place was at last made public. This, in theory, ought to be what really interested the reporters, but today, even the comparatively modest crowd from earlier in the week would not be coming around. The press conference at which the scientists would make their first disclosures had been called for 10 A.M. Pasadena time. Only hours after that schedule had been set, word had gone out to the space community that *Challenger,* which had been waiting out bad weather in Florida, had at last been cleared to fly and would be taking off at 8:38 A.M. If the liftoff had been set for later in the day, there was at least a chance that the reporters would take in the JPL conference first—if only by closed-circuit TV—and then turn their attention to the *Challenger* flight. With the space shuttle leaving when it was, however, virtually the entire press corps would be busy for virtually the entire day.

Nonetheless, Laeser once again decided to hold his conference as scheduled and hope for the best turnout possible. While waiting for the event to begin, he and his senior staff figured they might as well do what the reporters all over the country were doing, and gathered in the fourth-floor conference room of Building 264 to watch the *Challenger* launch on TV. By the time the liftoff was just a few minutes away, most of the seats at the room's thirty-person table were filled with *Voyager* scientists, all fixing their attention on a cluster of color TVs, which at the moment were showing an image of the bright white *Challenger* sitting on its pad, sweating vapor into the blue-gray Florida morning.

The men followed the countdown until the engines ignited, the shuttle shuddered, and the entire rattling assembly of spacecraft, fuel tank, and solid boosters muscled themselves into the sky. They watched the ship fly for fifteen seconds, thirty seconds, then nearly a minute. They heard the commander, Dick Scobee, announce, "Houston, we have roll program," as the spacecraft turned on its long axis to orient itself for its sprint to space. They heard Richard Covey, the spacecraft communicator in Mission Control, radio back, "Roger, roll, *Challenger.*" They heard the murmuring voice of the mission narrator confirming that the ship had three good fuel cells, three good engines, three good power-output units.

Finally, they heard Covey call out to Scobee, "*Challenger,* go with throttle up." They heard Scobee call back, "Roger, go with throttle up."

What they didn't hear a few seconds later—what nobody watching the launch on television heard—was the lone voice of a flight controller sitting at a console in Cape Canaveral who looked down at the monitor in front of him and saw, to his horror, that the dappling of numbers that usually filled the screen had suddenly vanished, the monitor going as blank as if the shuttle had suddenly come unplugged.

"Where in hell is the bird?" he called out to no one and everyone.

On the television screens in Building 264—and on millions of screens around the country—it became immediately clear where *Challenger* was, and where it was was nowhere at all. The *Voyager* scientists watched as the tiny image of the rising ship all at once transformed itself into a much bigger circle of smoke. The circle then swelled into a fireball and the fireball grew two ugly fingers.

"What *is* that?" someone in the room asked, knowing full well what it was.

"That isn't right," someone else said, knowing with certainty that it wasn't.

"Obviously a major malfunction," the voice from the television said self-evidently and absurdly.

Dick Laeser and the rest of the *Voyager* team watched slack-jawed as the death of the *Challenger* played out—as the ship blasted its vitals across the Florida sky, as those scorching remains fell sizzling into the ocean, as the crowd at Cape Canaveral pointed and gaped. When the drama was over, the Pasadena scientists stared stunned and red-eyed at one another, saying nothing because, after all, there was not a thing to say. Laeser, the Earth-based commander of a much smaller spacecraft that earlier in the week had had a much better day, then walked woozily back to his office, sat at his desk, and buried his head in his hands. After a moment he picked up the phone and called his director of public affairs.

"Cancel the press conference," he said hoarsely. "Just cancel it."

Nobody, least of all Laeser himself, cared a whit about Uranus this morning.

•   •   •

It would be several months before anyone in the larger space community thought about much beyond the loss of the *Challenger*. First there had to be the burials, then there had to be the wakes, then there had to be the inquests and the investigations and the hurled accusations. Then, finally, there was the inevitable resolution: the vow to build a better ship, a safer ship, one that would never die so gruesome a death again. With that promise, the scientists and reporters and public at large were at last ready to look skyward again, and when they looked, they found Uranus and its moons and the pictures *Voyager 2* had returned of both waiting for them. What they learned from the images was both more and less than what they'd hoped for.

Superficially, Uranus had promised to be something magnificent. With a diameter of 32,000 miles, it was the fourth largest body in the solar system, four times bigger than the modest Earth, and more than 10 times bigger than a nugget world like, say, Mercury. Uranus's vast, swampy atmosphere would almost certainly be a dynamic one. Its fields of radiated energy would almost certainly be crackling ones. It would, on the whole, be a planet well worth studying. When *Voyager 2* looked toward the planet, however, it essentially saw nothing at all.

Uranus, the JPL scientists discovered to their disappointment, was Titan all over again. Instead of Titan's characteristic orange, Uranus was a soft blue, but with that exception, it looked precisely like the far smaller Saturnian moon had looked. There was not a scrap of a cloud anywhere in its atmosphere, not a swirl or an eddy

breaking up the blankness of its face. There was only a vast, undifferentiated gas ball that gave up no visual clues to its composition or
its workings. Chemical sensors aboard the spacecraft *were* able to
sniff out the makeup of the planet's atmosphere and found it to be
nearly 83 percent hydrogen, 15 percent helium, and 2 percent
methane, plus some other stray hydrocarbons. Other instruments
were able to discern a curious, corkscrew magnetic trail following
Uranus as the planet simultaneously rotated on its side and revolved
around the sun. Apart from those barely illuminating findings, however, Uranus showed *Voyager* a face that was utterly expressionless.

But *Voyager* hadn't necessarily come to see the planet, it had
come to see the objects that circled the planet, and those objects delivered on their promise. The ring system that Earth-based astronomers thought they saw through their telescopes indeed existed,
and indeed orbited upright, circling the planet's sideways equator.
The rings were fine and faint formations, but not so fine and faint
that the cameras didn't reveal two additional ones that had never
been spotted from Earth, bringing the total number of the fairy-dust
bands to eleven. Like the rings around Jupiter and Saturn, the Uranian formations were probably the remains of a shattered moon, one
that previously circled the planet intact but at some cataclysmic moment in its history was reduced to rubble. And like the other planets'
rings, these probably would not last forever, decaying and crumbling
and eventually falling out of orbit altogether.

The death of this one moon and the ring it became didn't mean
that the planet couldn't be circled by a whole family of surviving
moons—and a big family at that. As poorly photographed as Uranus
had always been from Earth and as small as some moons could be,
JPL astronomers had little doubt they'd discover many more satellites orbiting the planet than the five that had already been spotted
by telescopes. In anticipation of this, Steve Synnott, who had hauled
in so many moons when the *Voyagers* flew by Jupiter, had developed
new satellite-hunting software that would make the job of spotting
undiscovered bodies far easier than it had ever been before. Rather
than assigning humans to the tedious squint-and-compare business
of studying stacks upon stacks of pictures, trying to find a single

moving dot in the vastness of Uranian space, Synnott instead automated the job. An imaging team member who found a single suspicious point in one picture could load that image—along with hundreds of subsequent images—into a computer and instruct the machine to scan the pictures and alert the scientists if the point seemed to move. If it did, the scientists could then calculate the nature of that motion to determine if it looked like orbital motion. If the motion was predictable enough, they would know they had themselves a moon.

Nonetheless, it was still up to human beings to tell the computer which little pinpoints to study, and that took work. The images that poured back from the spacecraft were messy ones, filled with noise and static and electromagnetic freckles, all of which showed up on the JPL television screens, and any of which could be a Uranian satellite. What's more, the screens themselves were imperfect, with blown pixels and other blemishes occasionally creating distracting dots of their own. Over time, the JPL researchers learned to recognize the unique flaws in each of their monitors, and took to drawing tiny grease pencil circles at random spots on the glass screens, reminding themselves that any dot they found inside a circle was probably not a moon.

Even with all of those distractions, the Uranian moon harvest turned out to be huge. The five known and relatively large moons had all been tracked in orbits between 80,500 and 362,000 miles high. When *Voyager 2* pointed its cameras closer to Uranus and focused them down to spot smaller objects, a whole school of tinier moons swam into view. As nearly as the scientists could make out, there were at least ten previously unknown, boulder-like bodies circling the planet, measuring between 16 and 95 miles in diameter. All of the moons were dark, sooty things, reflecting barely 7 percent of the light that struck them. Such floating lumps of coal looked to be exceedingly old and exceedingly rich in carbon—one more piece of proof that the solar system as a whole is fairly awash in the fundamental stuff of biology. The moons' reasonably orderly orbits suggested that they were not captured bodies snatched out of the skies by the Uranian gravity, but instead formed in place when the planet itself did.

Unremarkable as the moons looked, they were given some decidedly lyrical names. Synnott and the other JPL scientists largely ad-
hered to Uranus's Shakespearean naming tradition, christening the
first nine new moons Cordelia, Ophelia, Bianca, Cressida, Desde-
mona, Juliet, Portia, Rosalind, and Puck. The tenth one, Belinda,
was named after a character in Alexander Pope's *Rape of the Lock*.
Apart from Cordelia and Ophelia, which straddle and shepherd
Uranus's bright Epsilon ring, all of the small moons were found to
orbit essentially alone, having little gravitational influence on the
planet, the other rings, or even one another. As heavenly bodies
went, the ten new moons were thus little to speak of, significant
more for their very decorativeness than for any new scientific insights they offered.

The larger moons were a different matter. As a rule, the planets
of the inner solar system are denser than planets in the outer, mostly
because the blowtorch force of the nearby sun blasts away most
lighter gases, reducing Mercury, Venus, Mars, and even Earth to
small, relatively rocky pellets. More distant worlds like Jupiter and
Saturn, which feel far less of the sun's effects, retain their gases and
thus tend to be far larger and more vaporous. When it comes to the
moons that circle most planets, this density rule also applies, but for
different reasons.

Out in the solar system's successively more remote regions, temperatures fall lower and lower, with the thermometer in the Neptunian system approaching −400 degrees. The four large planets, which
generate their own gravitational and radioactive heat, are able to
compensate for this, keeping their insides hot and preventing their
atmospheres from freezing. The planets' moons have far weaker internal furnaces—and often have none at all. With no way to keep
themselves warm, they are thus completely exposed to the deep
freeze of deep space, growing steadily colder as they move farther
away. At such paralyzing temperatures, even light gases like carbon
monoxide and methane begin to solidify, settling down onto the surface of the world and becoming integrated into its ice-and-rock matrix. Hardened gases, however, are still just gases, and while they
may add to the overall weight of the body, they reduce its overall

density. The five large moons of Uranus appear to abide by this rule, appearing to be a bit less densely packed than Saturn's big moons and much less than Jupiter's.

So much volatile gas and ice chemistry, however, ought to make even relatively insubstantial moons active indeed, and at least some of the Uranian satellites fairly surge with energy. Ariel, measuring 718 miles across, is Uranus's brightest bauble, reflecting more than a third of the sunlight that strikes it. The moon's rhinestone reflectivity is a result of its bright, frozen surface, a surface that appears—by cosmic standards at least—to be relatively young. Though Ariel does have craters, many appear to be partially flooded with a now-hardened material, suggesting that something percolated up from underground and filled them after they formed. More curiously, Ariel, like Mars, appears to be scarred by extensive canyons and crevasses, some hundreds of miles long and more than six miles deep. Unlike Martian canyons, which are characterized by jagged walls and rubbly floors, Ariel's have been troweled—almost polished—until they are shiny and smooth. It's likely that some kind of flowing river of liquid was responsible for this natural buffing, but it's unlikely that the liquid was water, since water behaves like steel at Ariel's −335-degree temperature. Instead, whatever rivers existed on the moon were likely made of liquefied methane, ammonia, or carbon monoxide, driven to the surface by the heat of radioactive elements in the moon's core or by residual tidal pumping from the other Uranian moons.

Umbriel, a similarly sized sister moon located one orbit out from Ariel, is, superficially at least, everything its satellite sibling isn't. The darkest of the large Uranian moons, reflecting only 18 percent of the sunlight that strikes it, Umbriel also has one of the oldest surfaces in the Uranian system, pocked with ancient craters that have changed little since the moon's formation. This lack of surface activity is probably due to the moon's comparatively isolated orbit, which keeps it from getting tidally squeezed, but whatever the reason, Umbriel is clearly a static world, one that has experienced none of the upwellings that have so refreshed Ariel.

Oberon, the highest orbiting of the Uranian moons and, at 943 miles in diameter, the second largest, is similar to ancient Umbriel, with a dark, heavily pounded crust that has seen little resurfacing in the last 4.5 billion years. So uneroded is the face of Oberon that *Voyager 2* spotted a four-mile-high mountain on one of its horizons, the equivalent of a thirty-four-mile-high mountain on Earth. Titania, which measures 978 miles across and orbits one slot lower than Oberon, more closely resembles bright, white Ariel, with a newer, polished surface that has clearly benefited from some kind of periodic refinishing.

Similar as these four moons were to one another in size and composition, they were nothing like the smallest and the last of the main Uranian satellites: tiny, fractured Miranda. Measuring just 292 miles across, Miranda is the lowest flying of the five larger moons, and thus lies deeper than any of the others within the Uranian gravity well. The tidal squeezing that this has caused the moon to experience has evidently taken its toll. When *Voyager 2* made its initial Miranda flyby, the fractured surface of the world stopped the reporters in the JPL press room cold. With later refinement—allowing the scientists to make out features as small as a few hundred feet across—the images became more dramatic still. The scientists saw sheer scarps climbing three miles into the sky; cliff faces scarred by parallel, cat-like scratches, apparently created when huge fault blocks rubbed against one another; whole chunks of surface that appeared to be broken and tipped; vast fault lines running for miles through the crust; sudden changes in the moon's reflectivity, with dark, dull expanses lying in a quiltwork pattern next to bright, glassy ones.

Clearly, the *Voyager* team's initial analysis was a plausible one. This was a damaged world, a broken world, a world that had been busted to rubble as many as five times in its past only to shrug off the insult and gravitationally reassemble itself, fusing its shards together in an increasingly chaotic order. Had the breakup occurred only once, the *Voyager* scientists might have attributed it to a chance meteor hit. The fact that it appeared to have happened again and again almost certainly pointed the finger at gravitational pumping, a tidal violence caused by both the tug of giant Uranus lying beneath the moon and the intermittent turbulence caused by Ariel, Umbriel, Titania, and Oberon passing repeatedly overhead. These forces had helped break Miranda up in the past, and as long as they continued to exist, they would likely do so again.

If the moon was indeed going to crumble once more, however, it was going to crumble alone. On January 25, 1986, less than a day after *Voyager 2* arrived in the Uranian system and three days before seven luckless astronauts lost their lives on Earth, the now well-traveled interplanetary ship sped past Uranus, past its eleven rings, and past the last of its fifteen moons. *Voyager 2* would now have just three more years and one more planet ahead of it before it left the solar system altogether. After that, it would have forever.

Late 1986

It was long before *Voyager 2* reached its rendezvous with Neptune that JPL scientists began to realize it might be utterly pointless to make the trip. If the ship didn't collide with the planet, it would all but certainly fail to take any pictures of it; if it somehow took a few pictures, it would certainly fail to beam them home. No matter how rosily the researchers analyzed the journey that lay ahead of their spacecraft, even the most sanguine of them had a hard time believing that the ship's four-planet mission wouldn't end after three.

The biggest thing that stood in the way of *Voyager 2* successfully reconnoitering Neptune was the problem of light. Sitting nearly 2.8 billion miles from the center of the solar system, Neptune receives only the feeblest firefly illumination from the sun and remains lost in what, by Earthly standards at least, is a permanent night. With a little time, a lot of patience, and a lot of acclimation, it might be possible for a human eye to squint through the gloaming and make out a few details of the place. The less sharp eye of a camera would have a far harder time of things, particularly if that camera were mounted on a speeding spacecraft moving at 35,000 miles per hour over the blacked-out Neptunian landscape.

Making matters worse for scientists attempting such an improbable reconnaissance was the route they'd have to fly to get there. In addition to studying Neptune itself, the *Voyager* team was hoping to collect some pictures of the planet's vast, chemically rich moon, Triton. Discovered in 1846, only a few weeks after Neptune itself, the 1,674-mile-wide Triton had been extensively studied since with ground-based telescopes equipped with chemical spectrometers. Though the data such long-distance scanning could gather was imperfect, most astronomers had come to conclude that Triton is fairly drenched in organic chemicals, particularly ammonia, carbon, hydrogen, oxygen, and nitrogen. Circling Neptune in the same kind of highly inclined, backward orbit Phoebe inscribes around Saturn, Triton was almost certainly subject to all kinds of gravitational tugging from its parent planet, which could easily cause the moon's innards to heat up; this, in turn, could cook its organic chemicals together in any manner of interesting ways.

The problem was, trying to fly close enough to Triton to get a good look at it would take some doing. On its current trajectory, *Voyager 2* was following a path that took it just over the top of the solar system, buzzing the north pole of the planets at little better than crop duster altitude. At the moment, however, Triton's sharply tilted, 157-degree orbit placed it far below the Neptunian equator, and thus completely out of the ship's way. Reaching the moon was not impossible, but to get there *Voyager 2* would have to make the same kind of close pass over Neptune's north pole that *Voyager 1* had made over Titan's south pole when it passed through the Saturnian system. Just as the Titanian gravity had flipped the first ship upward, so too could the Neptunian gravity flip the second ship down toward Triton.

As nearly as the trajectory planners could figure it, managing this navigational stunt at Neptune would be even harder than it had been at Titan. In order to bend its trajectory enough to reach Triton, *Voyager 2* would have to pass barely 3,000 miles above the Neptunian cloudtops. When you're making a high-speed approach to a world whose equatorial diameter exceeds 30,000 miles, a margin of error of 3,000 miles is not much. When that world is shrouded in a dense atmosphere that swirls hundreds or thousands of miles into space, things get even trickier. The trajectory plotters at JPL might be able to map the kind of path *Voyager 2* would have to fly to reach Triton, but whether the ship could survive the maneuver was something else again.

Even if *Voyager 2* did manage to high-step its way around Neptune, fly past Triton, and somehow take pictures of both shadow worlds, there still was no guarantee those pictures would ever make it home. While the great, twelve-foot, parabolic antenna both *Voyagers* carried made it easy for the ships to beam information home to Earth, at some point even the best system reaches its operational limits. The strongest transmitter the spacecraft carried sent out a signal no more powerful than 23 watts—less electromagnetic oomph than a refrigerator lightbulb. Around Jupiter, Saturn, and Uranus, even so modest a signal was more than adequate. Around Neptune, however, the beacon became fainter still, spreading and fading until it was barely detectable at all. The answer to the problem lay not in making the transmissions from the spacecraft more powerful, but making the receivers on the ground listen harder.

Throughout the nine years the *Voyagers* had been flying, the JPL controllers had received data from the spacecraft through a trio of large Earth-based antennas arrayed at equal distances around the world. For the Neptune encounter, the JPL team hoped to add a lot of other ears. In Socorro, New Mexico, the space community maintained a very large array of deep-space antennas that had been named, prosaically enough, the Very Large Array. The VLA was made up of no fewer than twenty-eight antennas, each measuring no fewer than eighty-one feet in diameter, and all built to listen not just to the breathy piping of a single spacecraft barely 2.7 billion miles away in space, but to the electromagnetic roar of the cosmos as a whole, the accumulated emissions of countless billions of bodies located countless billions of miles away. If you were somehow able to aim this system to *Voyager 2,* you ought to be able to do a lot of listening, indeed.

The problem was, the *Voyager 2* scientists were not the only ones hoping to lay claim to the VLA. At any given time, cosmologists all over the world were regularly petitioning for a little time with the huge system. With the *Voyager* flyby of Neptune less than three years away and the waiting list a long one, the JPL bosses knew they would never get a turn with the antennas by the ship's 1989 rendezvous date unless they came up with some way to jump the line. They figured they knew just how.

For some time the engineers at the Pasadena lab had been studying the specs of the VLA system, and had come up with a way to redesign its programming guts and dramatically boost its power. If the government and university researchers who ran the VLA gave their approval, JPL would be happy to get to work straightaway on such an upgrade. All the lab asked in return was ninety days of exclusive access to the VLA in July, August, and September 1989, when *Voyager 2* would be approaching, reconnoitering, and leaving the Neptunian system. After the ship had passed through and the encounter was over, the upgrade would be the VLA's to keep. No fools, the managers of the antenna array saw the offer for the sweetheart deal it was and quickly accepted the arrangement.

Figuring out a way to take pictures that the beefed-up antennas could receive was a trickier business. Even in the dim conditions of the Neptunian system, it was possible for *Voyager 2*'s cameras to collect enough light to assemble an image, provided their shutters were left open long enough. In the early days of the MJS77 program the JPL designers had anticipated just this problem and had come up with the image motion compensation technique that allowed the spacecraft to pivot as it flew by a target planet, keeping its eye fixed on a single spot on the world so that the image wouldn't blur. Such a method was used successfully during the Saturn and Uranus encounters, but during those flybys the relatively bright light of the comparatively close sun had made the job easy. Out in the blackness where Neptune spins, the shutter would have to be left open for up to a full sixty seconds to gather enough light for a usable picture, a stretch of time that could challenge even the most deft motion-compensation strategy.

The answer was not to try to capture Neptune or Triton in a single photographic frame, but rather, in a lot of close-up frames. The greater the proximity of the ship to the planet and its big moon, the greater the detail it would be able to make out in the weak reflected sunlight. Taking such extreme close-ups made it impossible to frame either world in its entirety, but if the cameras took a patchwork of partial images—shifting their focus every sixty seconds to capture one part of the terrain, then another, then another—they ought to be able to assemble enough photographic panels for a full, panoramic portrait or two.

That was the theory anyway, and the *Voyager 2* imaging team was convinced it was a good one. The only remaining challenge, after the imagery and antenna problems had been worked out, was the ongoing trajectory problem, and the JPL engineers could not do a thing about that. If *Voyager 2* was going to fly by Triton, it was first going to have to pass within 3,000 miles of Neptune, facing the real possibility that it was going to fly *into* Neptune. The planet's atmosphere, the JPL scientists realized, would either reach out and claim their ship or it wouldn't. Either way, it would be another three years before they knew for sure.

August 25, 1989, 1 A.M.
Pasadena Time

From the surface of Triton you'd rarely get the chance to watch Neptune rise or set. Certainly, you'd *see* the planet. Indeed, it would be impossible not to see it: a dimly lit, ocean blue world measuring 31,000 miles across, lying in space just 220,000 miles away, it would cover fifteen times more sky than a harvest moon covers on an autumn night on Earth. Yet, unlike the moon, it would almost never rise or set.

What Neptune would appear to do is glide. Far ahead of you, hovering low above one of the vast, arctic ice plains that make up Triton, the planet would hang stationary in the sky, its blue disk barely grazing the white line of the horizon. Quickly, then—since Triton speeds around its parent planet at nearly six times the velocity Earth's own moon moves—Neptune would appear simply to drift right, or north, riding the ice prairie like a mammoth sailing ship, only vanishing from view when it had traveled into the distant north-ernmost latitude, where it would sink briefly out of sight. A short time later it would reappear to the left, or the south, once again hug-ging the horizon, and once again moving steadily to the right, with-out ever moving even incrementally up.

It's the sharply skewed angle at which Triton orbits Neptune that causes the planet to appear to follow such an unusual path, but from the surface of Triton it's Neptune that appears cockeyed, lying on its side with its south pole pointing left and its north pole pointing right. It's impossible to say what you'd have seen if you looked at that north Neptunian pole from your vantage point on Triton at about an hour after midnight California time on the morning of August 25, 1989. Certainly, the spacecraft that was soaring over the planet at that pre-cise instant was too small to see from so great a distance. Even if it *was* big enough, it was so close to Neptune that even a single sigh of atmosphere that happened to drift up at the wrong moment would have been more than enough to obscure it completely—and quite pos-sibly to bat it straight out of the sky. On the other hand, the ship was a bright thing—a crinkled clump of shiny gold metal with a higher re-flectivity than anything else in the local sky. It was at least theoreti-cally possible that a little of the feeble sunlight that makes it out this far might have been able to catch one of the facets of the spacecraft and flash a quick flicker down to you.

Regardless of whether a person standing on Triton would know the spacecraft barnstorming Neptune existed, countless people 2.7 billion miles away on Earth certainly did. As *Voyager 2* buzzed Nep-tune that Friday morning and six hours later buzzed Triton itself, it beamed a stream of 23-watt signals back toward terrestrial listeners at a speed of 186,000 miles per second. Even at that velocity the transmissions took four hours and eleven minutes to reach the wait-ing receivers in Socorro, New Mexico. From there they were relayed an additional 650 miles to the Jet Propulsion Laboratory in Pasadena, where they were translated into pictures. When they were, they told a complicated tale.

After *Voyager 2*'s rendezvous with Uranus three years earlier, JPL scientists did not necessarily expect much from the pictures the ship would return of Neptune itself. It was Uranus's fantastic distance from the sun that likely contributed the most to its unremarkable appearance, with the extreme cold of deep space helping to keep the planet still and quiet, and preventing so much as a breeze from disturbing its heavy quilt of atmosphere. Out around the vicinity of Neptune—a billion miles more distant than Uranus—things should be colder and quieter still.

But Neptune wasn't quiet. When *Voyager 2* engaged its thermal sensors and pointed them toward the planet, it discovered that Neptune actually radiates twice as much heat and nonthermal energy as it absorbs from the sun. Since even a perfectly polished mirror world could never reflect more energy than it receives, Neptune must be generating its own, likely from a dense mass of radioactive elements in its core. While the upper layers of the Neptunian atmosphere are a frigid −360 degrees, this sizzling planetary center causes the lower layers to soar above 900 degrees, and this stirs the atmosphere up pretty violently.

Around the Neptunian equator, gales howl at a supersonic 1,200 miles per hour. The blue face of the planet is streaked by white, wind-blown cirrus clouds, some of which move even faster, speeding around Neptune's entire 96,633-mile circumference in just sixteen hours. The southern hemisphere is stained by a great dark spot as wide as Earth itself, where anti-cyclonic winds blow westerly at nearly 700 miles per hour. In this meteorological maelstrom, atmospheric formations come and go so quickly that in just the thirty-six hours *Voyager 2* was watching, the shredded white clouds that surround the blue spot dissipated almost completely and then were quickly replaced by other ones.

The expanse of space surrounding the planet was just as complex. Earth-based astronomers had long observed the same kind of star-flickering phenomenon in the vicinity of Neptune that they had observed near the other outer planets, and concluded that this last of the giant worlds probably had at least a tenuous set of rings, too. In the case of Neptune, however, those rings appeared to be incomplete, since the stars seemed to wink off at some points near the planet but not others. This led the astronomers to conclude that Neptune is probably decorated not by complete bands, but by arcs, partial circles that simply do not contain enough material to stretch all the way around the world.

The theory was a sensible one—and indeed, judging by the visual evidence at least, was the *only* sensible one—but when *Voyager 2* arrived in the Neptunian system it found that it simply wasn't true. Neptune, it turned out, has no fewer than four concentric rings, all running a full 360 degrees around the planet's equator. While the rings are complete, they are surprisingly sloppy, dense and opaque in some points and transparently faint in others. The lack of tidiness is probably caused by a scattering of large boulder-like masses that are stirred in with the otherwise fine mash of the ring material, causing gravitational clumping that blocks out stars in some stretches of the rings, while allowing them to shine easily through elsewhere.

Equally striking were Neptune's moons. Before *Voyager 2* flew, the astronomical community knew of only two Neptunian satellites: the whale-sized Triton and the much smaller Nereid, a 210-mile moon discovered by Dutch-born astronomer Gerard Kuiper in 1949. As with Uranus, however, given the enormity and distance of Neptune, the JPL scientists always suspected there must be at least a few other small moons secretly circling the planet, and when *Voyager 2* arrived, they found out they were right. Flying through the Neptunian system, the spacecraft spotted no fewer than six previously undiscovered satellites, orbiting the planet from a low of just 30,000 miles to a slightly loftier 73,000 miles.

Of the six new moons, five—Naiad, Thalassa, Despina, Galatea, and Larissa—were little more than cosmic flotsam. Measuring between 36 and 119 miles in diameter, they fly low, stay close to Neptune's equatorial line, and speed around the planet in as little as seven hours. Clinging to Neptune's skirts this way, they have largely escaped bombardment by meteorites, since any incoming projectile would be drawn far more easily to the enormous, gravitationally heavy planet rotating slowly beneath the moons than to the darting, gravitationally light moons themselves. Just how the satellites came to fly in such low orbital slots was not clear, but the fact that their paths follow Neptune's center line so precisely suggests that, like the new Uranian moons, they hadn't just flown crazily in and gotten captured by the planet. Rather, they were probably the remains of a larger moon that accreted from the same primal cloud as Neptune and then broke apart as a result of a random collision. Much more than that, *Voyager 2* was not able to determine. As small as the moons are and as remotely as they were studied, the ship's cameras were able to gather up only a few fleeting pixels of visual information before the spacecraft raced on.

The last and highest flying of the new moons—Proteus—yielded at least a little more information, mostly because of its greater size and gravity. Measuring 257 miles across, it is roughly spherical in shape, but only *very* roughly. Compared to the neatly circular stars, planets, and moons that populate the cosmos, Proteus is an asymmetrical mess. A body this size, with a density as great as Proteus appears to have, almost certainly strains the very structure of the moon itself. Like Saturn's Hyperion, Proteus is probably only a few spoonfuls of matter away from being just large enough to collapse into a sphere. Indeed, with a diameter nearly eighty miles greater than Hyperion's, Proteus holds the distinction of being the largest irregular body, or the smallest near-regular one, in the solar system.

But it was Neptune's pair of better-known moons that yielded the most useful data during the *Voyager 2* flyby. Nereid, slightly more reflective than Earth's moon and slightly smaller than nearby Proteus, had already been known to be an erratically orbiting body, but the spacecraft showed how erratic. The moon sails around the planet in a nosebleed-high orbit, with an average altitude of nearly 3.4 million miles. Its average altitude, however, is not a reflection of how swooping that orbit is. At its lowest, Nereid comes within 841,000 miles of Neptune; at its most remote, it strays nearly 5.9 million miles away. Completing one of these huge, egg-shaped laps takes time, and while a moon like tiny Naiad can make nearly four circuits of the planet in the space of a single Earth day, Nereid takes nearly a year to complete just one. A satellite with so irregular an orbit was almost certainly imported from somewhere else in the solar system, and the JPL scientists were pretty sure they knew where.

In 1951, just two years after discovering Nereid, Gerard Kuiper was conducting some calculations about the total mass of the known solar system and came to the conclusion that at least some of it was missing. Forming bodies as big as the sun and the planets is not an efficient process, and for every clump of matter that gets gathered up and patted down into a star or a world, at least a few scraps get left behind. The overlooked material would likely be the most remote— the primal stuff that stayed out of the gravitational reach of the accreting bodies. Marshaling what he knew about the solar system's collective gravity and mass, Kuiper estimated how much lost matter there should be and where it should be located.

Surrounding the known solar system, he concluded, there ought to be a thin belt of matter reaching more than two and a half times farther into space than the 3.7-billion-mile orbit of Pluto. The matter should be made of the most ancient ingredients—rocky, carbon-rich lumps laced with organic gases and other volatiles, and occasionally covered by fluffy water ice and snow. There was no telling how many objects there might be in this river of debris, but Kuiper guessed that there ought to be tens of thousands, many as big as fifty or sixty miles across and a few much bigger.

More complex than this belt of theorized material—which was quickly dubbed the Kuiper Belt—was another hypothetical formation called the Oort Cloud. First proposed by Dutch astronomer Jan Oort in 1950, the Oort Cloud was essentially the same kind of formation as the Kuiper Belt, but fantastically bigger and fantastically farther away. Surrounding the solar system in a three-dimensional sphere instead of a mere two-dimensional disk, the Oort Cloud is thought to stretch eight hundred to one thousand times the distance of Pluto and, like the Kuiper Belt, is probably made of the icy, rocky residue left behind when the solar system formed. In the case of Oort Cloud objects, however, the material didn't simply hang at the fringes of the solar system while the sun, planets, and moons were accreting. Rather, it was actually in the thick of things, swirling among the nascent bodies as they were condensing. At some point, however, this loose detritus flew by one of the large outer planets at precisely the wrong angle and with precisely the wrong momentum and was gravitationally picked up and thrown. Propelled with all the force a gas giant could muster, the stray material found itself flung to the outermost frontiers of the solar system, where it was too close to escape the gravity of the sun completely, but much too far ever to be a true part of its family of planets and moons.

Not all Kuiper Belt and Oort Cloud objects would remain in such cosmic exile, however. Occasionally, one would stray from the distant swarm and tumble in toward the interior of the solar system. If the rogue was able to fall so far that it actually approached the sun, the solar energy would cause its gases to stream and flare, turning the frozen rock into a common comet, one that would fly in and out of the solar system again and again, as long as its new, irregular orbit survived. Other Kuiper Belt and Oort Cloud refugees would not be able to fly so free. Streaking in through the remote reaches of the solar system, they would have to tumble past Jupiter, Saturn, Uranus, and Neptune. If the trajectory an incoming rock was following brought it too close to one of the big worlds, there was always a chance it might get stuck on its gravitational flypaper, abandoning its original path and entering a permanent orbit around the planet.

That, the JPL astronomers figured, was probably what happened to Neptune's little Nereid, whose elliptical orbit looked like a last gravitational artifact of a violent capture. More important, it was also what probably happened to the planet's giant satellite Triton. The second most highly inclined moon in the solar system—behind Saturn's Phoebe—Triton not only circles Neptune at a 157-degree angle, but also clockwise, moving in a direction opposite almost all of the solar system's other worlds. An orbit like this suggests not only that Triton was captured by Neptune, but also that that capture was even more violent than Nereid's. The harder the gravitational brakes were slammed on, the more skewed Triton's orbit was likely to be, and the more skewed the orbit was, the more tidal energy the moon would likely absorb. This would have generated a lot of heat, meaning that Triton ought to be a warm place.

When *Voyager 2* turned its thermal sensors toward Triton, however, it found just the opposite. Whatever tidal energy the moon might once have radiated had apparently been largely spent, meaning that Triton was not only not warm, but paralyzingly cold—about −391 degrees, as far as the spacecraft could tell. This qualified Triton as the coldest object in the known solar system. Most of the frigid moon is covered by a solid permafrost of nitrogen, carbon dioxide, carbon monoxide, and methane—materials that would normally vaporize at the barest breath of solar warmth, but out in the Neptunian neighborhood, harden into icy armor. This frozen skin shines so snowy a shade of white that Triton as a whole reflects roughly 70 percent of the sunlight that strikes it. So far from the center of the solar system, that is not much sunlight at all, but if Triton were somehow made to orbit Earth—and were somehow able to stay frozen while it did—it would shine nearly six times brighter than a full moon.

Cold as Triton's exterior is, however, residual tidal forces do appear to have an influence on its innards. The moon as a whole is largely free of craters, indicating an outward-radiating heat that has repeatedly melted and resurfaced the crust. Fine cracks run like webwork through some parts of the surface, suggesting the same kind of thermal expansion and contraction that fractures Jupiter's Europa. Elsewhere, there are numerous glass-smooth plains, calling to mind the seas on Earth's moon. Unlike those lunar basins, however, which were smoothed over by molten rock seeping up from underground, Triton's seas were likely formed by a cold ammonia and water slurry that oozed slushily up from the warmer interior.

Of course, just when all this crustal sloshing and shifting took place is a mystery, and it's possible that recent epochs have gone by without much fresh resurfacing on Triton at all. But even if most of Triton's tidal warmth has dissipated, the Neptunian satellite is not a dormant place. As Voyager 2's initial images of Triton came back, JPL scientists saw features that they could swear looked like active volcanic geysers. Here and there over the ice white surface of the moon there appeared to be tall, black, feathery plumes, rising so high in the sky they even cast shadows on the ground below. Elsewhere, the plumes and shadows were gone but a distinctive black streak remained on the ice—the ghostly smoke of an earlier geyser that had settled back to the surface, leaving a telltale stain.

As tempting as it was to conclude that the thermal engine behind these blasts of icy exhaust lay in the tidal energy of the planet, the *Voyager* astronomers bet that the likeliest answer was the sun. Even in Triton's deep-space neighborhood, at least a shimmer of solar energy does manage to survive. When this faint illumination strikes Triton, it easily penetrates the nitrogen ice that covers its surface and bathes the black, carbon-rich layer underneath. Like all dark materials, carbon absorbs light and heat—even the impossibly weak light and heat from a sun more than 2.8 billion miles away. As the carbon grows incrementally warmer, the frozen nitrogen lying directly atop it flashes into gas and begins migrating beneath the overlying crust looking for a weak spot. When it finds one, it blows through the icy shell and explodes into the sky, carrying clouds of sooty carbon with it. After the blast of gas fully expends itself, the carbon settles back down to the ground and the crack closes over.

Though Triton's geysering has likely gone on for billions of years, it is unlikely to last billions more, largely because the moon itself is probably doomed. Planets and their satellites generally try to achieve a gravitational balance, and one of the things that throws that balance off is a retrograde orbit. When a moon orbits clockwise like Triton and a planet rotates counterclockwise like Neptune, the constant gravitational drag that results causes the moon to lose energy. As this happens, the satellite drops lower and lower in its orbit until—after millions upon millions of years—it simply falls out of the sky and collides with the planet below. Triton is now undergoing just such orbital decay and is all but certain to come to just such a calamitous end. The only thing that could save it will be if it plunges in at such an angle and such a speed that Neptune's gravity tears it apart before it completes its fall, turning it into a cloud of rock and ice that disperses around the world, forming a grand and gaudy set of rings. No matter how Triton meets its end, it is sure to die. And it will be Neptune itself, whose total number of known moons so recently grew to eight, that will be responsible for taking its first daughter's life.

•   •   •

On August 25, 1989, before the California morning had passed to afternoon, the Jet Propulsion Laboratory in Pasadena received a signal confirming that *Voyager 2* had passed the last of Neptune's moons, dropped out of the bottom of the solar system, and begun its infinite fall into interstellar space. In twenty-three years it would reach the heliopause, that point past which the effects of the charged particles in the solar wind are no longer felt. In 24,000 years it would reach the Oort Cloud. In 294,000 years it would make its first comparatively close approach to a star other than the sun, flying by the bright solar furnace of Sirius. Within a million years it would pass a dozen more stars.

Where *Voyager 2* will go after that is more difficult to chart, but long before it left the ground the JPL designers decided to plan for its journey. Attached to the side of both *Voyager*s is a twelve-inch gold-plated disk, etched with analog grooves like an ordinary LP record. Protecting the disk is an aluminum cover engraved with diagrams explaining in mathematical symbols the origin of each spacecraft, its basic trajectory, and a rough idea of when it was launched. Additionally, there are a few schematic scribbles explaining how to operate a tiny stylus stored beneath the cover.

Should anyone find the spacecraft, apply the stylus to the grooved disk, and somehow know to connect the whole apparatus to a turntable, a speaker, and a video system, the golden record would reveal a lot. Encoded in its grooves is a portfolio of color images from Earth, including seashores, snowflakes, dolphins, eagles, a mother nursing a baby, a house under construction, a fishing boat spreading its nets, and a gymnast spinning through her performance. Also included are sounds of the wind, the rain, and the surf; a cricket chirping, a frog croaking, and a horse clopping; a tractor, a riveter, a train, a bus, a Saturn 5 rocket leaving the ground, and spoken greetings in dozens of languages. Finally, the disk includes music—Mexican music, Javanese music, Peruvian music, Chinese music, Japanese music, Senegalese music; there is a little bit of Bach, a little bit of Mozart, a little bit of Beethoven, a little bit of Chuck Berry, and a complete recording of "Melancholy Blues," by Louis Armstrong and his Hot Seven Band. In all, there are 116 pictures, thirty-nine sounds, fifty-five samples of spoken languages, and twenty-seven selections of music.

It is virtually certain, of course, that the music will never be heard, the pictures will never be seen, the language will never be listened to or fathomed. Instead, the ships are probably destined to wander forever and alone, journeying indefinitely through the vast wilderness of the galaxy, without ever being noticed, much less intercepted. In the laughably improbable event that they are one day found, however—in the event that they are somehow captured and the code their records carry is cracked—the *Voyager* spacecraft could serve as one of humanity's most profound statements. There is some satisfaction, after all, in boasting to the cosmos that a civilization that could build a rocket and launch a space probe could also produce a Louis Armstrong.

7

The Deep Solar System

April 1998

If you wanted to hijack the *Voyager* spacecraft, you'd have a pretty easy time of it. Certainly, making contact with the two ships would be no simple matter; deciding what to say to them after you finally did get in touch would be a tricky business, too. But before you could communicate with the distant probes, you'd have to figure out where they were, and that at least would be a snap.

Where you'd find the two *Voyager* ships—or where you'd find the lonely listening post that would allow you to talk to them at least—would be in an unremarkable redbrick building on an unremarkable stretch of Madre Street, in Pasadena, California. Up the road a bit, just one or two clicks to the north, you'd find a little strip of shops—a grocery store, a coffeehouse, a magazine stand, a pet shop. Off in the other direction, about the same number of clicks to the south, you'd find the turnoff to Interstate 210, which would take you to the Jet Propulsion Laboratory. You'd know you were approaching JPL even if there weren't any signs to tell you so, mostly because of all the obstacles that prevent you from actually getting there. The sprawling facility is protected on one side by a forbidding stretch of patchy green mountains and on the other side by fences and guard booths and a series of round-the-clock checkpoints—just the kind of impenetrable perimeter you'd expect around so glamorous a lab doing such glamorous work.

There was a time when the *Voyager* listening post enjoyed the same kind of protection. But then, there was a time when the people behind the JPL barricades were fiercely concerned with what happened to the pair of ships. Every morning hundreds of scientists and support workers were waved past the guard stations and cleared into the buildings, and there spent the entirety of their days seeing to it that the ships kept flying, their instruments kept working, and the planets and moons they were approaching were the ones they were supposed to be approaching. But that was a lot of years and a lot of planets and a lot of cosmic miles ago. Things are different now.

If you wanted to reach the *Voyager* spacecraft in the last few years, what you'd have to do first is reach Richard Rudd. Rudd is the mission director of the *Voyager* program—maybe the eighth one, maybe the ninth one, maybe the eleventh one; people stopped paying close attention to the exact number after the whole project was bundled up and moved to its redbrick building on Madre Street. Not that the *Voyager*s don't still turn in a good day's work. *Voyager 1,* which hasn't been within hailing distance of a planet or moon since it soared up and out of the solar system in the autumn of 1980, is now 6.51 billion miles from Earth, or about twice as far from the sun as Pluto is and seventy times more distant than tiny Earth. *Voyager 2,* which fell out of the solar system after plunging by Neptune nearly a decade ago, is about 5.02 billion miles away, or fifty-four times the Earth-sun distance.

Out in that cosmic wilderness, the spacecraft's magnificent cameras don't get much of a workout, and it has been more than eight years since they sent home so much as a single grainy image. The final picture a *Voyager* did return, however, was a dandy. In 1991, eleven years after *Voyager 1* completed its primary mission, JPL scientists made an unexpected discovery. Calculating the positions of all nine planets and comparing them with the position of the spacecraft perched high above the solar plane, they realized that if the ship's cameras were aimed precisely enough and its shutters were held open long enough, it just might be possible to take the first-ever family portrait of the entire known solar system.

The imaging team told the engineers just how the picture should be taken, the engineers sent the instructions up to the ship, and over the next several days the ship did its best to execute the command, turning and firing, turning and firing, trying to herd all nine planets into a single photographic frame. As it turned out, it wasn't possible. Tiny Mercury, the innermost of the planets, was all but entirely washed out by the glare of the nearby sun. Mars, a bit larger and a bit farther from the solar system's center, was lost in shade. Pluto, spinning in its cosmic wilderness, was shrouded in shadow as well. Nonetheless, Venus, Earth, Jupiter, Saturn, Uranus, and Neptune were all visible, and all did sit still for the camera, appearing in the frame as six tiny points of light freckling a canvas that was otherwise a pure, glossy black.

The extraordinary photo was quickly processed into first dozens, then hundreds of copies and distributed around the JPL campus. Scientists tacked the picture up on their bulletin boards, taped it onto their office doors, took it home to show to their families. And then, almost by consensus, they forgot about it. There were other missions to be flown, after all, and newer ships to be built—ships that would have more serious work to do than fooling about with something as frivolous as a planetary family picture—and it was those projects that deserved the lab's attention now. By the time the second smallest of the points in that final *Voyager* photograph—the faintly blue one located not far from the sun—had completed one more rotation, the JPL engineers instructed the two ships to close their photographic eyes for good.

Nonetheless, even now, at the century's end, *Voyagers* 1 and 2 do still communicate with home—and quite a bit, in fact. Anywhere from ten to sixteen hours per day, the now-blind ships continue to speed through space, sniffing the void around them and sending home data on the fields and particles, solar wind, and magnetic fields they find there. In all, seven sensors continue to operate on both spacecraft, though the plasma detector on *Voyager 1* has worked only spottily for years, having been scalded by Jupiter's radiation fields when it flew by the planet in 1979. The transmissions that do come back from the *Voyagers* generally come back to Richard Rudd, who collects and collates them, and studies what they have to say to him. He knows, however, that what they do tell him is not likely to make many headlines. So routine are the ships' incoming signals and so unhurried are the doings in the lab that now and then the cubicle workstations outside Rudd's office door—the stations where a small team of flight controllers ought, in theory, to be sitting—are left completely unattended. To prevent the spacecraft from falling into careless or mischievous hands, some of the mission scientists have been careful to glue bright orange paper warnings to the keyboards and monitors of their unsupervised computers.

"CAUTION," the little stickers say. "THIS IS A LIVE *VOYAGER* CONSOLE. DO NOT TOUCH." It is here you could come to talk to the *Voyager*s if you ever had a mind to. And it is here, if you had the skills and the inclination and a slightly criminal bent, that you could try to shanghai the ships. The catch, of course—and this the JPL engineers know—is where could you try to send a spacecraft that's already nowhere at all?

The *Voyager*s thus continue to fly largely unremarked upon, telling their caretakers on the ground little that's scientifically newsworthy. Even if the control center on Madre Street is a sleepy one, however, that doesn't mean JPL engineers as a whole have nothing to keep them busy. Elsewhere in the cosmos, as it happens, other spacecraft are saying plenty.

For all the worlds the *Voyager*s toured, it was Jupiter that most demanded a return visit. It wasn't just the planet's vast size that captivated the scientists; it wasn't just its diaphanous rings or its swarm of moons. It was, ultimately, all of those things. A planetary system so dense and complex could not possibly be fathomed by a *Voyager* spacecraft blowing by at 35,000 miles per hour and then vanishing forever into deeper space. What was needed was a bigger and more versatile ship, one that wouldn't just reconnoiter the world but orbit it, survey it, dodge and dart among its moons, and perhaps even taste the atmosphere and chemistry of the planet itself. No sooner had the *Voyager*s completed their own Jovian flybys in 1979 than the JPL engineers began working on just such a souped-up space machine. They would call their spacecraft, fittingly enough, *Galileo,* and designing it would present them with challenges like none they had ever encountered before.

The problem with trying to put a spacecraft into orbit around a world is that that spacecraft has to be a big one. In order to leave the straight, ballistic path the *Voyager*s flew and instead inscribe an orbital circle, an interplanetary ship must first slow down, bleeding away enough speed and energy to allow the planet's gravity to catch and hold it. That kind of deceleration requires a powerful engine— one made up of a lot of machinery and carrying a lot of fuel; that, in turn, means a lot of weight. The *Voyager* spacecraft were relatively light, relatively lean machines, tipping the scales at just 1,793 pounds each. The *Pioneer*s were a flyweight 568. *Galileo*, the JPL engineers figured, would need to weigh close to three tons.

More difficult than the job of building so big a ship would be the job of flying it. If mission planners were serious about trying to sample the Jovian atmosphere, they were going to have to come up with a way to bring their spacecraft into direct contact with the planet's gases—and that could spell trouble. Flight planners designing trajectories for Earth-orbiting spacecraft know that even the most tenuous wisps of high-altitude air can be fatal to a ship, dragging on it until its attitude destabilizes, its orbit decays, and it goes plunging toward the ground like a brightly flaming meteor. A far faster interplanetary spacecraft encountering the far denser atmosphere of Jupiter would be annihilated on contact. In order to prevent this from happening, the JPL engineers would have to build a *Galileo* spacecraft that was essentially two spacecraft.

Stowed aboard the ship before it was launched would be a four-foot-wide, three-foot-long bullet-like probe, packed with half a dozen sensing instruments, including a wind-speed monitor, a chemical spectrometer, a lightning detector, a particle sensor, and a battery to power them all. During the years the spacecraft was making its transit to the Jovian system, the probe would do essentially nothing at all. Five months before arrival at Jupiter, however, *Galileo* would fire the little instrument package into space and then, with a puff from its thrusters, nudge itself off on a slightly different trajectory from that of its now-jettisoned cargo. Over the 150 days that followed, the main body of the *Galileo* ship would continue on a broad, arcing path that would take it obliquely toward the planet, where it would fire its engine and ease gently into Jovian orbit. The probe, by contrast, would aim itself for a bull's-eye strike just above the planet's equator.

When the ballistic projectile did hit Jupiter, it would hit hard, colliding with the atmosphere at nearly 100,000 miles per hour. Punching through the air, blunt end first, it would plunge for a full two minutes, until atmospheric friction had slowed it down to less than a hundredth of its original speed—slow enough to allow it to deploy a parachute. If the big cloth canopy successfully opened—by no means a sure thing—the probe would immediately switch on its sensors and begin sending what it learned about the atmosphere around it back to Earth. For the next hour and a quarter the probe would operate, dropping about 125 miles through the orange-red clouds, while the air pressure around it rose from almost nothing at all to twenty-five times the pressure found at sea level on Earth. Finally, the crush and heat of the atmosphere would grow so great that the probe would simply crumple, melt, and ultimately vaporize. Just over seventy-five minutes after arriving at Jupiter, the exquisitely engineered instrument package would be reduced to mere superheated vapor, adding its own tiny breath of largely metallic gases to the hydrogen and helium world it now inhabited. High above, the surviving mother ship would spend the next two years orbiting the planet alone.

That was the plan at least, but executing it—and paying for it—would be no small matter. It took more than a decade and more than a billion dollars before *Galileo* was actually engineered, built, and readied for flight, and it was not until October 1989 that it actually left its Florida launch pad—setting out on a slow, looping trajectory that would not get it to Jupiter until December 1995. But before *Galileo* could travel even a quarter of the way to the planet, it would become clear to JPL engineers that they were flying a badly snakebit ship.

The first of the spacecraft's problems concerned its high-gain antenna. Like the *Voyager* ships, *Galileo* was equipped with a sensitive antenna dish that would allow it to communicate with Earth from distances of hundreds of millions of miles. The *Voyager* antennas had been impressive assemblies, measuring about twelve feet across. For *Galileo*, the JPL engineers hoped to improve communications even further, equipping the new ship with a larger high-gain system measuring a formidable 15.75 feet. But while the bigger system was, indeed, a more powerful one, transporting it into space presented problems.

The *Voyager*s were fired on their interplanetary path by a Titan-Centaur booster, a stout, two-stage rocket with a roomy cargo bay more than big enough to handle the oversized high-gain antenna. *Galileo*, by contrast, was designed to be flown into Earth orbit inside the cargo bay of a space shuttle and then released into space to make its way to Jupiter on its own. A shuttle's cargo bay, however, is a cramped place, hardly big enough to accommodate the old *Voyager* antennas, much less *Galileo*'s larger one. The answer was to reengineer the *Galileo* antenna entirely. While the *Voyager* dishes were molded from epoxy into a single rigid piece, the *Galileo* high-gain antenna would be completely collapsible. Built from eighteen metal ribs and covered in a flexible gold mesh, it would fold up like an ordinary umbrella before being packed away in the shuttle. Later, after the ship had been set free from the orbiter and fired on its trans-Jupiter path, a signal would go up from the ground releasing the ribs, unfurling the antenna, and at last giving the spacecraft its electronic voice.

Precisely when that antenna deployment would take place presented problems of its own. In order to build up the speed it needed to fly all the way out to Jupiter, *Galileo* would not follow a straight, as-the-crow-flies path to the planet. Rather, it would be sent on a sort of corkscrew trajectory through the inner solar system, flying around and around the sun in an orbit that would take it once past Venus and twice past Earth. Each time the ship approached one of these large worlds, it would receive a gravitational kick that would add hundreds or thousands of miles per hour to its speed. Those velocity boosts would cause *Galileo*'s orbit to grow wider and wider, until finally, after six years, it would have spiraled all the way out to Jupiter.

Appealing as this fuel-efficient trajectory was, it presented some problems. If mission controllers unfurled the giant high-gain antenna when the ship was still moving through the inner solar system, there was a risk it could be damaged beyond repair as the reflective gold mesh absorbed the full, wilting heat of the nearby sun. To prevent such a mission-ending meltdown, JPL planners decided that *Galileo* would keep the skirts of its high-gain demurely furled for the first eighteen months it was in space, doing all its communicating with Earth through a far smaller low-gain antenna—one able to communicate at a rate of just eight data bits per second. Only when the ship was past the orbit of shadowy Mars would the far larger high-gain be brought on-line. Conversing with *Galileo* at the eight-bit rate the low-gain antenna would allow would be a painstaking business—a little like trying to have a conversation with a person who had all at once lost the ability to speak complete words and was instead reduced to spelling them out. But since such a plodding dialogue would last little more than a year and a half out of an eight-year mission, and since the alternative—exposing the high-gain antenna to the full ferocity of the close-up sun—was out of the question, mission designers figured they'd simply make do.

Even so redundant a pair of antennas, however, left at least a few engineers feeling skittish. While the low-gain antenna would be a perfectly adequate backup system during the lazy months *Galileo* was circling through the inner solar system, when the ship was actually orbiting Jupiter, things would be different. Pictures and other data coming back from the spacecraft would stream directly from its cameras and sensors, through its high-gain antenna, and back to Earth at a rate of thousands of bits per second. If the high gain should fail even briefly during a critical moment, the low-gain could never hope to keep up with this pace, and the data would simply be lost. In order to prevent this from happening, the designers decided to add one more piece of hardware to the ship: a tape recorder. In the event of a temporary high-gain breakdown, data would be stored on the recorder and later played slowly back to Earth through the far slower low-gain system. Conducting the entire mission this way, of

course, would be impossible, since even with the tape recorder helping out, the low-gain antenna could never transmit all the pictures
Galileo was built to collect. As a short-term fix, however, one that
would be used only until the high-gain antenna was working again,
this backup system was a good one.

With such triply redundant hardware built into their spacecraft,
mission planners could feel confident that they had designed a communications system that was almost completely resistant to failure,
and for the first eighteen months after *Galileo*'s 1989 launch, there
was no reason to assume anything *would* fail. Finally, on April 11,
1991, when the ship's trajectory had spiraled out as far as the edge of
the asteroid belt between Mars and Jupiter, JPL planners decided it
was at last probably safe to unlock the high-gain antenna and spread
its ribs. It was only then that they'd learn if triply redundant was redundant enough.

Though the deployment of the high-gain system was not a complicated exercise, it was a critical one, and for that reason the chief-
tains of the *Galileo* project made sure they were there to watch it
happen. On hand at the flight director's console that afternoon were
mission director Neal Ausman, deputy mission director Matt Landanow, and project manager Bill O'Neil. O'Neil and Ausman were
far and away the higher ranking of the three men, but Landanow,
they all knew, was far and away the most knowledgeable. As chief
engineer during the *Galileo* design phase, he had familiarized himself with every strut, nut, and rivet of the ship, and could practically
describe their placement and purpose from memory alone. If anything went wrong this afternoon, Landanow would likely be the
first person to recognize it—and the first person to come up with a
way to fix it.

For the first forty minutes or so after the deployment command
went up, O'Neil, Ausman, and Landanow had little to do. Like so
many other JPL controllers before them, they knew they would have
to tolerate the nonnegotiable limits of light speed, waiting twenty
minutes as their signal traveled from Pasadena to the spacecraft and
then another twenty minutes as it traveled back again. For that entire
time their screens told them nothing, flickering merely with the self-

evident information that their command had indeed been sent. Finally, after just over the anticipated forty minutes had elapsed, a column of numbers began to blink on the glass. Landanow gave the figures a quick scan and immediately noticed something amiss. He read them again—a bit more closely—and this time started to feel downright queasy. The antenna, from all indications, was pulling what the engineers called stall current. The motor was drawing power, the deployment gears were engaged, but the ribs of the umbrella appeared to be going nowhere at all.

"We're stuck," Landanow said flatly.

"How can you tell?" O'Neil asked.

"The current is saturated, something is jammed," Landanow said. "In any event, the antenna's not budging."

Ausman gave the numbers on the screen a read of his own, confirmed what Landanow was saying, and immediately called out to his flight controllers, instructing them to send a second deployment command up to the ship. The engineers complied, and forty minutes later another stall signal came down. A third command yielded a third signal, and a fourth a fourth. With each new report Landanow winced. If he knew this ship—and he surely did—he could all but guarantee that whatever was hanging up the antenna was not much: a single too-tight fitting, perhaps, a single protruding bolt, one that was situated in just such a way that it managed to jam all eighteen ribs. If it were somehow possible to transport the *Galileo* spacecraft to a hangar in Pasadena, Landanow knew he could probably roll over a stepladder, climb up to the antenna, and spring it free with his hands alone. But *Galileo* was not in a hangar in Pasadena; it was tens of millions of miles away, at the edge of the asteroid belt between Mars and Jupiter, and more elaborate measures would be necessary.

The first fix Landanow recommended was what the engineers referred to as hammering the motor. Sending signal after signal up to the ship, mission controllers would switch the antenna engine on and off at a speed calculated to cause the greatest vibrational resonance possible through the ribs. Anything that was stuck ought, in theory, to become easily unstuck. What the antenna ought to do and what it did do, however, turned out to be two different things, and no matter how many times the engineers hammered the engine throughout that day and the next one and the one that followed, nothing budged.

Making matters more frustrating still, as additional telemetry streamed down from the spacecraft and additional engineers had a chance to analyze it, it became clear that it wasn't the entire antenna that was jammed at all, but just a portion of it. On one side of the ship was a sun sensor that was bathed in solar light when the antenna was closed but ought to fall into shadow when it was opened; according to the data, the sensor was indeed now mostly shaded— but not entirely so. Working with a fully functional stand-in *Galileo* kept in a JPL clean room, the engineers positioned and repositioned the earthbound ship's antenna ribs, trying to reproduce the light and shade signals that were coming in from space. In the configuration that worked best, only three of the spacecraft's eighteen ribs were stuck in the closed position.

But a mostly open antenna was as bad as no antenna at all, and if the mission was going to proceed, the engineers would have to come up with a way to spring the rest of it free. If there was anything they had on their side, it was time. It would be another four years before *Galileo* finished inscribing its lazy arc to Jupiter, and as those years went by, flight controllers continued to try to fix what ailed their injured high-gain: They hammered and rehammered the balky deployment motor, studied and restudied the telemetry streaming back from the ship. Activating *Galileo*'s main engine, they shook the spacecraft this way and that, hoping to jolt the stuck ribs loose. Rotating the ship in toward the sun and then out toward space, they alternately heated and cooled the antenna, hoping the repeated expansion and contraction would pop something—anything—free. But month after month—and, ultimately, year after year—the condition of the high-gain antenna remained unchanged.

Finally, in the middle of 1995, as the spacecraft was approaching its early December arrival at Jupiter, the JPL engineers threw up their hands. The *Galileo* Jupiter ship—the billion-dollar machine the flight directors had spent more than a decade designing, building, and planning to fly—had been struck permanently mute. The ship's cameras and sensors were as alert as ever, its electronic brain was as sharp as ever, but the rare and perishable science that the instruments were built to collect would remain forever imprisoned within the ship.

Or at least it looked as if it would. If the booming voice of the high-gain antenna had been silenced, however, there was always the piping voice of the low-gain, and that provided some hope. As the plans for *Galileo* now stood, the orbiter portion of the spacecraft would be sending home a whopping load of photos—thousands, even tens of thousands, as it cruised around Jupiter and coasted among its moons. Indeed, so enormous would *Galileo*'s picture portfolio be that some of the images would even act as de facto movies—sequential, kinescopic snapshots that, when stacked up and riffled like a flip book, would reveal the motion of Jovian clouds, the pluming of volcanoes on the surface of Io, or maybe even the gush of a warm-water geyser as it hissed up through the ice of Europa.

Suppose, however, the flight controllers were willing to settle for fewer pictures? Suppose they instructed the cameras not to shoot indiscriminately at anything they saw, but to hold their fire until the most photographically opportune moments—when *Galileo* was cruising directly over a moon, say, or directly past the planet. The comparatively few pictures the spacecraft took could then be stored safely on the tape recorder, and later, in the weeks or months it would take the ship to make its next close pass by the next big moon or atmospheric formation, they could be played back to Earth through the slowpoke low-gain antenna. The process would be an excruciatingly tedious one, and it entirely ruled out luxuries like volcanic and atmospheric movies. What's more, it put a colossal burden on a fragile, fallible tape recorder that was intended only for the most occasional use and only in the most critical emergencies. But if the utter loss of the high-gain antenna did not constitute such an emergency, nothing did, and if the tape recorder was going to buckle under the weight of its new workload, that was simply a risk the engineers would have to take.

Their hands effectively tied, the *Galileo* flight directors approved the scaled-back mission, and as the spring and summer of 1995 gave way to autumn, and *Galileo* moved within eight weeks of its scheduled arrival at the planet, they prepared to implement the new plan. What nobody at JPL counted on was that the tape recorder was about to break down, too.

On October 11, 1995—four and a half years to the day after the *Galileo* antenna failed—Landanow, O'Neil, Ausman, and other mission supervisors were gathered in a meeting room in JPL's Building 264 reviewing the ship's upcoming insertion into Jovian orbit, when a member of the engineering team called on the phone. Strictly speaking, members of the engineering team should not be taking it upon themselves to phone any room in which so many project elders were gathered, and if one did, it could only mean that he was calling with very, very good news or very, very bad news. From the tone of the engineer's voice alone, Landanow could tell that this news was very, very bad, indeed.

"We've got a problem with the tape recorder," the engineer said.

"What kind of problem?" Landanow asked.

"It's spinning. We sent it a command instructing the reels to turn and they did turn, but the tape itself didn't go anywhere. Now the reels won't stop."

"Has the tape broken?"

"We don't know."

"Has it spewed off the reels?"

"We don't know."

Landanow rubbed his face. "Well," he said, "I guess we'd better find out."

Landanow adjourned the routine meeting that was under way and within the hour convened a far more urgent one, this time with the members of the flight engineering team. The group's general objective, of course, would be to diagnose the tape recorder's problem and see what they could do to fix it, but their more immediate concern was simply to bring the machine's madly whirling reels to a stop. As it turned out, they were able to accomplish this initial goal relatively easily. Sending up a series of override codes, they succeeded in disengaging and powering down the tape recorder entirely, rendering it temporarily useless, but at least keeping it from doing itself any further harm. Next, they set about analyzing the telemetry read-outs to see what had caused the breakdown in the first place.

From what the computer records of the past few hours indicated, the tape recorder's difficulties began when the ground controllers sent it a signal commanding the system to move in reverse, rewinding a short stretch of tape onto the feeder reel. The on-board machinery started to execute the simple command, but according to one curious data blip, it had to overcome a slight resistance first, almost as if the tape had become stuck on something and needed to be pulled free. When it finally *was* freed, the entire assembly simply went slack.

Without opening up the tape recorder and looking inside, there was no way of knowing what had caused the assembly to get hung up this way, but there were only three things that were likely to be re-sponsible. The recording tape itself was manufactured with a chemi-cal binder that helped give it its toughness and resistance to snapping. In the punishing temperatures of deep space, it was always possible that the chemicals on one patch of tape had undergone ad-hesive changes, causing them to become soft and gummy and to glue the tape in place. In the alternative, it was possible that the alu-minum dioxide that made up the machine's recording head was ac-cumulating an electrical charge, and as the tape passed by, it was getting stuck in place, much the way light, loose objects get stuck to a sweater crackling with static electricity. Finally—and more worri-somely—it was also possible that a small barb of metal had loosened itself somewhere in the guts of the machine and was jabbing at the tape as it spooled past, steadily weakening it and threatening to tear it in two. While this last alternative was clearly the most ominous, it was also the least likely, since if the tape was indeed being poked and damaged, it probably would have torn by now.

Whatever was causing the problem, the *Galileo* team eventually learned how to get around it. Putting the tape recorder through some careful, experimental spins, they discovered that—as the initial telemetry signals had suggested—the system almost always worked fine when the reels were moving forward; it was only backward that presented difficulties. In order to overcome this and get the reels going in reverse when they needed them to, the engineers discovered that all they had to do was advance them a little bit first. This somehow seemed to cock the spring-driven system and set it spinning in the other direction, much the way a smart tug downward on a common window blind gets the whole assembly flap-flap-flapping upward. Such a crude fix was not the solution the fussy engineers would have picked if they had had a choice, but the tape recorder had not offered them a choice, and if they wanted the balky system to function again, they would have to change the mission protocols to include this extra step. In the fall of 1995, the rules were indeed rewritten this way and the tape recorder went back into operation.

Several weeks later, on December 7, 1995, at 10:04 P.M. Pasadena time, the *Galileo* atmospheric probe, which had earlier been released from the main body of the ship, plunged into the rusty red Jovian clouds and began sending its stream of data back to the mother ship. Nearly two and a half hours later, at 12:27 on the morning of December 8, the spacecraft itself—its antenna arthritically frozen, its tape recorder cautiously turning—swung behind the great bulk of Jupiter and settled into orbit around it, adding a tiny metal moon to the sixteen natural ones that already circled the world.

For the probe, of course, the day it arrived at the planet was a short one. After entering the atmosphere and deploying its parachute, the 746-pound bullet died even faster than the mission designers had anticipated it would, surviving only fifty-seven minutes and thirty-six seconds before it was reduced to vapor by the crush of the Jovian air. In that brief time, however, it was able to switch on its instruments and drink in a gush of data from the environment around it, beaming it all up to the orbiter, which dutifully preserved it on its tape recorder until all transmissions from the probe stopped. Over the course of the next four months, the orbiter slowly dribbled the data back to Earth through its low-gain antenna. Though the probe's findings were a long time in coming, they turned out to be worth the wait.

Planetary scientists had always assumed that the uppermost reaches of Jupiter's atmosphere, where the air was vanishingly thin and the pressures were almost unmeasurably low, would also be paralyzingly cold—perhaps as low as −300 degrees. As it turned out, however, the temperatures in the region were about 100 degrees warmer than that. This suggested that the planet has some kind of internal heat source apart from its own intense gravitational energy, and that source, the scientists concluded, is probably a vast lode of radioactive material buried deep in the center of the world. Scientists studying data returned by *Voyager 2* had theorized just such a planetary oven at the heart of Neptune, and concluded that this could help explain the planet's high winds. Jupiter's winds were not quite so fast as Neptune's, but the planet did turn out to be a gusty place, with the *Galileo* probe detecting gales of 330 miles per hour in the upper atmosphere and 450 miles per hour down lower. If anything could drive those kinds of air bursts, it was likely buried radioactive heat.

The chemistry of the atmosphere was a bit less surprising. As scientists had anticipated, the Jovian air turned out to be about 13.6 percent helium and the remainder hydrogen—or about the same concentrations as the sun. This helped confirm the idea that Jupiter indeed swirled into existence from the same raw, gaseous material that formed the stars, and suggested that the planet might indeed have been a star itself if only it had been a little larger and a little denser. Simple organic elements like carbon, sulfur, and nitrogen are stirred into the atmospheric mix as well, but more complex hydrocarbons are largely lacking. This absence of preorganics is probably due to a paucity of water and oxygen in the Jovian stew, as well as the relative rarity of lightning, which can play a role in jolting hydrocarbons into existence. As the *Galileo* probe descended through the Jovian air, it did detect the occasional lightning bolt, but the closest one was a good 8,000 miles away.

Illuminating as the findings from the probe were, it was the data from the *Galileo* orbiter that provided the true Jupiter news. One of the least expected findings concerned not the planet itself but its rings. Ever since the *Voyagers'* Jovian flybys in 1979, planetary scientists had been assuming that the fine bands surrounding Jupiter, Uranus, and Neptune were the remains of small, annihilated moons, ones that were destroyed by collisions and then gravitationally prevented from pulling themselves back together. At least in the case of Jupiter, however, *Galileo* images revealed otherwise. Photographing the small inner satellites Metis, Adrastea, Amalthea, and Thebe, the spacecraft discovered that the Jovian moons appear not merely to orbit in the vicinity of the rings as scientists had previously thought, but rather to lead them, as if the fine, powdery material that makes up the bands were emerging from the rump ends of the little worlds themselves. Additionally, color enhancement of the images suggested that all of the rings and all four moons are made up of the same ruddy, rusty-looking material. This led the scientists to postulate a whole new theory of ring formation.

Incoming micrometeorites, they figured, probably bombard the four innermost moons almost constantly. When the rubble strikes the satellites, it blasts large clouds of dust from their surfaces. Unimpeded by atmosphere, which the moons don't have, or gravity, which they do have but only barely, the dust simply drifts into space where it follows behind the satellites in a ghostly wake. Though the rings are diffuse enough to disperse relatively quickly, no sooner do they start to dissipate than they are replenished by other micrometeorite hits, releasing other clouds of dust. As long as the tiny bits of cosmic gravel keep colliding with the satellites—which ought to be as long as the Jovian system exists—the rings will continue to circle the planet.

Orbiting just above the small, ring-producing moons, of course, are the large Jovian moons: Io, Europa, Ganymede, and Callisto—the satellites discovered by the eponymous Galileo himself. While the rings and the small moons earned some of the scientists' attention, it was those large moons they wanted their new Jupiter probe to study most closely. Once again, it was those large moons that wound up paying the true scientific freight of the mission.

Callisto, the outermost of the quartet of large satellites, appeared to the spacecraft to have a fine and faint atmosphere, made principally of hydrogen, oxygen, and carbon dioxide. Though such a mixture of gases sounds altogether terrestrial, Callisto's atmosphere is nothing like Earth's, partly because it lacks nitrogen—which makes up 79 percent of the Earthly air blanket—and partly because it is infinitely more tenuous. What's more, while Earth's atmosphere was created by a combination of gases—some left over from the planet's original formation, others imported by comets, still others discharged by volcanoes—Callisto's impossibly wispy air is produced principally by molecules of water ice that are regularly knocked loose from the surface by solar radiation and then scattered into the Callistan sky where they swirl and recombine.

Though the moon has only the faintest atmosphere, it did appear to the spacecraft that Callisto might—like its sister Europa—have another, far more dramatic feature: a hidden ocean. Images and data from the *Voyager* probes suggested that of the four Galilean moons, it was Callisto that was far and away the least dynamic—its innards inert, its surface frozen, its crust carrying the scars of meteorite hits billions of years old. During *Galileo*'s encounters with the 2,976-mile Callisto, however, the ship detected what seemed to be a distinct magnetic field surging around the world, one never noticed by the more primitive *Voyager* instruments. While magnetic fields on other bodies are often created by a mobile, metallic core, density studies of the Callistan interior pretty much ruled such a structure out. Instead, the JPL team came up with a far more imaginative explanation for the moon's curious magnetism.

When Jupiter's own magnetic field collides with Callisto, it ought to pass right through the moon, washing over rocks, ice, or any other materials that lie beneath its surface. Most of that subsurface matter would not interact with the field in any significant way, but some of it—most notably salty water—would. If Callisto had a briny ocean, the magnetic field would set up an electrical current in it, much like the chemically driven charge that crackles inside an automobile battery. When this current surges through a body the size of Callisto, it would set up an equally vast magnetic field that would surround the moon and reach miles into space, where it could be easily detected by a passing spacecraft.

Why, however, should Callisto have an ocean? Without the tidal pumping that close-up moons like Io and Europa undergo, this most remote of the Galilean satellites simply would not have enough subsurface motion to melt its ice. The only way Callisto *could* warm itself up would be if it, like Jupiter, was rich in heat-producing radioactive materials. These, combined with the moon's natural gravitational pressure, might be enough to stoke Callisto's internal oven and liquefy at least part of its mantle.

Such a thermal process, if it were taking place at all, might not be confined to Callisto. *Galileo* images of nearby Ganymede confirmed earlier *Voyager* findings that this largest of the Jovian satellites has a surprisingly young crust, one that appears to have been partially melted and remelted repeatedly over its history. Just what was behind this process had never been clear, but if cold Callisto has a radioactive core and a liquid ocean, there was every reason to believe that Ganymede—with its greater bulk, greater gravitational energy, and periodically refreshed face—would, too. Long thought of as ice worlds, Callisto and Ganymede had become, overnight, possible water worlds, mere moons that all at once seemed to have all the chemical promise of planets.

Dynamic as these outermost of the four large Jovian satellites suddenly seemed, it was the two inner moons—Io and Europa—that once again proved to be the truly seething satellites in Jupiter's system. Though the vagaries of *Galileo*'s orbit have allowed it to make few close passes by Io, the spacecraft's cameras and other instruments have nonetheless been able to conduct at least cursory surveys of the moon's surface and have detected dozens of new volcanic vents smoldering there, some of which approach a scalding 1,000 degrees Fahrenheit. A number of eruption sites spotted in 1979 were still simmering in 1997, and showed every sign of continuing to percolate for years—perhaps millennia—more. During one five-month stretch between April and September 1997, a 250-mile-wide volcanic field appeared around the well-photographed Pillan Patera peak, blackening an area larger than the state of New Mexico.

Europa's face also proved to be more active than it seemed when *Voyager* flew by a generation earlier. Skimming just 125 miles over the moon in March 1997, *Galileo* was able to spot objects on its surface as small as twenty feet across. With the help of such sharp, long-distance vision, the spacecraft photographed relatively small iceberg-like objects all over the moon's frozen crust, fractured and capsized like ordinary terrestrial icebergs dotting ordinary terrestrial oceans. Unlike Earthly bergs, however, which bob and float in liquid water, Europan bergs did not appear to be in motion. Rather they seemed to be locked in place, almost as if they had suddenly become trapped when the water around them cooled and refroze. Superficially, this would seem to argue against an active liquid ocean, but on a surface as apparently turbulent as this one, the ice almost certainly thaws and freezes in cycles, perhaps never getting a chance to grow more than 3,000 feet thick before warming and liquefying again.

Such water as there is on Europa is apparently just as chemically rich as the JPL scientists always assumed it was. All over the moon's surface, *Galileo* spotted vast stains that bore the chemical signature of salt—particularly natron and Epsom salts, which form only in the presence of liquid water. As tidal forces squeeze Europa, briny water apparently squirts out of the moon like juice from an overripe tomato. Though the water evaporates quickly in the vacuum of space, the salt settles back down to the surface and remains there. If the smudges the eruptions leave behind are indeed as salty as they seem, there is even greater reason to believe that the moon is home not merely to an ocean but to a *living* ocean. It is not in clean, pristine, inland waters that the majority of Earth's aquatic life thrives, after all, but in the great salty soup of the planet's vast oceans. If such a rich medium breeds organisms on Earth, there is no reason it couldn't do the same on Europa, too.

Galileo was given a bit of extra time to explore that question. The spacecraft's principal mission was planned to run only from December 1995 to December 1997. When that two-year tour was finished, however, mission managers decided to extend it, sending the still-chugging ship on eight more flybys of Europa, four more of Callisto, and up to two more of smoking Io. It is only after these encounters that the Jet Propulsion Laboratory scientists will finally be finished with the hobbled old spacecraft, turning their attention at last to other ships and other missions.

What will happen to *Galileo* then is not entirely clear. It's possible that without Earthly navigators continuously monitoring the ship, the swirling gravity of Jupiter and its moons will combine to fling it out of the Jovian system altogether. In the alternative, it's possible that the pinball complexity of *Galileo*'s orbit will cause it simply to collide with one of the moons, annihilating the ship like an ordinary meteor. It's even possible that JPL will not merely abandon the spacecraft, but instead will eventually cede command of it to a university lab where apprentice engineers will gun its engines and change its headings, fooling around with the physics of orbital flight until the systems of the spacecraft wink out completely. Whatever becomes of *Galileo*, its truly useful, truly scientific work will at last be done, leaving the JPL researchers a legacy of data to study long after the ship itself expires.

When *Galileo* does die, it will almost certainly have left at least one great Jovian question—the question of life—unanswered. Settling that issue will take other spacecraft making other voyages out to the Jupiter system—specifically to Europa—and JPL appears to be getting ready to make those trips. Already in development in the Pasadena labs is a Europa orbiter spacecraft that may be ready to leave Earth and head out to the Jovian system as early as November 2003.

Engineers hoping to study Europa up close have always found the idea of placing a spacecraft into orbit around the moon a daunting one—even more daunting than the *Galileo* engineers found the prospect of trying to put a spacecraft into orbit around Jupiter. Jupiter's gravity, after all, is far greater than little Europa's, and the planet is thus able to hold onto a ship that is still moving at a pretty fast clip. In order to orbit Europa, an interplanetary spacecraft has to slow down to a comparative crawl of just a few thousand miles per hour, and that requires it to carry a virtual bazooka of an engine.

The Europa orbiter now in development is little *but* engine. Weighing no more than three or four adult human beings, the compact spacecraft is about three-quarters combustion chamber, exhaust bell, fuel tank, and fuel. It is only the remaining 25 percent of the ship's weight that is set aside for the sensors, cameras, and other hardware it is being built to transport in the first place. Modest as the mass of those instruments will be, the work they do will be considerable.

When the Europa ship enters orbit around the moon, it will be flying at an altitude of just 124 miles—significantly lower than the space shuttle usually orbits Earth and a bit closer than *Galileo* has ever come to Europa on any of its flybys. With upgraded cameras orbiting at that close a proximity, the ship should get the best images yet of the moon, photographing surface features significantly smaller than the jagged icebergs *Galileo* spotted. If any of those icebergs happens to be floating, if the ice around them happens to be cracking, if subsurface water happens to be welling up through those cracks, the spacecraft ought to notice.

Even if nothing so aquatically obvious is happening on the moon, the Europa orbiter should still be able to determine if an ocean lies beneath the surface ice. Along with its battery of cameras, the ship will also be carrying a planetary radar, an instrument that's never flown out to the Jupiter neighborhood before. Soaring through its barnstorming orbit, the spacecraft will beam radar signals down to the surface, looking for the telltale echo of liquid water sloshing beneath the ice. If the ship indeed picks up such signature signals, it will map the spots where the underground water appears to be located, flagging the photographs of those areas for especially close study by the engineers on the ground.

Also assisting in the spacecraft's ocean hunt will be a more sophisticated device known as a laser altimeter. As Europa moves through its orbits around Jupiter, the tidal flexing caused by the planet and the surrounding moons subtly distorts the shape of the world. Just how much it distorts it depends on what the moon itself is made of. If Europa is solid ice all the way through, it should flex by no more than three to six feet during any one orbit. If, however, there's an ocean beneath the ice, things will be a good deal more malleable, with the moon changing shape by as much as twenty to thirty feet. Beaming a needle-thin laser signal down to the surface, the Europa orbiter should be able to measure the moon's shape to within just a few inches. Keeping track of the Europan flexing as the space-craft glides through its orbits—and correcting for surface irregulari-ties that could throw the readings off—the JPL scientists should be able to take some all but flawless measurement of the moon and draw some all but flawless inferences about its innards.

If those innards do look liquid, JPL has bigger plans still for Eu-ropa. Already being developed by the Pasadena engineers is another spacecraft that would not simply circle the Jovian moon, but land on it. Flying out to Europa and setting itself down on a patch of ice that the radar and laser suggested was especially thin, the ship would re-lease a four-foot cylindrical probe with a heated tip that would melt through the frozen crust looking for water. Should it find any—even a body as small as an underground lake or pond—it would release a smaller, one-foot, submarine-like probe into the water, which would propel itself around, sampling the marine chemistry and looking for signs of life.

Yet another, equally improbable spacecraft might be sent on a mission to bring a tiny bit of Europa home. Launched on a simple flyby trajectory, the sample-return ship would approach Europa and fire a twenty-pound, shot put–like sphere out into space. The cosmic cannonball would arc down toward the surface of Europa, strike the frozen crust with the force of a dynamite blast, and release a mush-room cloud of ice particles into space. The slower moving mother ship would then fly through this crystalline mist, collect a bit of it, and carry it back to Earth for analysis in the lab.

Whether any of these future missions will actually fly, of course, is by no means certain. If the tortured history of projects like *Voyager* proves anything, it's that a lot can happen between the time a space-craft is proposed and the time it's built, and no mission should be con-sidered a sure thing until the ship has actually left the pad. Even as the Europa probes are still being debated and designed, however, another JPL spacecraft—the *Cassini-Huygens* probe—is already speeding silently through the solar system, making its way toward another, even more distant destination: Saturn and its atmosphere-shrouded moon Titan.

Conceived shortly after the *Voyager* flybys of Saturn in 1980 and 1981 but not launched until October 1997, *Cassini* is modeled after its sister ship, *Galileo*—with a few decided improvements. There is no tangle-prone, snap-prone tape recorder on this spacecraft; any memories the ship needs to preserve are stored in a digital system that records data largely without moving parts. Similarly, there is no collapsible—and fallible—rib-and-mesh antenna. When *Cassini* left the ground, it left not inside the cramped cargo bay of a space shuttle, but inside the roomy payload area of a Titan-Centaur rocket. The more spacious missile meant more room for an antenna, and *Cassini* carried a dilly: a 13.1-foot hard-shell dish that was molded in a single unbreakable piece, in a single open position.

The similarities between *Cassini* and *Galileo*, however, outnumber their differences. Like its Jupiter-touring predecessor, the *Cassini-Huygens* ship is actually two ships: an orbiter and a probe. The main body of the spacecraft—named after Giovanni Cassini, the seventeenth-century astronomer who discovered four Saturnian moons and the gap in the planet's rings—was designed to orbit Saturn and spend at least four years photographing and studying it. Making no fewer than sixty laps around the world, the 2.5-ton *Cassini* will swoop to within 110,000 miles of Saturn's cloudtops, training its cameras and other instruments on the planet itself, its array of rings, and at least eight of its well-studied seventeen moons: Mimas, Enceladus, Tethys, Dione, Rhea, Titan, Hyperion, and Iapetus. The ship may also get at least a glimpse of a little-known eighteenth moon—tiny, twelve-mile-wide Pan, spotted on vintage *Voyager* photos in 1990.

Several months after *Cassini* goes into orbit around Saturn in January 2004, the *Huygens* probe attached to the exterior of the ship—named after Christiaan Huygens, who discovered Titan in 1655—will set to work. Measuring 8.9 feet across and weighing 770 pounds, *Huygens* is a blunt, bullet-shaped assembly modeled after the atmospheric instrument package carried aboard *Galileo*. Unlike the *Galileo* probe, however, this one was not built to study the chemistry of a planet; instead, it will take its plunge through the rust orange air of Titan.

Released from the speeding *Cassini* orbiter, the *Huygens* probe will navigate its way to the nearby moon and slam into its dense air blanket at a bone-jolting 13,750 miles per hour. Free-falling through the surrounding haze, the probe will encounter greater and greater atmospheric resistance until, after just three minutes, its velocity will have slowed to a comparatively sluggish 895 miles per hour. At that point, a 6.5-foot drogue parachute will pop from the pod, followed shortly after by a large, billowing twenty-seven-foot main chute. The *Huygens* probe will continue to descend through the opaque atmosphere for more than two and a half hours, its speed steadily slowing, the already dim light around it growing murkier still, until finally, when it is drifting downward at just 15 miles per hour, it will bump down on the Titanian surface. Just what the probe will find there depends on just where it lands.

Though no camera that processes images in visible light has ever been able to penetrate Titan's heavy atmosphere, cameras that work in the infrared can peer right through it. The Hubble space telescope, which has been orbiting Earth since 1990, is equipped with just such a camera, and not long ago it trained its gaze on Titan and spotted what appeared to be a curiously bright patch covering a portion of the moon's surface. If a formation that big were discovered on a world like Earth, scientists would guess it was an ocean or at least a large lake, and on the surface of Titan, it appears that that might be exactly what it is.

In the paralyzing cold that dominates Titan's climate, an ocean made of liquid water clearly could not exist. What some scientists think the Hubble telescope spotted instead is a rich mixture of liquid ethane and methane. Two of the signature chemicals of organic science, ethane and methane are the surest signs that Titan may indeed be a prebiological place. Should the *Huygens* probe land in this vast pool, accelerometers and float detectors attached to the buoyant machine will measure how much of it stays above the surface of the liquid and how much of it becomes submerged as it floats this way and that. Factoring this information together with the known mass of the probe should help scientists determine the density of the liquid, and that, in turn, should help them determine its precise composition.

Even before the buoyancy sensors can get to work, however, scientists should have a pretty good idea of whether *Huygens* touches down on land or in liquid. Built into the body of the probe is an ordinary microphone that will provide mission controllers with the decidedly *extra*ordinary ability to listen in on their ship. As the *Huygens* pod descends, JPL engineers will be able to hear the atmosphere screaming by, the chutes popping open, the winds pushing the little instrument package through the orange sky. If the probe indeed does land in the ethane-methane ocean, they should also hear the thump, the slosh, and then—implausibly—the soft slapping of the Titanian waves as they lap against the sides of the interplanetary buoy.

Wherever *Huygens* lands, it will have to do its work fast. With solar panels unworkable in the Titanian gloaming, radioactive generators producing too much heat for so small a craft, and fuel cells too unreliable, the only power source the probe can carry is a battery—one that will have a life span of just thirty to sixty minutes once the instrument package is on the surface. Throughout that brief period, the probe's numerous sensors will pull on that battery hard. Visual and infrared cameras will scan the landscape, peering out, down, and even up, in an attempt to determine how much feeble sunlight makes it through the blanket of clouds; a spectral radiometer and mass spectrometer will gather in sips of air and analyze its chemistry; a Doppler wind instrument will study the local breezes; an aerosol collector will examine liquid and solid particles suspended in the atmosphere. All of the findings from all of these sensors will be transmitted up to the *Cassini* mother ship orbiting overhead and then relayed the eight-tenths of a billion miles back to Earth. Less than an hour after the first data bit is sent out from the probe, the instruments as a group will flicker and die and the *Huygens* lander will fall forever silent. Unlike the *Galileo* probe, however, which was vaporized by the world it came to visit, the *Huygens* probe will endure for centuries, a metal mile-marker commemorating the farthest spot humanity has reached in its decades-long effort to place its machines on worlds not its own.

Once Titan is studied up close, JPL's preliminary look at the large bodies of the known solar system will at last be complete—or almost complete. While seven of the other eight planets and all of their large moons have at least been approached by Earthly machines, one planetary system—Pluto's—has gone utterly unexplored. It was in 1930 that Clyde Tombaugh, an astronomer at Lowell Observatory in Arizona, first spotted an impossibly faint dot lost among the far brighter stars of the wintertime sky. Tracking the dot for more than a year, he found to his surprise that it was moving—far more slowly than a speedy planet like, say, Mercury or Venus, but far faster than the stars themselves, which really oughtn't move at all. The dot, it turned out, was a ninth planet, a rocky, icy, 1,423-mile-wide world orbiting the sun nearly 3.7 billion miles from Earth. Nearly half a century later, in 1978, James Christy, an astronomer working at the United States Naval Observatory, discovered that tiny Pluto is itself circled by a tiny moon, a 737-mile ball, also made up principally of rock and water ice. The planet, sunk in the permanent darkness of deep space, was named Pluto, after the overseer of the underworld. Its moon was named Charon, the boatman said to ferry condemned souls across the River Styx.

By any measure, Pluto and Charon are an unusual pair. First of all, Charon is—in relative terms, at least—huge, measuring more than half the size of its parent world. Such a one-to-two moon-to-planet ratio is far and away the biggest in the solar system. Stranger still is the way the two bodies move. While Earth's moon orbits its parent planet at a mean distance of 238,000 miles and other large moons like Ganymede, Callisto, and Titan orbit at three or four times that altitude, Pluto and Charon spin in a close-up embrace just 12,000 miles apart, or only a bit more than the air-mile distance between Tokyo and Rio de Janeiro. To the untrained eye, Charon would not appear to be Pluto's satellite at all; rather, the two arm's-length worlds would seem to be orbiting each other, twirling about a common axis like the spherical weights at the opposite ends of a set of spinning barbells.

If the path Charon inscribes about Pluto is an uncommon one, the path both bodies inscribe around the sun is even more so. For the most part, the local planets maintain reasonably orderly solar orbits, moving about their anchor star in roughly circular paths that remain roughly in the plane of the solar system. Pluto's and Charon's orbit, by contrast, is an asymmetrical mess. Like the irregularly orbiting captured moons that circle the larger planets, the ninth planet and its companion satellite move about the sun at a dizzyingly inclined angle, rising 17.15 degrees above the solar equator and plunging 17.15 degrees below it. Additionally, the orbit is not even remotely circular. Though Pluto maintains an average distance of nearly 3.7 billion miles from the center of the solar system, at times it closes to within 2.8 billion miles, actually bringing it inside Neptune's orbit for twenty out of the 248 years it takes to travel around the sun.

For this reason more than any other, astronomers who have always thought of Pluto as a fully credentialed planet have lately begun to conclude that it might be nothing of the kind. Rather, like Neptune's Triton, Saturn's Phoebe, and all eight of Jupiter's outer satellites, Pluto and Charon might be nothing more than two bits of free-floating debris that went wandering through the solar system and became trapped in orbit around one of its bodies. If the planet and its moon are indeed nothing more than a pair of cosmic strays, it is almost certain that they came from the same rich river of detritus that gave birth to Triton, Nereid, and most of the solar system's comets: the Kuiper Belt. Superficially, Pluto and Charon certainly seem to be made of Kuiper Belt stuff—a bit of rock, lots of ice, a nimbus of haze surrounding them both. What's more, they move with Kuiper Belt motion. If their trajectories were to change by only a few degrees, the two bodies would not be orbiting the sun like planets at all, but swooping around it like comets, plunging toward the center of the solar system, flaring brightly for a few years, and then speeding back out again.

Formulating theories about the origin and composition of Pluto and Charon, of course, is not the same as proving those theories, and for years, astronomers have known that the only way to determine anything for sure is to fly out to the edge of the solar system and look up close. More than twenty years ago the outer planets offered to help out, moving into a tidy line that would have easily allowed a

Voyager spacecraft launched to Jupiter to make a sharp gravitational turn and speed out to Pluto. But the stewards of the ship decided not to make that trip, and the planets, indifferent to that choice, glided back out of position. Visiting Pluto now would require a whole new journey flown by a whole new kind of spacecraft. Even now, that spacecraft and that mission are being planned.

Sometime in the early winter of 2004, the way JPL envisions things, the lab will launch a small featherweight spacecraft dubbed the *Pluto-Kuiper Express*. Weighing as little as 500 pounds and carrying a dish antenna measuring no more than six feet across, *Pluto-Kuiper* will easily be the smallest deep-space machine ever built. Such a light ship can move at a considerable clip, and less than two years after its launch, it should have flown out as far as the neighborhood of Jupiter. Arcing partway around the planet, it will pick up a gravitational boost and fling itself in the direction of Pluto, arriving there sometime in early 2013 and flying directly through the 12,000-mile gap that separates the planet and its moon.

Assuming mission planners are able to pull off this feat of navigational needle-threading, the spacecraft should reward their efforts amply. Orbiting where Pluto does, the planet is 39 times farther from the sun than Earth is. Given the squared geometry of solar illumination, that means it receives 961 times less solar light, with the sun appearing to be little more than a very bright star, all but lost against a background curtain of other stars. Such a tiny stellar spangle does not provide much illumination, and taking pictures of a planet so feebly lit would not be easy. The *Voyager* probes overcame a similar problem in a relatively crude way—by leaving their cameras' shutters open, collecting as much light as they could, and pivoting the main body of the spacecraft in any way necessary to keep the picture from smearing.

Pluto-Kuiper Express will require no such complicated maneuvering. Built into the body of the ship will be a system that reads light not a beam or stream at a time, but a single photon at a time. Pouring into the imaging system, these massless subatomic units—essentially particles of light—strike electrons on a semiconductor, release a charge that excites a picture element, and then build up an image one fraction of a pixel after another. Systems like this—known as charge-coupled devices, or CCDs—work particularly well in extremely low light environments in which the detectors run little risk of being swamped by too much incoming illumination. They are also especially well adapted to cold, since heat can overexcite the hardware as easily as light, knocking loose electrons and creating false light and shadow where none exists. The deep black, bitterly cold Pluto environment, where temperatures approach −400 degrees, was practically made for a CCD.

When the *Pluto-Kuiper Express* arrives at its destination, its own CCD will not have much time to operate. The spacecraft will be able to photograph the tiny planet and moon for only about 3.2 days as it approaches them and another 3.2 days as it sails away. As it happens, however, this 6.4-day period is just a bit longer than a Pluto day, meaning that while the ship is in the vicinity of the Plutonian system, it will be able to watch most of the planet rotating beneath it, capturing images of both of its hemispheres as they turn slowly through space. Any feature on the face of the world that's at least as big as, say, a sports stadium will be faithfully recorded by the cameras.

Whatever pictures the *Pluto-Kuiper* ship does collect should be the last anyone sees of the surface of Pluto for a long time. For the last twenty years the planet has conveniently been in the portion of its orbit that brings it closest to the sun, having crossed inside the orbit of Neptune in January 1979 and crossed back out in February 1999. That relatively small change in proximity to the warmth of the solar fires is enough to heat up the Plutonian atmosphere slightly, keeping it gaseous and transparent. As Pluto moves farther away, however, once again becoming the ninth planet for the remaining 228 years of its orbit, the atmosphere will chill, freeze, and settle onto the surface in an opaque shell. When the *Pluto-Kuiper Express* reaches the planet, JPL scientists believe the air will still be vaporous enough for the spacecraft to see through. Any ships that are launched toward Pluto in the next two centuries, however, will have a far harder time of things, encountering only a flash-frozen globe, completely hidden by solidified gases from even the sharpest prying eyes.

Just where NASA is headed after it reconnoiters Pluto, after it returns to Europa, after it sails about Saturn and fires a probe into Titan, is unsettled for now. Those projects alone, of course, could keep the agency busy for the better part of a generation. Whatever ships will follow along after is for later generations to determine. But deep-space probes are not the only way to explore the cosmos, and even without sending a machine so much as a mile off Earth, scientists are discovering that the solar system still has new sides of itself to reveal.

Late in the summer of 1997, astronomers working in the mountainous Palomar Observatory in southern California decided to spend an evening or two studying the planet Uranus. Their telescope was a good one—a fat, fifteen-foot-wide reflector equipped with its own sensitive CCD. Such a sharp-eyed system was unlikely to miss any of the light Uranus reflected back to it, and indeed, the images the telescope collected were crisp ones. The astronomers could see the sea blue sphere of the planet itself—unchanged from the time eleven years earlier when *Voyager 2* glided serenely by. They could see the spangle of fifteen moons spinning like electrons around the world. They could watch as the moons orbited along in their neat ranks, all of them remaining in the equatorial plane of Uranus, all of them moving a uniform counterclockwise.

That, however, wasn't all the images revealed. Hovering far above Oberon, the highest flying of the Uranian moons, were two other shimmery points. Both of them appeared to be vaguely red in color; neither of them appeared to be even remotely round in shape. One of the two points—the lower-flying one—looked to be a good 4.5 million miles above the surface of Uranus; the other was a good 7.5 million. The astronomers took other pictures of the points and noticed that both of them appeared to be moving in an arc-like trajectory; more images still and the arcs became elongated circles, ones that took them in a high, elliptical orbit completely around the planet. That final finding, the scientists knew, changed things entirely. The two inconsequential points weren't just points at all, but two previously undiscovered moons, the first spotted anywhere in the solar system since the last image from the *Voyager*s came flickering back to Earth years earlier.

The new Uranian satellites—measuring 37 and 75 miles across—were not much to speak of: organics-rich cosmic BBs that probably flew in from the deeper cosmos and got caught in the planet's gravitational net. But while they might have been mere flying rocks once, they were undeniably moons now, and as such, they earned the right to be named. The smaller of the two satellites was accordingly dubbed Caliban, the sorrowful slave of the magician Prospero in Shakespeare's *The Tempest*. The larger one was named Sycorax, Caliban's sorceress mother.

With their discovery, Caliban and Sycorax brought the solar system's moon total to a robust sixty-three—or sixty-two more than the astronomical community ever imagined before Galileo Galilei began looking skyward nearly four centuries ago. If scientific history indicates anything at all, it is that even so great a satellite census could grow further still. Just who will count the new worlds, however, is impossible to say. It was the long-ago ships of the Jet Propulsion Laboratory that discovered so many of the sixty-three known moons. It was the long-ago scientists of earlier eras who discovered most of the rest. Now, it will be up to new ships—and to new observers peering up from the ground—to find any more.

Epilogue

The people in the room were trying hard not to spill food on all the spacecraft standing around them. It wasn't easy, what with hot hors d'oeuvres and cold hors d'oeuvres and little iced drinks to juggle. Already at least one person in the crowd had thumped into the side of a V-2 rocket, nearly sloshing a bit of soda down its metal skin. But while a V-2 could probably take the abuse, most of the rest of the machines—particularly the elegant *Ranger* moonship—were far more fragile pieces of hardware.

Or at least the people here today thought of them as fragile. But then, they *would* think of them that way. Three decades ago, deep in the early 1960s, the aristocratic-looking gentlemen in the room had had to build six of the spindly *Ranger*s, only to see them fall apart in their hands, before a seventh one finally operated the way it had been designed to operate. Jim Burke was among those engineers here today; so were Bud Schurmeier and Cliff Cummings, Burke's former bosses. Also here, arriving just a few minutes ago, was William Pickering, smiling, glad-handing, methodically working the room with a calm and clarity that belied his eighty-plus years.

The place all of these people had come today was the Smithsonian Institute's Air and Space Museum in Washington, D.C. And the reason they had come was to pay tribute to Jim Webb, the long-ago administrator of NASA. It had been more than two decades since Webb had left the agency, and in all that time nobody had thought to commemorate his tenure. In 1968 he had been allowed simply to empty out his desk, gather up his belongings, and slip quietly back into private life. Now, at last, someone in the protocol office had decided to rectify that oversight.

Hundreds of people from most of the NASA facilities had been invited here for the event today, including the men and women of the Jet Propulsion Laboratory in Pasadena. Pickering, who himself had left the agency years before, had been among the earliest of the former JPL scientists to answer the invitation, and to no one's surprise, he had answered yes. With word that the lab's first chief was going to attend, most of the other people who had worked for him decided to show up, too.

If JPL's scientists were well represented at the Air and Space Museum today, JPL's machines were even more so. Displayed around the place like so many animal heads from so many wilderness expeditions were the *Rangers*, *Surveyors*, *Vikings*, *Mariners*, *Voyagers*, and other ships that had, over the years, traveled to the moon, Mars, Jupiter, Saturn, Uranus, Neptune, and elsewhere. Standing nearby, dwarfing both these spacecraft and the men who built them, were the boosters that had started all the interplanetary ships on their journeys. Rising up toward the glass ceiling of the museum's atrium was an arsenal of Agenas, Atlases, Junos, Redstones, and Vanguards, all long since drained of their fuel and stripped of their vitals—all, essentially, taxidermized—built for flight but destined never to fly anywhere at all.

The event this evening played out over a few hours, and throughout that time Pickering had never had the opportunity to do more than briskly shake the hand of Webb, the man he was ostensibly here to honor. A long time ago, of course, he did not feel quite as warmly toward his onetime boss—not when all the *Rangers* were failing and the press was complaining and Webb was threatening to have Pickering's head if the faltering moonships didn't start flying right. But those years were past, and tonight Pickering would not mind finding a few minutes to reflect with Webb on happier times. Finally, toward the end of the night, as most of the other guests were gathering their coats and moving toward the exits, the former JPL chief noticed the former NASA chief standing quietly by himself near one of the old, hollow boosters. Pickering walked over and joined him.

"It was a pleasant evening," he said simply.

Webb nodded.

"Long overdue," Pickering added.

Webb shrugged, but with a small smile.

Pickering looked around the room slowly. Above his head, hung from cables attached to ceiling beams, were three of his spacecraft, suspended in eternal flight. Webb followed his gaze.

"Nice machines," the ex–NASA administrator said after a silence.

Now it was Pickering's turn to shrug. "They did the job," he said.

Pickering and Webb fell quiet again, and after a moment the muffled click of a circuit breaker sounded at the other end of the gallery and a bank of ceiling lights switched off, plunging part of the display into darkness. Another bank of lights followed, then another. The few people left in the museum started moving toward the exits. Wordlessly, the two scientists began to follow them. Before they had walked even a few steps, however, Webb stopped and turned to Pickering.

"You know," he said, "I may not have mentioned it at the time, but I always thought you fellows had a fine operation out there."

Pickering nodded.

"A real fine operation," Webb repeated. Pickering smiled and the two men walked on.

Overhead, a few more lights clicked softly off.

Appendix 1

The Sixty-three Moons

Name	Altitude (Miles)	Diameter (Miles)	Orbit (Days)	Discovery (Year)
Earth				
The Moon	238,000	2,155	27.3	—
Mars				
Phobos	5,815	.13.6	.32	1877
Deimos	14,545	8.7	1.3	1877
Jupiter				
Metis	79,360	25	.20	1979
Adrastea	79,967	12	.29	1979
Amalthea	112,406	117	.49	1892
Thebe	137,578	62	.67	1979
Io	261,392	2,251	1.8	1610
Europa	415,958	1,945	3.5	1610
Ganymede	663,400	3,262	7.1	1610
Callisto	1,167,469	2,976	16.7	1610
Leda	6,878,280	10	239	1974
Himalia	7,117,600	115	251	1904
Lysithea	7,266,400	22	259	1938
Elara	7,276,940	47	260	1905
Ananke	13,144,000	19	631	1951
Carme	14,012,000	25	692	1938
Pasiphae	14,570,000	31	735	1908
Sinope	14,694,000	22	758	1914

Name	Altitude (Miles)	Diameter (Miles)	Orbit (Days)	Discovery (Year)
Saturn				
Pan	82,820	12	.57	1990
Atlas	85,355	19	.60	1980
Prometheus	86,397	62	.61	1980
Pandora	87,854	56	.62	1980
Epimetheus	93,880	74	.69	1966
Janus	93,911	118	.69	1966
Mimas	115,022	243	.94	1789
Enceladus	147,572	310	1.4	1789
Tethys	182,689	657	1.89	1684
Telesto	182,689	19	1.89	1980
Calypso	182,689	16	1.89	1980
Dione	233,988	694	2.73	1684
Helene	233,988	20	2.73	1980
Rhea	326,764	949	4.5	1672
Titan	757,534	3,193	15.9	1655
Hyperion	918,282	180	21.2	1848
Iapetus	2,208,006	905	79.3	1671
Phoebe	8,030,240	136	550	1898
Uranus				
Cordelia	30,846	16	.34	1986
Ophelia	31,473	20	.38	1986
Bianca	36,682	27	.43	1986
Cressida	38,295	41	.46	1986
Desdemona	38,848	36	.47	1986
Juliet	39,901	52	.49	1986
Portia	40,980	68	.51	1986
Rosalind	43,354	33	.56	1986
Belinda	46,658	42	.62	1986
Puck	53,323	95	.76	1985
Miranda	80,505	292	1.4	1948
Ariel	118,375	718	2.5	1851
Umbriel	164,906	725	4.1	1851

Name	Altitude (Miles)	Diameter (Miles)	Orbit (Days)	Discovery (Year)
Titania	270,489	978	8.7	1787
Oberon	361,721	943	13.4	1787
Caliban	4,464,000	37	930	1997
Sycorax	7,564,000	75	1,280	1997
Neptune				
Naiad	29,902	36	.29	1989
Thalassa	31,043	50	.31	1989
Despina	32,586	92	.33	1989
Galatea	38,409	98	.42	1989
Larissa	45,601	119	.55	1989
Proteus	72,936	257	1.1	1989
Triton	219,976	1,674	5.9	1846
Nereid	3,418,060	210	360	1949
Pluto				
Charon	12,176	737	6.3	1978

Appendix 2

The Unmanned Missions to the Moons

Spacecraft	Date Launched	Destination	Results
Ranger 1	August 23, 1961	The moon	Failed to leave Earth orbit
Ranger 2	November 18, 1961	The moon	Failed to leave Earth orbit
Ranger 3	January 26, 1962	The moon	Missed moon by 23,000 miles
Ranger 4	April 23, 1962	The moon	Hit far side of moon; returned no pictures
Ranger 5	October 18, 1962	The moon	Missed moon due to on-board power failure
Ranger 6	January 30, 1964	The moon	Successfully crashed on lunar surface; camera blackout prevented return of pictures
Ranger 7	July 28, 1964	The moon	Successfully crashed in the Sea of Clouds; returned 4,316 pictures

Spacecraft	Date Launched	Destination	Results
Ranger 8	February 17, 1965	The moon	Successfully crashed in Sea of Tranquillity; returned 7,137 pictures
Ranger 9	March 21, 1965	The moon	Successfully crashed in crater Alphonsus; returned 5,814 pictures
Surveyor 1	May 30, 1966	The moon	Soft-landed in the Ocean of Storms; returned over 10,000 pictures
Lunar Orbiter 1	August 10, 1966	The moon	Entered lunar orbit; returned 207 pictures
Surveyor 2	September 20, 1966	The moon	Engine failure caused crash on lunar surface
Lunar Orbiter 2	November 6, 1966	The moon	Entered lunar orbit; returned 211 pictures
Lunar Orbiter 3	February 4, 1967	The moon	Entered lunar orbit; returned 211 pictures
Surveyor 3	April 17, 1967	The moon	Soft-landed in the Ocean of Storms; returned 6,315 pictures

Spacecraft	Date Launched	Destination	Results
Lunar Orbiter 4	May 4, 1967	The moon	Entered lunar orbit; flew over south pole; returned 193 pictures
Surveyor 4	July 14, 1967	The moon	Undetermined malfunction caused crash on lunar surface
Lunar Orbiter 5	August 1, 1967	The moon	Entered lunar orbit; returned 212 pictures
Surveyor 5	September 8, 1967	The moon	Soft-landed in the Sea of Tranquillity; returned over 18,000 pictures
Surveyor 6	November 7, 1967	The moon	Soft-landed in moon's Central Bay; returned over 29,500 pictures
Surveyor 7	January 7, 1968	The moon	Soft-landed in Tycho crater; returned 21,274 pictures
Mariner 9	May 30, 1971	Mars	Orbited Mars; flew by Phobos; returned 7,329 pictures
Pioneer 10	March 3, 1972	Jupiter	Successfully flew through Jovian system

Spacecraft	Date Launched	Destination	Results
Pioneer 11	April 5, 1973	Jupiter and Saturn	Successfully flew through Jovian and Saturnian systems
Voyager 2	August 20, 1977	Jupiter, Saturn, Uranus, and Neptune	Launched before *Voyager 1;* successfully flew through all four systems
Voyager 1	September 5, 1977	Jupiter and Saturn	Successfully flew through Jovian and Saturnian systems
Galileo	October 18, 1989	Jupiter	Currently orbiting planet
Cassini-Huygens	October 15, 1997	Saturn	Currently en route

Author's Note

Just as it takes a small army of willing people to send a spacecraft to its cosmic destination, so too does it take a smaller, more modest army of people to tell the tale of the mission that ship flies. I am deeply indebted to just such an army for making *Journey Beyond Selēnē* possible.

Most of the stories recounted here were reconstructed through interviews with the scientists, administrators, and engineers who have been flying the spacecraft of the Jet Propulsion Laboratory for the past forty years. The rest were researched with the help of thousands of pages of documents and news clips, all of which have been preserved in NASA's various libraries and all of which were graciously made available to me. Conversations that appear in quotes in the text were reconstructed through interviews with one—and often more than one—of the principals involved.

Among those scientists and engineers who gave most generously of their time to help me tell this tale, the one to whom I owe the greatest thanks is Dr. William Pickering—thinker, visionary, and New Zealand's gift to America. May we someday contrive a way to repay the debt. Also lending enormous help were the gifted and engaging Rich Terrile and Torrence Johnson of JPL; my thanks for all the times they took my calls and all the times they explained to me—patiently—how the solar system works. Appreciation goes as well to countless other folks at JPL and elsewhere in the space community, including Jim Burke, Karen Buxbaum, John Casani, Clifford Cummings, John Delaney, Eugene Giberson, Norm Haynes, Linda Morabito Hyder, Ray Heacock, Charles Kohlhase, Richard Laeser, Victoria Melikan Lairmore, Matt Landanow, Peter Lyman, Chris McKay, Dennis Madsen, Ellis Miner, Bill Muehlberger, Marcia Neugebauer, Bill O'Neil, Toby Owen, Robert Parks, Richard Rudd, Bud Schurmeier, Paul Spudis, Steven Squyres, Rob Staehle, David Stevenson, Steve Synnott, Ewen Whitaker, and Don Wilhelms. Special thanks are also owed to Dave Scott, a man uniquely qualified to describe any moon.

A number of fine authors and their fine books also assisted me in my research; among them: *Galileo's Commandment,* edited by Edmund Blair Boles; *Exploring Space,* by William E. Burrows; *A Man on the Moon,* by Andrew Chaikin; *Lunar Impact: A History of Project Ranger,* by R. Cargill Hall; *The Story of Astronomy,* by Lloyd Motz and Jefferson Hane Weaver; *Astronomy and Cosmology,* by John North; *The Once and Future Moon,* by Paul Spudis; *Voyager Tales,* by David W. Swift; *To a Rocky Moon,* by Don E. Wilhelms; and *Astronomy Through the Ages,* by Robert Wilson.

In addition to these people, numerous others in the publishing and journalism community lent their help. Thanks to Philip Elmer-DeWitt of *Time* magazine, for providing the kind of flexible environment in which it's possible to write both a news story and a book, and do a creditable job of both; thanks as well to Michael D. Lemonick, also of *Time,* for his counsel and commiseration. Much appreciation also goes to Alice Mayhew and Roger Labrie of Simon & Schuster, and to everyone at the Joy Harris Literary Agency, especially Leslie Daniels and the infinitely supportive, infinitely patient Joy Harris.

Finally, thanks and much love to Splash, Steve, Garry, and Bruce Kluger, Lori Oliwenstein, and Alene Hokenstad, the people with whom I have long shared my orbit; to Richard Kluger for setting an incandescent example; to Bridgette, Emily, Audrey, and Mateo, for reminding me that the universe can still be a place of enchantment; and finally, to my wife, Alejandra López Kluger, for getting up early to watch Venus rise with me—and for all the things that implies.

Index

About the Author

Jeffrey Kluger is a senior writer at *Time* magazine covering science and the space program. He is the coauthor, along with astronaut Jim Lovell, of *Lost Moon,* the book that served as the basis of the 1995 movie *Apollo 13.* He is also the author of *The Apollo Adventure,* which accompanied the release of the movie. His features and columns have appeared in *The New York Times Magazine, GQ,* and *Newsday,* among other publications. He lives in New York City.